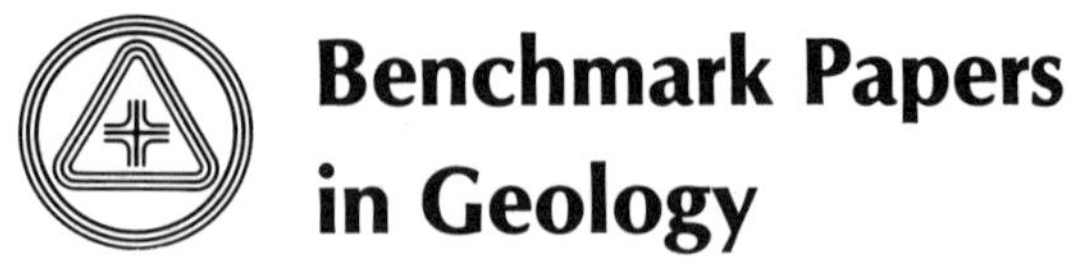

Benchmark Papers
in Geology

Series Editor: Rhodes W. Fairbridge
Columbia University

A selection from the published volumes in this series

Volume

 7 MARINE EVAPORITES: Origin, Diagenesis, and Geochemistry / *Douglas W. Kirkland and Robert Evans*

13 PHILOSOPHY OF GEOHISTORY: 1785-1970 / *Claude C. Albritton, Jr.*

14 GEOCHEMISTRY AND THE ORIGIN OF LIFE / *Keith A. Kvenvolden*

15 SEDIMENTARY ROCKS: Concepts and History / *Albert V. Carozzi*

20 PLAYAS AND DRIED LAKES: Occurrence and Development / *James T. Neal*

30 HOLOCENE TIDAL SEDIMENTATION / *George deVries Klein*

31 PALEOBIOGEOGRAPHY / *Charles A. Ross*

34 CRYSTAL FORM AND STRUCTURE / *Cecil J. Schneer*

37 STATISTICAL ANALYSIS IN GEOLOGY / *John M. Cubitt and Stephen Henley*

40 DIAGENESIS OF DEEP-SEA BIOGENIC SEDIMENTS / *Gerrit J. van der Lingen*

42 COASTAL SEDIMENTATION / *Donald J. P. Swift and Harold D. Palmer*

43 ANCIENT CONTINENTAL DEPOSITS / *Franklyn B. Van Houten*

45 SEA WATER: Cycles of the Major Elements / *James I. Drever*

65 DOLOMITIZATION / *Donald H. Zenger and S. J. Mazzullo*

73 CHEMICAL HYDROGEOLOGY / *William Back and R. Allan Freeze*

Related Titles

THE ENCYCLOPEDIA OF SEDIMENTOLOGY / *Rhodes W. Fairbridge and Joanne Bourgeois*

THE ENCYCLOPEDIA OF MINERALOGY / *Keith Frye*

A complete listing of volumes published in this series begins on p. 429.

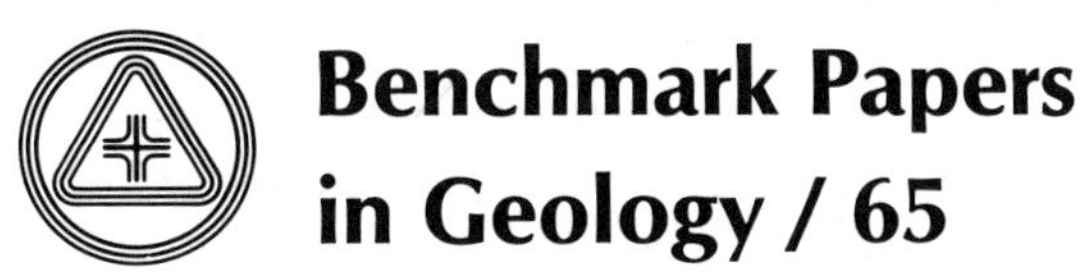

Benchmark Papers in Geology / 65

A BENCHMARK® Book Series

DOLOMITIZATION

Edited by

DONALD H. ZENGER

Pomona College, California

and

S. J. MAZZULLO

Geological Consultant

Hutchinson Ross Publishing Company

Stroudsburg, Pennsylvania

LIBRARY OF CONGRESS CATALOGING IN PUBLICATION DATA
Main entry under title:
Dolomitization.
   (Benchmark papers in geology; 65)
   Includes index.
   1. Dolomite.   I. Zenger, Donald H.   II. Mazzullo, S. J.   III. Series.
QE471.15.D6D64   552.5   81-7133
ISBN 0-87933-416-9   AACR2

Distributed worldwide by Van Nostrand Reinhold Company Inc.,
135 W. 50th St., New York, NY 10020

CONTENTS

Series Editor's Foreword   ix
Preface   xi
Contents by Author   xiii

Introduction   1

PART I: MAJOR PROBLEMS OF DOLOMITIZATION

Editors' Comments on Papers 1 Through 8   12

1   **CULLIS, C. G.:** The Mineralogical Changes Observed in the Cores of the Funafuti Borings   18
The Atoll of Funafuti. Boring into a Coral Reef and the Results, The Royal Society, London, 1904, pp. 392-393, 404-415, 420, Plate F

2   **VAN TUYL, F. M.:** New Points on the Origin of Dolomite   33
Am. Jour. Sci. **42**:249-260 (1916)

3   **PARSONS, L. M.:** Dolomitization and the Leicestershire Dolomites   45
Geol. Mag. **55**:246-258, Plate XI (1918)

4   **CHILINGAR, G. V.:** Relationship Between Ca/Mg Ratio and Geologic Age   59
Am. Assoc. Petroleum Geologists Bull. **40**:2256-2266 (1956)

5   **GRAF, D. L., and J. R. GOLDSMITH:** Some Hydrothermal Syntheses of Dolomite and Protodolomite   70
Jour. Geology **64**:173-186, Plate 1 (1956)

6   **DEGENS, E. T., and S. EPSTEIN:** Oxygen and Carbon Isotope Ratios in Coexisting Calcites and Dolomites from Recent and Ancient Sediments   85
Geochim. et Cosmochim. Acta **28**:23-44 (1964)

7   **FAIRBRIDGE, R. W.:** The Dolomite Question   107
Regional Aspects of Carbonate Deposition, ed. R. J. LeBlanc and J. G. Breeding, Soc. Econ. Paleontologists and Mineralogists Spec. Pub. 5, 1957, pp. 164-170

8   **INGERSON, E.:** Problems of the Geochemistry of Sedimentary Carbonate Rocks   116
Geochim. et Cosmochim. Acta **26**:830-837 (1962)

Contents

PART II: HOLOCENE DOLOMITE SEDIMENTOLOGY (MARGINAL MARINE)

Editors' Comments on Papers 9, 10, and 11   128

9   **ALDERMAN, A. R., and H. C. W. SKINNER:**   Dolomite Sedimentation in the South-East of South Australia   132
Am. Jour. Sci. **255**:561-567 (1957)

10   **CURTIS, R., G. EVANS, D. J. J. KINSMAN, AND D. J. SHEARMAN:** Association of Dolomite and Anhydrite in the Recent Sediments of the Persian Gulf   139
Nature **197**:679-680 (1963)

11   **SHINN, E. A., R. N. GINSBURG, and R. M. LLOYD:**   Recent Supratidal Dolomite from Andros Island, Bahamas   141
Dolomitization and Limestone Diagenesis, ed. L. C. Pray and R. C. Murray, Soc. Econ. Paleontologists and Mineralogists Spec. Pub. 13, 1965, pp. 112-123

PART III: MODELS FOR DOLOMITIZATION BY SURFACE-GENERATED HYPERSALINE SOLUTIONS

Editors' Comments on Papers 12 and 13   154

12   **ADAMS, J. E., AND M. L. RHODES:**   Dolomitization by Seepage Refluxion   157
Am. Assoc. Petroleum Geologists Bull. **44**:1912-1920 (1960)

13   **HSÜ, K. J., and C. SIEGENTHALER:**   Preliminary Experiments on Hydrodynamic Movement Induced by Evaporation and Their Bearing on the Dolomite Problem   166
Sedimentology **12**:11-25 (1969)

PART IV: ANCIENT ANALOGS OF PERITIDAL, EVAPORITIC DOLOMITIZATION

Editors' Comments on Papers 14, 15, and 16   182

14   **LAPORTE, L. F.:**   Carbonate Deposition Near Mean Sea-Level and Resultant Facies Mosaic: Manilus Formation (Lower Devonian) of New York State (Abstract)   185
Am. Assoc. Petroleum Geologists Bull. **51**:73 (1967)

15   **MATTER, A.:**   Tidal Flat Deposits in the Ordovician of Western Maryland   186
Jour. Sed. Petrology **37**:601-609 (1967)

16   **ZENGER, D.H.:**   Significance of Supratidal Dolomitization in the Geologic Record   195
Geol. Soc. America Bull. **83**:1-11 (1972)

PART V: DOLOMITIZATION IN A MIXING ZONE OF MARINE-METEORIC WATER

Editors' Comments on Papers 17 Through 20   208

17   **HANSHAW, B. B., W. BACK, and R. G. DEIKE:**   A Geochemical Hypothesis for Dolomitization by Ground Water   211
Econ. Geology **66**:710-724 (1971)

18	**BADIOZAMANI, K.:** The Dorag Dolomitization Model—Application to the Middle Ordovician of Wisconsin *Jour. Sed. Petrology* **43**:965-984 (1973)	227
19	**FOLK, R. L., and L. S. LAND:** Mg/Ca Ratio and Salinity: Two Controls over Crystallization of Dolomite *Am. Assoc. Petroleum Geologists Bull.* **59**:60-68 (1975)	247
20	**DUNHAM, J. B., and E. R. OLSON:** Diagenetic Dolomite Formation Related to Paleozoic Paleogeography of the Cordilleran Miogeocline in Nevada *Geology* **6**:556-559 (1978)	256

PART VI: LATE DIAGENETIC DOLOMITIZATION

Editors' Comments on Papers 21 and 22		262
21	**FREEMAN, T.:** Sedimentology and Dolomitization of Muschelkalk Carbonates (Triassic), Iberian Range, Spain *Am. Assoc. Petroleum Geologists Bull.* **56**:434-453 (1972)	268
22	**LOVERING, T. S.:** The Origin of Hydrothermal and Low Temperature Dolomite *Econ. Geology* **64**:743-754 (1969)	288

PART VII: OTHER MODELS AND CONCEPTS OF DOLOMITIZATION

Editors' Comments on Papers 23 Through 30		302
23	**HARRIS, L. D.:** Dolomitization Model for Upper Cambrian and Lower Ordovician Carbonate Rocks in the Eastern United States *U.S. Geol. Survey Jour. Res.* **1**:63-78 (1973)	312
24	**VON DER BORCH, C. C., D. E. LOCK, and D. SCHWEBEL:** Ground-Water Formation of Dolomite in the Coorong Region of South Australia *Geology* **3**:283-285 (1975)	328
25	**WOLFBAUER, C. A., and R. C. SURDAM:** Origin of Nonmarine Dolomite in Eocene Lake Gosiute, Green River Basin, Wyoming *Geol. Soc. America Bull.* **85**:1733-1740 (1974)	331
26	**SPOTTS, J. H., and S. R. SILVERMAN:** Organic Dolomite from Point Fermin, California *Am. Mineralogist* **51**:1144-1155 (1966)	339
27	**KAHLE, C. F.:** Possible Roles of Clay Minerals in the Formation of Dolomite *Jour. Sed. Petrology* **35**:448-453 (1965)	350
28	**GEBELEIN, C. D., and P. HOFFMAN:** Algal Origin of Dolomite Laminations in Stromatolitic Limestone *Jour. Sed. Petrology* **43**:603-613 (1973)	356
29	**LINDHOLM, R. C.:** Detrital Dolomite in Onondaga Limestone (Middle Devonian) of New York: Its Implications to the "Dolomite Question" *Am. Assoc. Petroleum Geologists Bull.* **53**:1035-1042 (1969)	367

Contents

30 SHINN, E. A.: Selective Dolomitization of Recent Sedimentary
Structures   375
Jour. Sed. Petrology **38**:612-616 (1968)

**PART VIII: SUMMARIES OF IMPORTANT CONTRIBUTIONS ON DOLOMITIZATION
FROM THE FRENCH, GERMAN, AND SOVIET LITERATURE**

Editors' Comments on Papers 31, 32, and 33   382

31 BOURROUILH-LE JAN, F. G.: Historical Résumé of French
Contributions to the Study of Dolomitization   383
Original article written expressly for this Benchmark volume

32 CHILINGAR, G. V.: Summary of Some Important Soviet Contributions
to the Field of Dolomitization   394
Original article written expressly for this Benchmark volume

33 FÜCHTBAUER, H.: Significant Contributions on Dolomitization from
the German Literature   403
Original article written expressly for this Benchmark volume

Author Citation Index   413
Subject Index   421
About the Editors   427

SERIES EDITOR'S FOREWORD

The philosophy behind the Benchmark Papers in Geology is one of collection, sifting, and rediffusion. Scientific literature today is so vast, so dispersed, and, in the case of old papers, so inaccessible for readers not in the immediate neighborhood of major libraries that much valuable information has been ignored by default. It has become just so difficult, or so time consuming, to search out the key papers in any basic area of research that one can hardly blame a busy person for skimping on some of his or her "homework."

This series of volumes has been devised, therefore, as a practical solution to this critical problem. The geologist, perhaps even more than any other scientist, often suffers from twin difficulties—isolation from central library resources and immensely diffused sources of material. New colleges and industrial libraries simply cannot afford to purchase complete runs of all the world's earth science literature. Specialists simply cannot locate reprints or copies of all their principal reference materials. So it is that we are now making a concerted effort to gather into single volumes the critical materials needed to reconstruct the background of any and every major topic of our discipline.

We are interpreting "geology" in its broadest sense: the fundamental science of the planet Earth, its materials, its history, and its dynamics. Because of training in "earthy" materials, we also take in astrogeology, the corresponding aspect of the planetary sciences. Besides the classical core disciplines such as mineralogy, petrology, structure, geomorphology, paleontology, and stratigraphy, we embrace the newer fields of geophysics and geochemistry, applied also to oceanography, geochronology, and paleoecology. We recognize the work of the mining geologists, the petroleum geologists, the hydrologists, and the engineering and environmental geologists. Each specialist needs a working library. We are endeavoring to make the task of compiling such a library a little easier.

Each volume in the series contains an introduction prepared by a specialist (the volume editor)—a "state of the art" opening or a summary of the object and content of the volume. The articles, usually some twenty to fifty reproduced either in their entirety or in significant extracts, are selected in an attempt to cover the field, from the key papers of the last century to fairly recent work. Where the original works are in foreign languages, we have endeavored to locate or commission translations. Geologists, because of their global subject, are often acutely aware of the oneness of our world. The selections cannot therefore be restricted to any one country, and whenever possible an attempt is made to scan the world literature.

To each article, or group of kindred articles, some sort of "highlight commentary" is usually supplied by the volume editor. This commentary should serve to bring that article into historical perspective and to emphasize its particular role in the growth of the field. References, or citations, wherever possible, will be reproduced in their entirety—for by this means the observant reader can assess the background material available to that particular author, or, if desired, he or she too can double check the earlier sources.

A "benchmark," in surveyor's terminology, is an established point on the ground that is recorded on our maps. It is usually anything that is a vantage point, from a modest hill to a mountain peak. From the historical viewpoint, these benchmarks are the bricks of our scientific edifice.

RHODES W. FAIRBRIDGE

PREFACE

Why a Benchmark volume on dolomitization? No other sedimentary rock has been more enigmatic and perplexing to geologists than dolomites, particularly with regard to their origin. In more practical terms, dolomites are of economic significance because of their common association with petroleum, base-metal ore deposits, and evaporites.

It has been difficult for us to make a selection of papers from a voluminous literature that would appropriately represent past research and advances in our understanding of dolomites and dolomitization. In an attempt to accomplish such a task, we polled approximately thirty-five carbonate petrologists for their suggestions after we had made an initial selection of papers. Although it was impossible to include every reference that was so recommended, we were gratified to know that we had, in fact, chosen all those papers suggested by two or more of those who responded to our questionnaire. Despite the rising costs of reprinting rights of some publishing companies, we were able to include essentially all of the papers that we had originally chosen. Of course, space and cost limitations necessitated the omission of certain related subjects, such as dedolomitization. So far as we are aware, only three of the selections reprinted here have appeared in previous Benchmark volumes. Papers 9 (Alderman and Skinner) and 10 (Curtis et al.) focus on dolomite and associated evaporites and were reprinted in *Marine Evaporites*, edited by Kirkland and Evans (1973). Paper 12, the classic article by Adams and Rhodes describing the model of seepage refluxion, had been included by A. V. Carozzi in his Benchmark volume *Sedimentary Rocks* (1975). Their inclusion in *Dolomitization* indicates their significance and our desire that this book stand independent of preceding ones.

Three of the papers in our introductory part are not reprinted in their entirety because of prohibitive length (Fairbridge, Paper 7) or their partial concern with subjects outside the realm of dolomitization (Cullis, Paper 1, and Ingerson, Paper 8). In Part IV (Paper 14), only the abstract of the article by Laporte is reprinted. The other twenty-six contributions are reproduced in their unabridged forms.

Part I includes some of the older papers on dolomites published in the early 1900s, as well as some later review papers. All of these papers deal with one or more of the major questions regarding dolomitization, such as the synthesis of dolomite, a primary versus replacement origin of the mineral, and the failure of uniformitarianism to account for the vast stratiform bodies of dolomite in the geologic record. Part II focuses on the fascinating discoveries of Holocene penecontemporaneous dolomite, particularly peritidal, that excited carbonate researchers in the late 1950s and early 1960s. Part III is concerned with various models that have been proposed to account for evaporative dolomitization, such as appears typical of many Holocene dolomite occurrences, whereas Part IV includes examples of the numerous publications that related Holocene peritidal dolomites to presumed ancient analogues. As a striking alternative to these hypersaline processes and products, the seawater-meteoric water-mixing model was popularized in the early 1970s (Part V). Late diagenetic dolomites and dolomitization provide the material for Part VI, which is followed by a discussion of yet other concepts of dolomitization (Part VII). It is hoped that, after considering Parts I through VII, the reader will appreciate the fact that dolomite does not have one unique origin. Finally, in what we consider to be a novel and useful approach in this series, we have devoted the last part to a very brief English-language summary of some of the important studies on dolomitization that were published originally by French, German, and Russian workers, primarily in their respective languages. The three selections in this part were very kindly authored by F. G. Bourrouilh-Le Jan, G. V. Chilingar, and Hans Füchtbauer, and we gratefully acknowledge their efforts and contributions.

Of course, the main purpose of any Benchmark volume is to provide highlights in the development of a particular field. Although these books are not primarily intended to be "state of the art" presentations, we hope that the papers presented here, coupled with our commentaries, will provide, in addition to an historical perspective, a gateway into modern researches on dolomitization.

We are grateful to the many persons who have helped us in the preparation of this book, particularly our foreign contributors and those authors and publishers who have granted us permission to use their material. Staff members in the Geology Department, Pomona College, and at Union Texas Petroleum Corporation's Stratigraphic Exploration Group were important in our support. S. J. Mazzullo wishes to acknowledge Fawn Meek for her typing duties and editorial assistance and Union Texas Petroleum Corporation for the time and resources necessary to complete this project. D. H. Zenger thanks Jean MacKay for her typing and Ann Zenger for editing early drafts of portions of the manuscript. William Back and Bruce Hanshaw kindly reviewed the commentary for Part V on mixed-water dolomitization.

D. H. ZENGER
S. J. MAZZULLO

xii

CONTENTS BY AUTHOR

Adams, J. E., 157
Alderman, A. R., 132
Back, W., 211
Badiozamani, K., 227
Bourrouilh-Le Jan, F. G., 383
Chilingar, G. V., 59, 394
Cullis, C. G., 18
Curtis, R., 139
Degens, E. T., 85
Deike, R. G., 211
Dunham, J. B., 256
Epstein, S., 85
Evans, G., 139
Fairbridge, R. W., 107
Folk, R. L., 247
Freeman, T., 268
Füchtbauer, H., 103
Gebelein, C. D., 356
Ginsburg, R. N., 141
Goldsmith, J. R., 70
Graf, D. L., 70
Hanshaw, B. B., 211
Harris, L. D., 312
Hoffman, P., 356
Hsü, K. J., 166

Ingerson, E., 116
Kahle, C. F., 350
Kinsman, D. J. J., 139
Land, L. S., 247
Laporte, L. F., 185
Lindholm, R. C., 367
Lloyd, R. M., 141
Lock, D. E., 328
Lovering, T. S., 288
Matter, A., 186
Olson, E. R., 256
Parsons, L. M., 45
Rhodes, M. L., 157
Schwebel, D., 328
Shearman, D. J., 139
Shinn, E. A., 141, 375
Siegenthaler, C., 166
Silverman, S. R., 339
Skinner, H. C. W., 132
Spotts, J. H., 339
Surdam, R. C., 331
Van Tuyl, F. M., 33
von der Borch, C. C., 328
Wolfbauer, C. A., 331
Zenger, D. H., 195

DOLOMITIZATION

INTRODUCTION

For a long time, my dear friend, I have realized that effervescence with acid is not always an essential characteristic of calcareous rocks, although all naturalists have indicated that this property is the best one by which this type of rock can be identified. I had observed that several rocks of this type were attacked by acid without producing the great release of air that causes effervescence. I have observed their solution proceeding quietly and reaching completion with the release of only a few large bubbles which rose slowly from the bottom of the solvent in which I had immersed them and eventually broke at its surface. Frequently I happened to spread acid on the surface of some rocks, which appeared to be calcareous by all other external characteristics, without producing the effervescence I had expected; and several minutes elapsed before seeing a very slight bubbly release that revealed the action of the acid. [de Dolomieu, 1791, pp. 3–4; English translation by Carozzi and Zenger, 1981.]

With these comments, Déodat de Dolomieu began his paper, a communication to his colleague, A. M. Picot, describing for the first time the rock that, a year later, was more fully described and named "dolomie" in his honor by N. T. de Saussure, the son of the more famous Horace-Bénédict de Saussure. (See partial translation of this paper by F. G. Bourrouilh-Le Jan in Part VIII.) The word *dolomite* appeared for the first time in the second edition of Kirwan's *Elements of Mineralogy* (1794) as the English equivalent of de Saussure's "dolomie."*

As is the case with most other subjects, one learns on examining

---

*Kirwan's description of dolomite (1794, pp. 111–112) was for the *mineral;* de Dolomieu (1791), on the other hand, was clearly referring to a rock (for example, "pierre," "marbre"). Subsequently, until Shrock (1948) proposed *dolostone, dolomite* was used by English-speaking geologists for both the rock and the mineral. Vatan (1958) showed that *dolostone* is etymologically improper, and, hence, many carbonate petrologists have favored the dual use of the term *dolomite.*

the history of the study of dolomitization that many ideas believed to have originated relatively recently were instead expressed either directly or peripherally at an earlier time.* Commencing with the studies of eighteenth-century workers (see Van Tuyl, Paper 2, for a concise coverage of this early work), we can clearly trace the evolution of the various concepts on the origin of dolomite that lead to what seems to be an unalterable fact: that there are "dolomites and dolomites" (refer to Chilingar et al., 1979).

Von Buch's early theory (Zirkel, 1866, pp. 251–252; see also Paper 2) on the origin of dolomite by the action of magnesium-rich volcanic vapors was early deemed untenable by Fournet (1849). However, Von Buch's ideas involved the dolomitization of a limestone precursor. Such a concept of replacement has been commonly credited to de Beaumont (1837) who proposed that dolomite resulted from the alteration of limestone in contact with magnesium bicarbonate solutions. Even earlier, Boué (1831) had advocated the direct precipitation of dolomite, and thus the stage was set for one of the continuing arguments: that of primary versus replacement origin. In short, with the exception of many Soviet carbonate petrologists who believe that primary precipitation of dolomite was a significant process in ancient rocks (G. V. Chilingar, oral communication, 1979; see also Paper 32), most geologists believe that field and petrographic evidence overwhelmingly indicate that most dolomites result from replacement. Related to this process is the matter of the timing of replacement, which can range from penecontemporaneous to epigenetic. Accordingly, each dolomite occurrence must be studied individually in order to understand the timing and mechanics of dolomitization. Although there appears to be a consensus of opinion among modern workers that early diagenetic, shallow-burial dolomitization has been more significant than other processes, it may be that late diagenetic dolomitization has so far been underestimated (Mattes and Mountjoy, 1980; see editors' comments for Part VI).

Compounding the problem of its origin has been the repeated failure to produce, by either precipitation or replacement, true (ordered and stoichiometric) dolomite under laboratory conditions at temperatures and pressures approaching those at the earth's surface (Berner, 1971, p. 149). By the early 1900s there was an extensive literature on the abortive attempts to synthesize dolomite under

*For reviews of dolomitization, some of which are too long for inclusion in this book, see Steidtmann (1911), Van Tuyl (1916; and a shorter summary published that same year, Paper 2), Fairbridge (Paper 7), Ingerson (Paper 8), Sonnenfeld (1964), Friedman and Sanders (1967), Zenger (1972), and Chilingar et al. (1979).

widely varying conditions (Van Tuyl, Paper 2); the situation has not improved much since. Sulfate and chloride salts of magnesium, ammonium carbonate, and organic matter have been the principal ingredients in many of the experiments. Generally the carbonate phase produced in such experimentation is not true dolomite, or if the product is bona fide, the experimental conditions under which it formed differed from those in the natural sedimentary environment. The term *protodolomite* was proposed by Graf and Goldsmith (Paper 5) for the calcium-rich, imperfectly ordered material produced artificially at low temperatures, in contrast to the true dolomite reportedly synthesized at temperatures over 200°C. In his review of the geochemistry of dolomite, Hsü (1967, p. 188) could still refer to "the unsatisfactory state of our knowledge in the chemistry of dolomite formation." English-language reviews of experimental studies on dolomitization include those of Fairbridge (Paper 7), Siegel (1961), Hsü (1967), and, for some more recent intriguing but still equivocal syntheses, Zenger (1972) and Chilingar et al. (1979). Currently we are uncertain about many aspects of dolomite chemistry such as the important matter of dolomite stability in association with various aqueous solutions (Carpenter, 1980; Zenger and Dunham, 1980, p. 5).

Intricately related to these questions of the origin of dolomite and dolomite synthesis is the apparent failure of uniformitarianism to provide the answer. Daly (1907, 1909) recognized the increase in Ca/Mg ratio in carbonate rocks with decreasing age; Chilingar (Paper 4), expanding on the earlier studies of Soviet workers, supported Daly's findings with many new data. Until about 1960 it was well known that while "marine" dolomites constitute a major portion of late Precambrian and Paleozoic carbonate sequences, they practically vanish by the Holocene, save for the limited occurrence in sediments of isolated rhombs and other grains that may well be detrital (Fairbridge, Paper 7). Despite the several surprising discoveries of Holocene protodolomites and dolomites since 1957, we feel that this discrepancy in temporal occurrence is still real.

The majority of reprinted papers here concern research since about 1955, which generally has built on that done during the preceding century and a half, albeit with an occasional lack of appreciation for some of the older work. Many of the papers published in the first half of the twentieth century are indeed noteworthy, although only a few, other than those included in this book are mentioned below. Skeats (1903) presented an exhaustive summary of dolomitization of uplifted Tertiary coral reefs in the Pacific. An excellent case for the widespread replacement origin of dolomite in the Bighorn Dolomite (Ordovician, Wyoming) was reported by

Blackwelder (1913). Parsons (1922) described the results of thorough field and petrographic studies of dolomitization in the Carboniferous Limestone (Mississippian) of the Midlands, England. Sander's classic monograph on fabrics in Alpine Triassic carbonates (1936; later translated by E. B. Knopf, 1951; also the contribution by Füchtbauer, Paper 33), included much on dolomitization. Laterally persistent dolomites in the Ellenburger Group (Lower Ordovician, Texas) were considered penecontemporaneous by Cloud and Barnes (1948), whereas those with abrupt latral changes to limestone were considered to be of late diagenetic origin.

The late 1950s and 1960s were highlighted by the startling discoveries of Holocene dolomite forming chiefly in marginal marine settings. First came the announcement of dolomitic sediments in an ephemeral lagoon (Coorong) and isolated lakes in southeastern Australia by Alderman and Skinner (Paper 9). This news was followed by reports of penecontemporaneous, evaporitic dolomite in peritidal sediments of Florida (Shinn and Ginsburg, 1964), the Bahamas (Shinn, Ginsburg, and Lloyd, Paper 11), the Persian Gulf (Wells, 1962; Curtis et al., Paper 10; Illing, Wells, and Taylor, 1965), and the island of Bonaire off the Venezuelan coast (Deffeyes, Lucia, and Weyl, 1965). The fluids necessary for dolomitization in these peritidal environments were attributed directly or indirectly by the authors to seawater that reached the site of dolomitization by capillary concentration (see Friedman and Sanders, 1967, p. 267), flooding, seepage refluxion (Adams and Rhodes, Paper 12), evaporative pumping (Hsü and Siegenthaler, Paper 13), or some combination of these processes. Subsequent to these early reports, by invoking uniformitarianism, geologists were quick to recognize ancient analogs—penecontemporaneous peritidal dolomite in the geologic record (for example, Laporte, Paper 14; Lucia, 1972; Matter, Paper 15; Roehl, 1967; Schenck, 1967; Textoris, 1969; Thompson, 1970; West, Brandon, and Smith, 1968; refer also to editors' comments for Part IV). The composite evidence in most of these particular instances seems valid. Nevertheless, the generalizations became overextended (Zenger, Paper 16), and a band-wagon trend developed, not an unfamiliar tendency in geology, particularly in those situations involving analogy between Holocene sediments and their ancient counterparts. As Carozzi aptly described (1975, pp. 5-6), we have witnessed the developments of the "Bahama" and "sabkha" syndromes.

With this situation, many geologists rested comfortably, content that the present indeed was *the* key to the past and that most ancient dolomites could now be accounted for. However, some objections to these analogies were raised by those who felt that much dolomite in the record is not of peritidal origin. For example, Freeman (1966) demonstrated that while two Ordovician carbonate units in Arkansas possessed depositional features characteristic of a peritidal origin, the dolomite present was not genetically related to features of desiccation but rather was of postlithification origin. Later he described a similar situation for the Triassic Muschelkalk of Spain (Freeman, Paper 21). The general significance of peritidal dolomitization in the geological record was questioned by Zenger (Paper 16), who pointed out that many ancient dolomites do not represent replacement of initial peritidal carbonates and even in those that apparently do, the dolomite may not have formed there penecontemporaneously; some undoubtedly were late diagenetic. (Aspects of late diagenetic dolomitization, including its possible underemphasis, are considered in Part VI.)

Although previously cited laboratory investigations and Holocene occurrences seemingly suggested that evaporative solutions were necessary in order to cause dolomitization, there was abundant evidence (primarily faunal), at times overlooked, that many dolomites in the geologic record may have formed within carbonates that were deposited in seawater of normal salinity (for example, Berry and Boucot, 1970, p. 87; Weller, 1911; Zenger, Paper 16). Furthermore, at the same time questions were raised concerning the inferred association and significance of the evaporitic waters in some of the Holocene peritidal occurrences (Atwood and Bubb, 1970; O. P. Bricker, written communication, 1970). In the early 1970s a new concept caught on—that of dolomitization by dilute waters characteristic of the groundwater zone intermediate between fresh and marine. Earlier work by Hsü (1963) and Runnells (1969) helped to establish a theoretical and experimental basis for subsequent, highly significant papers that produced such essentially synonymous terms as *schizohaline, dorag,* and *mixing-zone dolomitization:* Hanshaw, Back, and Deike (Paper 17), Land (1973a, 1973b), Badiozamani (Paper 18), Folk and Land (Paper 19), Land, Salem, and Morrow (1975), and others. In a sense, the carbonate world was set for a backlash to the peritidal concept and, ironically, a mixing-zone dolomitization syndrome developed. A significant question in the minds of many carbonate workers is whether such a model could account for dolomitization on a regional scale. Dunham and Olson (Paper 20) claimed to have affirmative evidence concerning this problem in Paleozoic carbonates in Nevada. As is evidenced by the papers in Zenger, Dunham, and Ethington (1980), it is clear that penecontemporaneous peritidal dolomitization, shallow-burial, mixing-zone dolomitization, and late diagenetic dolomitization have all been important processes in the geologic record.

Part VII includes a brief look into concepts and models of dolomitization not thoroughly covered in earlier papers, such as evaporative, subtidal dolomitization; lacustrine dolomitization; the possible role of organic matter on dolomitization; source of Mg^{2+} ions for dolomitization; terrigenous dolomite; and selective dolomitization. Space limitations preclude a more exhaustive treatment of all facets of the subject. For example, the classification of dolomites and dolomitic limestones has been only directly or partially stressed in several papers in this book (Papers 2, 3, 7, 18, and 21). Porosity of dolomites has been considered only indirectly despite its great interest to economic geologists because of the importance of dolomites as petroleum reservoirs and ore hosts (Murray, 1930; Bird and Jordan, 1977). Many studies, beginning with the early works of de Beaumont (1837) and de Morlot (1848) and continuing with Murray (1960) and Weyl (1960), attributed significant porosity development to mole-for-mole replacement of calcite and an accompanying 12 to 13 percent volume reduction, particularly if the carbonate source is local. However, little field evidence has been produced to substantiate the theoretical molecular replacement, as contrasted with isovolumetric replacement (Lindgren, 1912; Murray, 1930). Landes (1946) rejected the molecular replacement process, as well as the leaching and recrystallization theories, to account for local dolomitization, contending that excess solution over precipitation during the dolomitization process was responsible for porosity development. Certainly much porosity in dolomites develops after replacement.

The most perplexing problem remains unsolved: that of the great discrepancy between the impressive late Precambrian through early Paleozoic dolomite sequences, and despite the discovery of Holocene dolomite, the very minor amount existing today. Reviews of theories attempting to explain this anomaly are presented by Ingerson (Paper 8) and Zenger (1972). Little new has been published recently on this problem. Zenger (1972) and, in an overlooked proposal within a larger work, Boucot (1975, pp. 282–292) correlated widespread dolomites with extensive shelf seas. Boucot's suggestion of biotic dolomitization runs into difficulty because of the presence of many late Precambrian dolomites, as he now appreciates (oral communication, 1981). Mottl and Holland's (1975) novel idea of a Mg^{2+} flux into basalt, precluding its availability for dolomitization, at higher temperatures provided at spreading centers beginning with the present episode of continental drift in the Jurassic is based wholly on theory and laboratory work; also, it seems to ignore possible preceding periods of sea-floor spreading.

Recently Zenger and Dunham (1980, pp. 5–7) briefly summarized the present status of the study of dolomities and dolomitization. Although we have learned a great deal, much remains to be done in both the field and the laboratory. Nearly two centuries after de Dolomieu's introduction, we are at least able to appreciate the diversity of dolomite types and the complexities of dolomitization.

REFERENCES

Atwood, D. K., and J. N. Bubb, 1970, Distribution of Dolomite in a Tidal Flat Environment, Sugarloaf Key, Florida, *Jour. Geology* **78**:499–505.

Beaumont, E. de, 1837, Application du Calcul à l'Hypothèse de la Formation par Épigénie des Anhydrites, des Gypses, et des Dolomies, *Soc. Géol. France Bull.* **8**:174–177.

Berner, R. A., 1971, *Principles of Chemical Sedimentology*, McGraw-Hill, New York, 240p.

Berry, W. B. N., and A. J. Boucot, 1970, Correlation of the North American Silurian Rocks, *Geol. Soc. America Spec. Paper 102*, Boulder, Co., 189p.

Bird, K. J., and C. F. Jordan, 1977, Lisburne Group (Mississippian and Pennsylvanian), Potential Major Hydrocarbon Objective of Arctic Slope, Alaska, *Am. Assoc. Petroleum Geologists Bull.* **61**:1493–1512.

Blackwelder, E., 1913, Origin of the Bighorn Dolomite of Wyoming, *Geol. Soc. America Bull.* **24**:607–624.

Boucot, A. J., 1975, *Evolution and Extinction Rate Controls*, Elsevier, Amsterdam, 427p.

Boué, A., 1831, Compte Rendu des Progrès de la Géologie, *Soc. Géol. France Bull.* **1**:115.

Carozzi, A. V., ed., 1975, *Sedimentary Rocks: Concepts and History*, Benchmark Papers in Geology, vol. 15, Dowden, Hutchinson & Ross, Stroudsburg, Pa., 468p.

Carozzi, A. V., and D. H. Zenger, 1981, On a Type of Calcareous Rock That Reacts Very Slightly with Acid and That Phosphorescences on Being Struck (Sur un genre de Pierres calcaires très-peu effervescentes avec les Acides & phosphorescentes par la collision), translation with notes on de Dolomieu's paper reporting the discovery of dolomite, *Jour. Geol. Education* **29**:4–10.

Carpenter, A. B., 1980, The Chemistry of Dolomite Formation I: The Stability of Dolomite, in *Concepts and Models of Dolomitization*, ed. D. H. Zenger, J. B. Dunham, and R. L. Ethington, Soc. Econ. Paleontologists and Mineralogists Spec. Pub. 28, Tulsa, Ok., pp. 111–121.

Chilingar, G. V., D. H. Zenger, H. J. Bissell, and K. H. Wolf, 1979, Dolomites and Dolomitization, in *Diagenesis in Sediments and Sedimentary Rocks*, ed. G. Larsen and G. V. Chilingar, Elsevier, Amsterdam, pp. 423–536.

Cloud, P. E., Jr., and V. E. Barnes, 1948, The Ellenberger Group of Central Texas, *Univ. Texas Pub. 4621*, Austin, Tx., 473p.

Daly, R. A., 1907, The Limeless Ocean of Pre-Cambrian Time, *Am. Jour. Sci.* **23**:93–115.

Daly, R. A., 1909, First Calcareous Fossils and the Evolution of Limestone, *Geol. Soc. America Bull.* **20**:153–170.

Deffeyes, K. S., F. J. Lucia, and P. K. Weyl, 1965, Dolomitization of Recent and Plio-Pleistocene Sediments by Marine Evaporite Waters on Bonaire, Netherlands Antilles, in *Dolomitization and Limestone Diagenesis,* ed. L. C. Pray and R. C. Murray, Soc. Econ. Paleontologists and Mineralogists Spec. Pub. 13, Tulsa, Ok., pp. 71–88.

Dolomieu, D. de, 1791, Sur un Genre de Pierres Calcaires très-peu Effervescentes avec les Acides & Phosphorescentes par la Collision, *Jour. Physique* **39**:3–10.

Fournet, J., 1849, Apercus sur Diverses Questions Géologique, *Soc. Géol. France Bull.* (sér. 2) **11**:502–518.

Freeman, T., 1966, Post-lithification Dolomite in the Joachim and Plattin Formations, Northern Arkansas (Abstract), *Geol. Soc. America Spec. Paper 87,* p. 59.

Friedman, G. M., and J. E. Sanders, 1967, Origin and Occurrence of Dolostones, in *Carbonate Rocks: Origin, Occurrence, and Classification,* ed. G. V. Chilingar, H. J. Bissell, and R. W. Fairbridge, Elsevier, Amsterdam, pp. 267–348.

Hsü, K. J., 1963, Solubility of Dolomite and Composition of Florida Ground Waters, *Jour. Hydrology* **1**:288–310.

Hsü, K. J., 1967, Chemistry of Dolomite Formation, in *Carbonate Rocks: Origin, Occurrence, and Classification,* ed. G. V. Chilingar, H. J. Bissell, and R. W. Fairbridge, Elsevier, Amsterdam, pp. 169–191.

Illing, L. V., A. J. Wells, and J. C. M. Taylor, 1965, Penecontemporaneous Dolomite in the Persian Gulf, in *Dolomitization and Limestone Diagenesis,* ed. L. C. Pray and R. C. Murray, Soc. Econ. Paleontologists and Mineralogists Spec. Pub. 13, Tulsa, Ok., pp. 89–111.

Kirwan, R., 1794, *Elements of Mineralogy,* 2d ed., J. Nichols, London, 510p.

Land, L. S., 1973a, Contemporaneous Dolomitization of Middle Pleistocene Reefs by Meteoric Water, North Jamaica, *Bull. Marine Sci.* **23**:64–92.

Land, L. S., 1973b, Holocene Meteoric Dolomitization of Pleistocene Limestones, North Jamaica, *Sedimentology* **20**:411–424.

Land, L. S., M. R. I. Salem, and D. W. Morrow, 1975, Paleohydrology of Ancient Dolomites: Geochemical Evidence, *Am. Assoc. Petroleum Geologists Bull.* **59**:1602–1625.

Landes, K. K., 1946, Porosity Through Dolomitization, *Am. Assoc. Petroleum Geologists Bull.* **30**:305–318.

Lindgren, W., 1912, The Nature of Replacement, *Econ. Geology* **7**:521–535.

Lucia, F. J., 1972, Recognition of Evaporite-Carbonate Shoreline Sedimentation, in *Recognition of Ancient Sedimentary Environments,* ed. J. K. Rigby and W. K. Hamblin, Soc. Econ. Paleontologists and Mineralogists Spec. Pub. 16, Tulsa, Ok., pp. 160–191.

Mattes, B. W., and E. W. Mountjoy, 1980, Burial Dolomitization of the Upper Devonian Miette Buildup, Jasper National Park, Alberta, in *Concepts and Models of Dolomitization,* ed. D. H. Zenger, J. B. Dunham, and R. L. Ethington, Soc. Econ. Paleontologists and Mineralogists Spec. Pub. 28, Tulsa, Ok., pp. 259–297.

Morlot, A. de, 1848, Sur l'Origine de la Dolomie, *Acad. Sci. Comptes Rendus* **26**:311–315.

Mottl, M. J., and H. D. Holland, 1975, Basalt-Sea Water Interaction, Sea-Floor Spreading, and the Dolomite Problem (Abstract), *Am. Geophys. Union Trans.* **56**:1074.

Murray, A. N., 1930, Limestone Oil Reservoirs of the Northeastern United States and of Ontario, Canada, *Econ. Geology* **25**:452–469.

Murray, R. C., 1960, Origin of Porosity in Carbonate Rocks, *Jour. Sed. Petrology* **30**:59–84.

Parsons, L. M., 1922, Dolomitization in the Carboniferous Limestone of the Midlands, *Geol. Mag.* **59**:51–63, 104–117.

Roehl, P. O., 1967, Stony Mountain (Ordovician) and Interlake (Silurian) Facies Analogs of Recent Low-Energy Marine and Subaerial Carbonates, Bahamas, *Am. Assoc. Petroleum Geologists Bull.* **51**:1979–2032.

Runnells, D. D., 1969, Diagenesis, Chemical Sediments, and the Mixing of Natural Waters, *Jour. Sed. Petrology* **39**:1188–1201.

Sander, B., 1936, Beiträge zur Kenntniss der Anlagerungsgefüge (Rhythmische Kalke und Dolomite aus der Trias), *Tschermaks Mineralog. u. Petrog. Mitt.* **48**:27–139, 141–209. (Contributions to the Study of Depositional Fabrics [Rhythmically Deposited Triassic Limestones and Dolomites], trans. E. B. Knopf, *Am. Assoc. Petroleum Geologists Bull.*, 207p., 1951.)

Saussure, N. T. de, 1792, Analyse de la Dolomie, *Jour. Physique* **40**:161–173.

Schenck, P. E., 1967, The Macumber Formation of the Maritime Provinces, Canada—A Mississippian Analogue to Recent Strand-Line Carbonates of the Persian Gulf, *Jour. Sed. Petrology* **37**:365–376.

Shinn, E. A., and R. N. Ginsburg, 1964, Formation of Recent Dolomite in Florida and the Bahamas (Abstract), *Am. Assoc. Petroleum Geologists Bull.* **48**:547.

Shrock, R. R., 1948, A Classification of Sedimentary Rocks, *Jour. Geology* **56**:118–129.

Siegel, F. R., 1961, Factors Influencing the Precipitation of Dolomite Carbonate, *Kansas Geol. Survey Bull.* **152**:127–158.

Skeats, E. W., 1903, The Chemical Composition of Limestones from Upraised Coral Islands, with Notes on Their Microscopical Structures, *Harvard Museum Comp. Zoology Bull.* **42**:53–126.

Sonnenfeld, P., 1964, Dolomites and Dolomitization: A Review, *Bull. Canadian Petroleum Geology* **12**:101–132.

Steidtmann, E., 1911, Evolution of Limestone and Dolomite, *Jour. Geology* **19**:323–345, 392–428.

Textoris, D. A., 1969, Supratidal Origin of Appalachian Basin Dolostone (Abstract), *Geol. Soc. America Spec. Paper 121*, pp. 470–471.

Thompson, A. M., 1970, Tidal-Flat Deposition and Early Dolomitization in Upper Ordovician Rocks of Southern Appalachian Valley and Ridge, *Jour. Sed. Petrology* **40**:1271–1286.

Van Tuyl, F. M., 1916, The Origin of Dolomite, *Iowa Geol. Survey Ann. Report 1914* **25**:251–421.

Vatan, A., 1958, "Dolostone:" Discussion, *Jour. Sed. Petrology* **28**:514.

Weller, S., 1911, Are the Fossils of the Dolomites Indicative of Shallow, Highly Saline and Warm Water Seas? *Geol. Soc. America Bull.* **22**:227–231.

Wells, A. J., 1962, Recent Dolomites in the Persian Gulf, *Nature* **194**:274–275.

West, I. M., A. Brandon, and M. Smith, 1968, A Tidal Flat Evaporitic Facies in the Viséan of Ireland, *Jour. Sed. Petrology* **38**:1079–1093.

Weyl, P. K., 1960, Porosity Through Dolomitization: Conservation-of-Mass Requirements, *Jour. Sed. Petrology* **30**:85–90.

Zenger, D. H., 1972, Dolomitization and Uniformitarianism, *Jour. Geol. Education* **20**:107–124.

Zenger, D. H., and J. B. Dunham, 1980, Concepts and Models of Dolomitization—An Introduction, in *Concepts and Models of Dolomitization,* ed. D. H. Zenger, J. B. Dunham, and R. L. Ethington, Soc. Econ. Paleontologists and Mineralogists Spec. Pub. 28, Tulsa, Ok., pp. 1–9.

Zenger, D. H., J. B. Dunham, and R. L. Ethington, eds., 1980, Concepts and Models of Dolomitization, *Soc. Econ. Paleontologists and Mineralogists Spec. Pub. 28,* Tulsa, Ok., 320p.

Zirkel, F., 1866, *Lehrbuch der Petrographie,* vol. 1, Adolph Marcus, Bonn, Germany, 607p.

Part I

MAJOR PROBLEMS
OF DOLOMITIZATION

Editors' Comments on Papers 1 Through 8

1 CULLIS
Excerpts from *The Mineralogical Changes Observed in the Cores of the Funafuti Borings*

2 VAN TUYL
New Points on the Origin of Dolomite

3 PARSONS
Dolomitization and the Leicestershire Dolomites

4 CHILINGAR
Relationship Between Ca/Mg Ratio and Geologic Age

5 GRAF and GOLDSMITH
Some Hydrothermal Syntheses of Dolomite and Protodolomite

6 DEGENS and EPSTEIN
Oxygen and Carbon Isotope Ratios in Coexisting Calcites and Dolomites from Recent and Ancient Sediments

7 FAIRBRIDGE
Excerpt from *The Dolomite Question*

8 INGERSON
Excerpt from *Problems of the Geochemistry of Sedimentary Carbonate Rocks*

The following eight papers outline and discuss some of the major problems of dolomites and dolomitization. As such, they are arranged in quasi-historical perspective in order to illustrate the evolution of ideas related to the subject of dolomites. The abridged paper by Cullis (Paper 1) departs slightly from this theme; it was included as a classic, early petrographic investigation into complex patterns of dolomite microfabrics and diagenesis in Recent carbonates from Funafuti Atoll. The extensive formation of dolomite that was encountered below 637 feet in the Funafuti borings was believed to have originated as passively precipitated cements and replacements of calcitic skeletal fragments, fibrous (possibly marine) and scalenohedral cements, and mud matrix. Cullis not only recognized an initial fabric specificity (Shinn, Paper 30) and systematic progression of replacement (dolomitization, sensu stricto) but also the fact that dolomitization may either preserve or obliterate the microfabric detail of affected host grains, the latter process having been termed *recrystallization*. Although he suspected that the different modes of dolomite occurrence related somehow to ambient fluids, he had no supporting data with which to infer possible geochemical controls on dolomitization. In this regard it is interesting to speculate that, as in similar Pacific atoll-type settings (Reuling, 1934; Crickmay, 1945; Schlanger, 1963; Bathurst, 1975), dolomitization at Funafuti may, in fact, have resulted from exposure to "schizoid" fluids in a myriad of diagenetic environments.

The next two articles by Van Tuyl (Paper 2) and Parsons (Paper 3) are from the era of field rather than laboratory investigations of dolomitization. In his introduction, Van Tuyl (see also Van Tuyl, 1916) traced the history of dolomite studies since the early nineteenth century, remarking that the general consensus among such researchers as Von Buch, de Beaumont, Dana, and Sorby was that dolomite was of inorganic, replacement origin. However, the dolomite "problem" is readily apparent in these early studies, as evidenced by the arguments that dolomite also may be a primary precipitate from seawater, a residuum of the weathering of limestones and/or related to the weathering of ferromagnesian rocks and minerals, or of detrital origin (see Lindholm, Paper 29). Based on field and petrographic studies, both Van Tuyl and Parsons favored a replacement origin of dolomite, citing such evidence as the selective dolomitization of limestones and resultant formation of pseudo-brecciated and mottled textures. Parsons also attempted to correlate dolomite crystal size, clarity (for example, zoning), and degree of euhedrality with secondary dolomitization, a task still in vogue with some modern researchers. Parsons's studies of the Carboniferous of Great Britain resulted in the first rigid classification of dolomites according to timing of dolomitization, and he concluded that the majority of dolomites in this region are of early replacement origin ("Secondary: contemporaneous"). Subsequent classification schemes have been proposed more recently by Fairbridge (Paper 7), Friedman and Sanders (1967), and Badiozamani (Paper 18). These workers agreed that primary dolomite may occur in nature, although they stressed its limited significance in the geologic record and cited experimental syntheses that failed to produce dolomite. Steidtmann (1911, 1917) made significant contributions to thought on dolomitization; his 1917 paper was reprinted in *Sedimentary Rocks: Concepts and History,* edited by A. V. Carozzi (1975).

Research on dolomites waned from the early 1900s until the resurgence of interest in the 1950s in geochemical aspects of dolomitization. In Paper 4, Chilingar commented on possible reasons for the relative abundance of ancient dolomites and the nonlinear increase in the Ca/Mg ratio of carbonate rocks from Precambrian to Recent time. This work followed the previous studies by Daly (1907, 1909), Vinogradov, Ronov, and Ratynskii (1952), and others. Chilingar also suggested, based on his purportedly successful attempts at synthesizing dolomite, that some Precambrian and Paleozoic dolomites may have been precipitated from seawaters at high partial pressures of CO_2 ($>$4 atm.), as was advanced by Strakhov (1953).

Graf and Goldsmith (Paper 5) discussed in their contribution some possible reasons for the apparent reluctance of stoichiometric dolomite to precipitate in nature and under laboratory conditions (see also Berner, 1966). Their experimental results indicated that the process of achieving an ordered arrangement of Ca^{2+} and Mg^{2+} ions is extremely slow in aqueous (and dry) environments of surface-normal temperatures, whereas it best occurs by solid-state diffusion and/or ionic exchange in wet environments at temperatures greater than 500°C and 200°C–300°C, respectively. Therefore, because of the extremely slow rates of dolomite crystallization (Peterson, Bien, and Berner, 1963; Peterson, von der Borch, and Bien, 1966) and decreased mobility of Mg^{2+} at surface temperatures, the apparently intermediate mineral protodolomite forms instead. However, Hsü (1967) questioned whether protodolomite (or dolomite) is indeed the stable Mg carbonate at surface temperatures and pressures instead of, perhaps, hydromagnesite ($Mg_4(OH)_2(CO_3)_3 \cdot 3H_2O$) or nesquehonite ($MgCO_3 \cdot 3H_2O$). Although there have been numerous attempts at dolomite synthesis under normal surface conditions (Rivière, 1939; Oppenheimer and Master, 1963; Chazen and Ehrlich, 1973; some earlier studies reviewed in Paper 2), the question of stability in the natural system is as yet unresolved. The studies of the Recent ordered dolomites and protodolomites of the hypersaline Coorong region (Alderman and Skinner, Paper 9; von der Borch, 1965) and near-normal salinity Baffin Bay, Texas (Behrens and Land, 1972), may provide some insight into this problem. Also of interest is the article by Folk and Land (Paper 19) who suggested ordered dolomite may be the favored mineral phase under conditions of reduced salinity and Mg/Ca ratios of approximately one to one.

Paper 6 by Degens and Epstein represents an attempt at relating dolomitization to the environment of its formation through an analysis of carbon and oxygen isotopes in coexisting calcite and dolomite. Their study of continental to marine, synsedimentary to late diagenetic dolomites was predicated on the assumption that, at equilibrium, coexisting calcites and dolomites precipitated in the same environment should have a difference in $^{18}O/^{16}O$ values of approximately 6 to 10 per mil at room temperature. Based on their analyses, which indicated a similarity in isotope values for these minerals, they concluded that dolomite is an alteration product of calcite and that replacement occurs as a result of solid diffusion. However, most workers do not subscribe to this mechanism of solid state diffusion for the replacement process. For example, Liebermann (1967) insisted on the aqueous formation of dolomites, with coincident exchange of oxygen and carbon among the precursor phase, solute, and precipitate. Füchtbauer (Paper 33) refers to evidence by German researchers that argues against solid state replacement. Recent studies by Land (1980), aptly summarizing the state of the art of interpretation of stable isotope data with regard to dolomitization, also refuted the conclusions of Degens and Epstein; Land argued convincingly for an isotopic fractionation between calcite and dolomite.

The final two abridged papers by Fairbridge (Paper 7) and Ingerson (Paper 8) are in-depth synopses of then-current concepts of dolomitization; additional recent work addressing this subject includes Cloud and Barnes (1948), Sonnenfeld (1964), Pray and Murray (1965), Friedman and Sanders (1967), Lovering (Paper 22), Dott (1972), Zenger (1972), Chilingar et al. (1979), and most recently, Zenger, Dunham, and Ethington (1980). In their still-timely articles, Fairbridge and Ingerson discussed and summarized the "dolomite question:"

I. Primary Dolomites
   A. Although dolomite is stable in normal seawater at 25°C, it has never been unequivocally demonstrated to precipitate in rock-forming accumulations from seawater, nor artificially under "normal" conditions. Are magnesite and/or nesquehonite instead the stable Mg carbonates in seawater? Under what geochemical conditions may dolomite or protodolomite be directly precipitated from seawater? Is shallow or deep water the favored site of dolomite formation?
   B. If primary dolomites are indeed prevalent in certain Precambrian and Paleozoic shallow-marine strata, under what geochemical conditions did they form? Why is no shallow-marine dolomite found in Holocene deposits (prior to 1957)? Are some such dolomites of bacterial origin?

II. Secondary (Replacement) Dolomites
   A. What is the timing of dolomitization; is it penecontemporaneous (syndepositional) or post-burial?
   B. Under what conditions of depth, temperature, and pressure does dolomitization readily commence; is there a preferred site of dolomitization? Is time (age of the rocks) a consideration?
   C. What is the source of dolomitizing solutions? What are the mechanics of replacement? Does initial porosity, permeability and mineralogy of rocks affect the process of dolomitization?

These and other questions remain to be solved by petrologists and geochemists, and the following parts will address some of these specific problems. However, it is safe to say that we remain ignorant about many aspects of the geochemistry of dolomites, especially regarding its stability and synthesis (Zenger, Dunham, and Ethington, 1980, p. 5). Both Hsü (1967, p. 188) and Bathurst (1975, p. 541) have noted that we can say relatively little about the control by pH, Eh, or P_{CO_2} on dolomitization.

REFERENCES

Bathurst, R. G. C., 1975, *Carbonate Sediments and Their Diagenesis,* 2d ed., Elsevier, Amsterdam, 658p.

Behrens, E. W., and L. S. Land, 1972, Subtidal Holocene Dolomite, Baffin Bay, Texas, *Jour. Sed. Petrology* **42**:155–161.

Berner, R. A., 1966, Chemical Diagenesis of Some Modern Carbonate Sediments, *Am. Jour. Sci.* **264**:1–36.

Chazen, P. O., and R. Ehrlich, 1973, Low-Temperature Synthesis of Dolomite from Aragonite, *Geol. Soc. America Bull.* **84**:3627–3634.

Chilingar, G. V., D. H. Zenger, H. J. Bissell, and K. H. Wolf, 1979, Dolomites and Dolomitization, in *Diagenesis in Sediments and Sedimentary Rocks,* ed. G. Larsen and G. V. Chilingar, Elsevier, Amsterdam, pp. 423–526.

Cloud, P. E., Jr., and V. E. Barnes, 1948, The Ellenberger Group of Central Texas, *Univ. Texas Pub. 4621,* Austin, Tx., 473p.

Crickmay, G. W., 1945, Petrography of Limestones, in *Geology of Lau, Fiji,* ed. H. S. Ladd and J. E. Hoffmeister, Bernice P. Bishop Mus. Bull. 181, pp. 211–250.

Daly, R. A., 1907, The Limeless Ocean of Pre-Cambrian Time, *Am. Jour. Sci.* **23**:93–115.

Daly, R. A., 1909, First Calcareous Fossils and the Evolution of Limestone, *Geol. Soc. America Bull.* **20**:153–170.

Dott, R. H., Sr., ed., 1972, Carbonate Rocks I: Classifications-Dolomite-Dolomitization, *Am. Assoc. Petroleum Geologists Reprint Ser. 4,* Tulsa, Ok., 237p.

Friedman, G. M., and J. E. Sanders, 1967, Origin and Occurrence of Dolostones, in *Carbonate Rocks: Origin, Occurrence, and Classification,* ed. G. V. Chilingar, H. J. Bissell, and R. W. Fairbridge, Elsevier, Amsterdam, pp. 267–348.

Hsü, K. J., 1967, Chemistry of Dolomite Formation, in *Carbonate Rocks: Origin, Occurrence, and Classification,* ed. G. V. Chilingar, H. J. Bissell, and R. W. Fairbridge, Elsevier, Amsterdam, pp. 169–191.

Land, L. S., 1980, The Isotopic and Trace Element Geochemistry of Dolomite: The State of the Art, in *Concepts and Models of Dolomitization,* ed. D. H. Zenger, J. B. Dunham, and R. L. Ethington, Soc. Econ. Paleontologists and Mineralogists Spec. Pub. 28, Tulsa, Ok., pp. 87–110.

Liebermann, O., 1967, Synthesis of Dolomite, *Nature* **213**:241–245.

Oppenheimer, C. H., and I. M. Master, 1963, Transition of Silicate and Carbonate Crystal Structure by Photosynthesis and Metabolism (Abstract), *Geol. Soc. America Spec. Paper 76,* p. 125.

Peterson, M. N. A., G. S. Bien, and R. A. Berner, 1963, Radiocarbon Studies of Recent Dolomite from Deep Spring Lake, California, *Jour. Geophys. Research* **68**:6493–6505.

Peterson, M. N. A., C. C. von der Borch, and G. S. Bien, 1966, Growth of Dolomite Crystals, *Am. Jour. Sci.* **264**:257–272.

Pray, L. C., and R. C. Murray, eds., 1965, Dolomitization and Limestone Diagenesis, *Soc. Econ. Paleontologists and Mineralogists Spec. Paper 13,* Tulsa, Ok., 180p.

Reuling, H. T., 1934, Der Sitz der Dolomitisiering. Versuch einer Neuen Auswertung der Bohr-Ergebnisse von Funafuti, *Senckenberg. Naturforsch. Ges. Abh.* **428**:1–44.

Rivière, A., 1939, Observations Nouvelles sur la Mécanisme de Dolomitisation des Sediments Calcaires, *Acad. Sci. Comptes Rendus* **209**:691–692.

Schlanger, S. O., 1963, Subsurface Geology of Eniwetok Atoll, *U.S. Geol. Survey Prof. Paper 260 BB,* pp. 991–1066.

Sonnenfeld, P., 1964, Dolomites and Dolomitization: A Review, *Bull. Canadian Petroleum Geology* **12**:101–132.

Steidtmann, E., 1911, Evolution of Limestone and Dolomite, *Jour. Geology* **19**:323–345, 392–428.

Steidtmann, E., 1917, Origin of Dolomite as Disclosed by Stains and Other Methods, in *Sedimentary Rocks: Concepts and History,* ed. A. V. Carozzi, Benchmark Papers in Geology, vol. 15, Dowden, Hutchinson & Ross, Stroudsburg, Pa., pp. 359–385.

Strakhov, N. M., 1953, Diagenesis of Sediments and Its Significance for Sedimentary Ore Formation, *Akad. Nauk SSSR Izv. Ser. Geol.* **5**:12–49.

Van Tuyl, F. M., 1916, The Origin of Dolomite, *Iowa Geol. Survey Ann. Report 1914* **25**:251–422.

Vinogradov, A. P., A. B. Ronov, and V. M. Ratynskii, 1952, Variation in Chemical Composition of Carbonate Rocks of the Russian Platform, *Akad. Nauk SSSR Izv. Ser. Geol.* **1**:33–50. (Abstract trans. by G. V. Chilingar, *Geochim. et Cosmochim. Acta* **12**:273–276, 1957.)

von der Borch, C. C., 1965, The Distribution and Preliminary Geochemistry of Modern Carbonate Sediments of the Coorong Area, South Australia, *Geochim. et Cosmochim. Acta* **29**:781–799.

Zenger, D. H., 1972, Dolomitization and Uniformitarianism, *Jour. Geol. Education* **20**:107–124.

Zenger, D. H., J. B. Dunham, and R. L. Ethington, eds., 1980, Concepts and Models of Dolomitization, *Soc. Econ. Paleontologists and Mineralogists Spec. Pub. 28,* Tulsa, Ok., 320p.

THE MINERALOGICAL CHANGES OBSERVED IN THE CORES OF THE FUNAFUTI BORINGS.

By C. GILBERT CULLIS, *D.Sc., F.G.S.*

CONTENTS.

		Page
1.	Introduction	392
2.	The Mineralogical Characters of the Rocks occurring between the Surface and 637 feet.	393
	a. The Deposition of Secondary Calcite and Aragonite from Solution	394
	b. The Crystallisation of the fine Calcareous Detritus	398
	c. The Disappearance of Aragonite.	400
3.	The Mineralogical Characters of the Rocks from 637 feet down to 1114 feet	404
	a. The Replacement of Calcite and Aragonite by Dolomite, with the Formation of Casts.	405
	b. The Occurrence of Fibrous Deposits in the Cavities	409
	c. The Characters of the Lowest Rocks reached in the Borings	414

1. INTRODUCTION.

THE minerals which occur as the recognisable constituents of the Funafuti cores are calcite, aragonite and dolomite.

The small amount of calcium phosphate, which analysis shows to be present in all parts of the boring, is probably included within these, as an invisible impurity; it has not been detected as a distinct mineral. The possibility of the occurrence of magnesite, in certain of the rocks, has not been overlooked, but no evidence that magnesium carbonate exists in them, otherwise than in combination with calcium carbonate, has presented itself.

The discrimination between calcite, aragonite and dolomite, in thin sections of the rocks of modern coral reefs, which consist so largely of fresh calcareous organisms, and in which such crystals as occur are generally of microscopic dimensions, is not always a simple matter; it is most readily effected by micro-chemical methods. Of these there are two which are of special value, and which have been used with the most satisfactory results in the present investigation. The first is that of MEIGEN,*

* 'Centralblatt für Mineralogie,' 1901, pp. 577, 578.

by means of which aragonite, if present in a section, is stained, while calcite and dolomite remain unstained.   The second is that of LEMBERG,* by which calcite and aragonite are stained, while dolomite remains unstained.   The former method serves for the differentiation of aragonite from calcite and dolomite ; the latter for that of dolomite from calcite and aragonite (see Plate F).

The value of these two tests is increased by the fact that they provide a simple and trustworthy means of determining the mineralogical nature, not only of the well-crystallised inorganic portions of the rocks, but also of the original calcareous organisms. Concerning the organisms of the Funafuti reef—as exemplified in the rocks of the boring—the tests have given the following results.   Those composed of aragonite, are the calcareous alga *Halimeda*, the Madreporarian corals, the alcyonarian genus *Heliopora*, the hydrocorallines, the majority of the mollusca, and the minute spicules of tunicates.   The rest are of calcite, and comprise the calcareous algæ *Lithothamnion* and *Lithophyllum*, the foraminifera, the polyzoa, the echinodermata, and the spicules of alcyonaria.   These results, so far as they go, are in agreement with those obtained in the first instance by Dr. H. C. SORBY,† and by numerous investigators later.

The calcite and aragonite of which the various organisms are composed may be spoken of as *primary* calcite and aragonite ; that which makes its appearances in the rocks after their accumulation will be referred to as *secondary* calcite and aragonite. No such distinction is necessary in the case of the dolomite, since it is always of secondary origin.

The distribution of these three minerals in the boring is noteworthy.   Aragonite is confined to the upper cores, and dolomite to the lower, while calcite, which is the sole constituent of the middle cores, is also found above and below, in association with aragonite on the one hand and dolomite on the other.   Dolomite is not encountered in a clearly individualised form, until a depth of between 637 and 638 feet has been reached ; below this it continues, in varying amounts, to the bottom of the boring. It may be merely a coincidence, or it may be a fact of special significance, that in none of the cores has aragonite been found in association with dolomite.

[*Editors' Note:* Material has been omitted at this point.]

(3) THE MINERALOGICAL CHARACTERS OF THE ROCKS FROM 637 FEET DOWN TO 1114 FEET.

The rocks included between these limits are, for the most part, dolomites containing some 40 per cent. of magnesium carbonate, and composed of the one mineral dolomite

---

* ' Zeitsch. Deutsch. Geol. Gesellsch.,' 1888, vol. 40, p. 357.

† Presidential Address, ' Quart. Journ. Geol. Soc.,' 1879, vol. 35, Proc. pp. 58—66.

to the complete, or almost complete exclusion of recognisable calcite.   At three horizons, however, calcite is present in considerable proportions, causing the percentage of magnesium carbonate to fall below this maximum, and rendering it possible to divide the rocks into an alternating series of dolomites of fairly constant composition and dolomitic limestones* of variable composition.

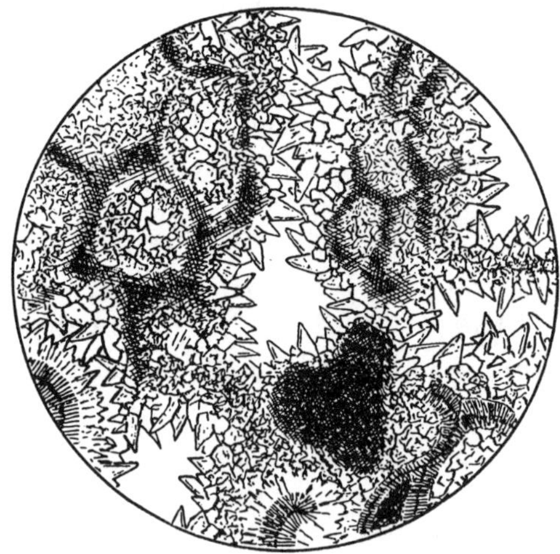

Fig. 39.—Main Boring. Core 313. Depth 637 feet.
× 200.

Section of the core immediately overlying that in which dolomite crystals first make their appearance. The rock is very porous. The cavities are lined with well-formed scalenohedral crystals of secondary calcite, which are often deposited in crystallographic continuity with the structural elements of the organisms. In the more solid parts of the mass the secondary calcite is granular in character.

Fig. 40.—Main Boring. Core 314. Depth 638 feet.
× 250.

Section of the highest core in which recognisable dolomite occurs. The organisms and their investment of acute crystals are of calcite. The rest of the mass consists of crystalline dolomite.

(a) *The Replacement of Calcite and Aragonite by Dolomite, with the Formation of Casts.*

The first core in which crystallised dolomite is seen comes from the depth of 638 feet, and contains 20·44 per cent. of magnesium carbonate. This, if assumed to be all combined with calcium carbonate to form the double carbonate of calcium and magnesium, indicates that nearly one-half of the rock here is composed of dolomite. The core above, from 637 feet, shows no dolomite, and contains only 2·44 per cent. of

* The term dolomitic limestone is here employed to denote a rock consisting of both calcite and dolomite, as distinguished from the dolomites which consist of the mineral dolomite only, though not necessarily of pure dolomite.

magnesium carbonate. The rise in the proportion of this constituent is therefore very sudden; but it is probably not quite so abrupt as these facts would imply, for number of small fragments, the precise position of which, in the boring, is uncertain but which are known to belong hereabouts, and most likely come from between the two cores in question, have given intermediate values. Sections of these fragments show some to consist of calcite only, others of calcite and dolomite. One of the former kind was analysed by Dr. SKEATS, and gave 3·82 per cent. of magnesium carbonate, while of three of the latter kind, two gave 11·67 per cent. and 16·21 per cent. respectively, and the third 29·72 per cent.

The cores following below that in which dolomite is first seen show a gradual increase in the proportion of magnesium carbonate, until a maximum of 40 per cent. or thereabouts is reached, at a depth of 650 feet.

When sections cut from a number of these cores are examined in descending order, this increase is seen to be represented by the gradual growth and extension of the areas composed of dolomite at the expense of those of calcite. So long as the proportion of magnesium carbonate falls short of its maximum, areas of unaffected calcite may still be discovered, but, as the maximum is approached, such areas become smaller and less numerous, and by the time it has been actually attained they have all disappeared, calcite having everywhere given place to dolomite. It is interesting to trace the various stages by which this progressive dolomitisation is effected. Sections taken from the depth of 637 feet consist entirely of calcite, partly of primary, partly of secondary origin. The rock here is very porous, and the interstices are lined with acute scalenohedral crystals of calcite, which are often disposed in continuity with the structural elements of the organisms by which the cavities are lined (fig. 39). In sections taken only a foot lower down dolomite is seen to be present as well as the primary and secondary calcite. The rock is less porous than that immediately above, the cavities being now in part occupied by crystals of dolomite which rest upon and enclose the projecting apices of the scalenohedra of secondary calcite (fig. 40; Plate F, fig. 3). If these dolomite crystals could be removed, the remainder would be essentially identical with the rock at 637 feet. It would appear, therefore, that the dolomite is, in part at least, an addition to the mass by direct deposition, and not always a substitution product for pre-existing calcite. As sections from successively deeper cores are examined, the lining of scalenohedral calcite, which intervenes between the unaltered organisms and the subsequently deposited dolomite, is seen to disappear by conversion into dolomite, which latter, by a process of inward invasion, ultimately comes into direct contact with the organisms themselves (Plate F, fig. 3), and the sites of aragonite organisms, which in the cores above are composed of secondary calcite, are also rapidly usurped by dolomite. At this stage the rock consists only of primary calcite and of dolomite, all secondary calcite having been changed into dolomite. It appears, therefore, that the secondary calcite of these rocks is more readily converted into dolomite than the primary calcite.

This fact is of interest, as an analogous difference in permanence may be observed in those parts of the boring where aragonite is altering to calcite ; it is the secondary aragonite which first undergoes the paramorphic change, the substance of the corals and other aragonite organisms preserving its integrity after the secondary aragonite of the cavities has been converted into calcite.

As still deeper cores are studied, the process of invasion by which the secondary calcite has been converted into dolomite is seen to attack the primary calcite of the organisms, with the ultimate result that they also become completely dolomitised. This change may or may not involve their re-crystallisation.   When it does, the

Fig. 41.—Main Boring. Core 342. Depth 660 feet. × 10.

Showing the characters of the "soft dolomite," as seen with low powers of the microscope.   In many cases the solid substance of the organisms has been entirely dissolved out, only the cells and tubules, filled with dolomitised "mud," remaining.   In others a certain proportion of the organic substance has withstood the solvent action, but has been converted into dolomite.

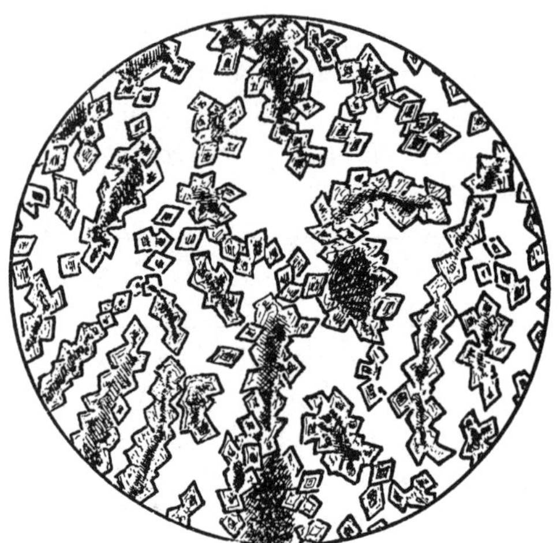

Fig. 42.—Main Boring. Core 366. Depth 698 feet. × 200.

A well-crystallised portion of the "soft dolomite," highly magnified. The mass is mainly composed of groups and strings of minute rhombohedra of dolomite, which define in a general way the outlines and major structural features of the organisms which were the original constituents of the rock. Here and there, in the midst of the larger groups of crystals, fragments of uneffaced organisms may still be recognised.

result is occasionally the almost entire obliteration of them, but much more often it is the destruction of their minor features of texture and structure only (fig. 43), their major features of structure·and form continuing to be more or less evident.   When, on the other hand, dolomitisation does not involve recrystallisation of the organisms, as is occasionally the case, their detailed structure may be preserved, and sometimes in a remarkably perfect manner.

The increase of dolomite, at the expense of calcite, continues until none of the latter mineral remains.   At the depth of 650 feet, or thereabouts, practically all calcite has disappeared, and thence, to the depth of 820 feet, with the rare exception of an incompletely converted organism, nothing but dolomite can be identified in the cores. Throughout the whole of this 170 feet the composition of the rocks is remarkably uniform ; the proportion of magnesium carbonate is about 40 per cent., sometimes a little less, occasionally a little more, but in no part of the boring, either here or elsewhere, does it exceed 43 per cent.

The proportion of magnesium carbonate in pure dolomite is 45·65 per cent. If 40 per cent. be taken as the average amount of magnesium carbonate in these most highly dolomitic rocks of the boring, and that be considered to be combined with calcium carbonate in the proportions represented by the formula $CaCO_3.MgCO_3$, then they must contain an excess of uncombined and uncrystallised calcium carbonate amounting to 13 per cent. or so. This may be included in the dolomite as an invisible impurity, but possibly it is this material which, in the condition of minute particles, renders a large proportion of the dolomite crystals more or less opaque, and often causes that zoned appearance which is so characteristic of dolomite in sedimentary rocks (figs. 42, 48).

The dolomites, which for 170 feet in this part of the boring maintain this peculiar constancy in chemical and mineralogical composition, are by no means uniform in their lithological character. Normally, the dolomites of the boring are dense and compact rocks, and a certain proportion of those of this 170 feet are of this nature, but for the 80 feet included between the depths of 660 feet and 740 feet they are of a totally different character, being so soft and friable as to crumble readily under the finger nail, and so porous as to feel, in the hand, scarcely heavier than pumice. The reason of these peculiar characters becomes evident under the microscope. When the rocks below 650 feet are examined in descending order, it is observed that many of the organisms show signs of corrosion by solution. The number of the organisms so affected increases with the depth, and also the proportionate part of each organism removed by solution ; and at the depth of 660 feet the organisms present the appearance shown in fig. 41. Solution has here removed much of their original substance, and especially certain parts which seem to be more readily dissolved than the rest, with the result that they are now little more than sponge-like skeletons of what they were originally. The undissolved parts are composed of dolomite, and each organism is invested by a very thin layer of tiny dolomite crystals. These are often very perfect in form, appearing in sections as minute isolated or nearly isolated rhombohedra. The majority of them contain a central zone or nucleus of the same form as the crystal itself, composed of a white opaque or semi-opaque material (fig. 42). The outer part of the crystal, which is transparent, is probably composed of pure or nearly pure dolomite. The central kernel, however, probably owes its opacity to the presence of calcareous impurity, which represents the excess of calcium carbonate which the rocks contain over and above that present in pure dolomite. As the middle of this 80 feet of soft dolomite is approached, signs of solution become more evident, and of such material as remains a larger proportion consists of the tiny zoned rhombohedra, and a smaller proportion of recognisable organisms (fig. 42); but with the exception of these variations in degree of solution and crystallisation, the rocks remain of the same essential nature through the whole of the interval between 660 and 740 feet. At about the latter level the cores begin to get heavier and harder once more, and within a very few feet they assume a dense and solid character. This change is denoted in thin sections by the gradual, though rapid disappearance of those evidences of solution which are so abundant in the rocks above.

Between the base of this soft dolomite and the point, another 80 feet lower down, at which the proportion of magnesium carbonate begins once more to lessen, the cores remain of the hard, dense and compact type (fig. 43), and such sections as have been cut from them reveal no important recurrence of those signs of excessive solution which are so distinctive of the soft dolomite above ; but the rocks exhibit all the various types of dolomitisation of the organisms which have already been described.

It may be mentioned here, that the phenomenon of reversal, already described on p. 403, is very characteristic of these lower parts of the boring. It is particularly well displayed by the corals, which are often preserved in the form of casts, the solid substance of the coral having been removed in solution, while the " mud " of the cavities has remained undissolved. The coral substance has not always been dissolved out, however ; sometimes it has been converted

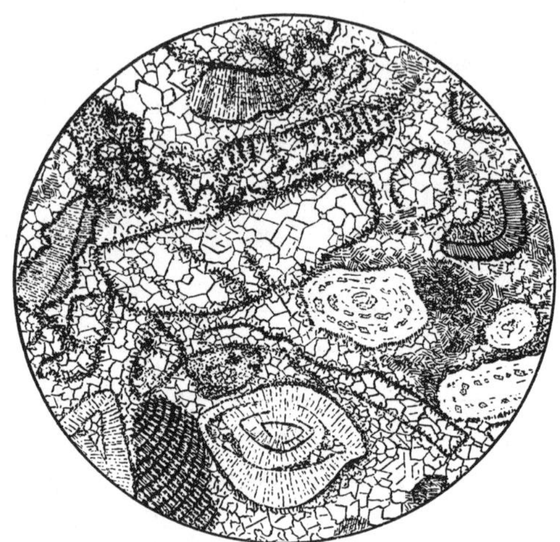

Fig. 43.—Main Boring. Core 51A. Depth about 760 feet. × 30.

Dolomitised " coral-reef sand." The calcite organisms, though dolomitised, show little sign of crystallisation ; the aragonite organisms, on the contrary, are almost completely recrystallised, and have consequently lost all their more delicate original characters. The granular cement probably represents, in large part, crystallised " mud," comparatively unaltered patches of which are seen here and there in the section. The dark line defining the original boundaries of the recrystallised organisms is well seen.

into secondary calcite, and this in its turn, at the horizons of complete dolomitisation, has been converted into dolomite. In such cases the corals are not preserved in the form of casts, but have simply undergone one or more changes of mineralisation. The mud of the cavities is usually dolomitised, but occasionally it still consists of calcite.

b. The Occurrence of Fibrous Deposits in the Cavities.

At the depth of about 815 feet the fibrous encrusting deposit which has already been stated to occur in the lower parts of the boring makes its appearance.

It is at first very thin, and constitutes only a minute proportion of the mass ; but traced downwards it increases rapidly in thickness and soon becomes an abundant and conspicuous element in the rocks (Plate F, figs. 4, 5, 6). With occasional interruptions and many variations in its detailed constitution, it continues to the bottom of the boring.

Mineralogically it is at first composed entirely of calcite. At greater depths it sometimes consists of calcite throughout (Plate F, figs. 4, 5), sometimes of regularly alternating layers of calcite and dolomite— the dolomite layers being thin in comparison with those of calcite, and destitute of the fibrous structure (figs. 44, 48 ; Plate F, fig. 6) — and sometimes of dolomite throughout (figs. 45, 49). Occasionally it may be uniformly clear and transparent, but usually it exhibits a very marked concentric structure produced by bands— from two or three to as many as fourteen or fifteen in number —of different degrees of transparency, which impart to it a markedly agate-like appearance. Associated with it, in the cavities in which it occurs, are numerous floors of "mud" (fig. 46), and there is repeated evidence that the darker layers of the deposit owe their comparative opacity to the inclusion of this finely divided silt within their substance. The deposit appears to have been formed in waters which were alternately clear and turbid with suspended matter, which settling slowly down accumulated in the irregularities of the cavity floors. Its occurrence as the innermost lining of the cavities proves it to be the latest addition to the rocks, by deposition from

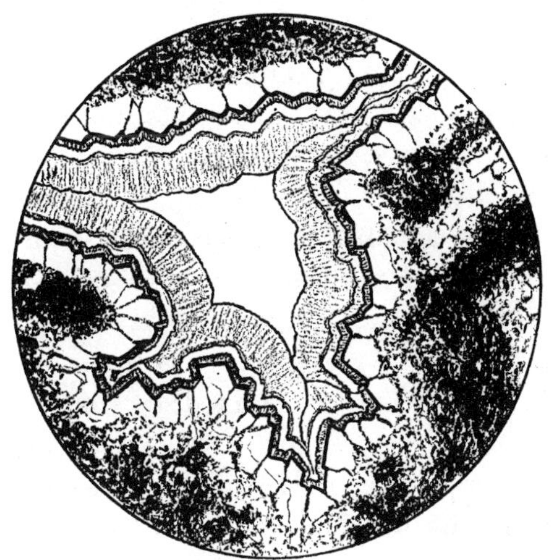

Fig. 44.—Main Boring.   Core 205A.   Depth 830 feet.   × 100.

Cavity, in dolomitised rock, containing banded encrusting deposit consisting of three layers, the first and third of which are of calcite, the second of dolomite. The dolomite layer may represent part of the original deposit converted into dolomite, or, as seems more probable, it may have been deposited directly as such. The encrusting material, the introduction of which was apparently subsequent to the dolomitisation of the rock, is separated from the boundaries of the original organisms by a layer of well-formed water-clear dolomite crystals (see also fig. 48 and Plate F, fig. 6).

solution ; its introduction, moreover, did not take place until after solvent action had affected many of the corals and other aragonite organisms, for it is frequently found lining or filling spaces which were formerly occupied by these.   In this respect it differs from the similar fibrous deposits seen in the first few feet of the boring, where the organisms of aragonite are as fresh and solid as those of calcite (Plate F, figs. 1 and 2).   Moreover, there can be little doubt that, in many cases, the rocks in which it occurs were at least partially, if not completely, dolomitised before its deposition in their cavities.

Below 820 feet and above 875 feet the rocks are dolomitic limestones containing a variable amount of recognisable calcite.   At the upper of these two limits they contain 39 per cent. of magnesium carbonate, and the proportion of calcite is very small indeed ; at the lower limit they contain 40·25 per cent. of magnesium carbonate, and there is no calcite at all, but as a descent is made from the upper limit, or an ascent from the lower, the relative amount of calcite rapidly increases, and for a considerable distance above and below a point situate mid-way between them, calcite constitutes not less than half the mass.   Such analyses as have been made from the cores of this part of the boring show a maximum of calcite at 826 feet, where the proportion of magnesium carbonate is less than 5 per cent. (4·83).   If it be assumed that this amount of magnesium carbonate is combined with calcium carbonate to form dolomite, and that all the rest of the calcium carbonate exists as visible calcite, then approximately nine-tenths of the mass must be composed of calcite and only one-tenth of dolomite, a conclusion which is quite in accord with what is observed in stained sections.

Starting from this point, at which so large a proportion of the mass consists of calcite, and working both upwards and downwards towards the condition of complete dolomitisation which exists above and below, it is interesting to note just the same series of changes as is to be observed in passing from the limestones at 637 feet into the dolomites below.   In the rocks which contain the smallest proportion of dolomite, and which, except for the presence of the fibrous encrusting deposit, may be compared with those at 638 feet, the calcite of the mass comprises the primary calcite of the unaltered organisms, the secondary calcite of the organisms that were originally composed of aragonite, and the secondary calcite deposited from solution, which last is of two kinds—an earlier-formed deposit of acute scalenohedral crystals investing the organisms, and the later-formed fibrous encrusting deposit.   The dolomite of the mass consists in the main of transparent crystals—there is also some dolomitised "mud"—which cover up and enclose the scalenohedra of calcite or invest the calcitised aragonite organisms (Plate F, fig. 4) in such a manner as to leave little doubt that they are the direct product of deposition from solution.   They are themselves covered up by the fibrous encrusting deposit of calcite.

A little higher up and a little lower down much of the secondary calcite gives place to dolomite, but the primary calcite of the organisms still remains unchanged, and so, in great measure, does the fibrous encrusting deposit.

Finally, as the distance from the starting point is gradually increased, and the horizons of complete dolomitisation, above and below, are more and more nearly approached, even the calcite organisms are attacked and pass over into dolomite ; the fibrous deposit too, when followed downwards, is seen to be dolomitised ; traced upwards it gradually diminishes in amount and finally disappears, so that at 820 feet on the one hand and 875 feet on the other, practically all calcite has disappeared.

Although the stages of this progressive dolomitisation, when studied in detail, are neither quite so simple nor quite so regular as here described, owing to local disturbing influences, it is believed that the general sequence of the changes here indicated is substantially correct.

From 875 to 1050 feet the rocks are again composed almost exclusively of dolomite,

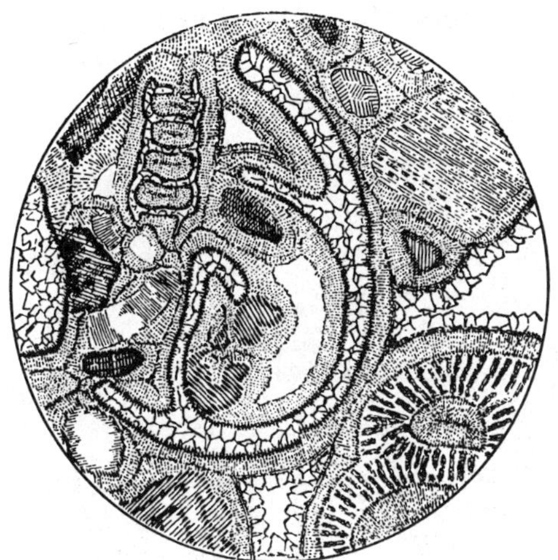

Fig. 45.—Main Boring.   Core 389A.   Depth 926–936 feet.   × 30.

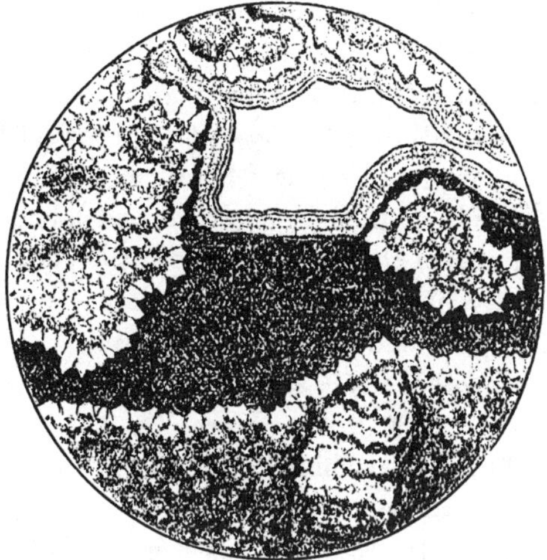

Fig. 46.—Main Boring.   Core 503A.   Depth 987–991 feet.   × 100.

Dolomitised " coral-reef sand " cemented by fibrous encrusting material. The fibrous deposit is in places coated with well defined water-clear crystals of dolomite. The rock, except for changes which have accompanied dolomitisation, is identical in character with that represented in fig. 27.

Dolomitised coral, with an irregular cavity containing dense homogeneous detritus and fibrous encrusting calcite. The fibrous deposit is the only calcite in the section. Neither the " mud " nor the fibrous material comes into direct contact with the coral. Each is separated from it by a layer of clear dolomite crystals.

visible calcite in the form of undolomitised structures being of rare occurrence. They contain, however, the same excess of uncombined calcium carbonate as the rocks occurring between 650 and 820 feet, the proportion of magnesium carbonate being approximately 40 per cent.

The most variable constituent in them is the fibrous encrusting deposit. This is sometimes entirely absent, sometimes scarce, and sometimes abundant ; it displays no constancy in its mineralogical composition, consisting most commonly of dolomite, but often of alternating layers of dolomite and calcite. Finely divided dolomitised detritus is common, and forms conspicuous floors to the cavities (fig. 46). At several horizons the rocks show signs, not only under the microscope, but also with the naked eye, of that solution which has produced such striking results in the soft dolomite occurring higher up in the boring (fig. 45).   In no case, however, are the effects of solution here so marked as at that horizon, nor have the rocks affected so great a vertical extent.   That solvent action did not come into operation until after the deposition of the fibrous encrusting material is proved by the fact that just as parts of the organisms have been dissolved away, so, occasionally, have the more soluble layers of this material itself been removed by solution—layers which are now represented by a series of roughly concentric or parallel vacuities.

The other characters of the dolomites at this depth, the state of preservation of the various organisms, their modes of mineralisation and so forth, are in all essentials identical with those of the rocks of the same composition which occur higher up in the boring.   Indeed, the lower dolomites differ from those above only in the presence of the fibrous encrusting material and the greater abundance of detrital mud.

Between 1050 and 1070 feet dolomitic limestones occur once more.   At the former depth, calcite, which in the dolomites immediately above is of very rare occurrence and constitutes, at most, but an insignificant proportion of the mass, begins to increase in quantity.   This increase continues until the depth of 1060 feet, or thereabouts, is reached, from which point a corresponding decrease ensues, with the result that, finally, at or about 1070 feet, dolomite is once more established in overwhelming preponderance, and so continues with little variation to the bottom of the boring.

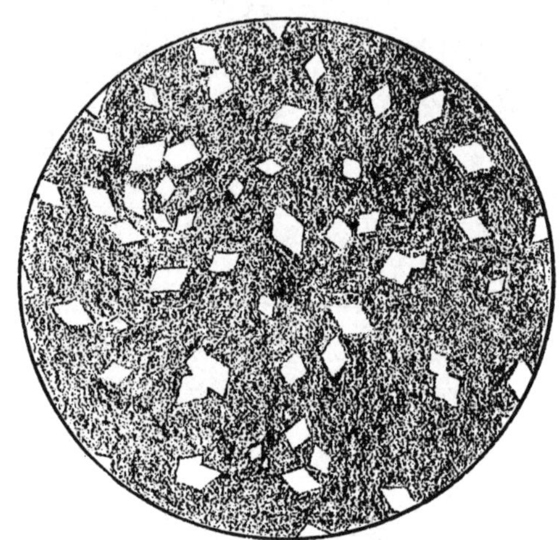

Fig. 47.—Main Boring.   Core 618A.   Depth 1060–1066 feet.   × 100.

Dolomitised " mud " containing well-formed dolomite crystals.

Starting from the level (1060 feet) at which the proportion of calcite is greatest—rather more than two-fifths of the mass—and working both upwards and downwards, the changes which present themselves are the same in kind and order as were observed in the dolomitic limestones occurring between the depths of 820 and 875 feet.   At first the calcite comprises not only the majority of the organisms which are primarily composed of it, but also a certain proportion of unaltered secondary calcite, whether it be that which in the form of scalenohedral crystals invests the calcite organisms, or that which represents the products of the paramorphism of the corals and other aragonite organisms, or that which as the fibrous deposit lines or fills the cavities of the mass.   The dolomite consists of crystals which may have been in part deposited directly as such, or may represent secondary calcite dolomitised ; a portion of the fibrous encrusting material also is generally composed of dolomite ; and when finely divided detritus occurs, it, too, is dolomitic almost without exception.   Higher up and lower down the secondary calcite more or less completely disappears, being gradually replaced by dolomite ; and, lastly, the calcite organisms are affected ; some alter earlier than others, but practically all have suffered conversion into dolomite when the upper limit of 1050 feet,

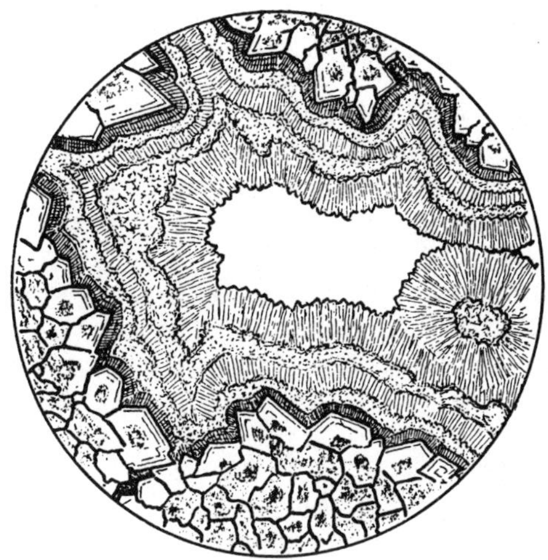

Fig. 48.—Main Boring.   Core 671A.   Depth
1090 feet.   × 200.

Fig. 49.—Main Boring.   Core 709A.   Depth
1114 feet.   × 35.

Partially dolomitic encrusting deposit occupying a cavity in dolomitised coral. The recrystallised coral substance is separated from the fibrous material by a layer of comparatively clear dolomite crystals. Of the five layers of the encrusting deposit the first, third, and fifth are of calcite, the second and fourth of dolomite.

Section of the deepest core. The organisms are bound together by a cement of three layers. The first layer is fibrous and somewhat opaque with included " mud " ; the second layer is a little more definitely crystallised and is less opaque with included matter ; the third layer is well crystallised and water-clear. Organisms and cement alike are composed of dolomite.

or the lower limit of 1070 feet, is reached.   None of the rocks included in this 20 feet show signs of solution.

c. The Characters of the Lowest Rocks reached in the Borings.

From 1070 feet to the bottom of the boring (1114 feet) the rocks consist once more of dolomite, to the complete or almost complete exclusion of visible and recognisable calcite.   What little calcite they do occasionally contain consists, for the most part, of undolomitised or incompletely dolomitised calcite organisms ; occasionally also the

fibrous encrusting deposit is partly composed of calcite, but by far the greater proportion of this material is of dolomite throughout.   From about 1100 feet downwards
the rocks show somewhat marked signs of solution.

The lowest core of the boring is composed of consolidated dolomitised " coral-reef
sand." It is made up of various small organisms, or fragments of organisms, which
are bound together by well-marked fibrous encrusting material, which more than
usually resembles that occurring near the top of the boring, in that it comes into
actual contact with the original bodies of the rock, and is not separated from them by
an intervening layer of dolomite crystals.

[*Editors' Note:* Material has been omitted at this point.]

DESCRIPTION OF PLATE F.

STAINED SECTIONS OF THE FUNAFUTI ROCKS.

(*Note.*—In figs. 1 and 2 the mineral which has taken the stain is aragonite, that which has remained unstained is calcite. In figs. 3, 4, 5 and 6 the coloured mineral is calcite, the uncoloured, dolomite.)

Fig. 1.—SOLLAS Boring No. 2.   Core D. 9.   Depth nearly 30 feet.   × 100.

To the right is a mass of coral with encrusting organisms.   To the left and below is a part of a *Halimeda* frond ; separating this from the encrusted coral is finely-divided calcareous detritus ("primary mud ").   Above is a large cavity lined with secondary calcite.   The cavities inside the coral and *Halimeda* frond are nearly all filled with secondary aragonite ; a few contain secondary calcite, which together with that in the cavity outside the organisms is of the fibrous encrusting type.

Fig. 2.—SOLLAS Boring No. 1.   Core C. 20.   Depth between 50 and 80 feet.   × 100.
"Coral-reef sand" with constituent fragments cemented together by secondary calcite of the fibrous encrusting type.

Fig. 3.—Main Boring.   Core 314 (lowest part).   Depth 638 feet.   × 200.

Partially dolomitised limestone.   In some parts the organisms are seen to be coated with acute crystals of secondary calcite, which either project into empty spaces or are embedded in granular dolomite ; in others these crystals have been dolomitised ; and in still others the organisms themselves have suffered the change, with a resulting partial or complete obliteration.   Some of the dolomite has possibly been deposited from solution, but the greater part is probably a substitute for pre-existing structures of calcite or aragonite.

Fig. 4.—Main Boring.   Core 225 A.   Depth 842 feet.   × 200.

The organisms are coated with secondary calcite in acute crystals.   These project into a layer of clear dolomite, the rhombohedral crystals of which are covered in turn by secondary calcite of the fibrous encrusting type.

Fig. 5.—Main Boring.   Core 237 A.   Depth 850 feet.   × 100.

The untinted part of the section represents a portion of a large mass of dolomitised coral, the cavities in and around which are lined with a thick deposit of fibrous encrusting calcite.

Fig. 6.—Main Boring.   Core 270 A.   Depth 868 feet.   × 200 (slightly diagrammatic).

The layer of clear and well-defined dolomite crystals rests upon and encloses a mass of coral, which has been converted into calcite.   This layer in turn is covered up by a thick deposit having a radiating structure, and consisting of bands of calcite and films of dolomite in regular alternation.

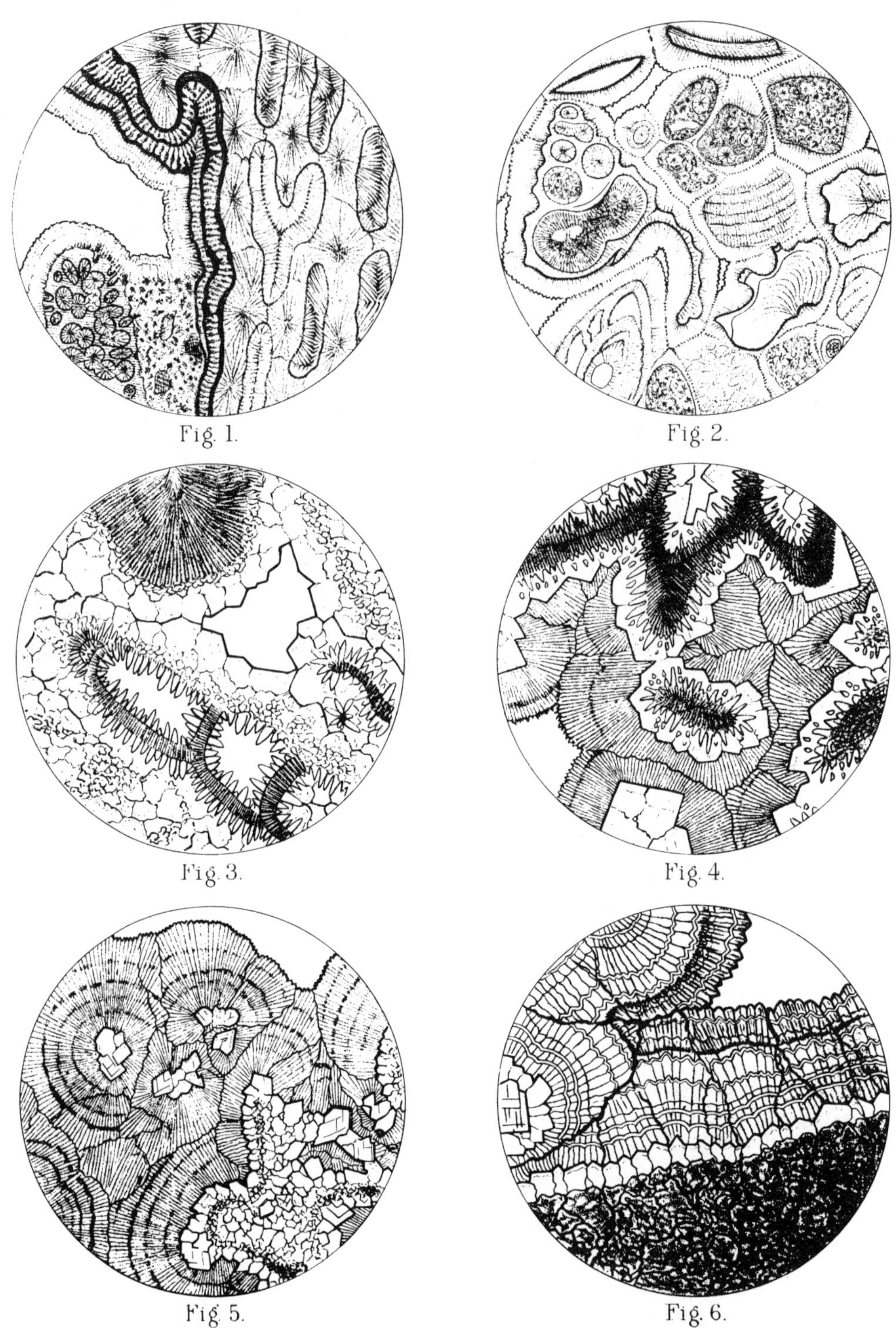

Fig. 1.

Fig. 2.

Fig. 3.

Fig. 4.

Fig. 5.

Fig. 6.

[*Editors' Note:* The original of Plate F is in color. In Figures 3 through 6 of this reproduction, dolomite crystals appear as unshaded, or white, in contrast to the darker calcite.]

2

*New Points on the Origin of Dolomite :** by
Francis M. VanTuyl.

Historical Review.

The problem of the origin of dolomite has long occupied the attention of geologists and many theories have been advanced for its formation, but no one of these theories has been widely accepted. Von Buch (1)† was the first to seriously attempt to explain the formation of the rock. As early as 1822 in his writings on the dolomite of the Tyrol, he ascribed its origin to the action of volcanic vapors, rich in magnesia, on limestone, and there was some basis for this belief, for the rocks are there penetrated by augite-porphyry. Frapolli (2) and Durocher (3) later expressed similar views upon the origin of the rock, and Favre (4), basing his supposition upon the conditions of the experimental production of dolomite by Marignac, concluded that the dolomite of the Tyrol was formed by the alteration of limestone beneath the sea at a temperature of 200° C. and at a pressure of 15 atmospheres, corresponding to a depth of 150 to 200 meters, by magnesium compounds furnished by the action of sulphurous and hydrochloric acids of volcanic origin on the lava of submarine melaphyr eruptions.

In 1834 Collegno (5) pointed out the frequent association of gypsum and dolomite in the St. Gothard region and regarded them both as transformation products resulting from the action of magnesium sulphate in surface waters or limestone. Morlot (6) also favored such a theory of origin.

As early as 1836 Beaumont (7) ascribed the origin of dolomite to the alteration of limestone by circulating solutions of magnesium bicarbonate and, assuming that the replacement was molecular, he calculated that the change should be accompanied by a decrease in volume of the original rock to the extent of about 12·1 per cent. Actual porosity determinations by Morlot (8) on a dolomite sample from the Alps later seemed to confirm this prediction.

In 1843, A. W. Jackson (9) suggested that ascending spring-water bearing magnesium bicarbonate might effect the change. Nauck, Haussmann, Bischof, Zirkel and others, however, subscribed to the view that ordinary circulating ground water bearing magnesium bicarbonate had attacked the limestone.

* The present article is based on a more extended paper which constitutes a portion of volume xxv of the Iowa Geological Survey. The reader is referred to this report for details.

† For numbered references to the literature, see the list at the end of this article.

33

Van Hise (10) also attaches much importance to dolomitization after the limestones emerge from the sea.

In 1846 Green (11) offered the suggestion that some dolomitic limestones might be formed by the decomposition 'of olivine sand incorporated in the original limestone and the recombination of the magnesia with the lime. He calls attention to the fact that olivine sand, derived from the action of the waves on lava, constitutes an important constituent of the coral-reef rock about the borders of the Hawaiian Islands and regards this as significant.

Dana (12) in 1843, attempting to account for the dolomite of the coral island of Metia, supposed that it had been formed by the action of magnesium salts of heated sea water on limestone. Twenty-nine years (13) later he expressed the view that the same dolomite had been formed in sea water at ordinary temperatures but perhaps in a contracting lagoon where magnesium and other salts were in a concentrated state. Sorby (14) likewise favored the theory of marine alteration and the same origin has been urged, either for dolomites in general or in special instances, by Von Richthofen, Doelter and Hoernes, Hoppe-Seyler, Mojsisovics, Murray, Skeats, and F. W. Pfaff. In support of this theory are also the observations of Weller (15) who, from a faunal study of the Galena and Niagara dolomites of the Upper Mississippi Valley, concludes that they were deposited originally as limestones and later metamorphosed. More recently, Blackwelder (16) has also advocated the replacement theory for the origin of the Big-horn dolomite of Wyoming, but owing to the very slight porosity of this rock he is led to suggest that the alteration proceeded contemporaneously with its deposition rather than subsequent to its consolidation.

F. W. Pfaff (17) believes that the alteration takes place at considerable depth and in concentrated seas, but Phillipi (18) vigorously controverts this view since he has good evidence that dolomitization may proceed in the open sea and at shallow depths. Skeats' (19) studies of the coral reefs of the Southern Pacific also seem to show that concentration and pressure are not important factors. On the other hand, both Nadson and Walther (20) have suggested that bacteria may play an important part in the alteration.

Still other geologists have supported the theory that dolomite represents a direct chemical precipitate from the ocean. Boné (21) as early as 1831 advocated this method of origin. Bertrand-Geslin (22) and Coquand (23) were also early supporters of this view. That dolomite can be formed as a chemical precipitate is pointed out by Zirkel (24) who shows that the occurrence of crystals of dolomite in veins and druses indicates its possible chemical deposition on a larger scale in nature. Fournet (25) regarded the dolomite beds interstratified with limestone in the Tyrol as original precipitates. His studies showed that the volcanic theory of Von Buch was no longer tenable. Others who have advocated the primary precipitation theory in one form or another are Loretz, Forchhammer, Hunt, Vogt, Daly, Linck, and Suess.

As to the nature and cause of the reactions which have been supposed to give rise to the chemical precipitation of dolomite, there have been differences of opinion. Forchhammer (26) attributed the reaction to the action of calcium carbonate of spring water on the magnesium salts of the sea, while Hunt, (27) basing his views on experimental evidence, regarded dolomite as the product of the action of sodium bicarbonate on the magnesium chloride and magnesium sulphate of the sea. Linck (28) and Daly (29), on the other hand, emphasize the importance of ammonium carbonate furnished by decaying organisms on the sea bottom as the precipitating agent.

Still another primary theory is that introduced by Lesler (30) to account for certain dolomitic layers in the " Calciferous " limestone near Harrisburg, Pa. These he believed to represent ordinary mechanical sediments which were deposited at the time the limestone was laid down. The clastic theory has been adopted more recently by Phillipi (31), who regards certain impure dolomites of the Muschelkalk of Germany as mechanical deposits possibly derived from the residuum of limestones low in magnesia. Grabau (32) has concluded that certain impure dolomitic limestones and waterlimes of the Salinan and Monroan series have had a similar origin.

An entirely different theory of origin is that which was introduced by Grandjean (33) in 1844, to explain the production of the dolomites of the Lahn district. He assumed that by the atmospheric leaching of the lime from an original limestone of low magnesia content, a true dolomite might in time result. Both Bischof and Hardman later demonstrated the plausibility of this theory experimentally, and Hardman (34) immediately accepted it to explain the origin of the Carboniferous dolomites of Ireland. In 1895 Hall and Sardeson (35) applied the same theory in interpreting the history of the Lower Magnesian series of the Upper Mississippi Valley.

Högbom (36) on the other hand, regards surface leaching as of minor importance and emphasizes the effect of marine leaching. He has proven the reality of this process, on a small scale at least, in the modern seas and concludes that some dolomites of former periods may have been formed in this manner. Judd (37) is of the opinion that the weakly dolomitic portions of the atoll of Funafuti may be explained upon the basis of this theory but regards the magnesia content of the more highly dolomitic portions as having been enriched by reaction with the magnesia of sea water.

Experimental Evidence.

On the experimental production of dolomite there is a voluminous literature. This has been well summarized by F. W. Pfaff (38), and later by Steidtmann (39). Dolomite has been frequently prepared artificially under conditions of high temperature or high pressure, or both, but it has been produced in the laboratory at ordinary temperatures and pressures only in rare instances, and then in minute amounts and under conditions which doubtfully operate in nature, at least on a large scale. It must be conceded then that these experiments furnish little evidence as to the actual conditions obtaining when extensive beds of dolomite are formed naturally. For the purpose of obtaining more accurate data on this point, a series of experiments was begun at ordinary temperatures and pressures early in 1912. In this series it was attempted to simulate natural conditions as near as they could be estimated, and to obtain some quantitative measurement of the effect of time and of concentration in the production of dolomite. In one set of experiments it was attempted to reproduce the conditions which exist in nature when limestone is altered to dolomite beneath the sea by solutions bearing magnesia. In these the effect of solutions of known concentration of $MgCl_2$ and $MgSO_4$, and of mixtures of the salts, both with and without the presence of $NaCl$, on powdered aragonite was tried. The concentration of the magnesium solutions used ranged from two to ten times the concentration of the magnesia in sea water. After a period of six months, residues from the experiments were thoroughly tested for dolomite. The results were entirely negative. No trace of dolomite could be found. Careful re-examination of the residues after a period of nearly three years still gave the same result. The analyses showed that the $CaCO_3$ had reacted slightly with the solutions, but no $MgCO_3$ had been deposited. Apparently the soluble trihydrate of $MgCO_3$ had been formed. It then appears that dolomite cannot be prepared artificially under these conditions.

In a second set of experiments it was attempted to obtain dolomite as a direct chemical precipitate at ordinary temperatures and pressures. First solutions of the bicarbonates of calcium and magnesium, after being standardized, were mixed in molecular equivalent proportions so as to give the same ratio of $CaCO_3$ to $MgCO_3$ as exists in normal dolomite. The solution was then allowed to evaporate spontaneously during a period of one month. It was noted that the carbonates came down separately with the $CaCO_3$ much in advance of the $MgCO_3$. The precipitate then contained only the mixed carbonates—no dolomite was formed. Scheerer (40) previously obtained the same results in a similar experiment. Negative results were still obtained when a solution prepared as above was inoculated with a crystal of dolomite and allowed to evaporate. Nor could the double carbonate be prepared upon evaporating spontaneously a solution of the two carbonates obtained by the action of carbonated waters on normal dolomite even when a dolomite crystal was introduced and a concentrated solution of sodium chloride and magnesium salts was added.

The experimental evidence so far obtained, therefore, does not suggest the circumstances under which large masses of dolomite can be formed in nature under ordinary conditions either by the alteration of limestone or by chemical precipitation. It is to be regretted that a careful study of the process of dolomitization where it is going on in the seas to-day has never been made. Such a study would doubtless throw much valuable light on the problem. It may well be that bacteria play an important rôle in the production of dolomite as suggested by Nadson.

Field Evidence.

Realizing the importance of careful field studies of dolomitic formations in interpreting the conditions of their origin, the writer undertook a study of the dolomites of the Upper Mississippi Valley under the auspices of the Iowa Geological Survey during the field season of 1912. More recently a grant from the Esther Herrman Research Fund of the New York Academy of Sciences has made possible much more extensive field studies in the eastern and central states. Dolomites ranging in age from the Cambrian to the Mississippian have now been examined and many samples collected for detailed chemical and petrographic study. It is possible to outline in this paper only some of the more important results obtained.

It should be pointed out that the term dolomite is used here in the broad sense to include both normal dolomite and dolomitic limestone. It is not necessary to differentiate between these in a discussion of their origin.

The field studies undertaken during the course of this investigation have alone furnished irrefutable evidence that most of the dolomites examined, regardless of their age, are replacement products. The following facts support this contention: (1) the lateral gradation of beds of dolomite into limestone, sometimes very abruptly; (2) the mottling of limestones by irregular patches of dolomite on the borders of dolomite masses ; (3) the existence of remnants of unaltered limestone in dolomite, and of nests of dolomite in limestone ; (4) the irregular boundaries between certain beds of limestone and dolomite ; (5) the presence of altered oolites or fossils in many dolomites ; (6) the protective effect of shale beds ; and (7) the obliteration of structures and textures.

In some instances the relationship of dolomite to limestone is such as to indicate that the alteration was accomplished by solutions which migrated from above downwards after the limestone was formed, or at least in the closing stages of its formation.

It is an interesting fact that certain layers have sometimes been passed over during the dolomitization of adjacent ones, and show little or no sign of alteration. The so-called inter-stratification of limestone and dolomite cited by some as evidence in favor of some primary theory of origin is then, in some cases at least, rather a pseudo-inter-stratification produced by the selective dolomitization of an original limestone. Some layers which have been passed over have been noted to be coarser grained than the adjacent layers which have been altered and this would seem to explain their greater resisting power. At times, however, the unaltered layers do not appear to differ markedly from the altered ones. The phenomenon is then difficult to account for. Normally the contact lines between such interbedded layers of limestone and dolomite are fairly regular and definite, but in some instances they are known to be very irregular and may even simulate irregular contacts produced by disconformity. A remarkable example of a pseudo-disconformity produced by uneven selective dolomitization has been observed in the St. Louis limestone near Farmington, Iowa. Here a bed of altered limestone is found resting very irregularly on a bed of dolomite. The two beds are very different physically and might readily be taken at first sight for two distinct formations, but when the contact is traced laterally for a short distance the lower bed loses its dolomitic character and passes into a limestone very similar to and continuous with the bed above.

Another striking relationship of limestone to dolomite is exhibited in a certain layer of an interbedded series of limestones and dolomites of the Beekmantown in the old Walton Quarry near Harrisburg, Pa. The beds dip south here at an angle of 30°. The layer in question is represented by dolomite six feet in thickness in the upper part of the quarry face and on each side of it appear good limestone layers. Now in the lower part of the quarry the lower half of this layer passes abruptly into limestone and continues to the quarry floor as two distinct layers each 3 feet thick. Samples of the dolomite at the point where it passes into limestone yielded 18·1 per cent of $MgCO_3$ while the limestone itself yielded only 0·83.

It will be noted that in the above instances the gradation of limestone into dolomite is abrupt, but in many cases the gradation takes place through transition zones of limestone mottled with dolomite. There can be no doubt but that these mottled limestones represent an incipient stage in the process of dolomitization and it is believed that many dolomites have passed through such a stage in the progress of their formation. In most cases the phenomenon of mottling appears to be of purely inorganic origin, having resulted from a process of dolomitization which began at certain favorable centers and spread outwards. In some cases, however, it has been produced by the selectine alteration of areas suggesting algæ and fucoids in the limestone first, and the spreading out of the dolomite from these as nuclei. The Tribes Hill limestone, as developed at Canajoharie, New York, furnishes an excellent illustration of the mottling produced by the latter method. All stages of mottling from altered fucoid-like markings to a rock uniformly dolomitic may be traced in this.

It has been observed that the spreading of dolomitization from certain centers in a limestone may give rise to mottling on a large scale if these centers be few and far apart. For example there is a conspicuous bed of dolomite pseudo-bowlders in the St. Louis limestone at Alton, Ill., which appears to have been formed entirely in this manner. These bowlder-like masses range from a few inches up to six feet in diameter and contain 32·39 per cent of $MgCO_3$ while the limestone matrix bears only 3·39. That they were formed in place is clearly indicated by the fact that the contact of the bowlders with the limestone matrix is occasionally gradational and that the stratification lines of the limestone may at times be traced directly through the bowlders. In a layer of limestone a few feet above the bowlder bed here a similar process of local dolomitization has given rise to the development of irregular lenses of dolomite.

If has often been noted during the course of the field studies that many dolomites known to be of secondary origin show little or no evidence of shrinkage and porosity determinations have since shown that the transformation of a limestone to dolomite, even subsequent to its deposition, need not necessarily be accompanied by a decrease in volume as pointed out by Beaumont and consistently adhered to by later writers on the subject. It seems probable, therefore, that the replacement may proceed at times according to the law of equal volumes as enunciated by Lindgren(41) and that the interchange need not be molecular. In view of this fact compact dolomites showing no shrinkage effects can no longer be regarded as primary.

Further studies will doubtless show that considerable shrinkage effects produced by dolomitization are not common. It is believed that many vesicular dolomites have resulted from atmospheric leaching long after their formation.

Petrographic Evidence.

The microscopic study of many thin sections of dolomitic limestone has not only further amplified and strengthened the field evidence but has also thrown new light upon the details of the process of alteration. By employing microchemical tests it has been possible to distinguish between calcite and dolomite in the sections and make clear the most intimate relationships of the two minerals. It should be stated, however, that these tests furnish no reliable guide to the exact amount of magnesia in the rock, for crystals containing less than 25 per cent of $MgCO_3$ may behave essentially like normal dolomite. But this is to be regarded in truth as a distinct advantage, for alterations of only a slight degree are indicated as well as the more marked ones.

It must be admitted that if dolomite has a diverse mode of origin the microscope fails to reveal it. Careful examination of every variety of dolomite fails to show any positive evidence in favor of either the primary chemical or the clastic theory of origin. On the other hand, there is abundant evidence in favor of the alteration theory. It is true that certain dolomites, whose origin is not certainly known from their field relations, possess an extremely fine and uniform texture, and this feature has in fact led Daly (42) to believe that these represent original chemical precipitates. In order to test the validity of this argument the finest grained dolomite of unknown origin encountered by the writer in these studies was compared with the finest grained dolomite of known secondary origin. For example the Jefferson City dolomite of the Ozark region, whose mode of formation is not definitely known, possesses unusually dense and compact layers which are seen under the microscope to be made up of minute granules the majority of which are below $\cdot003^{mm}$ in diameter, some measuring only $\cdot001^{mm}$. The granules are suggestive of an original chemical precipitate. The strength of this interpretation is weakened, however, by the fact that a dolomite of known secondary origin has been found in the Middle Devonian of Iowa which is equally as fine-grained. The latter dolomite has resulted from the alteration of a dense, lithographic-like limestone with the approximate retention of the original texture.

40

As regards the possibility that some dolomites may be of clastic origin, none has been found which exhibits any signs of clastic structure.   But that the original structure might have been obliterated during recrystallization is easily conceivable in rocks of this type.

Turning now to the dolomites which from their field relations are known to be secondary after limestone we have much more definite data.   Indeed in these, by virtue of the fact that the alteration has frequently been halted before it proceeded to completion, we are often able to trace all stages of dolomitization from a limestone showing only incipient alteration to a good dolomite.   Thus, it is possible to describe the steps normally passed through during the transformation of a limestone to dolomite.

So far as the testimony of the microscope goes the fine-grained limestones are more susceptible to alteration than the coarser-grained ones, a fact which is in keeping with the laws of chemistry.   The evidence also indicates that the alteration may not proceed in exactly the same manner in the two types of limestone.

The alteration of fine-grained compact limestones seems to be accompanied normally by a notable increase in size of grain. Usually the diameter of the dolomite crystals formed is many times as great as that of the original calcite grains.   But in rare cases, such as that of the dense Middle Devonian dolomite referred to above, the original structure and texture seems to be approximately retained.   In the dolomitization of such fine-grained limestones the replacement frequently begins at many centers throughout the rock and spreads outwards from these, or if the rock possesses fine stratification the replacement may follow closely these original lines of weakness in the early stages.   In those cases where the alteration begins at certain centers and spreads out from these, fucoid-like markings occasionally serve as the nuclei as in the case of the Tribes Hill limestone.   But as a general rule no organic influence is noted. Normally the limestone is altered uniformly in the process of spreading from the dolomite centers, but it cannot be said that it is altered completely, for the dolomite patches often possess less than twenty per cent of $MgCO_3$.   Small remnants of limestone, however, may occasionally escape alteration and become incorporated in the dolomite patches.   The boundary between the limestone and the spreading dolomite area may or may not be abrupt.   When abrupt, the rock may assume the appearance of a breccia and the term "pseudo-breccia" may aptly be applied.   When the boundary is gradational, on the other hand, rhombohedrons of dolomite, variable in size but usually nearly perfect in their development, are disseminated through the limestone a short distance in advance of the main dolomite area. As the replacement proceeds the dolomite areas grow larger and larger and eventually meet and become confluent thereby giving rise to a rock which is uniformly dolomitic. Further addition of magnesia may then take place by altering the rock more completely.

In the coarse-grained limestones, especially those which were originally coarse-grained, such as the crinoidal limestones, on the other hand, mottling does not seem to be the rule in the early stages of alteration. In these the replacement appears to affect the matrix first and to spread rapidly through the rock. The coarser grains are next affected, being broken down into aggregates of small dolomite grains. In the end a coarse-grained limestone may be changed over into a uniformly fine-grained one.

In the dolomitization of limestones of both types the calcareous skeletons of organisms appear in most cases to successfully withstand alteration and these, owing to their greater solubility than dolomite, are then removed to leave molds by a process of atmospheric leaching when the formation passes into the belt of weathering.

Conclusions.

Considering all the evidence, it seems probable that the great majority of our dolomites had their inception in the alteration of limestones. It will not be denied, however, that some dolomitic formations of minor importance may have had a different origin. For instance, some impure dolomitic limestones associated with shales very probably represent original clastic deposits which have not suffered any alteration whatever, and there is some reason for believing that certain dolomitic limestones high in siliceous material, such as the Silurian waterlimes of New York State, may have had a similar origin.

The importance of marine and surface leaching in increasing the magnesia content of limestones originally low in magnesia should likewise not be overlooked. There can be no doubt that this process has greatly enriched the more vesicular dolomitic limestones in magnesia. But the leaching theory does not explain the ultimate source of the magnesia. It merely shows how the magnesia content of a limestone originally low in this constituent can be enriched.

To return now to the dolomites which have resulted from the alteration of limestone, there are many reasons for believing that the more extensive of these have all been formed beneath the sea, and that dolomitization affected by ground water is only local and very imperfect. Some of the features which lend weight to this view are as follows: (1) The dolomite areas of mottled limestones are believed to have undergone recrystallization at the same time as the associated limestone areas as suggested by the occasional development of zonal growths of calcite and dolomite. (2) In imperfectly altered limestones the dolomite is seen to follow original lines of weakness rather than secondary structures such as joints or fractures. (3) In most cases of mottling the dolomitization appears to have progressed uniformly as we should expect it to in an uncrystallized rock, rather than to have progressed by forming veinlets and stringers in the early stages. (4) The existence of perfect rhombs of dolomite in many imperfectly altered limestones suggests that the latter had not yet solidified when the dolomite rhombs were formed. (5) The widespread extent and nearly uniform composition of many dolomites indicates that they must have been formed by an agent capable of operating uniformly over wide areas. (6) An adequate source of magnesia for transforming extensive limestone formations into dolomite is found only in the sea which contains many times as much of this constituent as ordinary ground water. (7) Many dolomites are directly and regularly overlain by pure limestone formations or by thick shale beds, proving that they must have been formed before these overlying beds were deposited.

Some dolomites of minor importance, such as those associated with ore deposits and probably most, if not all of those related to fractures (vein dolomites), must have been formed through the agency of ground water. But in general, ground water is incapable of carrying dolomitization far. Study of analyses of ground water, and of river water, shows these to be uniformly low in magnesia, this constituent normally being greatly exceeded in amount by lime. How, then, could such waters dolomitize limestone when they already contain lime far in excess of magnesia? The law of mass action speaks strongly against ordinary ground water being able to accomplish extensive dolomitization. In the case of mineral springs and the mineralizing solutions which are related to ore deposition, however, it is conceivable that magnesia might be present in sufficient proportions to accomplish local dolomitization and doubtless most vein dolomites have been so formed.

University of Illinois,
 Urbana, Ill.

REFERENCES TO THE LITERATURE.

1. Von Buch, quoted by Zirkel, Lehrbuch der Petrographie, 2d ed., vol. iii, p. 505.
2. Frapolli, Bull. Soc. géol. France (2), vol. iv, p. 832, 1847.
3. Durocher, Compt. Rend., vol. xxxiii, p. 64, 1851.
4. Favre, ibid., vol. xxviii, p. 364, 1849.

5. Collegno, Bull. Soc. géol. France, vol. vi, p. 110, 1834.
6. Morlot, Haidinger's Naturw. Abhandl., vol. i, p. 305, 1847.
7. Beaumont, Bull. Soc. géol. France, vol. viii, p. 174, 1836.
8. Morlot, Compt. Rend., vol. xxvi, p. 311, 1848.
9. Jackson, this Journal, vol. xlv. p. 140, 1843.
10. Van Hise, U. S. Geol. Surv., Mon. xlvii, p. 804 ff.
11. Green, Jour. Roy. Geol. Soc. Ireland, 2d ser., vol. iv (iii), pp. 140-143, 1846.
12. Dana, this Journal, vol. xlv, p. 120, 1843.
13. Dana, Corals and Coral Islands, p. 356.
14. Sorby, Rept. Brit. Assoc. Adv. Sci., p. 77, 1856.
15. Weller, Bull. Geol. Soc. America, vol. xxii, pp. 227-231, 1911.
16. Blackwelder, ibid., vol. xxiv, pp. 607-624, 1913.
17. F. W. Pfaff, Neues Jahrb., Beil. Bd.. xxiii, p. 529 ff., 1907.
18. Phillipi, Neues Jahrb., Festband 1907, i. p. 397 ff.
19. Skeats, Bull. Mus. Comp. Zool. Harvard College, vol. xlii, p. 52 ff., 1903.
20. Walther, Gesichte der Erde und des Lebens, p. 90.
21. Boué, Bull. Soc. géol. France, vol. i, p. 115, 1831.
22. Bertrand-Geslin, ibid.. vol. vi, p. 8, 1834.
23. Coquand, ibid., vol. xii, p. 314, 1841.
24. Zirkel, Lehrbuch der Petrographie, 2d ed., vol. iii, p. 503.
25. Fournet, Bull. Soc. géol. France (2), vol. vi, p. 502, 1849.
26. Forchhammer, Journal prakt. Chemie, vol. xlix, p. 52, 1850.
27. Hunt, this Journal (2). vol. xxviii, p. 382, 1859.
28. Linck, Monatsh. deutsch. geol. Gesellsch., 1909, p. 230.
29. Daly, this Journal (4), vol. xxiii, p. 93, 1907 ; also Bull. Geol. Soc. America, vol. xx, p. 153, 1909.
30. Lesler, Penn. Second Geol. Surv., Rept. M. M., p. 311 ff., 1879.
31. Phillipi, Frech's Lethæa Geognostica, vol. ii, p. 31, 1908.
32. Grabau, Bull. Geol. Soc.; America, vol. xxiv, pp. 524-526, 1913 ; also Principles of Stratigraphy, p. 760.
33. Grandjean, Neues Jahrb., 1844, p. 543.
34. Hardman, Proc. Roy. Irish Acad. (2), vol. ii, Science, p. 705 ff., 1875-77.
35. Hall and Sardeson, Bull. Geol. Soc. America. vol. vi, p. 167, 1895.
36. Högbom, Neues Jahrb.. 1894, vol. i, p. 262 ff.
37. Judd, The Atoll of Funafuti, p. 362 ff., 1904.
38. F. W. Pfaff, Neues Jahrb., Beil. Bd., xxiii, p. 529, 1907.
39. Steidtmann, Jour. Geol., vol. xix, p. 323, 1911.
40. Scheerer, Neues Jahrb., 1866, p. 1.
41. Lindgren, Econ. Geol., vol. vii, p. 521, 1912.
42. Daly, Bull. Geol. Soc. America, vol. xx, p. 153, 1909.

3

Reprinted from *Geol. Mag.* **55**:246–258 (1918), by permission of Cambridge University Press

DOLOMITIZATION AND THE LEICESTERSHIRE DOLOMITES.

By L. M. PARSONS, M.Sc. (Lond.), D.I.C., F.G.S.

(PLATE XI.)

PART I: EVIDENCES OF THE PERIOD OF DOLOMITIZATION.

CONTENTS.

1. Classification of Types.
2. Field Evidences.
3. Inherent Structural Evidences.
4. Selective Dolomitization.
5. The Absence or Presence of Fossils.
6. Classification of Petrological Types.
7. The Dense Yellow Dolomites of Breedon, etc.
8. The Red Ferruginous Dolomites of Breedon and Breedon Cloud.
9. The Barren Grey and Yellow Dolomites of Ticknall and Calke.
10. Fossiliferous Dolomitic Limestones of Ticknall and Calke.

IN the district north of Ashby-de-la-Zouch dolomitized Carboniferous Limestone crops out where the Leicestershire border adjoins that of South Derbyshire. The dolomites, which attain a thickness of nearly 900 feet in this area, have hitherto received very little attention. I therefore propose to give a short description of them with a view to ascertaining how far their mode of occurrence and structure afford additional examples of, or exceptions to, the usual conclusions adopted concerning the origin of dolomite. With this object in view I propose to give a brief resumé of the evidences generally relied upon to explain the origin of dolomite, before proceeding to describe the Leicestershire rocks.

1. CLASSIFICATION OF TYPES.

If dolomites are classified as simply as possible according to the period at which the dolomitization [1] took place, well-defined classes, as enumerated below, may be recognized:—

Primary dolomites.

1. Those deposited as *clastic* rocks derived from pre-existing dolomite.
2. Those *chemically precipitated* as dolomite, with or without the agency of organisms.

Secondary dolomites.

1. *Contemporaneous* dolomites, or those deposited as ordinary limestones which have been altered, soon after deposition, by the influence of magnesian salts in the sea in which the rocks were originally deposited.
2. *Subsequent* dolomites deposited as ordinary limestones which have been altered by the influence of waters belonging to some period later than that during which the rocks were originally deposited.

[1] Throughout this article the term dolomitization signifies the production of dolomite, either primary or secondary.

With regard to this classification it must be noted that as some confusion has existed concerning the significance of the term "contemporaneous", that name is used here strictly to denote dolomites of secondary origin.

The term "subsequent" has a wider significance than that of "vein" dolomitization, as certain leached dolomites and some other dolomitized rocks of undoubted subsequent origin cannot be described adequately as vein dolomites, since some of them do not occur in association with veins and channels. Evidently vein dolomites constitute a subdivision of subsequent dolomites. It is now generally admitted that the majority of dolomites are of secondary origin, the contemporaneous class probably being more numerous than the subsequent, but certain cases of dolomitization appear to be explained most satisfactorily by the theory of primary deposition. Many chemical experiments have been performed in the endeavour to produce dolomite at different pressures and temperatures, but with limited success.[1] In spite of the comparative failure of these experimental attempts to produce dolomite artificially, the fact that it does occur in nature as a chemical precipitate is shown by its occurrence in mineral veins. It is very doubtful whether any reliable evidence can be obtained from experiments performed under conditions which may be quite unlike the natural conditions which existed in the seas of remote geological periods.

In determining the class to which a dolomite belongs, one must rely upon the collective evidence afforded by : —

> Field relations,
> Inherent structural features,
> Selective dolomitization, and
> The absence or presence of fossils.

Space permits only an incomplete survey of the phenomena connected with dolomitization, and a short discussion of the more reliable sources of evidence is all that is attempted.

2. Field Evidences.

(*a*) The occurrence of truly bedded dolomites associated with beds of such deposits as gypsum or rock salt is considered to support the view that the dolomite was primarily precipitated.[2]

(*b*) The theory of primary deposition is also supported by the occurrence of genuine beds of dolomite alternating with beds of limestone. In this case it is inferred that the dolomite was either chemically precipitated,[3] or laid down as a clastic deposit derived from a source different from that of the non-dolomitic limestone. Pseudo-interbedding, characterized by the failure of the dolomitization to conform accurately to bedding planes, the lateral transition of dolomite into limestone, and sometimes by a streaky development of

---

[1] See F. W. Clarke, "Data of Geo-chemistry": Bull. Geol. Surv. U.S.A., No. 616, p. 559, 1916.

[2] See Weigelin, *Neues Jahrb.*, Beil. Bd. xxxv, p. 623, 1913.

[3] Suess, *The Face of the Earth*, English translation, vol. ii, p. 262, 1906.

dolomite in the intervening limestone, is in favour of subsequent dolomitization.[1]

(*c*) The presence of interbedded conglomerates containing fragments of dolomite derived from older dolomites below would suggest that the dolomitization of the older beds was not of subsequent origin, and that the dolomitic material of the conglomerate was a primary clastic deposit.[2]

(*d*) The persistence of uniform bedded dolomites over a wide area without lateral transition into unaltered or poorly magnesian limestones is considered to be one of the strongest evidences in favour of contemporaneous dolomitization.[3] This conclusion receives further support if such dolomites occur at the same stratigraphical horizons in different areas, but the presence of dolomite at a certain horizon in one area and its absence at the same horizon in another area, does not necessarily indicate dolomitization of subsequent origin. Dolomite may have been formed in shallow water, while poorly dolomitic or unaltered limestones were being deposited further from the shore-line.[4]

(*e*) Should beds of dolomite be found, when traced laterally, to pass into unaltered or poorly magnesian limestones, the inference is in favour of subsequent dolomitization,[5] provided that other more conclusive evidence of a different origin is not forthcoming.

(*f*) A patchy development of dolomite and limestone due to rapid lateral transition from one into the other is a modification of (*e*), and lends support to a similar conclusion.[6] A patchy development of dolomite and limestone on a small scale, known as pseudo-brecciation, is discussed later.

(*g*) The evidence in favour of subsequent dolomitization is much stronger when an irregular or patchy development of dolomite is associated with faulting or jointing. In such cases it is obvious that the fault planes and joints have probably served as channels for percolating magnesian waters of a period subsequent to that during which the rock was originally deposited.[7] Dolomites of this class are properly described as vein dolomites.

(*h*) Faulting associated with extensive and non-patchy dolomites (*d*) appear to indicate that the faulting occurred after dolomitization.

While discussing field relations we may observe that great thickness of bedded dolomite has been considered to support the view that the dolomite was of primary origin, but since a thick mass of limestone may be dolomitized during a long period of subsidence,[8] as in the case of certain coral reefs, it is questionable whether mere thickness

---

[1] Calvin, Iowa Geol. Surv., vol. vii, p. 151, 1896.

[2] See *Swansea* (Mem. Geol. Surv.), 1907, p. 13.

[3] Dixon, *Swansea* (Mem. Geol. Surv.), 1907, p. 13.

[4] Id., p. 15.

[5] Hardman, Proc. Roy. Irish Acad., ser. II, vol. ii, p. 728, 1875-7.

[6] F. M. Van Tuyl, " The Origin of Dolomite " : Iowa Geol. Surv., vol. xxv, p. 364, 1916.

[7] *The Geology of the South Wales Coal Fields* (Mem. Geol. Surv.), pt. ii, p. 33.

[8] Skeats, " On the Dolomites of the Southern Tyrol " : Q.J.G.S., vol. lxi, p. 97, 1905.

can be considered to yield any reliable evidence concerning dolomitization in general.

3. INHERENT STRUCTURAL EVIDENCES.

(*i*) The degree of idiomorphism of dolomite crystals may afford subsidiary evidence concerning the class to which the dolomite may be assigned.[1] In many cases where subsequent dolomitization is amply proved by other evidences, it is found that the rhombohedra are considerably more idiomorphic than those of most dolomites of undoubted contemporaneous or primary origin. In the latter cases there is a marked tendency of the crystals to interfere with one another, the resulting structure being a more or less granular mosaic.

(*j*) The size of the rhombohedra is usually larger in subsequent dolomites where the growth of crystals is less impeded than it is in contemporaneous and primary dolomites.[2]

(*k*) As in the case of idiomorphism and size, the degree of purity of a dolomite is suggestive, but by no means conclusive.[3]

Crystals of a primarily precipitated dolomite should, in general, be less contaminated with impurities than those of a dolomite of secondary origin, though a contemporaneous dolomite may attain a fair degree of purity. It is, perhaps, safer to place no reliance upon the degree of purity.

(*l*) The relation of iron oxides, particularly hæmatite, to the dolomite rhombohedra, affords one of the most reliable evidences concerning subsequent dolomitization.[4] When hæmatite is included either centrally or zonally in the rhombohedra the only possible conclusion is that the hæmatite was introduced at the time when the dolomitization took place. For instance, in a district where Trias comes above Carboniferous dolomites having zonal inclusions of hæmatite, the obvious inference is that the dolomitization was subsequent and associated with waters percolating through the Trias (see micro-photograph, Pl. XI, Fig. 1). On the other hand, should the hæmatite be only interstitial, the inference is that dolomitization took place before the introduction of iron oxide. In a case where Trias rests upon Carboniferous dolomites, the presence of only interstitial hæmatite in the dolomite would indicate that the dolomitization was certainly Pre-Triassic, perhaps contemporaneous. Inferences similar to those made from the presence of included hæmatite appear to be justified in cases where dolomitization is intimately associated with other ores such as galena and zinc blende.[5]

(*m*) The relation of rhombohedra to chert in cherty dolomites affords definite evidence with regard to the relative periods at which the dolomite and chert were respectively formed.[6] If rhombohedra occur enclosed by chert, the inference is that the dolomite was formed either before or simultaneously with the chert. On the

---

[1] See F. M. Van Tuyl, Iowa Geol. Surv., vol. xxv, p. 390 et seq., 1916.
[2] *Swansea* (Mem. Geol. Surv.), 1907, p. 16.
[3] F. M. Van Tuyl, Iowa Geol. Surv., vol. xxv, p. 319.
[4] *Swansea* (Mem. Geol. Surv.), 1907, pp. 15, 16.
[5] Schmidt, Trans. St. Louis Acad. Sci., vol. iii, No. 2, 1875, p. 246.
[6] H. H. Thomas, *Ammanford* (Mem. Geol. Surv.), 1907, p. 76.

other hand, should chert contain no included rhombohedra, the
formation of chert must have preceded that of dolomite, though this
does not necessarily prove subsequent dolomitization, since there is
no indication of any great period of time having elapsed between the
formation of chert and that of dolomite.

4. SELECTIVE DOLOMITIZATION.

The term "selective dolomitization" has been applied to certain
phenomena in which the formation of dolomite occurs more in certain
portions of rock, presumably the less coarsely crystalline and more
unstable parts, than in other more resistant portions. One result of
this differentiation is a rock of mottled appearance at times not unlike
a breccia. The relations between the dolomitic and non-dolomitic
portions of such rocks may afford evidence concerning the origin of
the dolomite.

(*n*) Pseudo-brecciation, in which a limestone assumes a mottled
appearance on account of a patchy development of dolomite on
a small scale, can scarcely be considered to yield evidence analagous
to that of patchy dolomitization on a larger scale (*f*). That
a pseudo-breccia is unlikely to be due to the formation of primary
dolomite appears to be a legitimate inference, but the question as
to whether the dolomitization of any particular pseudo-breccia is
contemporaneous or subsequent must be decided by inherent evidence
other than that of mere mottling. Thus a rock of this kind in which
hæmatite is merely interstitial may be referable to contemporaneous
dolomitization.[1]

Any particular pseudo-breccia may supply evidence of its own
origin, but cannot form the basis for a definite generalization
concerning the mottling of dolomitic limestones. Each case must
be considered on its own merits.

(*o*) A mottled appearance in which the dolomitic material is
worm-like or fucoid suggests dolomitization facilitated by the
presence of the remains of algæ or other organisms.[2]

Of particular interest are cases of fossiliferous dolomitic limestones
in which selective dolomitization has differentiated between the
matrix and organic structures. It frequently happens that in
dolomitic rocks proved by general evidence to be of subsequent
origin, the matrix, whether calcite or aragonite originally, has been
largely converted into recrystallized calcite prior to dolomitization.[3]
The recrystallised matrix, presumably on account of its relatively
coarse texture, is apparently more stable than the original calcareous
material of organic remains, consequently dolomite crystals have
developed more in fossil structures than in the recrystallized matrix.

(*p*) Hence it is inferred that differential dolomitization of this
kind may be an indication of subsequent alteration. That such
an inference may not always be legitimate is evident from the
occasional occurrence of dolomite crystals enclosed by recrystallized

---

[1] For instance, see *Swansea* (Mem. Geol. Surv.), 1907, pp. 14, 15.

[2] Peach & Horne, *North-West Highlands of Scotland* (Mem. Geol. Surv.),
1907, p. 366.

[3] *Swansea* (Mem. Geol. Surv.), 1907, p. 16.

calcite, suggesting that the dolomitization may have been contemporaneous.

It has been shown that in certain contemporaneous dolomites, fossil structures, including corals, have resisted dolomitization to a greater extent than the matrix has done owing to the non-recrystallized material of the matrix being more unstable than the more coarsely crystalline material of organic remains.

(*q*) From this it may be inferred that the greater development of dolomite in the matrix than in fossil structures, including corals, is in favour of the theory of contemporaneous dolomitization.[1] But while this may be true in many cases, it is by no means certain that selective phenomena of this kind should always be relied upon to furnish conclusive evidence of the period of dolomitization. In connexion with the question of the relative stability of matrix and fossil structures, it must be remembered that the calcareous contents of dolomitic limestones may have consisted originally of any or all of the following forms of calcium carbonate : aragonite mud, more coarsely crystalline aragonite, calcite mud, and more coarsely crystalline calcite. Of these, aragonite mud is certainly the most easily converted into dolomite, and coarsely crystalline calcite is the most stable, but whether the more coarsely crystalline aragonite of coral tissues is more unstable than calcite mud appears to be an open question. It is just this point which weakens evidence afforded by selective dolomitization. In the contemporaneous alteration of a coral limestone containing few other organic remains, the matrix consisting mostly of aragonite mud would certainly be more easily dolomitized than the coarser coral structures.

Most organic limestones, however, consist largely of calcitic remains as well as coral structures, so that calcite mud must have been present originally in the matrix, and may even have been in excess of aragonite mud. If calcite mud is more susceptible to alteration than coarser aragonite, in contemporaneous dolomitization the matrix would still be dolomitized in preference to coral structures. On the other hand, should calcite mud be more stable than coarser aragonite, contemporaneous alteration of a mixed organic limestone would result in the development of dolomite more in coral structures than in the matrix. Again, there appears to be no means by which the proportions of aragonite and calcite muds in the original matrix may be ascertained. Considerations of this kind suggest that the phenomena of the selective dolomitization of organic rocks can scarcely be relied upon to supply sound evidence, particularly in cases where such phenomena appear to contradict the evidence afforded by field relations or other reliable features.

Selective alteration of oolitic limestones may be more reliable, since the differentiation in such cases is between coarser calcite and finer calcite. If dolomitization has attacked ooliths in preference to a recrystallized matrix, the alteration is probably subsequent; on the other hand, the development of dolomite more in a non-recrystallized matrix than in ooliths may indicate contemporaneous dolomitization.

[1] *Swansea* (Mem. Geol. Surv.), p. 15.

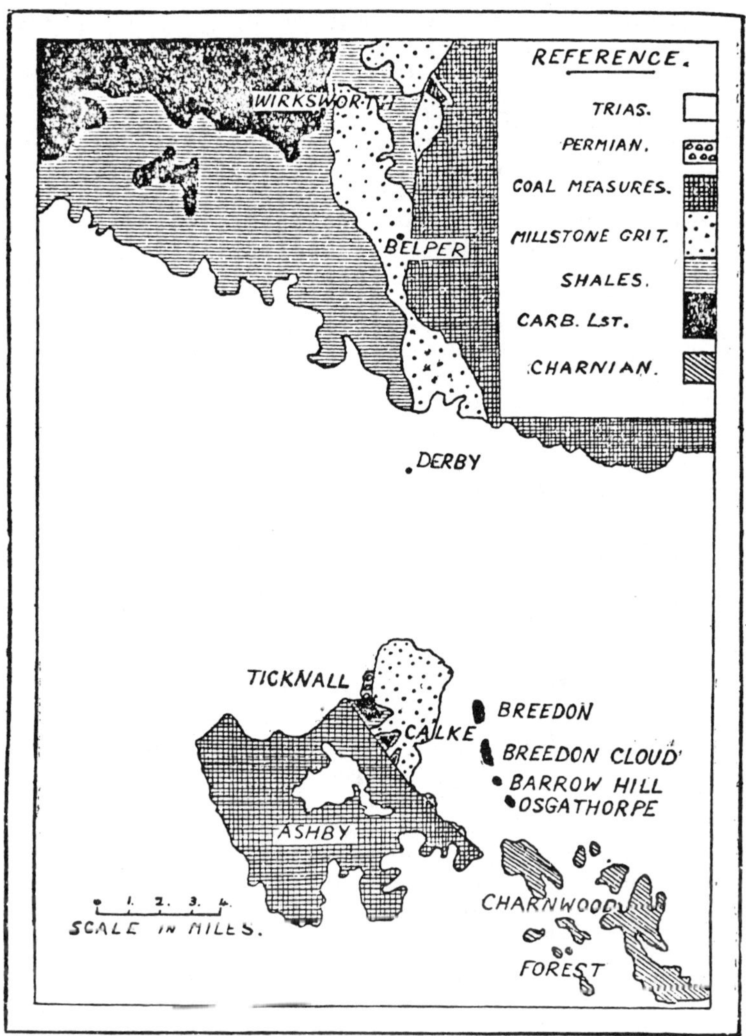

Map showing the position of the Leicestershire Dolomites compared with those of the Carboniferous Limestone of Derbyshire and of the Charnwood Pre-Cambrian.

5. The Absence or Presence of Fossils in Dolomites.

It is doubtful whether any reliable evidence of the period of dolomitization can be obtained from the absence of organic remains. A primarily precipitated dolomite would probably be deposited under conditions unfavourable to life, but it is also conceivable that either contemporaneous or subsequent dolomitization could be so complete as to obliterate all traces of organic structures that may have been present originally. In other words, the final stage of secondary dolomitization might produce the complete alteration of a rock which may have exhibited selective phenomena in its early stages of alteration.

(*r*) The presence of fossils in a dolomite is a little more significant, particularly if they are at all numerous. The argument is then against the theory of primary precipitation. In a rock showing no selective features the condition of fossil structures as casts or as replacements in dolomite, appears to yield very little evidence of contemporaneous or subsequent alteration since both casts and replacements of the same class of organisms may occur in the same bed of dolomite.

The perfect replacement by dolomite of coral structures is certainly *suggestive* of contemporaneous alteration, since such replacements have been found to occur in modern reefs before calcitic recrystallization of coral tissues took place.[1]

Part II: The Leicestershire Dolomites.

A series of faulted inliers of Carboniferous dolomites extends in a north-westerly direction from the edge of Charnwood, commencing with small patches of dolomite at the village of Osgathorpe, and ending to the north in the mass forming Breedon Hill, a landmark for many miles. Between these limits of the series are situated the dolomite hill known as Breedon Cloud and the much smaller though petrologically similar inlier called Barrow Hill. In each of these cases the Carboniferous Limestone is surrounded by unconformable Keuper. A few miles to the west of Breedon there are small valley inliers of Lower Carboniferous rocks containing bedded dolomites and dolomitic limestones at Ticknall and Calke Park. At these localities the Lower Carboniferous is succeeded conformably by Millstone Grit, which is overstepped at one or two places by Trias.

6. Classification of Petrological Types.

Disregarding details of stratigraphy and palæontology which I have described in another paper,[2] and considering the formations purely from a petrological point of view without assuming the mode of origin of the dolomite, we may distinguish in the area different types of dolomitic rocks as follows:—

Dolomites proper, containing a proportion of magnesium carbonate approaching 40%
- Dense yellow dolomites of Breedon and Breedon Cloud (1).[3]
- Red ferruginous dolomites of Breedon and Breedon Cloud (2).
- Barren grey and yellow dolomites of Ticknall (4).

Dolomitic limestones containing a relatively small percentage of magnesium carbonate
- Fossiliferous dolomitic limestones of Ticknall and Calke (3).

[1] See Cullis, "The Atoll of Funafuti": Report of Coral Reef Committee, Royal Society, 1904, section xiv, p. 407.

[2] See Abstracts of the Proc. Geol. Soc. of London, No. 1004, March 14, 1917.

[3] The numbers in parentheses indicate relative stratigraphical positions in ascending order.

In this table the types predominating in the district are placed higher in the list.

7. The Dense Yellow Dolomites of Breedon, Breedon Cloud, etc.

The bulk of the dolomites of the two Breedons, Barrow Hill, and Osgathorpe consists of dense yellow material having a specific gravity and chemical composition approaching those of a pure dolomite. The proportion of Magnesium Carbonate varies slightly in different beds, but averages nearly 40 per cent, while iron compounds and insoluble residues are present in small amounts.

The field relations of these rocks are studied best at Breedon-on-the-Hill, where quarries are being worked in a direction at right angles to the strike. At this locality more than 800 feet of fairly thick-bedded dolomites succeed one another without any marked variation in petrological characters and without any apparent complications due to faulting. Though the chemical composition and texture of the material forming one stratum may be slightly different from that of another, the inherent characters of any particular bed appear to be uniform. The dolomitization is in no case patchy (*f*).[1] Laterally the beds do not pass into unaltered or poorly dolomitic limestones (*d*), though this fact would have greater significance if the Carboniferous Limestone of the area had a larger outcrop. The absence of faulting at Breedon suggests the improbability of subsequent vein dolomitization associated with dislocation (*g*). At Breedon Cloud strike faulting does occur, but is there associated with non-patchy yellow dolomites which do not pass laterally into unaltered limestone (*h*).

Conglomerates and pseudo-breccias are not present. Microscopic sections of the yellow dolomite of Breedon, Breedon Cloud, and Barrow Hill show a fine-grained crystalline structure composed mainly of small grains more or less allotriomorphic, though some rhombohedral outlines may be seen (*j* and *i*). The degree of purity is not high since minute dusky inclusions of insoluble matter are very numerous (*k*). There are no zonal or central inclusions of hæmatite, and what little iron oxide does occur, mainly limonite, is interstitial (*l*). Chert is absent from the material exposed in the workings of Breedon-on-the-Hill, though it occurs in the yellow dolomites of a higher horizon at Breedon Cloud. Sections do not show any dolomite rhombohedra in the chert (*m*). Fossils are not numerous at Breedon, Barrow Hill, and Osgathorpe; those that do occur at these localities consist chiefly of dolomite casts of Brachiopoda and a few corals. At Breedon Cloud, however, higher beds are exposed and fossils are more plentiful. Corals are preserved as dolomite replacements and as casts (*r*). *Syringopora* is usually found as casts, but *Michelinia*, *Campophyllum*, and other genera exhibit septa, tabulæ, and other structures beautifully preserved in minutely crystalline dolomite. Goniatites (*Glyphioceras*) also occur as dolomite replacements, but Brachiopoda are present as casts.

The conclusion to be drawn from the collective evidence concerning

[1] Italic letters in parentheses refer to corresponding evidences mentioned in the earlier part of the article.

the yellow dolomite of Breedon, etc., is undoubtedly in favour of contemporaneous dolomitization.

8. The Red Ferruginous Dolomites of Breedon and Breedon Cloud.

The uppermost portion of the Carboniferous Limestone sequence seen at Breedon and Breedon Cloud is composed of several feet of thinly bedded red dolomites quite distinct from the more massive yellow dolomites below. These red dolomites cannot be seen to pass laterally into unaltered limestone (*d*), but it must be remembered that the outcrop is not very extensive. The rock is patchy in the sense that the proportion of magnesium carbonate varies in any particular stratum, but there are no external appearances analogous to those of pseudo-brecciation (*n*). Faulting occurs at Breedon Cloud, though I infer that dislocation has not been a factor of the dolomitization for the following reasons: (1) no faulting occurs at Breedon-on-the-Hill where these red dolomites are otherwise precisely similar to the corresponding rocks at Breedon Cloud, (2) the faulting at Breedon Cloud is also associated with the yellow dolomites yielding very strong evidence of contemporaneous dolomitization (*h*).

Microscopic sections exhibit a structure characterized by idiomorphic, fairly large, and very impure rhombohedra (*i*, *j*, and *k*). Well-marked central and zonal inclusions of hæmatite occur in the crystals of dolomite forming these beds (*l*) (micro-photo, Pl. XI, Fig. 1). The outer zones of the crystals are free from hæmatite, but contain other inclusions similar to those in the rhombohedra of the yellow dolomites. A few streaks and patches of recrystallized calcite are present, and it is probable that this recrystallization took place prior to dolomitization.

It has yet to be proved whether coarsely crystalline calcite, either original or recrystallized, is under any particular conditions immune from alteration to dolomite. In connexion with this question, the condition of crinoid stems and ossicles in these red dolomites is interesting. Micro-photo, Pl. XI, Fig. 2, taken from a specimen of the Breedon rock, shows a crinoid ossicle invaded by hæmatite-bearing dolomite near the central passage and around the external margin, which is badly corroded by the alteration.

Other organic structures are obscure. That other fossils were present in these rocks originally, is shown by the occasional occurrence of coral and Brachiopod casts. It appears that we have here an instance of undoubted subsequent dolomitization in which both the matrix and organic structures, with the exception of encrinites, have been completely altered. Even crinoid remains, a most stable form of calcite, have been altered to some extent.. With regard to the matrix, there is no way of ascertaining its original condition. It may have been calcite or aragonite, or a mixture of the two; or it may have been calcite recrystallized prior to dolomitization. Evidently this rock affords an illustration of the fact that conclusions are not easily made from phenomena connected with selective dolomitization.

The evidences concerning the origin of the Breedon and Breedon Cloud red dolomites indicate quite definitely that the dolomitization was subsequent and associated with waters percolating through the Trias, which formation rests upon the upturned edges of the red dolomites at these localities. It seems fairly evident that this subsequent dolomitization has attacked previously unaltered limestones (2)[1] lying between apparently contemporaneous dolomites stratigraphically lower (1) at the Breedons and higher (4) at Ticknall.

9. The Barren Grey and Yellow Dolomites of Ticknall and Calke.

About ten feet of bedded dolomites, yellow below but grey above, occur at the very top of the Carboniferous Limestone at Ticknall and Calke. They are succeeded by dark shales which pass up conformably into Millstone Grit. The chemical composition of these dolomites is similar to that of certain rocks described by Professor Skeats as " Dolomites of theoretical composition ".[2] A comparison of the results of analysis makes this evident.

	Dolomite of theoretical composition (Skeats).	Ticknall grey dolomite.
Calcium carbonate	54·7	57·1
Magnesium carbonate	45·3	38·3
Iron compounds	—	2·7
Insoluble residue	·033	1·5

In the Ticknall dolomite a small amount of free calcite is present, as shown by slides stained with Lemberg's solution.

These rocks are seen best in the old lime works of Ticknall, but the exposures are very limited in extent, so that the bedded nature of these dolomites and their apparent non-passage into unaltered limestone cannot be used as reliable evidence of their origin (*d*). There is, on the other hand, not the slightest visible development of patchy dolomitization (*f*).

A small fault occurs, but this shows no definite connexion with the origin of the dolomite. The yellow and grey varieties of the rock are very similar in their microscopic characters. Sections show a crystalline mass in which a large number of crystals have been rounded off, presumably by simultaneous development (*i*), but some rhombohedral outlines are retained, particularly in the case of some of the larger crystals (micro-photo, Pl. XI, Fig. 3). There are very numerous inclusions, consisting mainly of minute particles of insoluble matter incorporated during crystallization (*k*), but there are no central nor zonal inclusions of hæmatite (*l*), iron oxide in the form of limonite being interstitial. This feature has special significance in view of the fact that reddish rocks of Permian and Triassic ages rest upon the limestone in the north-west corner of the Ticknall exposures. Chert is absent, and organic remains, if present originally, have been completely obliterated (*r*). According to the

---

[1] Numbers refer to stratigraphical positions given in the classification of petrological types.

[2] Q.J.G.S., 1905, p. 105.

evidence it seems reasonable to infer that the Ticknall dolomites are of contemporaneous origin.

10. Fossiliferous Dolomitic Limestones of Ticknall and Calke.

Certain limestones occurring at a slightly lower horizon than that of the barren dolomites of Ticknall, are interesting mainly on account of their representing an incomplete stage in the process of dolomitization. These limestones are best studied at Ticknall, but some of the Calke specimens yield very fine slides showing "selective" phenomena. The amount of magnesium carbonate does not exceed 16 per cent. As in the case of the barren dolomites above, these rocks occur apparently in definite beds and are not associated with faulting of any importance. Chert is absent. Microscopic sections show idiomorphic crystals of dolomite having a fair degree of purity (k), and devoid of zonal hæmatite inclusions (l). The rhombohedra show a decided preference for organic structures (p), both coralline and brachiopod, though dolomitization occurs to some extent in the matrix. Micro-photo, Pl. XI, Fig. 4, shows rhombohedra developed in organic structures, the matrix consisting of recrystallized calcite in places. Though the features shown by this photograph are typical, some individual crystals are developed partly in an organism and partly in the matrix. The rock being of a mixed organic nature, there is no knowledge of the relative proportions of aragonite and calcite muds in the original matrix, and even if this point could be decided there would still be the question of the comparative stability of coarser aragonite and of calcite mud to be determined. In one or two cases dolomite crystals are entirely surrounded by recrystallized calcite, from which it appears that either the recrystallization took place after the formation of dolomite or recrystallized calcite has been converted into dolomite. The inference that dolomitization was prior to recrystallization would tend to support the theory of contemporaneous origin. The selective phenomena exhibited by the Ticknall and Calke dolomitic limestones do not, in my opinion, yield any conclusive evidence concerning the period of dolomitization, and though the balance of evidence derived from other sources may be slightly in favour of contemporaneous alteration, it may be wiser to consider the matter "not proven" since some of the evidence is of a conflicting nature.

Having arrived at the general conclusion that most of the dolomites of the Leicestershire area are of contemporaneous origin, may I suggest that they appear to have been formed in shallow portions of the Carboniferous sea situated in the Charnwood region. The dolomitic nature of these rocks compared with that of the more normal limestones occurring at the same horizons (D_1 and D_2) in Derbyshire may thus be explained. The case may be somewhat similar to that of the *laminosa* dolomites of the Bristol area, which are situated about twenty-five miles distant from the more normal limestones of the Mendips. The presence of a rich Lamellibranch fauna in the D_2 subzone in Leicestershire, and the presence of only a few feet of Carboniferous Limestone at places south of Osgathorpe as shown by a boring at Desford where the Carboniferous Limestone was found to rest on Pre-Cambrian, supply evidence that shallow-water conditions existed in the area during Carboniferous times.

EXPLANATION OF PLATE XI.

MICROPHOTOGRAPHS OF THE LEICESTERSHIRE DOLOMITES.

FIG. 1.—Red dolomite, Breedon, Leicestershire. Idiomorphic rhombohedra, having central zonal inclusions of hæmatite and fairly clear outer zones. × 25.

,, 2.—Red dolomite, Breedon. Calcite of crinoid ossicle invaded by fine-grained hæmatite-bearing dolomite. Matrix completely dolomitized. × 25.

,, 3.—Grey dolomite, Ticknall, south - eastern border of Derbyshire. A mosaic of rather allotriomorphic grains devoid of zonal hæmatite but having many dusky inclusions. A few rhombohedral outlines are shown. A little limonite is interstitial. × 25.

,, 4.—Fossiliferous dolomitic limestone, Calke Park, Derbyshire. Typical section showing preference of dolomite for organic structures. The matrix is partly composed of recrystallized calcite. × 25.

GEOL. MAG.. 1918.

PLATE XI.

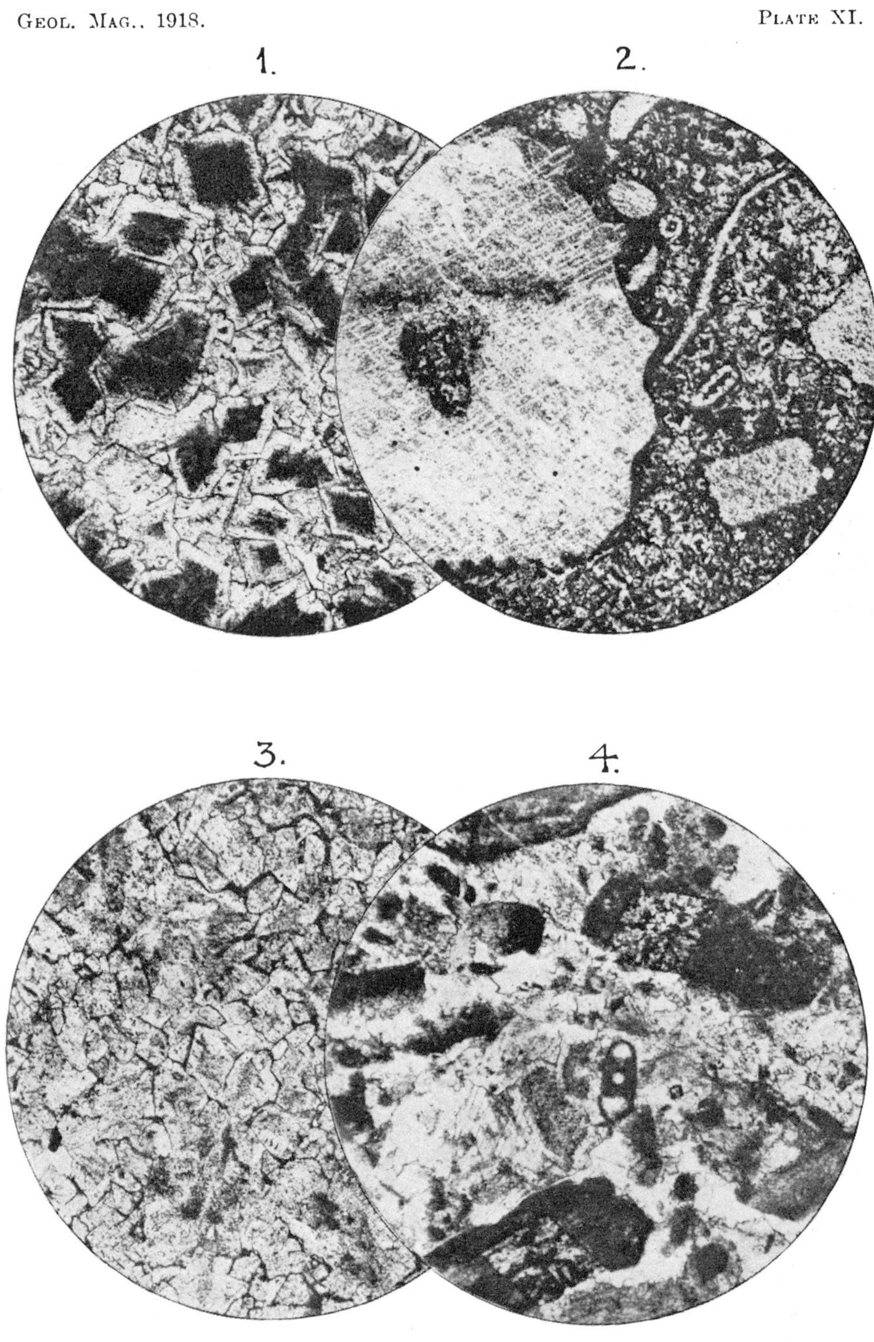

G. S. Sweeting, photo.

Bale, imp

4

RELATIONSHIP BETWEEN Ca/Mg RATIO AND GEOLOGIC AGE[1]

GEORGE V. CHILINGAR[2]
Los Angeles, California

ABSTRACT

The analyses of numerous limestone samples showed that there is a general increase in the average Ca/Mg ratio in going up the geologic column with superimposed periodic fluctuations. A general decline in abundance of dolomites in going up the geologic column was previously shown by R. A. Daly (1910). The writer, however, failed to establish a simple relationship between the Ca/Mg ratio and the age of carbonate rocks. The cyclic occurrence of dolomitic and calcitic limestones, observed by the writer, suggests that the formation of dolomites was occurring in cycles with a gradual accumulation of magnesium in sea water until favorable conditions were established for the formation of dolomites.

Inasmuch as a high CO_2 pressure of the atmosphere during pre-Cambrian and Paleozoic time is advanced by Russian geologists as a possible explanation for the abundance of dolomites, an attempt was made to precipitate dolomite directly out of sea water at high CO_2 pressure. This attempt was successful and indicated that some pre-Cambrian and Paleozoic dolomites may have been formed by direct chemical precipitation out of sea water.

EVOLUTION OF DOLOMITES

The Ca/Mg ratios[3] of limestones and dolomites are not erratic and unpredictable as might appear at first glance. Daly (1910, p. 165) and Chilingar (1953, p. 202) demonstrated that the average Ca/Mg ratio of limestones and dolomites in the United States and Canada decreases with age of sediments. Figure 1 and Table I, which demonstrate this relationship, are based on the results obtained by Daly (61 per cent) and the present investigator (39 per cent). The Ca/Mg ratios of some limestones, dolomites, and recent calcareous sediments are presented in Table II.

In explaining this gradual diminution in number of dolomites in younger rocks, Daly (1907, p. 96) assumed that the bottom temperatures of the pre-Cambrian seas were relatively high. Animal remains falling to the sea bottom were undergoing putrefaction and the resulting ammonium carbonate increased the pH and carbonate ion concentration, thus precipitating $CaCO_3$. This immediate precipitation, therefore, made the calcium salts unavailable for incorporation into the carbonate shells. Daly failed, however, to mention that organic decay will also generate CO_2 and fatty and other acids, which tend to keep

[1] Manuscript received, May 16, 1956.

[2] University of Southern California.

This writer is greatly indebted to T. Clements, W. H. Easton, K. O. Emery, O. L. Bandy, C. M. Beeson, and R. H. Merriam, of the University of Southern California, under whose direction this study was undertaken, and whose help and critical advice were invaluable in carrying it to completion.

The investigator would also like to express appreciation to J. E. Lamar of the Illinois Geological Survey, H. Gould of the University of Washington, G. Kanakoff of the Los Angeles County Museum, and H. J. Bissell and D. Clark of Brigham Young University for supplying numerous samples of limestones, dolomites, and recent calcareous sediments.

Some chemical analyses of carbonate rocks have been done by Griffin-Hasson Laboratories, Los Angeles, California. The help extended by A. Dollar, the chief chemist of the Griffin-Hasson Laboratories, and Wilson Orr of the University of Southern California, is also greatly appreciated.

[3] The Ca/Mg ratios given in the present work are weight ratios.

TABLE I. EVOLUTION OF DOLOMITES

Age	No. of Samples Analyzed	Average Ca/Mg Ratio
Pre-Cambrian	70	4.0 : 1
Cambrian	40	4.2 : 1
Ordovician	100	3.5 : 1
Silurian	250	3.0 : 1
Devonian	160	7.0 : 1
Mississippian ⎫ Pennsylvanian ⎬ Permian ⎭	400	16 : 1
Cretaceous	85	56 : 1
Tertiary	50	53 : 1
Quaternary and Recent	250	40 : 1

TABLE II. Ca/Mg RATIO OF SOME LIMESTONES, DOLOMITES, AND RECENT CALCAREOUS SEDIMENTS*

Age, Name, and Locality	Ca/Mg Ratio
Recent	
Calcareous sediments on Great Bahama Bank, Bahamas (Ca/Mg ratio increases with greater distance from shore)	19–170
Shore material, west side of Andros Island, Bahamas	8.9
Calcareous shallow-water deposits, off Florida	30–140
Beach sand, east side of Sands Key, Florida	35
Tortugas, Florida (depth of 60 feet)	64
Australian algal limestone, southeastern corner of Australia (Limestone "Biscuit")	34
Honeycombed algal limestone, Karatta Lake	10
Coral sand, Achang Reef, Guam	27– 28
Coral sand, Pago Reef, Guam	14– 23
Agana Beach, Guam	36
Fine calcareous sand and silt, Cocos Lagoon, Guam	18– 30
Globigerina ooze, 11° 16′ N., 165° 30′ E., depth of 2,010 fathoms (core)	105
Globigerina ooze, 11° 22.5′ N., 165° 31′ E., depth of 1,550 fathoms (core)	77
Globigerina ooze, 11° 59.5′ N., 164° 44′ E., 840 fathoms (dredge)	306
Great Isaac Bank, Caribbean Sea	30– 45
Persian Gulf sediments	3– 12
Pleistocene	
Fresh-water marl, Confusion Range, Utah	31
Lomita marl member of San Pedro formation, San Pedro, California	44– 54
Eocene	
Sables dolomitiques de la Forêt de Saint-Gobain, Aisne, France	17– 19
Glauconitic limestone, Conflans-Sainte-Honorine, Seine-et-Oise, France	26
Triassic	
Schlern dolomite, southern Tyrol, Austria	1.63–3
St. Cassian limestone, Sett Pass, Austria	9– 21
Wengen limestone, southern Tyrol	19
Buchenstein limestone, southern Tyrol	13– 36
Giau Pass limestone, southern Tyrol	17
Lower Muschelkalk dolomite, southern Tyrol	6
Jurassic	
Hydraulic limestone, Wassy, Yonne, France	50

* The chemical analyses of Illinois and Indiana limestones and dolomites are in close accord with those reported by Krey and Lamar (1925) and Patton (1910).

TABLE II (*continued*)

Age, Name, and Locality	Ca/Mg Ratio
Argillaceous limestone, Wassy, Yonne, France	40
Saint-Girons dolomite, south of Saint Girons, Ariege, France	1.7
Cretaceous	
Sables de Dizy-le-Gros, Aisne, Paris basin	2– 3
Sables de Beynes, Seine-et-Oise, France	2
Craies dolomitique de Beynes, Seine-et-Oise, France	4
Craies magnesiennes de Bimont, Oise, France	70– 88
Permian	
Arcturus formation (crinoidal lower part), Illipah Quadrangle, Nevada	84
Arcturus formation (arenaceous), Moorman Ranch, Illipah Quadrangle, Nevada	29– 31
Middle of Permian section near Mitchell's Caverns, California	18– 33
Pennsylvanian	
La Salle limestone, La Salle County, Illinois	6–120 (Avg. 30)
McLeansboro limestone, Illinois	4–120 (Avg. 49)
Ely limestone, Illipah Quadrangle, Nevada	70–132
Pontiac limestone, Illinois	70–300
Lonsdale limestone, Peoria County, Illinois	30– 90 (Avg. 58)
Bird Spring limestone, Nopah Range, California	28– 37
Maria Creek limestone, Freelandville, Sullivan County, Indiana	18
Minchall limestone, Portersville, Daviess County, Indiana (Ca/Mg ratio decreases from bottom to top of outcrop)	16– 44
St. David limestone, St. Clair County, Illinois	89
Seahorne limestone, Schuyler County, Illinois	143
Knobby limestone, Scott County, Illinois	211
Mississippian	
Kinkaid limestone, Harrisburg, Illinois	41
Clore limestone, Illinois	31– 57
Menard limestone, Illinois	13– 56 (Avg. 39)
Vienna limestone, Illinois	14
Okaw limestone, Illinois	30– 70 (Avg. 40)
Glen Dean limestone, Martin County, Indiana	7– 25
Golconda limestone, Illinois	35– 64 (Avg. 40)
Beach Creek limestone, Owen County, Indiana	57– 72
Beaver Bend limestone, Monroe County, Indiana	142
Paoli limestone, Indiana	95–165
Renault limestone, Illinois	80– 90
Aux Vases sandstone, Indiana	2– 35
Ste. Genevieve limestone, Indiana	15–250 (Avg. 45)
Ste. Genevieve limestone, Illinois	20–200
Ste. Genevieve limestone (Fredonia oölite member), Harrison County, Indiana (Ca/Mg ratio decreases from bottom to top of outcrop)	18–160
Ste. Genevieve limestone (Levias limestone member), Owen County, Indiana (Ca/Mg ratio increases from bottom to top of outcrop)	86–130
St. Louis limestone, Indiana	6– 60 (Avg. 20)
St. Louis limestone, Illinois	2–309 (Avg. 40)
Salem limestone, Indiana	6–141
Salem limestone, Illinois	4–157 (Avg. 30)
Harrodsburg limestone, Bloomington, Monroe County, Indiana	4– 56
Keokuk limestone, Illinois	6–136 (Avg. 30)
Bullion limestone member of Monte Cristo formation, near Mitchell's Caverns, California	16– 17
Burlington limestone, Illinois	19–162 (Avg. 60)
Anchor limestone member of Monte Cristo formation, near Mitchell's Caverns, California	11– 18
Chouteau limestone, Calhoun County, Illinois	40

TABLE II (*continued*)

Age, Name, and Locality	*Ca/Mg Ratio*
Lodgepole limestone, Madison group, Fergus County, Montana	21
Dinantian dolomite, Flaumont-Waudrechies, Nord, France	25
Devonian	
Dolomie des Noces du Boulonnais, France	2– 3
Beechwood limestone, Blocher, Scott County, Indiana	7– 8
Silver Creek limestone, Indiana	2– 5
Speed limestone, Speed, Clark County, Indiana	23
Cedar Valley limestone, Jersey County, Illinois	117
Jeffersonville limestone, Bartholomew County, Indiana	2– 29
Devils Gate formation (uppermost beds), near Hamilton, Nevada	55– 59
Hamilton limestone, Illinois	5–300 (Avg. 28)
Grand Tower limestone, Jackson County, Illinois	48
Clear Creek formation, Alexander County, Illinois	21
Geneva dolomite, Indiana	1.8– 3
Bailey limestone, Union County, Illinois	11– 15
Silurian	
Kenneth limestone, Kokomo, Howard County, Indiana	126
Kokomo limestone, Kokomo, Howard County, Indiana	8
Niagaran dolomite, Chicago, Illinois	1.63– 5 (Avg. 1.7)
Racine dolomite, Illinois	1.66–1.7
Joliet dolomite, Illinois	1.7
Waukesha dolomite, Illinois	1.7
Port Byron dolomite, Illinois	1.68
Huntington dolomite (reef), Monon, Indiana	1.63–1.9
Liston Creek formation, Sweetser, Grant County, Indiana	24
Mississinewa shale, Sweetser, Grant County, Indiana	1.7
Louisville limestone, Clark County, Indiana	6
Laurel limestone, Indiana	2.5– 8
Osgood formation, New Point, Decatur County, Indiana	5.6
Brassfield limestone, Decatur County, Indiana	124
Kankakee limestone, Illinois	1.7– 26
Sexton Creek limestone, Jersey County, Illinois	106
Edgewood limestone, Illinois	2– 26
Girardeau limestone, Alexander County, Illinois	126
Ordovician	
Maquoketa formation, Illinois	1.63– 13
Kimmswick limestone, Illinois	60–270 (Avg. 70)
Platteville limestone, Illinois	1.8– 86 (Avg. 4.7)
Galena dolomite, Illinois	1.7–1.8
Elkhorn formation, Richmond, Wayne County, Indiana (Ca/Mg ratio decreases from bottom to top of outcrop)	2– 18
Plattin limestone, Calhoun County, Illinois	10
Joachim limestone, Calhoun County, Illinois	2.5
Whitewater formation, Indiana	11– 37
Shakopee dolomite, Illinois	1.7–3.2
Saluda limestone, Dearborn County, Indiana	2.8– 46
Cambrian	
Algal limestone in Pioche shale, Pioche, Nevada	94

$Ca^{··}$ and $Mg^{··}$ in solution. A limited decomposition, mainly anaerobic, of the organic matter may, however, temporarily create a high pH necessary for the precipitation of $CaCO_3$. Many carbonate concretions in Cretaceous shales of the Magdalena Valley of Colombia, South America, contain such marine organisms

62

as ammonites and fish (Weeks, 1953). On putting a dead fish into a beaker with sea water for several days, the present investigator obtained a white film of $CaCO_3$ precipitate on the body of the fish.

The Ca/Mg ratio of the pre-Cambrian limestones (4.0:1) is very close to the Ca/Mg ratio of the present-day rivers draining terranes, which mostly consist of granite, phyllite, mica-schist, and basalt (Clarke, 1924, pp. 63–121). For example, the Ca/Mg ratio of Ottawa River (above Ottawa city), which drains Archean igneous and metamorphic terranes in Canada, is 3.69:1 (Daly, 1910, p. 159). This appears to indicate that the river-borne calcium and magnesium were wholly precipitated upon entering the sea.

A possible explanation for the evolution of dolomites is the selective return of calcium to the lands. There is a selective weathering of calcium over magnesium in the sediments, and a gradual increase with time of Ca/Mg ratio in solutions contributing to the sea. Steidtmann (1911, p. 394) showed that Ca/Mg ratio of fresh limestones is many times higher than that of their altered equivalents. He also showed that the percentage loss of calcium through weathering of acid igneous rocks averages higher than magnesium, and that the Ca/Mg ratios of streams are usually higher than the Ca/Mg ratios of terranes through which they flow. The writer also observed that the weathered limestone samples have lower Ca/Mg ratios than their fresh equivalents (Chilingar, 1953, p. 307).

Permanent loss of muds, which have very low Ca/Mg ratios, from lands is another reason for the selective and permanent loss of magnesium from the continents.

A possible reason for the decrease in Ca/Mg ratio since the Cretaceous is the fact that pelagic foraminifera started to extract great quantities of calcium out of the sea water and deposit it in the oceans during and after Cretaceous. This calcium is thus withdrawn from the cycle and never returned to the lands. If this removal of calcium to the deep seas should continue, the Ca/Mg ratio of sediments might drop down to the value of pre-Cambrian time (4.0:1). The extrapolation of the curve of evolution of dolomites by the writer shows that the Ca/Mg ratio might drop down to 3:1 in about 100 million years (Fig. 1).

One objection against the removal of calcium from the cycle by foraminifera is the fact that the magnesium in pelagic Foraminifera is also removed simultaneously. The average Ca/Mg ratio of globigerina ooze, according to Kuenen (1941, p. 165), is 24.8, and several samples of globigerina ooze (depth of 840–2,010 fathoms) analyzed by the writer showed a Ca/Mg ratio range of 70–300. Consequently, more calcium than magnesium is being removed from the cycle, and the former objection appears to be invalid.

Direct precipitation of dolomites out of sea water during pre-Cambrian and Paleozoic time.—Soviet scientists believe that high P_{CO_2} existed during the pre-Cambrian time. If all the carbon of shales, coals, and limestones were returned to the atmosphere, the amount of CO_2 would be increased several thousand times. The formation of great masses of iron carbonate during pre-Cambrian time also supports the former belief. According to Vinogradov (1940, p. 235), the unsaturation of sea water with $CaCO_3$ due to the excess of CO_2 also prevented invertebrates from accumulating $CaCO_3$ in their tissues. Noting, however, that the total Ca content of present-day oceans is 552,800,000,000,000 metric tons and its yearly increment is 557,670,000 tons (Clarke, 1924, p. 138), it would take only about 2 million years to supersaturate the oceans with Ca. In order to determine the validity of Vinogradov's explanation, therefore, it is necessary to know the CO_2 pressure of the pre-Cambrian atmosphere, pH of the sea water, and the corresponding $CO_3'' : HCO_3'$ ratio.

Strakhov (1953, p. 24) believes that due to the high content of CO_2 in the atmosphere and the resulting high concentration of $Ca\cdot\cdot$ and $Mg\cdot\cdot$ in sea water during pre-Cambrian and Paleozoic time, the dolomites were precipitated directly

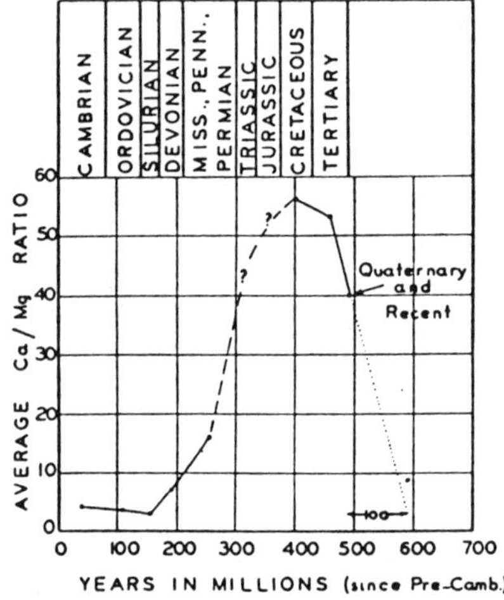

Fig. 1.—Changes in average Ca/Mg ratios with time.

out of sea water. Strakhov points out that direct precipitation of dolomite is taking place even at the present time in the eastern parts of Lake Balkhash and in its shallow-water embayments, upon evaporation in the arid climatic conditions. The water in Lake Balkhash is of a hydrocarbonate type with the pH varying from 81. to 9.3. Strakhov also believes that most diagenetic dolomites are post-Paleozoic.

In testing Strakhov's explanation for the direct precipitation of dolomites, the writer successfully precipitated dolomite out of sea water. One liter of sea water was placed in a container provided with an electrical stirrer. After adding 21.3 grams of $MgCO_3$ and 1.97 grams of $CaCO_3$ to this sea water (maximum (?) solubilities of $MgCO_3$ and $CaCO_3$ in sea water at CO_2 pressure of 4 atmospheres),[4]

---

[4] Possibly, these are not the absolute solubilities, because the solid and liquid phases were kept in contact only for a period of one week.

the container was placed in an autoclave with CO_2 pressure of 4 atmospheres. The Ca/Mg ratio of the precipitate, which formed at the end of 2 weeks, was equal to 1.8. Although the precipitated particles were too small for determination of their optical properties, the refractive index was that of a dolomite (1.7). The X-ray analysis also established the presence of dolomite.

All previous attempts of geologists and chemists in precipitating dolomite out of sea water failed because none of the investigators used high CO_2 pressure.

Lack of evolution of dolomites in Europe.—As indicated previously, Fig. 1 is entirely based on the chemical analyses of limestones and dolomites in Canada and the United States. An attempt to show the existence of a similar evolution in Europe was unsuccessful largely because of the abundance of dolomites in the Triassic and Jurassic sediments of Europe. These dolomites rival the pre-Cambrian and Cambrian dolomites of America. The lack of evolution of dolomites in Europe, however, can be determined definitely only by a field geologist working in Europe.

CYCLIC OCCURRENCE OF DOLOMITES OR HIGHLY DOLOMITIC LIMESTONES

The writer has observed a cyclic occurrence of highly dolomitic limestones. For example, Figure 2 shows the periodic occurrence of dolomitic and calcitic lime-

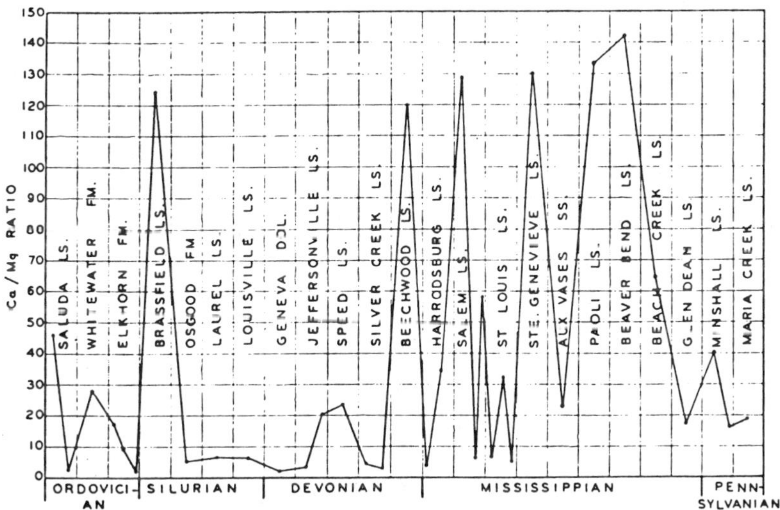

FIG. 2.—Ca/Mg ratio in limestones and dolomites of central and southern Indiana.

stones in central and southern Indiana. These periodic fluctuations are superimposed on the general decline in number of dolomites (or increase in average Ca/Mg ratio) in going up the geologic column.

A possible explanation of the cyclic occurrence may be the periodic formation of dolomites. The first stage of the cycle would involve accumulation of Mg in the oceans and a simultaneous deposition of calcitic limestones. The second stage would be the formation of dolomites and dolomitic limestones when the Mg content of the oceans reaches a certain limit. The formation of dolomites would eventually impoverish the sea water of Mg and a new cycle, with accumulation of Mg, would start.

Other possible explanations for the periodic formation of highly dolomitic limestones include variations in depth and temperatures of basins of deposition. Numerous samples from each limestone have been analyzed in order to preclude sampling errors, such as reporting the Ca/Mg ratio of a locally dolomitized area.

Accumulation of magnesium in oceans at present time.—The Ca and Mg content of river waters is 20.39 per cent and 3.41 per cent of the total dissolved solids, respectively, and of the oceans is 1.15 per cent and 3.69 per cent. As 2.735×10^9 tons of dissolved material are brought by the rivers to the oceans every year, the annual increment of magnesium is equal to $(2.735 \times 10^9) \times 3.41$ per cent $= .934 \times 10^8$ tons. The amount of Ca which is contributed to sea water annually is about $(2.735 \times 10^9) \times (20.39 - 1.15)$ per cent $= 5.26 \times 10^8$ tons, which under steady-state conditions would be precipitated annually. Due to the fact that the average Ca/Mg ratio of recent sediments is around 40:1, the annual precipitation of magnesium amounts to $(5.26 \times 10^8)/40 = .13 \times 10^8$ tons. With the total Mg content of 1.721×10^{15} tons in the oceans (Clarke, 1924, p. 138) and the annual increment of $(.934 \times 10^8) - (.13 \times 10^8) = .804 \times 10^8$ tons, it would, therefore, take about 21 million years to accumulate the present-day content of Mg in the oceans $[(1.721 \times 10^{15})/(.804 \times 10^8) \cong 21]$. It should be also remembered, however, that to-day's rate of inflow of Ca and Mg is probably much greater than during much of the geologic time.

This continuous increase in the Mg content of the oceans may eventually again produce favorable conditions for the extensive formation of dolomites.

Frequency distribution of Ca/Mg ratios in limestones and dolomites.—The frequency distribution of Ca/Mg ratios in limestones and dolomites is shown in Figure 3. This figure is based on chemical analyses of 3,500 limestone samples, of which 1,000 samples have been analyzed by the present investigator and the Griffin-Hasson Laboratories, Los Angeles, California. The remainder of the chemical analyses were obtained from the American and European literature. The predominance of dolomites with Ca/Mg < 2 and limestones with Ca/Mg > 32 is clearly shown in Figure 3. This predominance of pure limestones and dolomites possibly indicates that the margin between the formation of limestones and dolomites is a very narrow one. Probably small changes in the chemical composition of sea water may bring about drastic changes in the composition of limestones.

The more recent limestones are largely calcitic, whereas the older ones comprise both limestones and dolomites. Abundant calcitic limestones can be also found in the pre-Cambrian sediments.

RELATIONSHIP BETWEEN Ca/Mg AND Sr/Ca RATIOS

The Sr/Ca ratio is of growing importance in age determinations; therefore, an attempt was made to correlate it with the Ca/Mg ratio. According to Kulp, Turekian, and Boyd (1952), the Sr/Ca ratio of fossils depends on salinity and composition of the original ocean environment, the ratio of aragonite to calcite in the original shell, and the subsequent recrystallization history. The tempera-

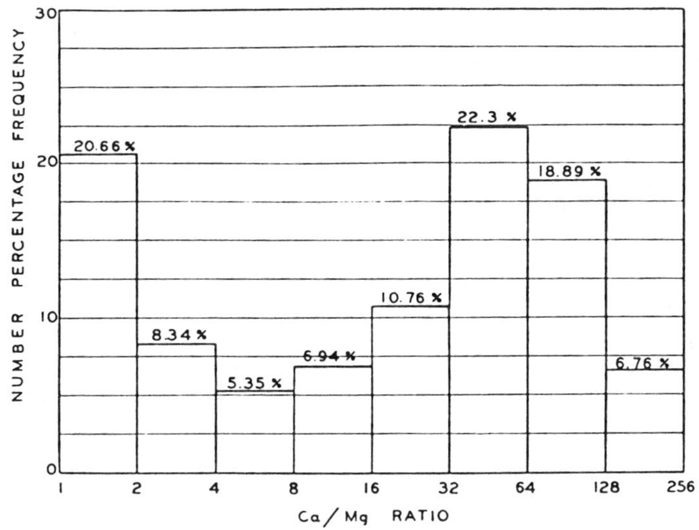

FIG. 3.—Frequency distribution of Ca/Mg ratios in limestones and dolomites.

ture has a very minor effect on the Sr/Ca ratio, whereas the Ca/Mg ratio is largely dependent on temperature.

Inasmuch as the strontium is assimilated more readily by aragonite than by calcite, whereas the reverse is true of magnesium, some relationship between Sr/Ca and Ca/Mg ratios was expected. The tabulation of the two ratios as determined for a considerable number of limestones (Table III), however, fails to show any correlation. These results are in close accord with those of Kulp, Turekian, and Boyd (1952).

USE OF Ca/Mg RATIO FOR IDENTIFICATION OF LIMESTONES AND DOLOMITES

A general relationship between the Ca/Mg ratio and age became evident during the process of the present investigation. There is a decline in abundance of dolomites (or increase in the average Ca/Mg ratio) in going up the geologic column, with superimposed periodic fluctuations of calcitic and dolomitic limestones. Experimental results obtained by the writer also indicate that direct precipitation of dolomite out of the sea water was probably possible during pre-Cambrian and Paleozoic time when the CO_2 pressure in the atmosphere was

TABLE III. RELATIONSHIP BETWEEN Ca/Mg AND Sr/Ca RATIOS

System and Formation*	Ca/Mg	Sr/1,000 Ca
Recent		
Globigerina ooze	70–110	9–15
Pennsylvanian		
Maria Creek ls.	18.7	0.91
Minshall ls.	43.5	0.70
Mississippian		
Glen Dean ls.	23.0	0.55
Beach Creek ls.	71.5	0.53
Beaver Bend ls.	142.0	0.47
Paoli ls.	164.5	0.56
Aux Vases ss.	36.0	0.48
Ste. Genevieve ls.	130.1	0.30
St. Louis ls.	29.0	0.36
Salem ls.	127.2	0.29
Devonian		
Beachwood ls.	7.9	1.85
Silver Creek ls.	4.4	3.90
Speed ls.	23.5	2.91
Jeffersonville ls.	28.7	2.20
Geneva dol.	2.0	1.01
Silurian		
Kenneth ls.	126.5	1.70
Kokomo ls.	8.8	1.72
Huntington dol.	1.9	1.00
Louisville ls.	6.6	1.62
Liston Creek fm.	24.3	1.02
Laurel ls.	2.8	1.50
Mississinewa sh.	1.7	2.60
Osgood fm.	5.6	1.55
Brassfield ls.	124.0	0.83
Ordovician		
Elkhorn fm.	9.5	0.81
Whitewater fm.	36.5	1.73
Saluda ls.	45.9	0.21

* *Globigerina* ooze is also included here.

probably higher than it is to-day. There is no simple relation between the Ca/Mg ratio and the age of carbonate rocks, however, and the writer failed to find a method for determining the age of limestones from the average Ca/Mg ratios.

If the Ca/Mg ratios of all limestones and dolomites in a region are known, a field geologist unfamiliar with the area and fossils may be able to identify a limestone from its Ca/Mg ratio. As one analysis by the Versenate technique takes only about 25 minutes (Chilingar and Terry, 1954, p. 368), this method seems to be quite practicable.

CONCLUSIONS

The findings of the present study are summarized in the order of their presentation in the text.

1. There is no simple relation between the Ca/Mg ratio and the age of carbonate rocks. There is, however, a general decline in abundance of dolomites (or increase in the average Ca/Mg ratio) in going up the geologic column, with superimposed periodic fluctuations of calcitic and dolomitic limestones.

2. Dolomite can precipitate directly out of sea water upon evaporation at high CO_2 pressure.

3. The chemical analyses of 3,500 samples of limestones and dolomites show the predominance of dolomites with Ca/Mg $<$ 2 and limestones having Ca/Mg $>$ 32. The dominance of pure limestones and dolomites has been previously brought out by Steidtmann (1917, p. 450).

4. There seems to be no relationship between the Sr/Ca and Ca/Mg ratios.

The future line of research suggested by this work is the establishment of the upper and lower limits of CO_2 pressure at which dolomite can precipitate directly out of sea water. This might shed some light upon CO_2 pressure of atmosphere during the pre-Cambrian and Paleozoic time.

BIBLIOGRAPHY

CHILINGAR, G. V., 1953, "Use of Ca/Mg Ratio in Limestones as a Geologic Tool," *Compass*, Vol. 30, No. 4, pp. 202–09.

———, AND TERRY, R. D., 1954, "Simplified Techniques of Determining Calcium and Magnesium Content of Carbonate Rocks," *Petrol. Eng.*, Vol. 26, No. 12, pp. 368–70.

CLARKE, F. W., 1924, "The Data of Geochemistry," *U. S. Geol. Survey Bull. 770.* 841 pp.

DALY, R. A., 1907, "The Limeless Ocean of Pre-Cambrian Time," *Amer. Jour. Sci.*, Vol. 23, pp. 93–115.

———, 1910, "First Calcareous Fossils and the Evolution of Limestone," *Bull. Geol. Soc. America*, Vol. 20, pp. 153–70.

KREY, F., AND LAMAR, J. E., 1925, "Limestone Resources of Illinois," *Bull. Illinois Geol. Survey 46*, pp. 311–33.

KUENEN, P. H., 1941, "Geochemical Calculations Concerning the Total Mass of Sediments in the Earth," *Amer. Jour. Sci.*, Vol. 239, pp. 161–90.

KULP, J. L., TUREKIAN, K., AND BOYD, D. W., 1952, "Strontium Content of Limestones and Fossils," *Bull. Geol. Soc. America*, Vol. 63, pp. 701–16.

PATTON, J. B., 1949, "Crushed Stone in Indiana," *Indiana Dept. Conserv., Geol. Div., Prog. Rept. 3.* 47 pp.

STEIDTMANN, E., 1911, "Evolution of Limestone and Dolomite," *Jour. Geol.*, Vol. 19, pp. 323–45, 393–428.

———, 1917, "Origin of Dolomite as Disclosed by Stains and Other Methods," *Bull. Geol. Soc. America*, Vol. 28, pp. 431–50.

STRAKHOV, N. M., 1953, "Diagenesis of Sediments and Its Significance for Sedimentary Ore Formation," *Akad. Nauk SSSR Izv.*, Ser. Geol., No. 5, pp. 12–49.

VINOGRADOV, A. P., 1940, "On the Cause of the Absence of Calcareous ($CaCO_3$) Skeletons in Pre-Cambrian Invertebrates," *Akad. Nauk SSSR Doklady*, Tom. 27, No. 3, pp. 232–35.

WEEKS, L. G., 1953, "Environment and Mode of Origin and Facies Relationships of Carbonate Concretions in Shales," *Jour. Sed. Petrology*, Vol. 23, No. 3, pp. 162–73.

5

SOME HYDROTHERMAL SYNTHESES OF DOLOMITE AND PROTODOLOMITE[1]

DONALD L. GRAF AND JULIAN R. GOLDSMITH
Illinois Geological Survey and University of Chicago

ABSTRACT

The formation of dolomite from a number of metastable carbonate materials has been investigated. Aragonite–basic magnesium carbonate mixtures, Ca-Mg carbonate gels, and magnesian calcites produced by calcareous algae and echinoids were treated in the presence of H_2O and $H_2O + CO_2$, in the temperature range from 25° C. to 450° C. The dolomite produced at the higher temperatures is the stable material of 1:1 composition, ordered with respect to Ca and Mg. The dolomite-like materials, or *protodolomites*, produced in some runs at lower temperatures are somewhat calcium-rich and do not show superstructure reflections. They appear to be intermediate stages in the development of the stable configuration.

We believe that the necessity of attaining an ordered arrangement of Ca and Mg ions at rapid rates of crystallization is responsible for the difficulty of forming dolomite at low temperatures. At temperatures above about 500° C., ionic mobility is great enough for dolomite to form, even by dry, solid-state cation diffusion, in reasonable experimental times. Diffusion rates at earth-surface temperatures are, however, so greatly reduced that an aqueous environment appears to be necessary. Even then, cation ordering must take place essentially by surface diffusion rather than by internal rearrangement. This greatly increases the difficulty of both dolomite nucleation and subsequent growth upon either ordered or simple rhombohedral carbonate substrates. As a result, a metastable phase or phase assemblage is the most probable initial precipitate.

INTRODUCTION

The manner in which dolomite is formed in nature, particularly in sedimentary environments, has been the subject of much speculation. Various authors have advocated a variety of origins, including primary precipitation from sea water, authigenesis or penecontemporaneous alteration of $CaCO_3$ by sea water, and postdiagenetic secondary replacement of $CaCO_3$ by ground water, connate water, or other solutions. Field and petrographic studies have failed to yield generally accepted conclusions. A considerable variety of laboratory experimentation also has been concerned with the dolomite problem, but no integrated picture has emerged.

Prior to 1924 the crystal structure of dolomite was unknown. Inasmuch as interest in the dolomite problem dates back more than a century, the great bulk of field and laboratory observations was not obtained or evaluated from a crystal-chemical viewpoint. More recently, the concepts of order-disorder and substitutional solid solution have been applied in studies of a number of mineral groups.[2]

Graf and Goldsmith (1955) experimentally determined the equilibrium thermal decomposition curve of dolomite, and also the stability relations of magnesian calcites in equilibrium with dolomite and in equilibrium with periclase, at temperatures from about 500°

---

[1] A portion of this material was presented at a Conference on Carbonate Rocks held at the University of Michigan on February 12, 1955. Manuscript received June 10, 1955.

[2] With specific reference to the rhombohedral carbonates, calcite may be considered as made up of alternating layers of Ca^{++} and CO_3^- ions. The Mg^{++} ions in magnesian calcites, found in nature in some marbles and invertebrate skeletal remains (Chave, 1954a; Goldsmith, Graf, and Joensuu, 1955) apparently are randomly substituted for Ca^{++} ions. However, the Ca^{++} and Mg^{++} ions in dolomite are not only present in the ratio of 1 to 1, but also are ordered, i.e., cation planes populated entirely by Mg^{++} alternate with those populated entirely by Ca^{++}. The relative difficulty of achieving this ordered arrangement during crystal growth should be highly significant to the problem of dolomite genesis.

70

to 850° C. and CO_2 pressures from about 15 lbs/in² to 20,000 lbs/in². In this temperature interval, dolomite can be synthesized under appropriate CO_2 pressures by reaction between $MgCO_3$ and $CaCO_3$, between MgO and $CaCO_3$, or by rearrangement of the cations in magnesian calcite to form dolomite plus a calcite poorer in Mg than the starting material. The latter process requires no transport of material between distinct crystalline entities but only cation rearrangement by dry, solid-state diffusion within small crystalline units.

The production of dolomite by these dry, solid-state reactions is readily accomplished in the laboratory at temperatures above approximately 500° C. Many metamorphic carbonate rocks have attained temperatures high enough for the dolomite observed to have formed in a relatively short time by similar reactions within the system CaO-MgO-CO_2. Although it is not generally possible to specify the mechanism by which a metamorphic dolomite has formed, a few marbles have been studied (Goldsmith, Graf, and Joensuu, 1955) in which dolomite was produced as an exsolution product from a magnesian calcite, stable at elevated temperatures, upon cooling of the rock.

The significant point is that *at elevated temperatures ionic mobility is great enough for cation ordering to take place* and thus develop the stable, ordered dolomite. At temperatures below about 400° C., the rate of reaction for dry dolomite syntheses becomes so slow that no dolomite can be detected by X-ray diffraction after hundreds of hours.

In this paper a series of hydrothermal experiments at lower temperatures are described and interpreted. It would be expected that the reaction rate in a hydrothermal environment would be increased relative to that in a dry system. These experiments do not represent an attempt to delineate the equilibrium relations in even a small part of the complex system CaO-MgO-H_2O-CO_2. Our intention has been merely to obtain some information on the rate and the structural mechanisms of dolomite formation under a limited number of conditions.

FORMATION OF DOLOMITE AT LOW TEMPERATURES IN THE PRESENCE OF WATER

Dolomite is more stable in sea water at earth-surface temperatures than the calcite-magnesite assemblage, as indicated both by the widespread occurrence of dolomite in sedimentary rocks and by limited thermochemical data (Halla and Ritter, 1936). However, we are not aware that any of the numerous attempts to precipitate dolomite from solution at comparable temperatures have been successful. The development of a metastable phase or phase assemblage at low temperatures, which apparently takes place in this case, is well known in many systems. Thus one principal concern is the rate of dolomite formation from an initially produced metastable assemblage. In this study, a number of experiments were carried out in which several metastable Ca-Mg carbonate assemblages were hydrothermally treated at temperatures ranging from 450° to 25° C. This temperature range extends up to the region in which dry synthesis of dolomite is observable in the laboratory.

EXPERIMENTAL PROCEDURE

The experiments were carried out in the presence of distilled water in modified 18-ml. and 1-ml. Morey-type pressure vessels (Morey and Ingerson, 1937). The degree of water filling varied from 50 to 80 per cent by volume. Above the critical temperature of water, the pressure varies with the degree of filling, but no attempt was made to control this pressure.

The following metastable starting materials were used:

a) Powdered calcitic skeletal parts of marine invertebrates with relatively high metastable Mg contents, as follows:

Sample	Description	Mole per cent $MgCO_3$ (see Goldsmith, Graf, and Joensuu, 1955)
I	Calcareous algal material	17
II	Calcareous algal material	21
III	Calcareous algal material	17.5
IV	Echinoid test..........	12
V	Echinoid test..........	8
VI	Echinoid spines.......	6
VII	Echinoid test..........	7

Approximately 10 mg. samples were used, enclosed in Pt foil.

b) Ca-Mg carbonate gels, produced by pouring a 5N solution of K_2CO_3 into a 5N solution of mixed Ca and Mg chlorides. Gels were precipitated directly in the pressure vessels from solutions having Ca:Mg ratios of approximately 48:52 and 22:78.[3] No additional liquid was added to these saline solutions. Upon completion of a run, the products were washed by repeated decantation before examination.

c) Intimate fine-grained mixtures of aragonite and basic Mg carbonate, produced by repeated washings of gels like those in (*b*), and dried before use. Materials having three different Ca:Mg ratios were prepared, 86:14, 54:46, and 28:72. Approximately 0.3-gm. samples were used.

In addition, a number of runs were made with powdered natural dolomites from Monroe County, N.Y., and Serra das Eguas, Brazil, as starting materials, using approximately 10-mg. samples.

Approximately 0.3 gm. of CO_2 in the form of dry ice was added to a number of runs of dolomite and the invertebrate skeletal materials. This was done to evaluate the effect of CO_2 in a qualitative way, the partial pressure of the CO_2 being unknown.

The synthetic products were very fine-grained, and interpretations were neces-sarily based upon the results of X-ray powder diffraction. X-ray photographs of all samples were made with 114.6-mm. Philips powder cameras using Mn-filtered Fe radiation.

HYDROTHERMAL TREATMENT OF
NATURAL DOLOMITE

The runs in table 1 were made to obtain an approximate upper temperature limit of stability of dolomite in water. This information was necessary for subsequent experiments with materials (*a*), (*b*), and (*c*) to determine whether dolomite was stable under the conditions of the particular experiment.

The data from these runs indicate that dolomite is stable under the conditions of the present experiments below approximately 350° C. Schloemer (1952) obtained an upper limit of about 390° C. at water pressures up to 1,000 atm. However, it should be stressed that these values are strongly dependent upon the partial pressure of CO_2 in the system, as well as upon the total pressure. CO_2 is present in varying amounts in the air and water contained in the bomb, and, in addition, a small amount of decomposition of the dolomite yields CO_2 to the system. If a very small sample of dolomite is used, as was the case in these experiments, the production of sufficient CO_2 by dolomite dissociation to equilibrate the system may require the breakdown of a large enough fraction of the sample so that the breakdown products, calcite and brucite, are detectable by X-ray diffraction. If a large sample is used, higher temperatures may be necessary before the percentage decomposition reaches the limit of detection.

---

[3] Solutions prepared from weighed amounts of hydrous Ca and Mg chlorides, in the approximate molar ratios 1:1 and 1:3, proved upon analysis to have the compositions 48:52 and 22:78.

Runs 7 and 8 in table 1 illustrate the fact that the introduction of CO_2 into the system raises the temperature of detectable decomposition above 350° C.

HYDROTHERMAL TREATMENT OF NATURAL MAGNESIAN CALCITES

Runs using the calcareous algal and echinoidal materials described earlier in the section on experimental procedure where $x > y$, and where the Mg content of the calcite reaction product is neglected. In several runs at 300° and 350° C., however, a small amount of a poorly crystallized dolomite-like material was developed. Only the strongest reflection, (112),[4] is visible in the X-ray diagrams, and determination of composition from this single, rather diffuse reflection (Goldsmith, Graf, and Joensuu, 1955) is subject to error. Nevertheless, it appears

TABLE 1

DATA FOR HYDROTHERMAL RUNS USING DOLOMITE AS STARTING MATERIAL

Run	Starting Materials	Temp. (° C.)	Time (Hrs.)	Products
1	Dolomite, Monroe Co., N.Y.	300	1,150	Unchanged
2	Dolomite, Serra das Eguas, Brazil	300	1,150	Unchanged
3	Dolomite, Serra das Eguas, Brazil	350	360	D, (C*), (B)
4	Dolomite, Monroe Co., N.Y.	350	360	D, (C*), (B)
5	Dolomite, Serra das Eguas, Brazil	400	140	C_3, B, (D)
6	Dolomite. Monroe Co., N.Y.	400	140	C_4, D, B
7	Dolomite, Serra das Eguas, Brazil, +CO_2	350	360	Unchanged
8	Dolomite, Monroe Co., N.Y., +CO_2	350	360	Unchanged

KEY TO SYMBOLS USED IN TABLES 1–4

A............aragonite
B............brucite
bmc..........basic magnesium carbonate
N............nesquehonite
M............magnesite, no detectable Ca substitution
M_3, etc........magnesite, 3 mole per cent Ca substitution, based upon measurement of (112) spacing, Straumanis-type correction for shrinkage, and assumption of linear change of (112) spacing for compositions between dolomite and magnesite
C............calcite, no detectable Mg substitution
C_5, etc........calcite, 5 mole per cent Mg substitution, based upon measurement of (521) and (633)(552) spacings, shrinkage correction from Nies scales, and assumption of linear change of these spacings for compositions between calcite and dolomite
C*...........calcite, showing poor reflections only, Mg substitution between 0 and ∼20 mole per cent, but not measured
D............dolomite, no detectable deviation from 1:1 composition
D*...........protodolomite, state of order and composition in mole per cent (as, $Ca_{45}Mg_{55}$) given under "Remarks." Composition determined from measurement of (112) spacing, Straumanis-type correction for shrinkage, and assumption of linear change of (112) spacing for compositions between calcite and magnesite
( )..........quantity near the limit of detection by X-ray powder diffraction methods
The products are listed in order of estimated abundance.

are listed in table 2. At 200° C., only one of the four materials showed any change in 10 days, the formation of a small amount of brucite being noted. In those runs in which no CO_2 was added, the general tendency, at temperatures of 300° C. and above, was to form brucite plus calcite poorer in Mg than the starting material. The reaction can be written as:

that an excess of Ca over the 1:1 Ca:Mg ratio of dolomite is present.

In those runs with added CO_2, no change of any kind was observable in 11 days at 300° C. The absence of brucite apparently is a consequence of the added CO_2 pressure (see preceding equation).

[4] The indices used in this paper are based upon the true rhombohedral structure cell (*see* Bradley, Burst, and Graf, 1953).

$$\overset{\text{Magnesian Calcite}}{Ca_xMg_y(CO_3)_{x+y}} + yH_2O \rightleftharpoons \overset{\text{Calcite}}{xCaCO_3} + \overset{\text{Brucite}}{yMg(OH)_2} + yCO_2$$

As in the runs at 300° C. without CO_2, even incipient equilibrium development of dolomite cannot be detected in the length of time of the experiments. No brucite was formed in the runs with added CO_2 at 350° C. and 400° C.; in some of the samples, however, a small amount of the dolomite-like material described was produced. Therefore at 400° C. the added CO_2 pressure was sufficient to make dolomite stable in the presence of H_2O, and the rather high temperature resulted in a reaction rate sufficient to produce a detectable amount of dolomite-like material in 13 days.

The amount of Mg that is stable in the

TABLE 2

DATA FOR HYDROTHERMAL RUNS USING NATURAL MAGNESIAN CALCITES AS STARTING MATERIAL

Run	Starting Materials	Temp. (° C.)	Time (Hrs.)	Products	Remarks
9	I†	200	240	Unchanged	
10	II†	200	240	Unchanged	
11	III†	200	240	C_{15}, (B)	III retains virtually all its Mg
12	IV†	200	240	Unchanged	
13	I	300	240	C_{15}, $C_{<15}$, (B)	Retains virtually all its Mg; two magnesian calcites
14	II	300	240	C_{17}, $C_{<17}$, (B)	
15	III	300	240	C_{15}, (B)	
16	IV	300	240	C_8, C_4, (B)	Two magnesian calcites
17	I	300	1,150	C_{15}, (B)	Retains most of its Mg
18	II	300	1,150	C_{16}, (B)	
19	III	300	1,150	C_{16}, $C_{<16}$, (B)	
20	IV	300	1,150	C_4, (B), (D*)	Retains most of its Mg; only (112) of D* visible { $\sim Ca_{60}Mg_{40}$
21	V†	300	1,150	C_4, (D*)	$\sim Ca_{58}Mg_{42}$
22	VI†	300	1,150	Unchanged	
23	I	350	360	C_{14}, C_4, (B)	Two magnesian calcites
24	II	350	360	C_4, $C_{>4}$, (B)	
25	III	350	360	C*, C*, (B)	
26	IV	350	360	C_3, (B)	IV has lost most of its Mg
27	VII†	350	504	C_5, (D*)	VII retains considerable Mg; (D*) is $\sim Ca_{55}Mg_{45}$
28	I	400	384	C_4, B	Has lost most of its Mg
29	II	400	384	C_3, B	
30	III	400	384	C_4, B	
31	IV	400	384	C, (B)	
32	V	400	384	C, C*, (B)	Some coarse-grained V left
33	VI	400	384	C*, C, (B)	Most of VI has lost its Mg
34	I +CO_2	300	264	Unchanged	
35	II +CO_2	300	264	Unchanged	
36	III +CO_2	300	264	Unchanged	
37	IV +CO_2	300	264	Unchanged	
38	V +CO_2	300	264	Unchanged	
39	VI +CO_2	300	264	Unchanged	
40	I +CO_2	350	360	Unchanged	
41	IV +CO_2	350	360	C_{10}, (D*)	(D*) is $\sim Ca_{54}Mg_{46}$
42	V +CO_2	350	360	C_6, (D*)	(D*) is $\sim Ca_{54}Mg_{46}$
43	I +CO_2	400	312	Unchanged	
44	II +CO_2	400	312	C*, C*	Two magnesian calcites
45	III+CO_2	400	312	C_{15}	III retains most of its Mg
46	IV +CO_2	400	312	C_7, (D*)	(D*) is $\sim Ca_{55}Mg_{45}$; IV retains considerable Mg
47	V +CO_2	400	312	C_5, (D*)	(D*) is $\sim Ca_{55}Mg_{45}$; V retains most of its Mg
48	VI +CO_2	400	312	Unchanged	

† Mg substitutions, in mole per cent, are I, 17; II, 21; III, 17.5; IV, 12; V, 8; VI, 6; and VII, 7. These materials are described more fully in the section on "Experimental Procedure."

calcite structure as a function of temperature has not been determined under hydrothermal conditions. This equilibrium is more complex than that in the dry system (Graf and Goldsmith, 1955), as variation in the relative H_2O and CO_2 pressures produces variation in the phase

HYDROTHERMAL CRYSTALLIZATION OF GELS

Table 3 presents the results obtained with gels having Ca:Mg ratios of about 48:52 and 22:78. The first of these materials, which has approximately the same Ca:Mg ratio as dolomite, yielded a carbonate having a Ca content some-

TABLE 3†

DATA FOR HYDROTHERMAL RUNS USING CARBONATE GEL AS STARTING MATERIAL

Run	Starting Materials	Temp. (°C.)	Time (Hrs.)	Products	Remarks
49		25	1,940	D*	Very diffuse X-ray reflections, no superstructure reflections, (222) position abnormal, $\sim Ca_{59}Mg_{41}$
50		72	46	D*	Diffuse X-ray reflections, no superstructure reflections, (222) position abnormal, $\sim Ca_{58}Mg_{42}$
51		72	470	D*	$\sim Ca_{57}Mg_{43}$
52		80	2,000	D*	$\sim Ca_{55}Mg_{45}$
53	$Ca_{48}Mg_{52}$ gel	100	276	D*	$\sim Ca_{58}Mg_{42}$
54		200	46	D, (M_7)	Superstructure reflections of D very weak
55		200	477	D, M_7	Superstructure reflections of D broadened, appear weak
56		200–210	2,010	D, (M)	Sharp X-ray reflections
57		280–300	207	D, M	Fairly diffuse X-ray reflections
58		387	101	D, M, (B)	
59		25	1,940	N, bmc	
60		72	46	bmc, D*	D* reflections broad, partly masked by bmc reflections, center around $Ca_{50}Mg_{50}$
61		72	470	D*, M	M reflections sharp; D* diffuse, no superstructure lines, $\sim Ca_{55}Mg_{45}$
62		80	2,000	D*, M	M reflections sharp; D* diffuse, no superstructure lines, centers around $Ca_{49}Mg_{51}$
63	$Ca_{22}Mg_{78}$ gel	100	276	D, M	M reflections sharp; D* diffuse, no superstructure lines
64		200	46	M, D*	D* reflections diffuse, no superstructure lines, (222) weak
65		200	477	M, D	D reflections diffuse, superstructure lines very weak
66		200–210	2,010	M, D	D more coarsely crystalline than M (spotty lines)
67		280–300	207	M, D	Superstructure reflections of D weak
68		387	101	M, D, (B)	

† The symbols $Ca_{48}Mg_{52}$ and $Ca_{22}Mg_{78}$ indicate accurately the molar composition of the chloride solutions used, but errors inherent in measuring out the small quantities of liquid needed for preparation of gels before each run may have resulted in deviation of some gel compositions from the values stated.

assemblage, with an attendant difference in the amount of Mg that is stable in the calcite which may be in equilibrium with either periclase, brucite, or dolomite. From these data, however, it would appear that the amount of Mg originally present in the algal and echinoidal materials is in excess of that stable in solid solution under the conditions of the present experiments.

what greater than that of dolomite[5] when run at temperatures of 100° C. or less for periods up to 83 days. It should be noted that this phase is not a true dolomite, inasmuch as no superstructure reflections

[5] Any Mg not present in this phase may have been held in solution owing to the presence of Cl^- in this series of runs. As in all the experiments, other phases may not be detected by the X-ray techniques used, either because they are not well crystallized or not present in sufficient amount.

can be detected, excess Ca appears to be present, and the (222) spacings are abnormal, a point to be discussed later. This product is a single phase material yielding an X-ray diffraction diagram having high-angle reflections which, although diffuse, clearly indicate a compositional deviation from that of dolomite. At temperatures from 200° to 387° C. a material was produced which appears to be true dolomite, although the superstructure reflections are somewhat weak in the X-ray patterns of samples from the two shorter runs at 200° C.

The second of the gels, with a Ca:Mg ratio of 22:78, produced dolomite-like material in three runs between 70° and 80° C. In two of the three runs, measurement of the diffuse (112) reflection indicated a composition centering around $Ca_{50}Mg_{50}$. Runs at temperatures ranging from 200° to 387° C. produced normal dolomite, although superstructure reflections were weak or missing in the shorter runs at 200° C. The excess Mg present in the system appeared as magnesite, nesquehonite, basic magnesium carbonate, or brucite, depending upon the temperature.

HYDROTHERMAL TREATMENT OF PRECIPI-
TATED CARBONATES

The runs using precipitated washed carbonates of three different compositions are listed in table 4. The $Ca_{86}Mg_{14}$ material yielded a metastable assemblage of brucite and magnesian calcite in the 215° to 450° C. range. No dolomite was observed.

The $Ca_{54}Mg_{46}$ material yielded a metastable assemblage of aragonite and brucite after 20 days at 220° C., the aragonite of the starting material having persisted. At 300° C., an increasing amount of a dolomite-like material, associated with metastable phases, was

observed with increased time. At 348°–450° C., dolomite giving superstructure reflections was produced, although these reflections are weak in some cases.

The $Ca_{28}Mg_{72}$ material yielded at least a trace of dolomite or dolomite-like material at temperatures from 215° to 450° C. Superstructure reflections are present for all dolomites produced at 300° C. or above. The formation of dolomite in runs at 400° and 450° C., when the $Ca_{54}Mg_{46}$ and $Ca_{28}Mg_{72}$ materials were used, probably resulted from increase in the temperature at which dolomite breaks down, owing to the relatively high partial pressure of CO_2 produced by the extensive carbonate dissociation.

NATURE OF THE SYNTHETIC DOLOMITES

All Fe- and Mn-free natural dolomites that the authors have examined show superstructure reflections. In addition, X-ray spacings indicate that there is no detectable deviation from a 1:1 Ca:Mg ratio.[6] Synthetic dolomites (Graf and Goldsmith, 1955) produced dry under equilibrium conditions in the temperature range from 500° to 800° C. appear identical to the natural dolomites. Only in a few runs at somewhat higher temperatures is a deviation in composition of 1 or 2 per cent in the direction of excess Ca noted, but superstructure reflections are still present. Therefore, dolomite that is stable at earth surface conditions must be an ordered compound with a 1:1 Ca:Mg ratio.

The dolomites produced in the higher temperature hydrothermal experiments (see, for example, runs 56–58 and 68, table 3; runs 85 and 95, table 4) are

[6] While this paper was in press, the authors examined a partially dolomitized Tertiary limestone, the dolomite of which was ordered but deviated some 10 per cent from the 1:1 Ca:Mg ratio in the direction of excess $CaCO_3$. This material will be discussed further in a later publication.

76

normal in the sense described in the preceding paragraph. Dolomites produced at somewhat lower temperatures or in shorter times have weakened superstructure reflections, although they are of ideal 1:1 composition (see, for example, runs 54–55, 63–65, table 3; runs 83 and 84, table 4). In addition, the time and temperature required to produce normal dolomite vary with the type of starting material used. Other dolomite-like materials formed at still lower temperatures or in shorter times produce no visible superstructure reflections, and their X-ray spacings are indicative of compositions that may deviate as far as $\sim Ca_{67}Mg_{33}$, although generally not more calcian than $Ca_{60}Mg_{40}$ (see runs 49–53, table 3; runs 70–72, table 4). These dolomite-like materials produce diffuse X-ray reflections, and the over-all sharpness of the diffraction pattern can in general be correlated with the relative ideality of the dolomite, with regard to both composition and degree of order. Similar dolomite-like materials were produced during the rapid cooling of dry periclase-calcite assemblages through the dolomite stability field, from temperatures above 500° C. (Graf and Goldsmith, 1955).

TABLE 4

DATA FOR HYDROTHERMAL RUNS USING COPRECIPITATED ARAGONITE AND
BASIC MAGNESIUM CARBONATE AS STARTING MATERIAL

Run	Starting Materials (Ppt.)	Temp. (° C.)	Time (Hrs.)	Products	Remarks
69		215	498	C_8, (B)	C_8 has diffuse X-ray reflections
70		300	72	C_5, (B), (D*)	D* is $\sim Ca_{67}Mg_{33}$; C_5
71		300	480	C_5, (B), (D*)	D* is $\sim Ca_{64}Mg_{36}$; C_5 have sharp X-ray reflections
72		300	2,424	C_4, (B), (D*)	D* is $\sim Ca_{61}Mg_{39}$; C_4
73	$Ca_{86}Mg_{14}$	348	571	C_3, (B)	Fairly sharp X-ray reflections
74		375	498	C_2, B	
75		400	72	C_2, B	Sharp X-ray reflections
76		400	480	C_2, B	
77		450	504	C_1, B	Very sharp X-ray reflections
78		220	479	A, B	
79		300	72	C_{13}, M_{12}, B, (D*)	The only D* reflection visible, (112), is broad and centers around $Ca_{50}Mg_{50}$
80		300	504	C_{11}, M_{15}, B, (D*)	The only D* reflection visible, (112), is broad and centers around $Ca_{54}Mg_{46}$
81		300	2,712	D*, C_{17}, B	D* is $\sim Ca_{60}Mg_{40}$
82	$Ca_{54}Mg_{46}$	348	571	D*, B, (C)	D* is $\sim Ca_{60}Mg_{40}$
83		375	498	D, C_2, B	Superstructure reflections weak
84		400	72	D, C*, B	Superstructure reflections weak, (222) broad and diffuse
85		400	504	D, C_3, B	
86		450	504	D, C, B	Superstructure reflections and (222) weak
87		215	498	M_{15}, C*, B, (D*)	Only fuzzy (112) of D* visible, $\sim Ca_{57}Mg_{43}$; (222) of M in abnormal position
88		300	72	M_{11}, C_{13}, B, (D*)	Only fuzzy (112) visible for D*, $\sim Ca_{58}Mg_{42}$
89		300	504	D, M_{12}, B	M_{12} and D have fairly sharp X-ray reflections
90		300	2,712	D, M_{11}, B	
91	$Ca_{28}Mg_{72}$	340	550	D, B, M	X-ray reflections of D rather diffuse
92		375	498	D, B, M	
93		400	72	B, C_4, D	
94		400	504	B, C_1	
95		450	480	D, B, C	

There are several types of evidence that help further to characterize the dolomite-like materials:

1. Even though the experimental hydrothermal environments utilized had a threefold variation in Mg concentration, the great majority of the dolomite-like materials produced have compositions between $Ca_{60}Mg_{40}$ and $Ca_{50}Mg_{50}$. Even if runs 70 and 71, table 4, which yielded $\sim Ca_{67}Mg_{33}$ and $\sim Ca_{64}Mg_{36}$ compositions, are considered, there is still a distinct compositional break between these materials and the magnesian calcites produced in these experiments, which contain less than 20 mole per cent $MgCO_3$. Furthermore, the dolomite-like materials have been produced from three quite different experimental environments—dry periclase–calcite mixtures, gels, and small samples of magnesian calcite in large volumes of water. The dolomite-like materials cannot, therefore, be merely the result of crystal-growth processes sensitive to particular environmental compositions or states of aggregation of reactants.

2. In spite of the absence of superstructure reflections, these materials cannot be merely magnesian calcites very rich in Mg. Highly magnesian calcites are unstable in both dry and hydrothermal systems at the temperatures of this study. One would not expect dolomite formation to proceed by way of such highly unstable states as disordered $Ca_{55}Mg_{45}$ and $Ca_{60}Mg_{40}$ phases (tables 2–4). It therefore appears necessary to postulate for the dolomite-like materials a rather high degree of short-range Ca-Mg order, a structural arrangement which would be relatively more stable in the experimental environments studied than would complete disorder.

3. The structural imperfection of the dolomite-like material from some samples (see runs 49–53, table 3, for example) is manifested not only by the absence of superstructure reflections but also by the character of the (222) reflection produced by planes parallel to the carbonate groups.[7] The position of the (222) reflection, relative to other reflections, is variable for the materials from runs 49–53, even though they have essentially the same composition. This variable displacement is in the direction of the (222) position appropriate for calcite. In some cases the (222) reflection is more diffuse than neighboring reflections in the diffraction diagram.

The absence of superstructure reflections indicates that the simple ordered succession of cation planes, Ca-Mg-Ca-Mg, is not consistently maintained throughout discrete crystalline entities. The aberrant nature of the (222) reflection indicates that the distances between successive pairs of cation planes are not the same—which means that the ideally disordered succession in which all cation planes are occupied by Ca and Mg ions in equal numbers does not exist. The cations may tend to segregate into more or less Ca-rich and Mg-rich planes, which do not occur in the simple succession Ca-Mg-Ca-Mg. Another possibility is the existence of small domains, some of which may approach ideal dolomite in composition and ordering, but with the composition of a given cation plane changing from domain to domain. Thus the nature of the superstructure and (222) reflections suggests at least two possible structural arrangements which

[7] The higher-order (333) and (444) reflections are generally too poorly resolved in these materials to be of use, and the lower-order (111) reflection is not observed in ordinary powder photographs, even for well-crystallized dolomites.

could have the relatively high degree of short-range order postulated in (2).

If the structural arrangement of the dolomite-like materials is actually more stable under the given experimental conditions than that of obviously metastable assemblages, then these materials are intermediate steps leading to stable 1:1 ordered dolomite. A gradual rearrangement of the whole mass of dolomite-like material to dolomite might be expected in time, without the intermediate formation of metastable assemblages. This behavior would contrast with that, for example, of low-Mg calcites, which have been observed in this study to break down to calcite plus brucite with the development of only a small amount of dolomite or dolomite-like material during the same time. In order to evaluate this hypothesis, a number of hydrothermal runs were made in which small amounts of dolomite-like material were removed at successively greater elapsed times from the same charge, maintained at a given temperature. The results of these experiments are summarized in table 5.

The data of table 5 clearly show that the dolomite-like materials are gradually converted to dolomite, without intermediate breakdown to metastable assemblages. The transition is best shown by samples 20–22, which were prepared in the temperature range between 190° and 230° C. The X-ray reflections of sample 20 have intensity maxima at positions corresponding to a composition more calcium-rich than that of dolomite. These reflections persist with diminished intensity in the diagram of sample 21, together with a new set of intensity maxima at positions characteristic of the true 1:1 dolomite composition. Weak superstructure reflections are present for the 1:1 material. Sample 22 has been

completely converted to dolomite; the first set of intensity maxima has disappeared in the X-ray diagram of this material, and the reflections no longer have a bimodal intensity distribution as they did for sample 21. The transition is partially completed in samples 12 and 13, prepared in the temperature range from 160° to 190° C., and it may be supposed that with adequate experimental times even lower temperatures would suffice for the transition.[8]

Although the dolomite produced in the longest runs at the highest temperatures listed in table 5 appear to have slightly more diffuse reflections than the best natural dolomites, this effect is a minor one and may be due solely to small particle size. In practice, no difference would be apparent if an X-ray diagram of one of these best synthetic dolomites were reproduced in plate 1, alongside the natural dolomite shown there.

The data in table 5 clarify the behavior of the dolomite-like materials to the extent that a general definition for

[8] It might have been expected that the transition from dolomite-like material to dolomite could be followed in tables 3 and 4, where there are several series of runs carried out at the same temperature for different lengths of time but with a new batch of gel prepared for each run. The X-ray diagrams of these dolomite-like materials each show a single set of reflections, and in some instances (runs 50–52, table 3, and runs 70–72 and 88–89, table 4) this set of reflections appears to shift toward 1:1 dolomite composition with increased time. However, there are other instances (runs 60–61, table 3, and 79–81, table 4) where the apparent shift is in the opposite direction. Actually, this "shift" of unimodal intensity maxima is inconsistent with the bimodal intensity distributions described in table 5, and some of these supposed reactions in tables 3 and 4 occur at temperatures lower than those for which no reaction is recorded in comparable times in table 5. The supposed compositional changes in tables 3 and 4 must result from the inherent inaccuracy in mixing small volumes of liquids used in preparing successive batches of gel, yielding dolomite-like materials of slightly different composition which persisted metastably.

TABLE 5†

DATA FOR HYDROTHERMAL RUNS, USING CaMg CARBONATE GEL OF APPROXI-
MATELY $Ca_{50}Mg_{50}$ COMPOSITION AS STARTING MATERIAL

Experiment No.	Experimental Conditions	Character of Product, from Powder X-ray Diffraction Diagrams
1	5 hrs. at 160° C.	Single, poorly crystallized phase; no super-structure reflections; greatest (112) intensity at a position somewhat more Ca-rich than dolomite, but tailing-off toward ideal dolomite composition; no change with time
2	18 hrs. at 160° C.	
3	48 hrs. at 160° C.	
4	108 hrs. at 160° C.	
5	274 hrs. at 160° C.	
6	5 hrs. at 190° C.	Same as for samples 1–5
7	17 hrs. at 190° C.	
8	90 hrs. at 190° C.	
9	351 hrs. at 190° C.	
10	274 hrs. at 160° C.+ 17 hrs. at 190° C.	Same as for samples 1–5
11	274 hrs. at 160° C.+ 42 hrs. at 190° C.	
12	274 hrs. at 160° C.+ 90 hrs. at 190° C.	All reflections becoming bimodal, with new intensity at ideal dolomite composition
13	274 hrs. at 160° C.+351 hrs. at 190° C.	
14	5 hrs. at 230° C.	Same as for samples 1–5; small amount of calcian magnesite
15	20 hrs. at 230° C.	
16	48 hrs. at 230° C.	
17	228 hrs. at 230° C.	Ideal dolomite composition; superstructure reflections present, but weak and diffuse; (222) as sharp as other reflections; (444) weaker and more diffuse than $(11\bar{2})$, although these two reflections are comparable for well-crystallized dolomite
18	780 hrs. at 230° C.	All reflections somewhat sharper than in sample 17, especially superstructure reflections and (444), but the latter are not yet quite ideal
19	351 hrs. at 190° C.+ 5 hrs. at 230° C.	Same as for samples 1–5
20	351 hrs. at 190° C.+ 20 hrs. at 230° C.	
21	351 hrs. at 190° C.+ 48 hrs. at 230° C.	All reflections becoming bimodal, with new intensity at ideal dolomite composition; very weak superstructure reflections; trace calcian magnesite
22	351 hrs. at 190° C.+228 hrs. at 230° C.	Ideal composition; all reflections present but still somewhat diffuse; no magnesite
23	351 hrs. at 190° C.+780 hrs. at 230° C.	Reflections sharper than for sample 22, but still not quite ideal; (444) and superstructure reflections as good as others
24	9 hrs. at 275° C.	Ideal dolomite composition; (222) as sharp as the other reflections, which are still diffuse; superstructure and (444) reflections very broad and weak; small amount of calcian magnesite
25	32 hrs. at 275° C.	No change from sample 24
26	99 hrs. at 275° C.	Definite improvement in sharpness of (444); other reflections slightly sharper
27	200 hrs. at 275° C.	No change from sample 26, except magnesite now is Ca-free
28	536 hrs. at 275° C.	No significant change from sample 27; superstructure and (444) reflections still noticeably diffuse

† Repeated samplings of several initial charges were made in order to follow the conversion of protodolomite to dolomite.

80

this class of materials becomes possible. The term *protodolomite* is suggested for single-phase rhombohedral carbonates which deviate from the composition of the dolomite that is stable in a given environment, or are imperfectly ordered, or both, but which would transform to dolomite if equilibrium were established.

INTERPRETATION AND DISCUSSION

It is apparent from this investigation and from previous experimental studies by others that it is difficult to crystallize dolomite at low temperatures. No unequivocal laboratory precipitation of dolomite at earth-surface temperatures is known to the authors. Not until temperatures of approximately 500° C. in the dry state and 200°–300° C. in the wet state were reached under the rather specialized conditions of our experiments did dolomite formation proceed in convenient laboratory times.

The general difficulty involved in the production of certain types of ordered compounds is discussed by Goldsmith (1953) in terms of the "simplexity" of the phases involved. The minerals most difficult to synthesize generally are those in which there are two or more cations occupying nonequivalent positions in the structure, particularly if the cations are crystallochemically similar and occupy energetically similar positions. During rapid crystallization, or at low temperatures, the probability of ideal cation sorting into ordered positions is greatly decreased. Instead of the stable, ordered modification, a disordered form, or even an entirely different metastable phase assemblage, may be produced.

Failure to produce the stable modification should be considered in terms of both nucleation and subsequent growth.

The probability of forming a nucleus of the critical size necessary for survival is less for the ordered form than for the disordered form or forms, because of the more complex arrangement of the ordered nucleus. If the several different types of cation sites do not greatly differ structurally and energetically, cations arriving at improper sites are more likely to be accepted. Once accepted, they are less likely to become ordered by either surface or bulk diffusion, inasmuch as the difference in energy levels of the different positions is slight. Similarly, an assemblage of metastable phases might be produced at the expense of the more stable, ordered modification, particularly if the free-energy difference between the assemblage and the stable modification were small.

The difficulty with which dolomite is crystallized at low temperatures, as evidenced by experimental data, is considered a consequence of the necessity of attaining an ordered Ca-Mg configuration. The crystallochemical differences between Ca^{++} and Mg^{++} are great enough for an ordered Ca-Mg compound to be stable over a very large temperature range. At somewhat elevated temperatures there is sufficient thermal motion to allow cation ordering to proceed readily. However, at earth-surface temperatures the differences between Ca^{++} and Mg^{++} are not as easily expressed; the thermal motion is not sufficient to produce ordered dolomite readily, either by original crystallization or by rearrangement or recrystallization of a metastable assemblage. The ordering requirement severely decreases reaction rate whether dolomite is being formed by dry, solid-state diffusion, or in the presence of water. The effect of the ordering requirement presumably is felt in both decreased nucleation rate and decreased growth rate upon already formed nuclei.

This discussion has offered an explanation in terms of the "simplexity principle" as to why dolomite does not form more readily, and has put forth the related protodolomite concept, which appears to offer an explanation of how the required cation ordering is achieved through the initial crystallization of imperfectly organized assemblages. Neither of these principles gives insight at present as to which natural environments are most favorable for dolomite formation. The detailed mechanisms by which dolomite is formed in sedimentary environments are unknown. Indeed, very little is known even of the primary phase assemblages from which dolomite is produced. The intermediate protodolomite stages that some of the synthetic materials pass through in the process of reorganization to dolomite at somewhat elevated temperatures may or may not be realized in nature. Sufficiently slow crystallization might produce dolomite directly, as a primary phase. The virtual absence of dry diffusion in carbonates at earth-surface temperatures is shown by the persistence of metastable magnesian calcites in fossils as old as Eocene (Chave, 1954b) that were embedded in relatively impervious shales. By contrast, well-crystallized dolomite is found in Pleistocene formations. Thus the amount of dolomite formed at low temperatures by dry diffusion must be insignificant compared with that formed in the presence of water. Solution and recrystallization, if slow enough to produce dolomite, would still be much more rapid than reorganization by solid-state diffusion. However, the persistence of calcitic material for long periods in contact with sea water, at least in the deeper oceanic basins, shows that the intimate contact of solid $CaCO_3$ and Mg^{++} in solution is not necessarily sufficient for dolomite formation.

Ultimately, an understanding of dolomite formation at earth-surface temperatures must include the mechanisms by which ordered dolomite nuclei are formed, either in solution or as two- or three-dimensional nuclei upon substrates of other materials such as calcite.[9] More complex types of nucleation at or near structural defects within carbonate crystals also may be significant. Compositionally zoned, unit-extinguishing rhombohedra that have been observed in sedimentary carbonate rocks indicate that an ordered carbonate, such as ankerite or dolomite, may plate out upon a simple substrate, such as calcite, and vice versa.

A knowledge of the ionic composition of the solution also is needed, but this implies much more than mere solubility determinations. It is the concentration of specific types of ions at the crystal-solution interface, and their acceptance, rejection, or lateral transport along the growing crystal face, which is significant. Thus it is practically impossible at present to make useful statements about the effect of such variables as salinity, pH, and Fe^{++} concentration upon the rate of dolomite formation. The complex interrelationships of the chemical and physical conditions of the environment in which dolomite is produced are largely unknown. It is evident also that attempts that have been made to define conditions of equal Ca^{++} and Mg^{++}

---

[9] The case of nucleation within a carbonate crystal by dry cation diffusion at elevated temperatures is relevant to dolomite formation in metamorphic rocks but would not appear to be significant in sedimentary rocks.

concentration as a guide to dolomite formation represent an oversimplified approach; for a growing dolomite crystal could surely select a 1:1 Ca:Mg population from solutions having a range of Ca:Mg ratios.

REFERENCES CITED

BRADLEY, W. F., BURST, J. F., and GRAF, D. L., 1953, Crystal chemistry and differential thermal effects of dolomite: Am. Mineralogist, v. 38, p. 207–217.

CHAVE, K. E., 1954a, Aspects of the biogeochemistry of magnesium. 1. Calcareous marine organisms: Jour. Geology, v. 62, p. 266–283.

———— 1954b, Aspects of the biogeochemistry of magnesium. 2. Calcareous sediments and rocks: *ibid.*, p. 587–599.

GOLDSMITH, J. R., 1953, A "simplexity principle" and its relation to "ease" of crystallization: Jour. Geology, v. 61, p. 439–451.

———— GRAF, D. L., and JOENSUU, O. I., 1955, The occurrence of magnesian calcites in nature: Geochim. et Cosmochim. Acta, v. 7, p. 212–230.

GRAF, D. L., and GOLDSMITH, J. R., 1955, Dolomite-magnesian calcite relations at elevated temperatures and CO_2 pressures: Geochim. et Cosmochim. Acta, v. 7, p. 109–128.

HALLA, F., and RITTER, F., 1936, Eine Methode zur Bestimmung der Änderung der freien Energie bei Reaktionen der Typus $A(s) + B(s) = AB(s)$ und ihre Anwendung auf das Dolomitproblem: Zeitschr. physikal. Chemie, v. 175A, p. 63–82.

MOREY, G. W., and INGERSON, E., 1937, A bomb for use in hydrothermal experimentation: Am. Mineralogist, v. 22, p. 1121–1122.

SCHLOEMER, H., 1952, Hydrothermal Untersuchungen über das System $CaO\text{-}MgO\text{-}CO_2\text{-}H_2O$: Neues Jahrb. Mineralogie, p. 129–135.

PLATE 1

a, calcite.

b, dolomite; the principal superstructure reflections, (100), (221), and (11$\bar{1}$), are indicated by arrows.

c, protodolomite from run 51, table 3; note the absence of superstructure reflections, the general diffuseness of all reflections, and the enrichment in Ca indicated by the shift of the back reflections marked by a diagonal line.

d, protodolomite and magnesite from run 61, table 3; the magnesite reflections are much sharper than those of protodolomite, indicating the relative perfection of crystallization.

e, the development of a small amount of protodolomite from magnesian calcite echinoidal material, run 27, table 2. The (112) reflection of protodolomite is indicated by an arrow. All reflections are spotty, indicative of rather coarse crystal size. The spottiness of the (112) reflection of protodolomite suggests that the relatively large crystallite size of the calcite was preserved during the transition to protodolomite.

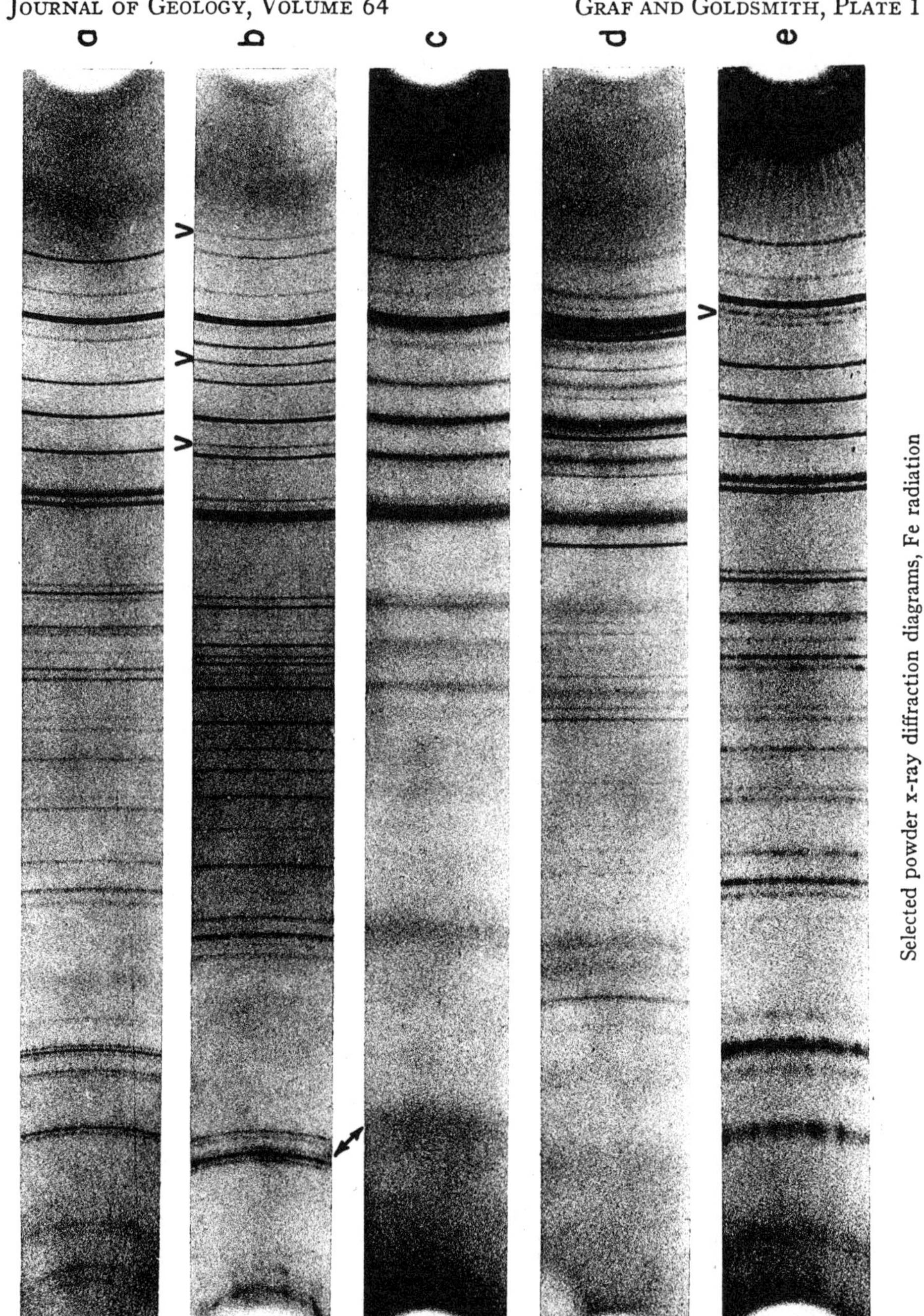

Selected powder x-ray diffraction diagrams, Fe radiation

6

Oxygen and carbon isotope ratios in coexisting calcites and dolomites from recent and ancient sediments

Egon T. Degens and Samuel Epstein
California Institute of Technology*

Abstract—The O^{18}/O^{16} and C^{13}/C^{12} ratios of coexisting sedimentary dolomites and calcium carbonates from a variety of environments, modes of formation and geological ages were measured. The data reflect the processes of dolomitization of calcium carbonates and their subsequent history. Dolomites usually have δO^{18} values characteristic of marine calcium carbonates or values of diagenetically altered marine limestones. Invariably the δO^{18} of dolomite are the same or greater than that of the coexisting calcite. This is in accordance with previously postulated mechanisms based on isotope measurements of hydrothermal and of laboratory synthesized cogenetic calcites and dolomites. The carbon isotope values of the measured carbonates are usually within the range found for marine limestones.

Introduction

The wide distribution of dolomite deposits in the stratigraphic column and the relatively limited extent to which dolomite is deposited at present is a well known geological feature. The reason for this phenomenon is not well understood in terms of conditions of carbonate formation. For example, is a particular environment necessary for dolomite to form, or are most sedimentary dolomites a result of alteration of pre-existing limestones?

Although a great deal of work has been done in this connection, employing geological, petrographical, geochemical and physical–chemical tools, an isotope study seemed pertinent. For general references on the origin of dolomites, see Fairbridge (1957), Cloud and Barnes (1957), and Ingerson (1962).

Some previous data on the O^{18}/O^{16} ratios indicate that at equilibrium the dolomite value should be greater by about 8 per mil at room temperature relative to that of calcite. The above relationship was obtained indirectly by comparing the O^{18}/O^{16} ratio between quartz and dolomite and between quartz and calcite of hydrothermal origin (Clayton and Epstein, 1958; Engel *et al.*, 1958). This inference was strongly supported by the results of studies on high temperature synthetic carbonates in the 400°–600°C range (Epstein *et al.*, 1963). The isotope values obtained from these synthetic dolomites and calcites are, in terms of their O^{18}/O^{16} difference, in accordance with values that are inferred from naturally occurring hydrothermal carbonates formed at about the same temperature level. In view of the considerable isotope fractionation between calcites and dolomites, it might be expected that dolomites and calcites precipitated in the environment of deposition will also have a difference in δ-values af obout 6–10 per mil at room temperature.

The similarity in δ-values for coexisting dolomites and calcites in recent sedimentary deposits and in some incompletely equilibrated synthetic high-temperature carbonates suggested that dolomite is an alteration product of crystalline

---

* Division of Geological Sciences, Contr. No. 1169.

85

calcium carbonate. The present study attempts to apply the above information on the isotope relationship between coexisting dolomite and calcite to the history of alteration of dolomites and limestones in the stratigraphic column.

ANALYTICAL PROCEDURES

Both in the case of dolomite and calcite, the CO_2 for mass spectrometer analysis is obtained by reacting the carbonate, ground to <200 mesh, with 100% phosphoric acid at 25°C (McCREA, 1950).

Carbon dioxide was analysed for carbon and oxygen isotopes by a 6-in., 60° sector type mass spectrometer using a double collecting system.

The data are reported as per mil deviation relative to the PDB_I Chicago belemnite Standard (CRAIG, 1953, 1957):

$$\delta C^{13} = \left(\frac{R}{R_{standard}} - 1\right) \times 1000$$

$$R = C^{13}/C^{12} \text{ ratio in the sample}$$

$$R_{standard} = C^{13}/C^{12} \text{ ratio in the standard.}$$

δO^{18} is defined in similar terms of O^{18}/O^{16} ratios. Appropriate correction factors described by CRAIG (1957) were applied. The data for oxygen δ-values reported in this paper will be lower by about $29.5\%_0$ when compared to the data of EPSTEIN *et al.* (1963). The relationship is:

$$\delta_{S.M.O.W.,*} = (\delta_{P.D.B_I}) : 1.03 + 29.5.$$

Since most samples investigated did not permit a physical separation of the dolomite from the calcite or vice versa, a chemical separation technique was used which allowed CO_2 developed from calcite and dolomite to be collected separately (EPSTEIN *et al.*, 1963). This method is principally based on the marked difference in the relative rates of reaction which are established between dolomite and calcite when treated with phosphoric acid.

All samples were analysed by X-ray diffraction and the dolomite/calcite or dolomite/aragonite ratio was estimated semiquantitatively (± 10 per cent). Calcium and magnesium analyses were made for most of the ancient carbonates and the results are reported elsewhere (DEGENS *et al.*, 1960; CHILINGAR, 1956). By using the calibration curve (EPSTEIN *et al.*, 1963, Fig. 1) and the total volume of CO_2 evolved during the first 60 min and during the period of 4 and 70 hr of reaction, the ratio of dolomite and calcite of a carbonate mixture could be determined at the ± 2 per cent level. A check on the reliability of this procedure was made possible by collecting the total CO_2 of a carbonate mixture (Table 2, No. 38–47), measuring its isotope ratio, and comparing the resulting δ-values with those obtained from the pure calcite and dolomite. This then allows the calculation of the absolute amount of dolomite and calcite in the studied sample.

SAMPLE MATERIAL AND GEOLOGICAL BACKGROUND INFORMATION

A wide range of carbonate specimens representing a variety of genetic types of coexisting dolomites and calcites was analysed isotopically in the present investigation. Care was taken that the selection included representative dolomites of

---

* Standard mean ocean water.

recent and ancient, continential and marine, and synsedimentary and late diagenetic origin.

Some pertinent information as to type, age, collection site, and geological
background of the samples are presented below. For our case it is obvious that
the best approach to the problem of dolomite formation should involve a study
of dolomite–calcite syngenetic pairs. This type of approach was deemed advan-
tageous, because under such circumstances some of the knowledge available from
previous isotope studies on calcium carbonates could be applied to the solution of
the dolomite problem in hand.

(1) *Nos. 1 and 2, Bahamas* (MILLER, 1961). Clear, pale pink dolomite rhombohedra up to 0·14 mm long are intricately associated with recent marine evaporite
minerals, including aragonite and calcite. The textural relationship suggest that
dolomites are younger than the remainder of the minerals which recently formed
in evaporation pans of Great Inagua, southernmost of the Bahama Islands in the
West Indies. Both samples investigated are derived from the same locality. The
radiocarbon age of the dolomite is about 2000 years, as kindly checked at the Shell
Laboratories at Houston, Texas (TAFT, personal communication).

(2) *Nos. 3–5, Florida Bay* (TAFT, 1961). Dolomite crystals (euhedral rhombohedron; $1–60\mu$ in size) occur in carbonate sediments now accumulating in shallow
sea water along the Southern coast of Florida. These dolomites are associated with
shell fragments. The salinity in some of the dolomite-containing muds which
occasionally run dry can amount to 70 per mil (TAFT, personal communication).
Preliminary C^{14} analyses on a small size dolomite mineral concentrate show no
indication for the presence of radiocarbon, which means that the samples are
probably older than ~40,000 years. At present, these Florida dolomites are being
rechecked for their radiocarbon age, this time using a different preparation and
concentration technique.

(3) *Nos. 6–9, Coorong Lagoon, South Australia* (ALDERMAN and SKINNER, 1957;
ALDERMAN and VON DER BORCH, 1960; SKINNER, 1960; WELLS, 1962; SKINNER *et
al.*, 1963). Microcrystalline dolomite sediments with all the characteristics of
ancient evaporite dolomites are present in restricted marine environments of
South Australian lagoons. These dolomitic sediments apparently resulted from
post-depositional alteration of an original aragonite precipitate in an annually
desiccated restricted marine environment. Based on various geological criteria,
the penecontemporaneous dolomitization has gone to completion within a time
span of 4000–8000 years. The ratio of dolomite/$CaCO_3$ in the studied carbonate
mixtures is approximately 60 to 40; according to X-ray diffraction analysis, the
dolomites are well ordered and not of a "proto" dolomite type, as has been suggested earlier. Radiocarbon data (SKINNER *et al.*, 1963) give ages for total carbonate
of a few hundred to a few thousand years.

(4) *Nos. 10–18, Lake Bonneville, Knolls, Utah* (GRAF *et al.*, 1959, 1961; BISSELL
and CHILINGAR, 1962). At two collection sites, 1·61 and 1·60 miles west of Knolls,
Utah, two of each sample were taken from the lake beds at a depth of about 25 cm
below the present water sediment interface (samples Nos. 10 and 11; and 12 and
13, respectively). The sediment is laminated; gray and white layers alternate.
Thin section studies and X-ray data reveal that dolomite euhedra are virtually

restricted to the gray layers and only traces of dolomites are present in the white calcareous portion of the hand-specimens. Only one 2 mm gray layer each was taken from samples 10 and 12, whereas a total of six individual layers were taken from sample 13, all of which were analysed separately (Nos. 13–18). Sample 13 is the top and sample 18 is the bottom layer; the thickness of the whole sample is 4 cm. The rocks can be classified petrographically as argillaceous lime and calcareous–dolomitic clay, respectively, depending on the carbonate concentration which may range from 30 to 70 per cent. The radiocarbon age of the dolomite-bearing strata is 11,300 ± 25 years (GRAF *et al.*, 1961). This C^{14} data correlates with the beginning of a period of relatively greater aridity, namely the disappearance of ancient Lake Bonneville and the appearance of the present Great Salt Lake. Dolomitization probably started at the time the Great Salt Lake incident took place.

(5) *Nos. 13–23, sediment in Experimental Mohole, near Guadalupe Island, Mexico* (RIEDEL *et al.*, 1961; RITTENBERG *et al.*, 1963). Dolomite beds occur at a depth of 170 m below the water–sediment interface; two of these were isotopically studied (Nos. 22 and 23). Inasmuch as the strata did not contain $CaCO_3$, for comparison calcite-containing beds at a depth of 83, 110 and 138 m, respectively, were studied (Nos. 19, 20 and 21). On the basis of X-ray diffraction analysis, the calcites from sample 21 contain a few per cent of $MgCO_3$ in solid solution.

(6) *Nos. 24 and 25, Steinheim Basin, Southwest Germany, Miocene* (BAJOR, 1963; MENSINK, 1963). Steinheim Basin sediments contain multitudes of well-preserved gastropod shells of *planorbis sp.* This species changes morphologically from the bottom to the top of the exposed sediment profile (about 100 m) in such an obvious and systematical manner that it has received great attention from paleontologists over the last 100 years. The question was whether these changes in morphology were a result of evolution or environment, or both. Recently the Steinheim sediments and in particular their fossil content have been carefully re-examined both paleontologically and geochemically by MENSINK (1963) and BAJOR (1963).

According to BAJOR, Steinheim Basin sediments are deposited in a continental evaporation lake environment. Most of the limestone beds are thought to be derived from reworked gastropods, calcareous algae, ostracodes, and other lime-secreting organisms. Dolomites occur as massive blocks in the lower portion of the lake strata. The structural–textural relationship with the surrounding bedrock suggests a penecontemporaneous origin for the dolomite.

(7) *Nos. 26–32, Alpine Dolomites, Austria, Triassic* (SANDER, 1937). Based on petrographical evidence, these dolomites are thought to be of syngenetic origin. Most of the dolomites are finely laminated or interbedded with calcareous limestones. Samples Nos. 26 and 32 are from Sauwand, near Gusswerk, Steiermark; Nos. 27, 29, 30 and 31 from Bleiberg, Kärnten; and No. 28 is from Presegger See, Kärnten (FLÜGEL and SCHULTZ, personal communication).

(8) *Nos. 33–47, Buntsandstein drill-core, Goldenstedt, northwest Germany, Triassic* (TRUSHEIM, 1961; REUTER, 1961). The sediments are taken from a drill-core at a depth of 2755–2760 m below the present surface. The specimens are spaced at intervals of approximately 50 cm. Samples Nos. 33 and 38–47 each represent 20 cm core sections. From one 20 cm core, 4 subsamples, spaced in 5 cm intervals,

were taken (Nos. 34–37). On petrographical and geochemical grounds it is assumed that the dolomites, which occur as finely laminated bands, are of synsedimentary origin. They are deposited in a marine environment close to the ancient shoreline. Thin-section studies reveal the presence of alternating layers of dolomite, calcite, and occasionally anhydrite.

(9) *Nos. 48 and 49, Caliche deposits of arid region, Coburg, Bavaria, Triassic (Keuper)* (KNETSCH *et al.*, 1961; WELTE, 1962). Subcrustal carbonate caliche

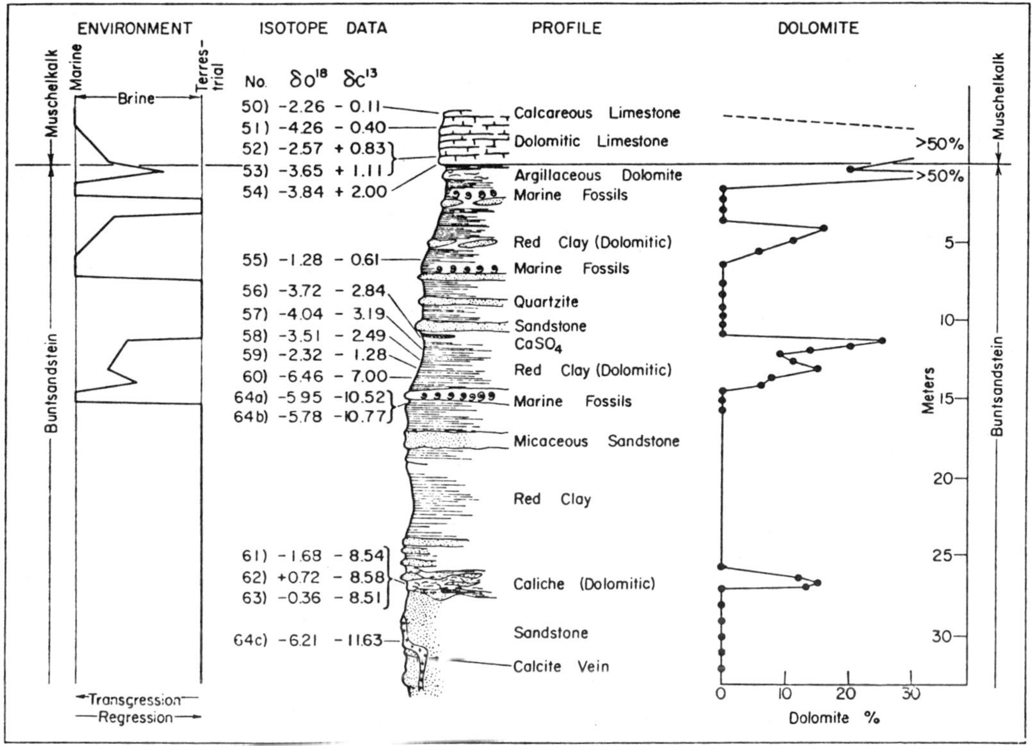

Fig. 1. The δO^{18} and δC^{13} along the Buntsandstein–Muschelkalk profile at Dallau, southwest Germany.

deposits contain syngenetic calcites and dolomites. The dolomites are believed to have formed from pre-existing $CaCO_3$ under extreme climatical conditions probably at a time the climate changed from semiarid to arid.

(10) *Nos. 50–64, Various carbonates from Dallau profile, southwest Germany, Triassic* (DEGENS *et al.*, 1960). Different types of dolomites are present across the profile, ranging from truly marine (50–55) to continental (61–63). Samples 56–59 may be termed intermediate dolomites, since they were deposited under brackish water or lagoonal conditions. All are of penecontemporaneous origin and were formed most likely at a time the sediments were emerged and exposed to the atmosphere. Calcite of syngenetic origin appears higher in the section (No. 50 and 51) probably a result of greater water depth, which prohibited seasonal desiccation as observed in the lower portion of the profile, where dolomite is the only synsedimentary carbonate species. Anhydrite is associated with some of the dolomite

beds. Samples Nos. 61–63 are dolomitic caliche deposits; they were formed at a time of extensive arid to semiarid weathering. Samples 64 a, b and c are secondary calcites which were epigenetically introduced into the strata, probably in the course of Tertiary weathering. Some additional information may be obtained from Fig. 1.

The sedimentological history of the profile can be summarized as follows: Semiarid to arid climatical conditions prevailed for the most part of the continental Buntsandstein. Caliche deposits occasionally developed in fossil soil profiles. The first marine transgression at profil-meter 15 (Fig. 1) can be regarded as the overture to the coming Muschelkalk sea. However, before the marine Muschelkalk finally could overcome the terrestrial Buntsandstein, transgressions and regressions rapidly alternated. Sea waters left after the retreat of the ocean were evaporated or exchanged for fresh water. The sediments frequently run dry as can be inferred from mud cracks, NaCl pseudomorphs and other structural–textural features recorded in the sediments.

The dolomite at the basis of the Muschelkalk was deposited under shallow marine conditions. Moving to the north along the Buntsandstein/Muschelkalk outcrop belt into more open marine conditions (deeper water), a limestone facies gradually replaced the dolomite.

(11) *Nos. 65–70, Cretaceous strata of the Western Interior* (SABINS, 1962). The dolomites are single rhombic crystals abraded to various degrees of roundness and sorted to the size of associated clastic sand grains. The fabric relationships are of the depositional type, which eliminates a diagenetic origin. SABINS (1962), on the basis of petrographical criteria, terms these dolomites "primary." Fossil relicts, *inoceramus sp.*, are sporadically present; secondary calcite is highly abundant. In other words, three different genetic types of carbonates are present in the sediments. All samples are from the New Mexico–Colorado area: Nos. 65 and 66, Montezuma Co., Colorado; Nos. 67 and 68, McKinly Co., New Mexico; and Nos. 69 and 70, San Juan Co., New Mexico.

(12) *Nos. 71–85, Late diagenetic–epigenetic dolomites from the Precambrian to the Permian* (CHILINGAR, 1956). Based on the distribution of Ca/Mg ratio in sedimentary carbonates with time, DALY (1907, 1909) proposed a model of the "evolution of dolomite." Recently, CHILINGAR (1956) confirmed Daly's observation and the progressive decrease in the occurrence of dolomites from the late Precambrian to the Cretaceous is now widely accepted. The slight inconsistency, i.e. an increase in Ca/Mg ratio from the Cretaceous to the Present, has been accounted for by the enforced extraction of lime from the sea by pelagic calcareous foraminifera and coccoliths.

A number of Precambrian to Permian marine carbonates, on which CHILINGAR outlined his concept of the geochemical evolution of the ocean and the dolomites with time, were included in the present investigation. Stratigraphically younger samples of CHILINGAR's collection were too small in dolomite to allow an isotope analysis. Thin-section studies reveal that all 15 specimens analysed are recrystallized. Calcite and dolomite fissures and capillary veins occur frequently in all of the carbonates. No original depositional sedimentary structures can be recognized. Based on these petrographical features, the carbonates are late-diagenetic–epigenetic in character.

The samples are from the following location sites: No. 71, middle of Permian section near Mitchell's caverns, California; No. 72, Anchor limestone (Carboniferous), member of Monte Christo Formation, near Mitchell's caverns, California; No. 73, Salem limestone (Carboniferous), Indiana; No. 74, Burlington limestone (Carboniferous), Illinois; Nos. 77 and 78, Gardner formation (Carboniferous), Utah; No. 79, Devils Gate formation, uppermost beds (Devonian), near Hamilton, Nevada; No. 80, Niagaran dolomite (Silurian), Chicago, Illinois; No. 81, Springfield dolomite (Silurian), Lewisburg, Ohio; No. 82, Platteville limestone (Ordovician), Illinois; Kimmswick limestone (Ordovician), Illinois; and Nos. 84 and 85, Noonday dolomite (Precambrian), Nevada.

RESULTS

The δ-values of the studied dolomites and calcites are presented in Tables 1–5 and Figs. 1–5. The tables are arranged according to age of the specimen, degree of consolidation, and genetic type in terms of penecontemporaneous vs. diagenetic-epigenetic. The differences between the δ-values of coexisting dolomites and calcites are expressed as ΔO^{18} and ΔC^{13}, respectively. The percentages of dolomite and calcite are determined for most of the sediments and are included in the tables.

The results show that dolomites from recent and unconsolidated marine sediments range in δO^{18} from -0.8 to $+4.9$ (average $+2.1$) and in δC^{13} from -1.2 to $+1.4$ (average -0.2). The corresponding δ-values for the associated calcites are -0.4 to $+4.6$ (average $+1.8$) for oxygen and -0.8 to $+1.2$ (average -0.4) for carbon, respectively.

Ancient marine dolomites of penecontemporaneous origin show a spread in δO^{18} from -4.3 to $+2.6$ (average -1.3) and in δC^{13} from -0.6 to $+3.6$ (average $+1.9$). The data for coexisting calcites are -7.2 to $+1.4$ (average -4.8) for oxygen and -3.5 to $+2.6$ (mean $+0.6$) for carbon, respectively.

Diagenetic–epigenetic dolomites of marine facies vary in δO^{18} from -8.9 to -3.2 (average -5.8), and in δC^{13} from -5.0 to $+4.4$ (average -0.1). Their accompanying calcites range in δO^{18} from -18.6 to -3.8 (average -7.9) and in δC^{13} from -4.7 to $+3.7$ (average -0.8).

Terrestrial synsedimentary dolomites of all ages are mostly considerably lighter in δO^{18} than their time equivalent marine dolomites, whereas the δC^{13} values do not in general reflect environmental differences in terms of marine vs. continental.

The upper Cretaceous samples from the New Mexico–Colorado area are characterized by their extreme low δO^{18}-values for the "primary" as well as secondary carbonates. This feature is uncommon for normal marine carbonates. The δC^{13} of the shell debris, however, corresponds to δ-values generally reported for marine limestones.

In summary, Recent syngenetic dolomites and coexisting calcites are isotopically not widely remote from each other. The reported values fall well within the range expected for calcareous limestones formed under marine or continental conditions. Exceptions are the δC^{13} values in the studied fresh-water carbonates which are considerably heavier than the range normally found in terrestrial limestones not associated with dolomites (CLAYTON and DEGENS, 1959).

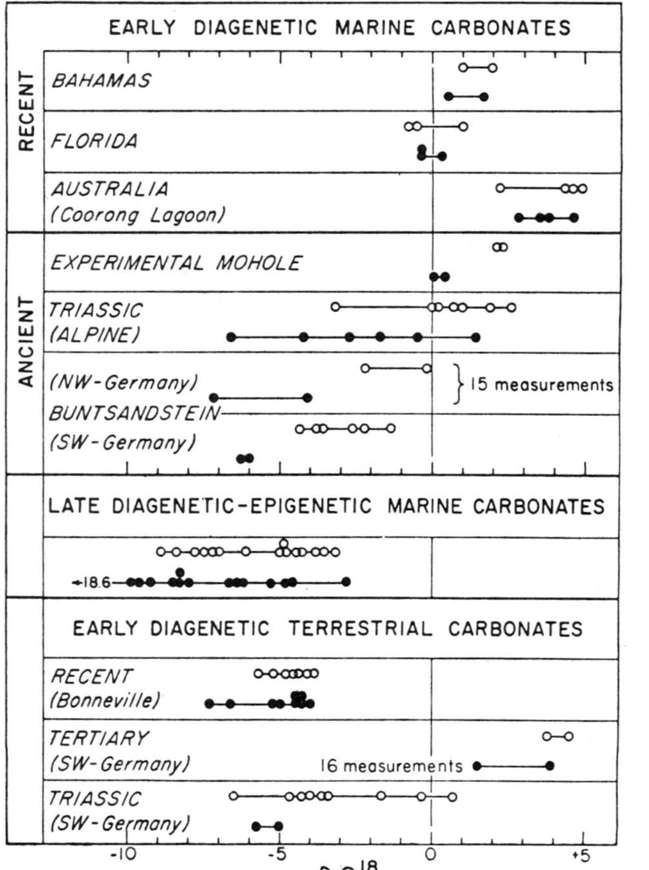

Fig. 2. The δO^{18} values of coexisting dolomite–calcite pairs of various origins and geologic ages. Open circles represent calcite values and solid dots the dolomites.

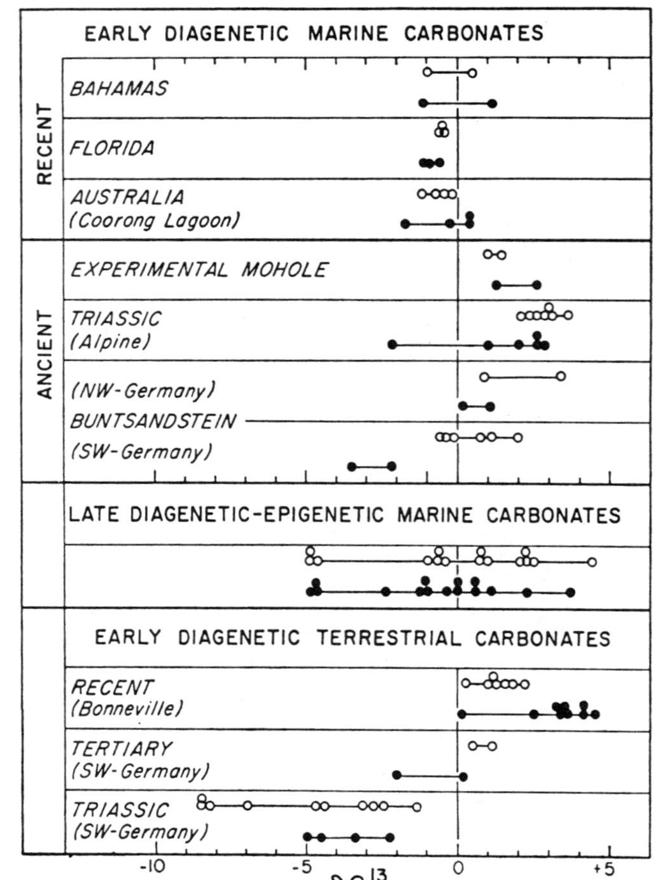

Fig. 3. The δC^{13} values of coexisting dolomite–calcite pairs of various origins and geologic ages. Open circles represent calcite values and solid dots the dolomites.

The isotope data for synsedimentary ancient marine dolomites and associated calcites are identical in δC^{13} with the corresponding recent phases. Ancient dolomites vary only moderately in δO^{18} from recent materials, whereas pronounced differences in δO^{18} between modern and fossil calcites are established. Ancient calcites are consistently lighter.

Independent of the type of carbonate species, late-diagenetic marine limestones are always light in their oxygen isotope distribution; the carbon isotope distribution, on the other hand, is identical to that observed in facies equivalent synsedimentary carbonates. It seems, therefore, that at least for our samples the marine calcites and dolomites are isotopically similar in δC^{13} values and are not significantly altered by diagenetic recrystallization processes.

DISCUSSION

1. *Recent dolomite–calcite pairs of marine origin*

The geological field criteria and the petrographical relationships of the coexisting carbonates from the Bahama evaporite ponds, the Florida Bay area, and the South Australian lagoons are in support of the penecontemporaneous nature of dolomite and calcite (MILLER, 1961; TAFT, 1961; ALDERMAN and SKINNER, 1957; SKINNER, 1960; ALDERMAN and VON DER BORCH, 1960; WELLS, 1962; and SKINNER *et al.* 1963). Most significant for the interpretation of the mechanism of how these dolomites formed at present is the result that the ΔO^{18} values of all studied dolomite–calcite pairs do not exceed ± 1 per mil (Table 1, Nos. 1–9). Namely, coexisting dolomites and calcites are isotopically similar.

If equilibrium conditions existed in the dolomite–calcite pairs, and if the proposed mechanism is correct, then dolomites precipitated in an aqueous environment at room temperature should be heavier by 6–10 per mil in δO^{18} over cogenetic calcites and aragonites (CLAYTON and EPSTEIN, 1958; ENGEL *et al.*, 1958; and EPSTEIN *et al.*, 1963). The lack of such a relationship in the investigated sedimentary dolomite–calcite pairs and the consistent Δ dolomite–calcite values of about zero suggests that dolomitization must have proceeded without significantly altering the O^{18}/O^{16} record of the precursor carbonate and that dolomite must have been derived by way of metasomatism of calcite or aragonite; namely the transformation of the original calcareous ooze to dolomite must have taken place without chemically effecting the CO_3 unit. Consequently, it can be inferred that the growth of dolomite did take place under solid-state conditions from crystalline calcium carbonate.

In this connection, the data of FAIRBRIDGE (1957) on the occurrence and origin of Funafuti dolomite are noteworthy. According to his observation, calcareous algae enriched in $MgCO_3$ are geochemically reorganized into low-Mg calcite and dolomite during diagenesis. Sea water in this case presumably does not serve as a potential Mg source for the newly-formed dolomite. This viewpoint on the thermodynamical instability of $MgCO_3$ that is present in solid solution in $CaCO_3$ is also supported by work of CHAVE (1954), LOWENSTAM (1961), CHAVE *et al.* (1962), SEIBOLD (1962), and CHILINGAR (1962). According to the isotope data by Epstein *et al.* (1963), this reorganization into calcite and dolomite seems to have no significant isotope effect on the Funafuti dolomite, and the δ-values of the dolomites seem to correspond to those originally fixed in the calcium carbonate at the time of its formation.

93

Table 1. The δO^{18} and δC^{13} values of syngenetic dolomites and coexisting calcites from unconsolidated sediments

No.	Geologic age	Facies	Dolomite δO^{18}	Calcite δO^{18}	Dolomite δC^{13}	Calcite δC^{13}	Dolomite–Calcite ΔO^{18}	Dolomite–Calcite ΔC^{13}	Dolomite %*	Calcite %*
1	Recent	marine	+1·04	+0·46	+0·54	+1·19	+0·58	−0·65	80	20
2	(Bahama)		+1·88	+1·68	−0·95	−1·12	+0·20	+0·17	15	85
3	Recent		−0·79	−0·42	−0·53	−0·97	−0·37	+0·44	4	96
4	(Florida)	marine	−0·60	−0·43	−0·54	+0·55	−0·17	+0·01	9	91
5			+0·96	+0·32	−0·59	−1·05	+0·64	+0·46	14	86
6			+2·17	+2·78	−0·80	−0·33	−0·61	−0·47		
7			−4·43	+3·54	−0·25	+0·35	+0·89	−0·60	n.d.	n.d.
8	Recent	marine +	+4·58	+3·76	−1·17	−1·78	+0·82	+0·61		
9	(South Australia)	lagoonal	+4·94	+4·61	−0·45	+0·44	+0·33	−0·99		
10			−4·32	−4·51	+1·76	+3·59	+0·19	−1·83	12	88
11			−4·43	−4·49	+2·18	+4·50	+0·06	−2·32	10	90
12			−4·75	−6·59	+1·30	+2·51	+1·84	−1·21	31	69
13	10,000 years	continental-	−5·29	−7·27	+1·02	+0·12	+1·98	+0·90	48	52
14	(Utah)	brine	n.d.	−4·49	n.d.	+4·19	—	—	3	97
15			−3·89	−5·01	+1·16	+3·47	+1·12	+2·31	28	72
16			n.d.	+4·04	n.d.	+4·19	—	—	2	98
17			−4·11	−5·13	+1·52	+3·33	+1·02	−1·81	33	62
18			−5·77	−4·38	+0·26	+3·47	−1·39	−3·21	6	94
19			—	−0·04	—	+1·30	—	—	—	100
20	Tertiary	marine	—	+0·04	—	+1·55	—	—	—	100
21	(Experimental		—	+0·43	—	+2·56	(+1·66†)	(−1·21†)	—	100
22	Mohole)		+2·09	—	+1·35	—·	—	—	100	—
23			+2·27	—	+1·05	—	—	—	100	—

* Per cent of carbonate fraction.
† Δ-values of closest spaced carbonates.

2. *Dolomite–calcite pairs in unconsolidated continental sediments*

The ΔO^{18} and ΔC^{13} values of continental dolomite–calcite pairs from the Lake Bonneville area (Table 1, Nos. 10–18) are somewhat higher than those found in their recent marine equivalents (Table 1, Nos. 1–9). These differences, however, are small compared to the Δ-values one would expect if dolomites and calcites would precipitate in isotopic equilibrium with each other. Considering the fact that continental environments of this type fluctuate strongly in water temperatures and O^{18}/O^{16} ratios, which might cause large differences in the expected δO^{18} of calcite and dolomite, the observed differences in δO^{18} between both carbonates are low. This again suggests that, similar to recent marine dolomite–calcite pairs, the studied continental dolomites are derived from a $CaCO_3$ precursor material rather than being directly precipitated from an aqueous solution.

Noteworthy are the positive δC^{13}-values for both calcite and dolomite. The carbonates of the lower δC^{13}-values are related to the dark bands of the formation which might indicate presence of some biological activity and addition of lighter carbon to the depositional environment. The δC^{13}-values of both carbonates, however, seem to be too positive for a fresh-water environment (CLAYTON and DEGENS, 1959). A relatively sterile environment limiting biological growth and permitting establishment of isotopic equilibrium with surrounding atmospheric CO_2 could account for these values.

3. *Tertiary carbonates from Experimental Mohole (Guadalupe Island, Mexico)*

From Tertiary sediments in Experimental Mohole, RIEDEL *et al.* (1961) report the occurrence of dolomite bands at 170 m depth of burial. The sediment–water interface is located at a water depth of about 4000 m. The dolomites are underlain by beds of basalt of unknown thickness. Studies by RITTENBERG *et al.* (1963) indicate that interstitial waters throughout the 170 m sediment profile are with the exception of calcium geochemically not significantly different from mean ocean water.*

Inasmuch as the Mohole dolomite beds so far as we know are only found at the bases of the section (170 m) and, furthermore, do not contain measurable amounts of $CaCO_3$, for comparison the latter carbonate was taken from limestone beds at 138, 110, and 83 m, respectively.

The isotope data are reported in Table 1 (Nos. 18–23). The O^{18}/O^{16} and C^{13}/C^{12} ratios are in accordance with those expected for carbonates formed in a normal marine environment. On the basis of the paleotemperature scale by EPSTEIN *et al.* (1953), the dolomites yield water temperatures of around $+5°C$, and the calcites somewhat higher in the section record temperatures of $+13°$ to $15°C$.

Interstitial water in the dolomite beds still contains the same Mg-level as that of the modern sea (RITTENBERG *et al.*, 1963). This, perhaps, may serve as an indication that sea water entrapped in the sediment is not a significant source for magnesium in the dolomite. Part of the magnesium may have been derived from the magnesium carbonate present in solid solution in the original limestone material, the rest has to come from below. The occurrence of basalt in the immediate environs is interesting, since Mg may have originated from here. In spite of this,

---

* Relative to mean ocean water.

the formation of dolomite has not resulted in significant Δ-values. This adds further to the contention that the formation of the dolomite is a secondary process unaccompanied by changes in δO^{18} and δC^{13}.

4. Lake-water dolomite–calcite pairs of Tertiary age, Steinheim Basin, south-west Germany

In the work of BAJOR (1963), the presence of stratified dolomite in the lake beds was established. The environment of deposition in terms of O^{18}/O^{16} ratio of the water can be reasonably estimated from the δO^{18} values of the aragonitic gastropod shell fragments. The isotope data indicate (Table 2, Nos. 24 and 25) that the former lake was greatly enriched in O^{18} by at least 10–15 per mil over the fresh-water one would expect at this locality. This increase in the heavy isotope can be accounted for only by severe evaporation processes. The δC^{13} values suggest that the dissolved carbonate phases in the former lake environment may have been isotopically equilibrated with the CO_2 in the atmosphere or with underlying marine limestones.

Magnesium-rich calcite is abundant particularly in the lower calcareous portion of the profile where the dolomite also occurs. This feature is largely a result of biological activity, i.e. calcite-secreting organisms, incorporating a few per cent $MgCO_3$ into their skeletons. The instability of $MgCO_3$ in solid solution resulted in the penecontemporaneous formation of dolomite and low-magnesium calcite. Calcareous algae, because of their high $MgCO_3$ concentration (up to 30 per cent), can be regarded as the most likely precursor material for the dolomite, and the "reef"-type structure of the dolomite beds is in support of this inference.

In this case, the Δ-dolomite–calcite values are also close to zero. Here there is another type of dolomitization environment which still gives the same results.

5. Dolomite–calcite pairs of Triassic age

The majority of the sample material (Tables 2 and 3, Nos. 26–63) was obtained from the Alpine and German Triassic. The samples were selected from here, because a great deal of general geologic, petrographic, and geochemical information is available on the "dolomite problem."

It is interesting to note that among the 22 dolomites of apparently marine facies (based on sedimentological criteria) (Table 2, Nos. 26–47), only one (No. 31) stands isotopically somewhat outside the normal range of marine carbonates. For the others the δO^{18} values would correspond to paleotemperatures between $+5°$ and $+25°C$. The associated calcites, however, are consistently lower in δO^{18}, which is in accordance with the experimental data of EPSTEIN et al. (1963). These show that the rate of O^{18}/O^{16} exchange between calcite and water exceeds the rate of dolomite equilibration with the surrounding water phase. In other words, if the penecontemporaneous nature of a marine dolomite can be established, its O^{18}/O^{16} ratio (in principle) may eventually serve as material for paleotemperature determinations, particularly in such instances where calcareous limestone or fossils have already exchanged.

Samples Nos. 50–63 are of particular interest because they include dolomites of terrestrial, lagoonal, and marine origin, occurring within the same vertical

Table 2. The δO^{18} and δC^{13} values of syngenetic dolomites and coexisting calcites from Tertiary and Triassic sediments

No. (cont.)	Geologic age	Facies	Dolomite	Calcite	Dolomite	Calcite	Dolomite–Calcite		Total Carbonate		Dolomite	Calcite
			δO^{18}		δC^{13}		ΔO^{18}	ΔC^{13}	δO^{18}	δC^{13}	(wt. %)	
24	Tertiary	fresh water	+4·45	+2·36*	+1·11	−0·96	+2·09	+2·07	n.d.	n.d.	n.d.	n.d.
25	(Southwest Germany)	(evap. lake)	+3·86	+2·36*	+0·52	−0·96*	+1·50	+1·48				
26	Rhät		+1·04	—	+3·57	—	—	—			100	
27			+0·74	−0·52	+2·05	+2·02	+1·26	+0·03				
28	Nor	marine	+2·57	+1·41	+2·60	+2·56	+1·16	+0·04				
29	(Austria)		+0·18	−1·72	+3·10	2·59	+1·90	+0·51	n.d.	n.d.	n.d.	n.d.
30			−0·02	−2·74	+2·96	+1·01	+2·72	+1·95				
31	Karn		−3·20	−4·21	+2·86	+2·60	+1·01	+0·26				
32			+1·90	−6·58	+2·34	−2·21	+8·48	+4·55				
33			−1·13	−4·49	+1·98	−0·73	+3·36	+1·25	−2·04	+1·46	16·5	12.8
34			+1·47	+4·19	+1·80	+1·01	+2·72	+0·79				
35			−1·61	−4·63	+1·95	+0·89	+3·02	+1·06	n.d.	n.d.	n.d.	n.d.
36			−1·73	−4·15	+1·76	+0·98	+2·42	+0·78				
37			−1·79	−4·42	+0·94	+0·96	+2·63	−0·02				
38			−0·77	−5·40	+2·26	+0·69	+4·63	+1·57	−2·03	+1·76	15·0	10·2
39	Buntsandstein	marine	−0·76	−6·10	+1·88	+0·18	+4·34	+1·70	−2·62	+1·52	23·4	7·8
40	(Northwest Germany)		−1·23	−6·38	+2·00	+0·09	+5·15	+1·91	−2·39	+1·40	13·9	7·3
41			−2·13	−7·21	+2·40	+0·30	+5·08	+2·10	−2·86	+1·74	12·0	6·5
42			−1·61	−6·58	+2·55	+0·37	+4·97	+2·18	−2·72	+1·80	13·2	6·1
43			−0·16	−5·90	+3·36	+0·71	+5·74	+2·65	−2·00	+2·15	14·1	6·6
44			−1·38	−5·90	+2·34	+0·55	+4·52	+1·79	−2·40	+1·96	19·4	7·5
45			−1·94	−5·74	+1·87	+0·81	+3·80	+1·06	−2·33	+1·70	25·8	2·8
46			−1·42	−5·49	+2·24	+0·78	+4·07	+1·46	−3·00	+1·68	14·4	15·5
47			−1·54	−5·68	+2·42	+0·24	+4·14	+2·18	−3·00	+1·42	10·5	14·3

* Mean value of 16 gastropod samples ranging from: $\delta O^{18} = +0.52$ to $+3.85$; $\delta C^{13} = -2.00$ to $+0.13$.

Table 3. The δO^{18} and δC^{13} values of syngenetic dolomites and coexisting calcites of the German Triassic

No. (cont.)	Geologic age	Facies	Dolomite	Calcite	Dolomite	Calcite	Dolomite–Calcite	
			δO^{18}		δC^{13}		ΔO^{18}	ΔC^{13}
48	Keuper	Continental	−3·96	−5·84	−4·61	−4·95	+1·88	+0·34
49	(Bavaria)	(caliche)	−4·08	−5·06	−4·44	−4·63	+0·98	+0·19
50	Muschelkalk		−2·26	−6·19	−0·11	−3·47	+3·93	+3·36
51	(Southwest		−4·26	−6·05	−0·40	−2·24	+1·79	+1·84
52	Germany)	Marine	−2·57	—	+0·83	—	—	—
53			−3·65	—	+1·11	—	—	—
54			−3·84	—	+2·00	—	—	—
55			−1·28	—	−0·61	—	—	—
56			−3·72	—	−2·84	—	—	—
57			−4·04	—	−3·19	—	—	—
58		Fresh water	−3·51	—	−2·49	—	—	—
59	Buntsandstein	lagoonal	−2·32	—	−1·28	—	—	—
60	(Southwest		−6·46	—	−7·00	—	—	—
61	Germany)		−1·68	—	−8·54	—	—	—
62		Continental	+0·72	—	−8·58	—	—	—
63		(caliche)	−0·36	—	−8·51	—	—	—
64a *			—	−5·95	—	−10·52	—	—
64b *		Continental	—	−5·78	—	−10·77	—	—
64c *			—	−6·21	—	−11·63	—	—

* Epigenetic calcites.

cross-section of less than 30 m. The facies relationships throughout the sediment profile are well understood and the penecontemporaneous character of all dolomites is established (DEGENS *et al.*, 1960).

In Fig. 1, the stratigraphic position of the individual carbonates and a few basic details on the geology and petrography are graphically presented.

Dolomite constitutes the only synsedimentary carbonate phase in the Buntsandstein; marine limestones appear higher in the Muschelkalk section. Secondary calcites of presumably Tertiary age are found as fillings in Buntsandstein clay concentrations or along joints.

The isotope data support the results of previous geological and geochemical studies (DEGENS *et al.*, 1960). Marine dolomites have δC^{13} from -0.6 to $+2.0$ per mil, and their δO^{18} values indicate paleotemperatures between $+20°$ and $+35°C$. The fresh water and fresh water lagoonal dolomites show also a reasonable range in δO^{18} from -6.5 to -2.3 and in δC^{13} from -7.0 to -1.3. Continental dolomitic caliche deposits (Nos. 61–63), although high in δO^{18}, are in environmental accordance with regard to C^{13}/C^{12} ratio ($\delta C^{13} = -8.5$) (CLAYTON and DEGENS, 1959). It should be pointed out that recent calcareous caliche deposits from Greece yield δ-values of comparable magnitude ($\delta O^{18} = -1.5$; $\delta C^{13} = -10.6$). The high δO^{18}-values in both cases can be attributed to extensive evaporation. which is known to be effective during caliche formation (KNETSCH, 1937; RUTTE, 1958). In the case where dolomites and calcites coexist—two caliche deposits from the Keuper (Nos. 48 and 49)—the coexisting dolomite and calcite show no significant isotope difference from each other for both oxygen and carbon.

The secondary vein and vug calcites present in the Buntsandstein contain less than 100 ppm in magnesium (Nos. 64 a, b and c). They probably have formed during Tertiary and later time weathering and denudation of the overlaying Muschelkalk plateau. Their fresh water character is isotopically well-established ($\delta O^{18} \sim -0$; $\delta C^{13} \sim -11$). The limestone material in the Muschelkalk formation and in the secondary calcites resemble each other in δO^{18}. In view of the light δC^{13} values of the calcareous portion of the Muschelkalk, the isotope data suggest that some secondary calcites low in δC^{13} are associated with the Muschelkalk and, furthermore, that the original synsedimentary Muschelkalk calcites are equilibrated in terms of O^{18}/O^{16} ratio with meteoric or surface waters. Dolomites, on the other hand, seem to be not affected by any of these exchange processes. Thus, in the Muschelkalk, the δ-values of the dolomites are probably the δ-values of the original marine limestone (the dolomite precursor) and the calcareous portion we now find has been re-equilibrated with meteoric waters.

6. *Dolomite–calcite pairs of the Cretaceous of the Western Interior*

Certain types of dolomites which occur in the Cretaceous strata of the Western Interior are interpreted as being of primary origin; that is, they are formed concurrently with deposition. The inference is based on the size, fabric and facies distribution of the dolomites. A secondary formation or a detrital origin is ruled out on various geological and petrographical grounds (SABINS, 1962).

Marine fossils, *inoceramus sp.*, associated with the sediments yield δC^{13} values which are in accordance with a marine environment. The extremely light O^{18}/O^{16}

Table 4. The δO^{18} and δC^{13} values in coexisting carbonates of the Upper Cretaceous

No. (cont.)	Geologic age Facies	Dolomite Inoceramus Calcite*			Dolomite Inoceramus Calcite*			Dolomite–Inoceramus		Dolomite–Calcite*	
		δO^{18}			δC^{13}			ΔO^{18}	ΔC^{13}	ΔO^{18}	ΔC^{13}
65		-9.81	-9.58	-10.49	-4.18	$+0.67$	-1.09	-0.23	-4.85	$+0.68$	-3.09
66		-6.91	-9.68	-10.98	-0.81	$+3.61$	-1.42	$+2.77$	-4.42	$+4.07$	$+0.61$
67	Upper Cretaceous Marine	-9.36	-9.65	-7.46	-3.11	$+1.10$	-1.94	$+0.29$	-4.21	-1.90	-1.17
68	(New Mexico–Colorado)	-10.26	-9.58	-12.28	-2.32	$+0.86$	-8.05	-0.68	-3.18	$+2.02$	$+5.73$
69		-9.32	n.d.	-17.74	-1.76	n.d.	-4.35	—	—	$+8.42$	$+2.59$
70		-9.38	n.d.	-15.65	-4.07	n.d.	-9.47	—	—	$+6.27$	$+5.40$

* Secondary calcite.

Table 5. The δO^{18} and δC^{13} values of diagenetic dolomites and coexisting calcites from the Precambrian to the Permian

No. (cont.)	Geologic age	Dolomite δO^{18}	Calcite δO^{18}	Dolomite δC^{13}	Calcite δC^{13}	Dolomite–Calcite ΔO^{18}	Dolomite–Calcite ΔO^{13}
71	Permian	−7·17	−9·64	+2·01	+0·57	+2·47	+1·94
72		−4·83	−18·57	+0·72	−0·08	+14·74	+0·81
73		−4·94	−6·23	−0·65	−0·99	+1·29	+0·34
74	Carboniferous	−7·00	−8·26	−0·97	−2·29	+1·26	+1·32
75		−3·15	−3·78	+2·46	+2·32	+0·63	+0·14
76		−3·45	−6·71	+0·95	+0·25	+3·26	+1·20
77		−8·92	−8·53	+0·86	+0·62	−0·39	+0·24
78		−4·34	−4·61	+4·38	+3·74	+0·27	+0·64
79	Devonian	−8·37	−9·29	−4·86	−4·68	+0·92	−0·18
80	Silurian	−4·73	−5·29	+2·32	+1·11	+0·56	+1·21
81		−6·10	−9·99	−0·64	−1·09	+3·89	+0·45
82		−4·36	−4·73	−0·62	−1·01	+0·37	+0·39
83	Cambro–Ordovician	−4·93	+6·45	+2·18	+0·06	+1·52	+2·12
84		−7·77	−8·03	−4·99	−4·66	+0·26	−0·33
85	Precambrian	−7·46	−8·29	−5·00	+4·55	+0·83	−0·45

ratios on the other hand, however, can only reasonably be accounted for by post-depositional isotopic equilibration with meteoric or connate waters. Assuming that connate waters represented the aqueous phase, during diagenesis, the δO^{18} values in the shell carbonate would correspond to temperatures in the neighbourhood between 60° and 70°C.

The dolomites have the same δO^{18} values as their coexisting marine fossils and are lighter by about 4 per mil in δC^{13}. This would exclude a primary origin, as has been proposed by SABINS (1962). Although we presently do not know how much time and temperature is required for dolomite to attain isotopic equilibrium with the interstitial water in terms of O^{18}/O^{16} ratio, based on the experimental work of EPSTEIN *et al.* (1963), it is less probable that the light δO^{18} values in the Cretaceous dolomite is a result of isotopic equilibration with the surrounding aqueous phase. A secondary origin for the studied dolomite seems more likely, in view of the small differences in δO^{18} between fossil shells and dolomite.

Secondary calcites are relatively abundant and account for a great percentage of the carbonate material present in the studied sandstones. In samples Nos. 65–67 (Table 4), they correspond isotopically with their coexisting dolomite and fossil shell material. Regarding the comparatively light δC^{13} and δO^{18} values in samples Nos. 68–70, bicarbonate contributions from biogenic sources and the presence of meteoric waters at the time the latter three calcites formed must be assumed.

7. *Late diagenetic–epigenetic dolomite–calcite pairs*

The δO^{18}-values for this group of dolomite–calcite pairs fall outside the range for marine limestones. The δC^{13} would correspond to marine conditions, with the exception of one Devonian and two Precambrian samples. The light δO^{18}-values are caused by isotopic exchange of the limestones with meteoric or connate waters. The fact that Δ-dolomite–calcite is small suggests a sequence of events undergone by the carbonate formations. Provided the previously proposed fractionation

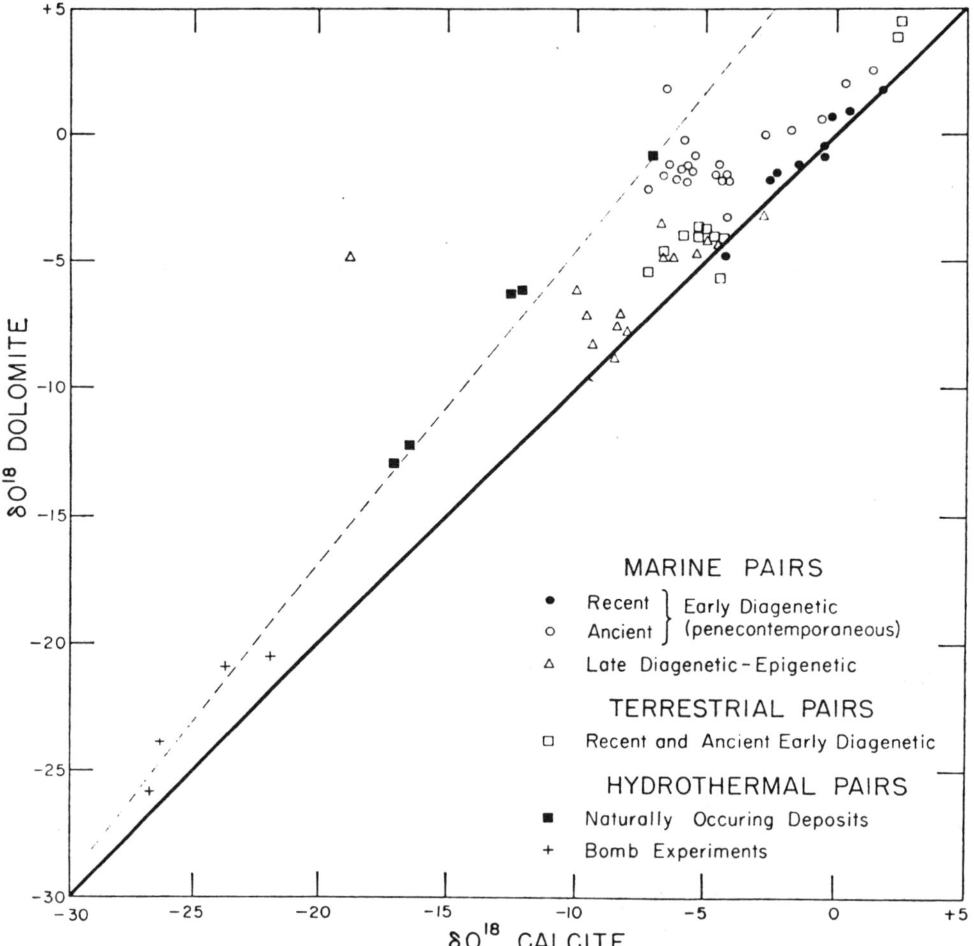

Fig. 4. The δO^{18} relationship between coexisting dolomites and calcites. Black line: Δ-dolomite–calcite equals zero; dashed line: the dolomite–calcite relationship obtained by Clayton and Epstein, 1958 (the assumed equilibrium curve between dolomite and calcite and a large reservoir of hydrothermal fluids).

factor at room temperature is correct, the above data indicate that the limestone re-equilibrated to the negative δ-values and that dolomitization proceeded after this event with different degrees of re-equilibration of the dolomite to give Δ-values from zero to $+4\cdot0$. If dolomitization took place when the δ-values of the calcite were zero, then the present Δ-dolomite–calcite data would have to be greater than the equilibrium Δ-values, which is not the case.

It is interesting to note that recrystallization and dolomitization even at a late diagenetic stage do not significantly alter the C^{13}/C^{12} record originally laid down by the synsedimentary calcium carbonate precursor.

SUMMARY AND CONCLUSIONS

The summary of the relationship of the δ-values of the dolomites and calcites is shown in Figs. 4 and 5. Based on these graphs, the following statements can be made:

1. Nearly all the δO^{18}-values lie between ΔO^{18} = zero and ΔO^{18} as determined by the hydrothermal samples previously reported (ENGEL *et al.*, 1958; CLAYTON and EPSTEIN, 1958.)

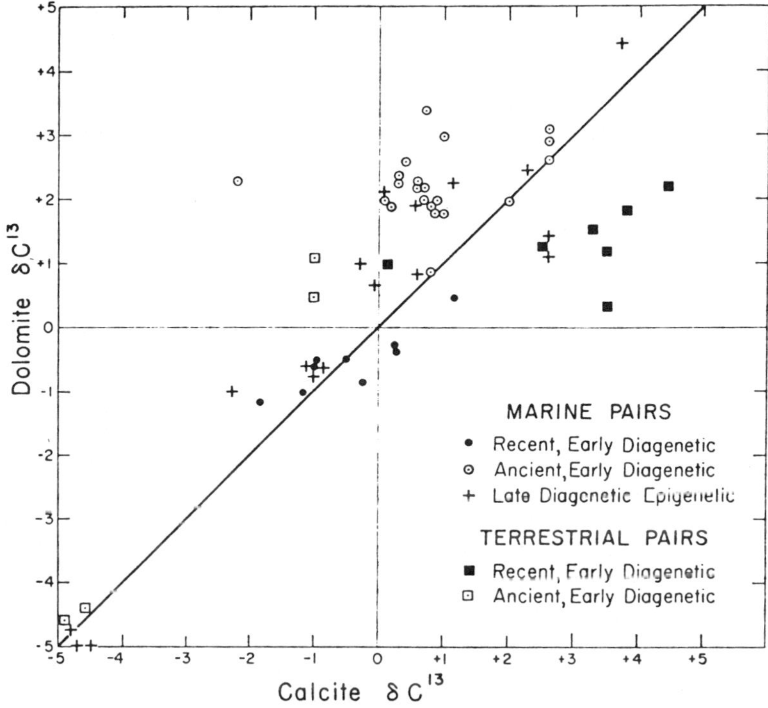

Fig. 5. The δC^{13} relationship between coexisting dolomites and calcites. Black line: Δ-dolomite–calcite equals zero.

2. Recent sedimentary dolomite–calcite pairs formed at moderate temperatures under surface or subsurface conditions are isotopically similar in terms of their δO^{18}- and δC^{13}-values. This feature holds true for all pairs independent of facies relationships or mode of formation. The oxygen and carbon isotope data recorded in recent dolomites correspond to those expected for calcite precipitated in isotopic equilibrium with the environment of deposition.

3. The isotope data for recent coexisting dolomites and calcites speak against a primary origin for the dolomite. Recent dolomites are not precipitated from an aqueous solution but are formed by way of metasomatism from pre-existing

103

crystalline calcium carbonate. Geological and C^{14} data indicate that dolomitization may already start shortly after deposition.

4. Inasmuch as no isotope fractionation for both the carbon and oxygen occurs during the alteration of the $CaCO_3$ precursor material to dolomite, dolomitization has to proceed under solid-state conditions without chemically altering the CO_3^{--} unit.

5. Ancient dolomites of penecontemporaneous origin show the same isotope characteristics as observed in present-day dolomites of analogous facies. However, their associated calcites are frequently depleted in O^{18}. This relationship can be easily accounted for by oxygen-isotope equilibration of the calcite with subsurface waters (DEGENS and EPSTEIN, 1962). The reluctance of dolomite to adjust isotopically to changes in temperature and O^{18}/O^{16} ratio of formation waters has been effectively demonstrated by EPSTEIN *et al.* (1963). This characteristic in principle makes dolomites a possible tool for the evaluation of paleotemperatures in the ancient sea. The δO^{18} of synsedimentary Triassic marine dolomites would correspond to temperatures between 5° and 35°C.

6. Late diagenetic–epigenetic dolomites of marine facies are low in δO^{18}, whereas their δC^{13} apparently is not significantly affected by dolomitization processes. Calcites are isotopically similar to the coexisting dolomites. This characteristic may serve as criteria that recrystallization of the limestone occurred before dolomitization took place. In other words, secondary calcites rather than the original limestones were metasomatically replaced by dolomite.

7. It is at present not fully understood what kind of environmental conditions in terms of alkalinity, organic activity, chemistry, or other parameters have to be met in order to form dolomite. Assuming the 6–10‰ room-temperature-fractionation between dolomite and calcite is correct, the isotope data indicate that sedimentary dolomites, independent of whether they are penecontemporaneous or late diagenetic, and marine or terrestrial, are similar in their mode of formation. All are derived by metasomatic replacement of calcite or aragonite. It seems most probable that the $CaCO_3$ polymorphs are metastable in certain diagenetic environments and that the calcium carbonates are transformed into the more stable dolomite by straightforward solid diffusion mechanism.

Acknowledgements—The writers are indebted to M. BAJOR (Bonn, Germany), H. J. BISSELL (Provo, Utah), G. V. CHILINGAR (Los Angeles), K. O. EMERY (Woods Hole, Mass.), E. FLÜGEL (Darmstadt, Germany), H. A. LOWENSTAM (Pasadena), W. R. RIEDEL (La Jolla, California), H. C. W. SKINNER (Washington, D.C.), W. TAFT (Palo Alto, California), F. TRUSHEIM (Hannover, Germany) and D. WELTE (Würzburg, Germany) for kindly contributing sample material.

The aid of the American Chemical Society and Atomic Energy Commission AT(04-3)-427 is gratefully acknowledged.

REFERENCES

ALDERMAN A. R. and VON DER BORCH, (1960) Occurrence of hydromagnesite in sediments in South Australia. *Nature, Lond.* **188,** 931.

ALDERMAN A. R. and SKINNER H. C. W. (1957) Dolomite sedimentation in the South-East of Australia. *Amer J. Sci.* **255,** 561–567.

BAJOR M. (1963) Unpublished Ph.D. thesis, Bonn University, Germany.

BISSELL H. J. and CHILINGAR G. V. (1962) Evaporite dolomite in salt flats of Western Utah. *Sedimentology* **1,** 200–210.

CHAVE K. E. (1954) Aspects of the biochemistry of magnesium—I: Calcareous marine organisms. *J. Geol.* **62**, 266–283.

CHAVE K. E., DEFFEYES K. S., WEYL P. K., GARRELS R. M. and THOMPSON M. E. (1962) Observations on the solubility of skeletal carbonates in aqueous solution. *Science* **137**, 33–34.

CHILINGAR G. V. (1956) Relationship between Ca/Mg ratios and geologic age. *Bull. Amer. Ass. Petrol. Geol.* **67**, 1559–1562.

CHILINGAR G. V. (1962) Possible loss of magnesium from fossils to the surrounding environment. *J. Sed. Petr.* **32**, 136–139.

CLAYTON R. N. and EPSTEIN S. (1958) The relationship between O^{18}/O^{16} ratios in coexisting quartz, carbonate and iron oxides from various geological deposits. *J. Geol.* **66**, 352–373.

CLAYTON R. N. and DEGENS E. T. (1959) Use of carbon-isotope analyses for differentiating fresh-water and marine sediments. *Bull. Amer. Ass. Petrol Geol.* **43**, 890–897.

CLOUD P. E. and BARNES V. E. (1957) Early Ordovician sea in Central Texas. *Geol. Soc. Amer., Memoir* **67**, 163–214.

CRAIG H. (1953) The geochemistry of the stable carbon isotopes. *Geochim. et Cosmochim. Acta* **3**, 53–92.

CRAIG H. (1957) Isotopic standards for carbon and oxygen and correction factors for mass spectrometric analysis of carbon dioxide. *Geochim. et Cosmochim. Acta* **12**, 133–149.

DALY R. A. (1907) The limeless ocean of Pre-cambrian time. *Amer. J. Sci.* **23**, 93–115.

DALY R. A. (1909) First calcareous fossils and the evolution of limestone. *Geol. Soc. Amer. Bull.* **20**, 153–170.

DEGENS E. T., KNETSCH, G. and REUTER H. (1960) Ein geochemisches Buntsandstein-Profil von Schwarzwald bis zur Rhön. *N. Jb. Geol. Pal. Abh.* **111**, 181–233.

DEGENS E. T. and EPSTEIN S. (1962) Relationship between O^{18}/O^{16} ratios in coexisting carbonates, cherts and diatomites. *Amer. Assoc. Petrol. Geol. Bull.* **46**, 534–542.

ENGEL A. E. J., CLAYTON R. N. and EPSTEIN S. (1958) Variations in isotopic composition of Oxygen and carbon in Leadville limestone and its hydrothermal and metamorphic phases. *J. Geol.* **66**, 374–393.

EPSTEIN S., BUCHSBAUM R., LOWENSTAM H. A. and UREY H. C. (1953) Revised carbonate-water isotopic temperature scale. *Bull. Geol. Soc. Amer.* **64**, 1315–1326.

EPSTEIN S., GRAF D. L. and DEGENS E. T. (1963) Oxygen isotope studies on the origin of dolomites. *Isotopic and Cosmic Chemistry*, dedicated to H. C. Urey 70th Birthday (eds. H. Craig, S. L. Miller and G. T. Wasserburg), p. 188. North Holland, Amsterdam.

FAIRBRIDGE R. W. (1957) The Dolomite Question: Symp. Regional Aspects of Carbonate Deposition, edited by R. J. LE BLANC and J. G. BREEDING. *Soc Econ. Paleo. Min.* Special Publication No. 5, 125–178.

GRAF D. L., EARDLEY A. L. and SHIMP N. F. (1961) A preliminary report on magnesium carbonate formation in glacial Lake Bonneville. *J. Geol.* **69**, 219–223.

GRAF D. L., EARDLEY A. J. and SHIMP N. F. (1959) Dolomite formation in Lake Bonneville, Utah (Abstract). *Bull. Amer. Geol. Soc.* **70**, 1610.

INGERSON E. (1962) Problems of the geochemistry of sedimentary carbonate rocks. *Geochim. et Cosmochim. Acta* **26**, 815–847.

KNETSCH G. (1937) Beiträge zur Kenntnis von Krustenbildungen. *Ztschr. deutsch. geol. Ges.* **89**, 177–192.

KNETSCH G., DEGENS E. T., WELTE D. and REUTER H. (1960) Untersuchungen und Schlüsse zur Verteilung von Strahlungsträgern in Sedimenten Frankens. *Glückauf* **96**, 172–183.

LOWENSTAM H. A. (1961) Mineralogy, O^{18}/O^{16} ratios, and strontium and magnesium contents of recent and fossil brachiopods and their bearing on the history of the oceans. *J. Geol.* **69**, 241–260.

McCREA J. M. (1950) On the isotopic chemistry of carbonates and a paleotemperature scale. *J. Chem. Phys.* **18**, 849–857.

MENSINK H. (1963) Habilitation-thesis, Bonn University, Germany. To be published.

MILLER D. N. (1961) Early diagenetic dolomite associated with salt extraction process, Inagua, Bahamas. *J. Sed. Petrol.* **31**, 473–476.

Reuter H. (1961) Geochemische Untersuchungen einer radioaktiven Anomalie in einem norddeutschen Buntsandsteinprofil. *Erdöl und Kohle,* **14,** 802–803.

Riedel W. R., Ladd H. S., Tracey J. I., Jr., and Bramlette M. N., 1961, Preliminary drilling phase of Mohole Project, II. Summary of coring operations (Guadelupe Site). *Bull. Amer. Ass. Petrol. Geol.* **45,** 1793–1798.

Rittenberg S. C., Emery K. O., Hülsemann J., Degens E. T., Fay R. C., Reuter J. H., Grady J. R., Richardson S. H. and Bray E. E. (1963) Biogeochemistry of sediments in Experimental Mohole. *J. Sed. Petrol.* **33,**

Rutte E. (1958) Kalkkrusten in Spanien. *N. Jb. Geol. Pal. Abh.* **106,** 52–138.

Sabins R. R., Jr. (1962) Grains of Detrital, Secondary, and Primary Dolomite from Cretaceous Strata of the Western Interior. *Bull. Geol. Soc. Amer.* **73,** 1183–1196.

Sander B. (1937) Beiträge zur Kenntnis der Anlagerungsgefüge: Rhytmische Kalke und Dolomite der Trias. *Min. Petr. Mitt.* **48,** 27–139.

Seibold E. (1962) Kalk-Konkretionen und karbonatisch gebundenes Magnesium. *Geochim. et Cosmochim. Acta* **26,** 899–909.

Skinner H. C. W. (1960) Formation of modern dolomite sediments in South Australian Lagoons (Abstract). *Bull. Geol. Soc. Amer.* **71,** 1976.

Skinner H. C. W., Skinner B. J. and Rubin M. (1963) Age and accumulation rate of dolomite-bearing carbonate sediments in South Australia. *Science* **139,** 335–336.

Taft W. H. (1961) Authigenic dolomite in modern carbonate sediments along the Southern coast of Florida. *Science,* **134,** 561–562,

Trusheim F. (1961) Über radioaktive Leithorizonte im Buntsandstein Norddeutschlands zwischen Ems und Weser. *Erdöl und Kohle,* **14,** 797–802.

Wells A. J. (1962) Recent dolomite in the Persian Gulf. *Nature, Lond.* **194,** 274–275.

Welte D. H. (1962) Sedimentologische Untersuchungen uranhaltiger Keupersedimente aus der Umgebung von Lichtenfels bei Coburg. *Geol. Bavar.* **49,** 91–123.

7

THE DOLOMITE QUESTION

RHODES W. FAIRBRIDGE
Columbia University, New York, New York

[*Editors' Note:* In the original, material precedes this excerpt.]

VII. SUMMARY ON DOLOMITIZATION

1. The mineral dolomite, $CaMg(CO_3)_2$, is not found in rock-forming accumulations on the sea floor today, or at shallow depth in Recent marine sediments. Isolated euhedral rhombohedra of dolomite are apparently authigenic, since some contain inclusions of glauconite, another authigenic mineral formed by chemical alteration on the sea floor; these authigenic dolomite crystals are rather uniformly the same size (10-20 microns) and restricted to fairly deep sea environments (500-2000 fathoms) within 300 miles of the coast. Traces of dolomite are suspected from X-ray work on Recent intertidal deposits. In any case, a direct precipitation of the double carbonate is not proven.

2. The mineral magnesite, $MgCO_3$, is not found in the strictly marine environment, but only in intermediate facies, in the lagoonal evaporites. The hydrated form, nesquehonite, $MgCO_3 \cdot 3H_2O$, is generally restricted to mineralized areas.

3. Calcite, containing up to 30 percent magnesium carbonate in solid solution, is fixed organically on a large scale in shallow warm marine environments. Experimentally it has been fixed inorganically with up to 8 percent magnesium carbonate and this may take place in nature on coral reef flats, intertidal areas, etc. Aragonite carries little magnesium, but often appreciable strontium carbonate in solid solution.

4. Both high-magnesium calcite and aragonite are unstable or metastable. At normal temperatures (up to 80°F) and pressures (up to say 20 atmospheres), aragonite may, under favorable conditions, invert to calcite almost completely

107

in the course of 10,000 years. High-magnesium calcite breakdown is slower but is accelerated by pressures of the order of 20 atmospheres (corresponding to a depth of about 100 fathoms or 600 feet). With this instability the $MgCO_3$ is available for conversion to the mineral dolomite, which is 12 percent denser than calcite.

5. The formation of the mineral dolomite, if not by direct precipitation in the marine realm, must be by metasomatic replacement, in which an increase of the Mg^{++} ions takes place, either at the expense of some of the Ca^{++} ions or by filling primary cavities and space made available by the leaching of the more soluble aragonite component.

6. The source of the Mg^{++} ions is by diffusion from the sea, connate water, ground water, and so on.

a) In the marine realm Mg^{++} ions are provided by sea water. This contains 1.3 parts per thousand of Mg^{++}, and 0.4 part of Ca^{++}. However, according to present beliefs regarding the solubility of these ions in sea water, magnesium is undersaturated, while calcium in warm surface waters is supersaturated. Mg^{++} and Ca^{++} solubility curves in sea water intersect under conditions such as increased alkalinity (pH 9 to 10), elevated temperature, elevated pressure. The alkalinity may be raised by various means such as removal of CO_2 and breakdown of bicarbonates by bacterial action.

b) Further supply of Mg^{++} ions for late-stage diagenesis, postorogenic and continental dolomitization comes from connate water and ground water. Leaching, dedolomitization, and selective precipitation render many underground waters extremely high in magnesium.

7. The chemical, experimental data on dolomitization show that:

a) Under normal temperatures and pressures:

  i) $CaCO_3$ and $MgCO_3$ may be precipitated from sea water by increasing the alkalinity.

 ii) As $CaCO_3$ is less soluble in sea water than $MgCO_3$, but $Ca(OH)_2$ is more soluble than $Mg(OH)_2$, the solubility curves intersect under increasing alkalinity, permitting both carbonates to precipitate together. With the pH at about 8.5, $CaCO_3$ precipitates, and at 9-10 $MgCO_3$ falls freely as well.

iii) Calcite and magnesite, however, form an unstable association, and tend to dissociate (Baer, 1924), and with excess magnesium ions available should theoretically (not proven) tend to recombine as dolomite (Halla, 1935, 1936), even without added temperature or pressure.

 iv) Pure calcite, treated with $MgSO_4$, $MgCl_2$, or merely sea water (kept at a high pH by removal of CO_2), produces a high-magnesium calcite, without recourse to increased temperature or pressure (Rivière, 1939a,b).

  v) Analyses of marine carbonate-secreting plants and the more primitive invertebrates (together with the clastic sedimentary derivatives) suggest that biogenic fixation of high-magnesium calcite is far more important in nature than the inorganic compound.

b) Under increased temperature and pressure:

  i) With considerable heat or pressure on a wet mixture of $CaCO_3$ and $MgCO_3$, the double carbonate, dolomite, may be synthesized (Sterry Hunt, 1859, 1866).

 ii) With considerable heat and pressure, $CaCO_3$, treated with $MgSO_4$, $MgCl_2$, and other salts, especially in the presence of NaCl, produces solid solutions of $CaCO_3$ and $MgCO_3$ and ultimately dolomite (theory of Haidinger; see von Morlot, 1847; Klement, 1895).

iii) High-magnesium calcite is unstable and with time and only very mild heat or pressure inverts to normal calcite, liberating $MgCO_3$, but as magnesite and calcite are unstable partners (see "(a) (iii)"), in the presence of Mg^{++} saturation, dolomite will form, leaving a residue of low-magnesium calcite if the Mg^{++} is insufficient to convert the entire carbonate available.

c) Under increased CO_2 pressure:
The solubility curves of $CaCO_3$ and $MgCO_3$ also intersect with increased CO_2 pressure, at which point dolomite is theoretically precipitated (Baer, 1924, 1932). This was recently repeated by Chilingar (1956a) and confirmed by X-ray analysis. Increased CO_2 may be obtained in the ocean by

i) Increase in depth, by a simple hydrostatic pressure increase;

ii) Bottom circulation, from cold CO_2-rich latitudes;

iii) Organic activity, at depth, by benthonic organisms, including certain bacteria.

d) Other special environments:

i) Decay environments in general, rich in CO_2, $(NH_4)_2CO_3$, $(NH_4)_2S$, H_2S, bituminous hydrocarbons, seem to favor dolomite formation (theory of Pfaff; also Udluft, 1929).

ii) In the presence of small amount of ammonia salts, such as $(NH_4)_2SO_4$, NH_4Cl, etc. (liberated by bacteria in some marine sediments), it is said that dolomite can be precipitated directly from $MgSO_4$ and $MgCl_2$ (Koehler, 1931; Linck, 1937); it may be, however, that this is via the intermediary of magnesium-rich calcite, by rather rapid contemporary metasomatism.

8. Dolomitization by metasomatic replacement of $CaCO_3$ by Mg^{++} is therefore possible in the following environments (see fig. 14):

a) On the sea floor or sea shore:
No important dolomitization occurs at or near the surface of the sea floor, but considerable fixation of Mg-rich calcite occurs, that may become mobilized later (see (b)). However, two exceptional cases of probable dolomitization do occur in nature that may well be called "contemporaneous":

i) In deep-sea environments, isolated crystals of authigenic dolomite are found at the surface or at shallow depth in Recent sediments that are difficult to explain. They may be derived from Mg-rich calcite fixed by floating (pelagic) algae or foraminifera or by neritic benthonic algae carried seaward after storms; or transported by turbidity flows; or alternatively from chemically (inorganically) fixed Mg-rich calcite, formed *in situ* from isolated bottom water characterized by low CO_2 and high pH. Metasomatism on the surface would be assisted by slow sedimentation. A direct precipitation of dolomite in this environment (increased pressure) does not seem to occur.

ii) In the intertidal belt of warm shallow-sea environments, unconfirmed indications of isolated dolomite crystals suggest that organically fixed Mg-rich calcite (as grains in newly formed clastics) may become unstable in the presence of heat (from the sun) and high concentration of Mg^{++} owing to evaporation or other alkalinization of sea water (introduced by spray or diffusion from tide pools), leading to dolomite metasomatism.

b) In soft sediments (early diagenetic stage):
Metastable aragonite and Mg-rich calcite buried under a blanket of youthful sediments and subjected to increasing pressure, with adequate supplies of Mg^{++} ions from sea-water diffusion, seem to provide a primary

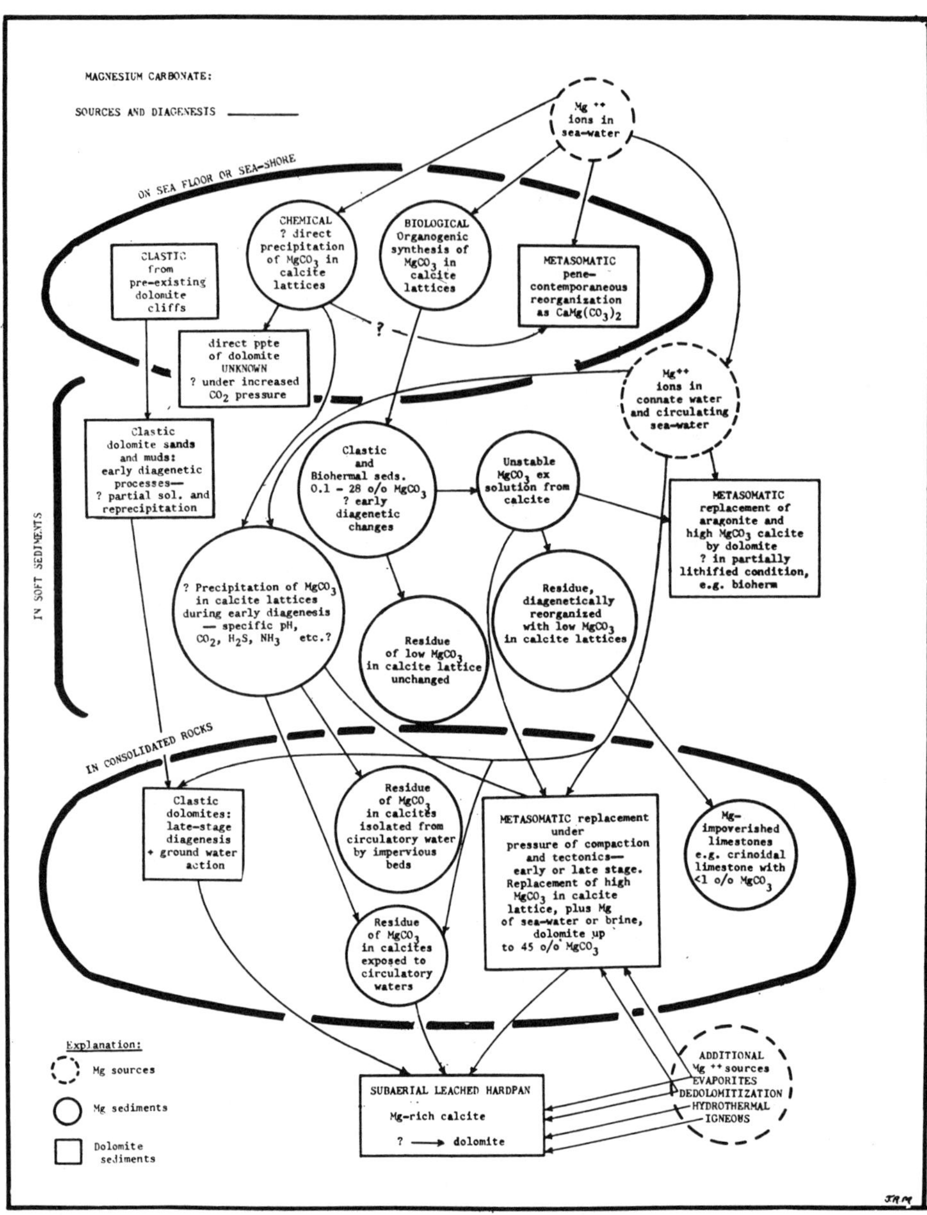

FIG. 14. Four locations in the evolution of marine magnesian sediments to dolomite are: (i) on the sea floor (strictly marine), or on the seashore (and evaporite lagoons, i.e., Intermediate Class); (ii) in soft unconsolidated sediments (still marine, i.e., penecontemporaneous metasomatism or "early diagenetic"); (iii) in consolidated rocks (subsequent metasomatism or "late diagenetic"—may be pre- or postorogenic); (iv) continental (subaerial, leached "hard pan," a soil formation or ground water phenomenon). Not shown are special locations—evaporite lagoons, magnesian lakes, hydrothermal springs, etc.

raw material for dolomite metasomatism. Adequate supply of Mg^{++} depends on grain size or porosity of the initial formation.

Porous or clastic sediments are thus a *sine qua non*. However, the dolomitization process generally shows an initial preference for the finer-grained clastics.

An ideal site, providing a fairly rigid, but extremely porous and cavernous framework, is a coral reef foundation or similar biohermal structure. Within the core of a pervious atoll the following conditions may be provided: unstable raw material, unlimited Mg^{++} solutions in a saturated state owing in part to an increased pressure but also to high alkalinity, and reducing condition owing to removal of oxygen from the incoming water by organic activity in the outer atoll wall; bacterial action will remove further oxygen from the sulfates and bicarbonates of the sea water, further increasing the alkalinity. After the initial nuclei of dolomite develop from the Mg-rich calcites, complete dolomitization of the entire calcite formation may follow, provided the permeability is maintained.

As regards the impervious or extremely fine-grained sediments, it is difficult to see how the necessary supply of Mg^{++} ions could enter into such a formation as one of the pelagic ooze class, say, a globigerina ooze, Kullenberg core samples, even of great length, of globigerina oozes from many parts of the deep ocean have not disclosed the development of dolomite. With advanced diagenesis, elevation above sea level, jointing, and the availability of ground water, a totally different condition is developed (see under (c)).

   c) In consolidated sediments (late diagenetic stages):
   It is difficult to determine precise stages during later diagenesis, but broadly there may be distinguished pre- and postorogenic stages, and if there have been multiple movements then of course more.

   i) "Preorogenic stage" implies that the sediment has become well cemented and probably elevated but not seriously folded or sheared. Since no orogenic disturbance may have occurred, this class is generally referred to merely as "late diagenetic." Joint planes will have developed giving connate water mobility and ground water access to otherwise less pervious formations. In carbonate rocks solution effects provide additional access to waters possibly saturated in Mg^{++}. Dolomitization possibly initiated at stage (b) may now proceed. However, it appears unlikely that limestones originally very low in magnesium, such as those of cold deep-water facies, would become dolomitized; the nuclei provided by the unstable high-magnesium calcites would be absent and the supply of additional Mg^{++} in solution would not be assisted by leaching, unless located in the same stratigraphic column as magnesium-rich evaporites or neritic formations.

   ii) "Postorogenic stage" implies that advanced deformation has led to much more extreme jointing, shearing, and faulting, thus opening up far more of the fine-grained sediments to artesian and ground-water circulation. Pressure effects, particularly shearing pressures, have been demonstrated to materially assist dolomitization. Thus, for any one formation of fairly uniform constituents the degree of dolomitization tends to be less away from any orogenic zone (theory of van Hise). Regional pressure often seems to have instituted a regional dolomitization that transcends the original high-magnesium limestone beds.

In the most advanced stage where actual metamorphism takes place, recrystallization tends to inhibit further ground-water movement and thus dolomitization ceases.

9. Continental dolomitization is distinct from the marine class which has been the main subject of inquiry, but the Mg^{++} for epigene continental dolomites may be derived by leaching or dedolomitization of those dolomites due to diagenesis of marine sediments. Thus there are dolomitic "hard pans," which are developed during soil formation over peneplaned land areas containing magnesium-rich rocks, probably in the same way as laterite is formed under semitropical climatic conditions. Magnesium-rich terrains also lead to lake and swamp dolomites, as well as certain hydrothermal dolomites.

10. Ca/Mg ratio and paleogeographic implications. Dolomitization in stages (see "8," (a), (b), (c)) may explain in part Daly's (1907, 1909) fundamental statistics of the magnesium geoeconomy, i.e., that the oldest periods contain the most dolomite in proportion to limestone. The oldest rocks have had the greatest opportunity to suffer late diagenetic and postorogenic metasomatism. However, it should be added that paleogeographic studies show conclusively that the Proterozoic and Paleozoic epochs were marked by far wider epicontinental seas than exist today, and that calcareous algae, the most active secreters of magnesium-rich calcite, were infinitely more widespread. Both Chave and Chilingar have demonstrated that in marine organisms, for each particular class, the higher the temperature of the water, the higher the magnesium in solid solution in the skeletal calcite. Thus regional dolomitization is favored by the warm epicontinental seas of thalassocratic periods.

11. Unsatisfied problems regarding dolomite are still numerous; they include:

a) The hydrologic question: Was the salinity of sea water the same during the Proterozoic and Paleozoic as now? And was not the CO_2 pressure higher, as suspected by a number of workers (Strakhov, 1953; Chilingar, 1956a,b; Cooper, 1956)? In any case a contemporaneous metasomatism of unstable high-magnesium algal debris would be possible if the marine magnesium salts were near saturation close to the bottom.

b) The contemporary deep-sea problem: Weynschenk and some others suggest that dolomitization is going on today at some depth in deep-sea carbonate oozes. The overall depth satisfies the pressure requirement, but the fine-grained character of the sediment will not favor unlimited access by Mg^{++}-rich solutions. No deep-sea core of sufficient length has yet been able to test this hypothesis.

c) The problem of the primary precipitate: So far there is very little evidence to support the idea of any primary precipitation of dolomite in the marine realm, either now or in the past. All cases ultimately seem to come down to magnesium enrichment and metasomatic replacement. Modern spectrometric work on the carbon isotopes may ultimately solve the problem. Wickman, Blix, and Ubisch (1951) have determined from the C^{12}/C^{13} ratios of 45 carbonate samples that all the dolomites show an enrichment of the heavy carbon isotope (see discussion also in Craig, 1953). Paleotemperature work by means of oxygen isotopes may also assist the matter (Rankama, 1954; Urey, Lowenstam, *et al.*, 1950; Teis, 1951). Determination of Mg^{24}/Mg^{25} ratios may further indicate the extent of metasomatism (Ingerson, 1953).

d) What is the nature of the apparently inorganic enrichment of magnesium that seems to occur in organic debris after death in relatively shallow water (Mägdefrau's evidence), or in relatively deep-sea calcareous oozes (over 1000 fathoms)? Most recent deep-sea expeditions have ignored $MgCO_3$ analyses; it is possibly not too late for many core samples to be examined for $MgCO_3$. Does the Mg/Ca ratio rise as one goes down in each core? Or is the ratio increase merely a function of water depth? Cayeux (1935, p. 414) has suggested that the organically fixed magnesium-rich calcite nuclei form "centers of attraction" for further (inorganic) magnesium growth. Beales (1953) speaks of their "trigger effect."

e) More information is needed on the factors—especially pH, pressure, time —that control the stability/instability of aragonite and high-magnesium calcite.

f) If aragonite carries relatively large quantities of $SrCO_3$ in solid solution, as calcite carries $MgCO_3$, what becomes of the strontium when the aragonite is dolomitized? Only sparse traces of strontium are reported from dolomites (Odum, 1950). No significant trend is noted by Chilingar (1956a) in determining the Sr/Ca ratio of many ancient limestones and dolomites.

12. As promised at the beginning, an attempt should now be made to answer a number of questions on various aspects of dolomitization which were asked of different speakers at this symposium.

i) "Do we know of any calcium magnesium carbonates or dolomite being deposited in the oceans today?" . . . *Yes*, but in small quantities only. The isolated crystals are idiomorphic and may be primary precipitates, but are most likely formed by contemporary metasomatism, i.e., *before* burial.

ii) "What is the validity of the concept of 'penecontemporaneous' dolomitization?" . . . *True*, most metasomatism probably begins soon after burial at a critical depth in the ocean (several hundred feet below sea level), but *not all* dolomitization is of this sort, and in many formations several stages may be recognized (see (iii)).

iii) "At what phase of diagenesis does dolomitization begin?" . . . In some cases *soon after burial;* in others *after consolidation;* or, *after emergence and leaching;* again, in others they are *postorogenic.*

iv) "Are the modern 'dolomitized reefs' actually dolomite in the same sense as the ancient dolomite?" . . . Mineralogically and chemically, *yes.* Stratigraphically, *no.* Modern atoll reefs represent a special environment; the Paleozoic facies were probably neritic. Both cases involve essentially the metasomatism of organically fixed high-magnesium calcites.

[*Editors' Note:* Only those references cited in the preceding excerpt have been reproduced here.]

REFERENCES

BAER, O., 1924, Versuch einer Lösung des Dolomitproblems auf phasentheoretischen Grundlage: Senckenberg. Nat. Gesell., v. 6, pp. 116-118.

——, 1932, Beitrag zum Thema Dolomitentstehung: Centralbl. f. Min. Geol. Pal., Abt. A, pp. 46-62.

BEALES, F. W., 1953, Dolomitic mottling in Pallisser (Devonian) limestone, Banff and Jasper National Parks, Alberta: Bull. Amer. Assoc. Petr. Geol., v. 37, pp. 2281-2293.

CAYEUX, L., 1935, Les roches sédimentaires de France. Roches carbonatées (calcaires et dolomies): Paris (Masson), 463pp.

CHAVE, K. E., 1954, Aspects of the biochemistry of magnesium: Journ. Geol., v. 62, Pt. I. Calcareous marine organisms, pp. 266-283; Pt. 2. Calcareous sediments and rocks, pp. 587-599.

CHILINGAR, G., 1956a, Relationship between Ca/Mg ratio and geologic age: Bull. Amer. Assoc. Petr. Geol., v. 40, pp. 2256-2266.

——, 1956b, Use of Ca/Mg ratio in porosity studies: Bull. Amer. Assoc. Petr. Geol., v. 40, pp. 2489-2493.

COOPER, Byron N., 1956, Primary dolomite?: Min. Ind. Journ. (Va. Poly. Inst.), v. 3, pp. 5-7.

CRAIG, H., 1953, The geochemistry of the stable carbon isotopes: Geochim. et Cosmochim. Acta, v. 3, pp. 53-92.

DALY, R. A., 1907, The limeless ocean of Pre-Cambrian time: Amer. Journ. Sci., v. 23, pp. 93-115.

——, 1909, First calcareous fossils and the evolution of limestone: Bull. Geol. Soc. Amer., v. 20, pp. 153-170.

HALLA, F., 1935, Eine Methode zur Bestimmung der Änderung freien Energie bei Reaktionen des Typus A(s) + B(s) = AB(s) und ihre Anwendung auf das Dolomitproblem: Zeitschr. Phys. Chemie, Abt. A, v. 175, pp. 63-82. (Experimental part with F. Ritter.)

——, 1936, Bemerkungen zur kongruenten Löslichkeit des Dolomits: Nachtrag zu einer vorangegangenen Arbeit: Zietschr. Phys. Chemie, Abt. A, v. 175, pp. 396-399.

HUNT, T. S., 1859, On some reactions of the salts of lime and magnesia and on the formation of gypsums and magnesian rocks: Amer. Journ. Sci., ser. 2, v. 28, pp. 170-187, 365-383.

——, 1866, Further contributions to the history of lime and magnesia salts: Amer. Journ. Sci., ser. 2, v. 42, pp. 49-67.

INGERSON, E., 1953, Nonradiogenic isotopes in geology; a review: Bull. Geol. Soc. Amer., v. 64, pp. 301-374.

KLEMENT, C., 1895, Sur l'origine de la dolomie dans les formations sédimentaires: Bull. Soc. Belg. de Géol., mém. v. 9, pp. 3-23 (see also: Tschermak's Min. und Petr. Mitth., n.s., v. 14, 1894, pp. 531 et seq.).

KOEHLER, E., 1931, Ueber die Entstehung von Schaumspat und Dolomit: Chemie der Erde, v. 6, pp. 257-268.

LINCK, G., 1937, Bildung des Dolomits und Dolomitisierung: Chemie der Erde, v. 11, pp. 278-286.

MÄGDEFRAU, K., 1933, Ueber die Ca-und Mg-Ablagerung bei den Corallinaceen des Golfs von Neapel: Flora, n.s., v. 28, pp. 50-57.

MORLOT, A. von, 1847, Ueber die Dolomit und seine künstliche Darstellung aus Kalkstein: Haidinger, Nat. Abh., v. 1, p. 305. (Also: Sur l'origine de la dolomie: C. R. Acad. Sci., Paris, v. 26 (1848), pp. 311-315.)

ODUM, H. T., 1950, Biogeochemistry of strontium: Yale Univ., Ph.D. Thesis. (See also: 1950. Strontium biogeochemistry, ecosystems, and paleoecological tools: Rept. Comm. Mar. Ecol. and Paleoecol., no. 10 (1949-50), N.R.C., Washington, pp. 55-58. Also: 1951. Notes on the strontium content of sea water: Science, v. 114, p. 214.)

PFAFF, F. W., 1895, Beiträge zur Erklärung über die Entstehung des Magnesits und Dolomits: Neues Jahrb. f. Min. Geol. Pal., Beil.-Bd. 9, pp. 485-507.

____, 1907, Ueber Dolomit und seine Entstehung: Neues Jahrb. f. Min. Geol. Pal., Beil.-Bd., 23, no. 3, pp. 529-580.

RANKAMA, K., 1954, Isotope geology: London (Pergamon Press), 536pp.

RIVIÈRE, A., 1939a, Sur la dolomitisation des sédiments calcaires: C. R. Acad. Sci., Paris, v. 209, pp. 597-599.

____, 1939b, Observations nouvelles sur le mécanisme de dolomitisation des sédiments calcaires: C. R. Acad. Sci., Paris, v. 209, pp. 691-692.

STRAKHOV, N. M., 1953, Diagenesis of sediments and its significance for sedimentary ore formation: Akad. Nauk. SSSR Izv., Ser. Geol., no. 5, pp. 12-49.

TEIS, R. V., 1951, Isotopic method of determining the temperatures of formation of carbonate minerals: Doklady (C. R. Acad. Sci. U.S.S.R.), Moscow, v. 79, pp. 291-294.

UDLUFT, H., 1929, Die Genesis der flächenhaft verbreiteten Dolomite des mitteldevonischen Massenkalkes: Jahrb. Preus. Geol. Landesanst. (Berlin), v. 50, pp. 396-436.

UREY, H. C., LOWENSTAM, H. A., EPSTEIN, S., and McKINNEY, C. R., 1950, Measurement of paleotemperatures and temperature of the Upper Cretaceous of England, Denmark, and southeastern United States: Bull. Geol. Soc. Amer., v. 62, pp. 399-416.

VAN HISE, C. R., 1904, A treatise on metamorphism: U.S. Geol. Surv., Mon. 47.

WEYNSCHENK, R., 1951a, The problem of dolomite formation considered in the light of research on dolomites in the Sonnwend mountains (Tirol): Journ. Sed. Petr., v. 21, pp. 28-31.

WICKMAN, F. E., BLIX, R., and UBISCH, H., von, 1951, On the variations in the relative abundance of the carbon isotopes in carbonate minerals: Journ. Geol., v. 59, pp. 142-150.

8

Problems of the geochemistry of sedimentary carbonate rocks

Earl Ingerson

[*Editors' Note:* In the original, material precedes this excerpt.]

Formation of Dolomites—Dolomitization

The problem of the origin of dolomite is one of the most fascinating as well as one of the most important in all of geochemistry or in sedimentary petrology. As indicated above, thermodynamic considerations show that dolomite is stable in normal sea water at 25°C with respect to $CaCO_3$ or mixtures of $CaCO_3 + MgCO_3$.

Yet it has never been demonstrated unequivocally that dolomite has ever been precipitated in nature directly from solution, nor has dolomite ever been formed in the laboratory under conditions similar to normal marine, lagoonal, or lacustrine environments. An extensive and varied literature has grown up around this intriguing problem. It has been summarized from time to time, some of the most comprehensive reviews being those by Cloud and Barnes (1948) and by Fairbridge (1957).

Many theories have been proposed to explain the formation of particular dolomites, but a discussion of these theories and their variants is far beyond the scope of this report. If we eliminate hydrothermal dolomite and the small amounts that may be formed by leaching of magnesian limestone or precipitated by bacteria (see p. 28, *Minor and Trace Elements*), the problems of the sedimentary dolomites centre chiefly around (1) primary vs. secondary dolomite, (2) possible conditions of precipitation of primary dolomite, (3) source of solutions to form secondary (replacement) dolomite, and (4) questions of timing and conditions of replacement.

Primary dolomite

Although the existence of primary dolomite has not been proved beyond question, there are certain situations where direct precipitation is the most logical explanation. Fairbridge (1957) mentions isolated dolomite rhombs in deep sea sediments that have been interpreted as primary. Strakhov (1953) reports similar dolomite crystals, but in greater abundance, in the sediments of Lake Balkash. Graf *et al.* (1959) report a thin bed of nearly pure unconsolidated dolomite about a foot under the surface of the desert west of Great Salt Lake, Utah. Shoreward from the dolomite the same horizon contains aragonite and a "magnesite-like" material (hydromagnesite?).

Sander (1936, p. 123 of 1951 translation) lists three kinds of primary dolomite but none of them corresponds to precipitation of dolomite on the sea floor like $CaCO_3$ is precipitating on the Bahama Banks.

The most thoroughly documented case of a dolomite-bearing sediment where the dolomite appears to be primary is that described by Alderman and Skinner (1957) and by Alderman (1959), in southeast South Australia. In the shallow end of the Coorong, in other shallow land-locked lagoons and in many interdunal lakes

116

of the area the sediments contain dolomite, locally up to almost 50 per cent. The lakes are remnants of former lagoons, so the dissolved salts are present in more or less the same proportions as in sea water, but salinities of the bodies in which dolomite-bearing sediments have been observed vary from about half to almost four times that of normal sea water. Precipitation takes place in the spring and summer and is most abundant where plants are most active, with consequent elevation of the pH to 8·5 to 9·2.

The association of shallow water, plant growth, elevation of pH and dolomite-bearing sediments is undoubted. There are two observations, however, that suggest another possibility than direct precipitation:

1. The dolomite always has calcite associated with it, but the reverse is not true. The irregular variation in the composition of the sediments from almost pure calcium carbonate to mixtures of subequal parts dolomite and $CaCO_3$ might indicate the derivation of the dolomite by varying amounts of replacement. If all of the carbonate is being precipitated from solution, one would expect a little more consistency in the calcite–dolomite ratio. Perhaps a study of the relation between salinity and calcite–dolomite ratio would be instructive. A really positive demonstration could be made, if the periods of precipitation of calcite and dolomite are not always *identical*, by placing vessels on the bottom to catch the crystals as they form and sink.

2. ALDERMAN and VON DU BORCH (1960) have described an occurrence in one of the lagoons where the surface layer consists of approximately equal parts of aragonite and hydromagnesite. The hydromagnesite decreases downward and disappears at a depth of 9 in., at which level the sediment is essentially aragonite. Farther down calcite appears, and aragonite decreases. Aragonite disappears at a depth of about 2 ft, below which level the sediment is essentially calcite and dolomite.

This situation reminds one of the Great Salt Lake occurrence described by GRAF *et al.*, but the details of the latter have not yet appeared in print. In either case the dolomite *could* have been formed by reaction of the calcite and/or aragonite with the residual high-Mg solution. Since aragonite is more soluble and reactive than calcite, it is possible that the calcite–aragonite ratios in the original sediment and rate of inversion of aragonite to calcite have something to do with the observed calcite–dolomite ratios.

The fact that the dolomite occurs in discrete crystals does not preclude the possibility that it was formed by reaction or replacement. Dolomite has a greater tendency to form idiomorphic grains than does calcite and would be expected to take its crystal form against soft unconsolidated $CaCO_3$ sediments. This would also apply, of course, to the dolomite rhombs observed in deep-sea sediments and those of Lake Balkash, the Dead Sea, etc. FAIRBRIDGE (1957) mentions the possibility that the dolomite rhombs of deep sea sediments might be formed by replacement of calcite ooze.

KHODAK (1956) describes an extensive formation of lower Cambrian dolomite of the Aldan region of the Yakut which he believes to be primary. His principal arguments are that *no* relicts of calcite could be found although the formation is several hundred meters thick, that fine sedimentary structures indicating shallow water deposition are preserved and that there is a very sharp contact with the underlying limestone.

Regardless of the verity of these and other examples that might be primary, the current consensus is that quantitatively primary dolomite is unimportant. KRYNINE (1958), for example, states that, although primary dolomite is a possibility, the great bulk of it is formed "penecontemporaneously in plastic carbonate oozes."

Secondary dolomites

Most recent writers agree with KRYNINE on the origin of the bulk of sedimentary dolomites. See, for example, BROWN (1959), FAIRBRIDGE (1950, 1957), ILLING (1959), MENNIG and VATAN (1959), NEWELL et al. (1953), NOLAN (1935), STRAKHOV (1958), and TEODOROVICH (1959). There is by no means complete agreement on details, but there is a surprising degree of unanimity about many of the important factors involved in the formation of dolomite.

Source of solutions

Since important amounts of Mg have to be added to calcite, even magnesian varieties, to form pure dolomite rocks and since ground water is low in Mg, the only logical source of Mg-bearing solutions is ocean water itself either with normal salinity or a modified residual solution (connate water). FAIRBRIDGE (1950) thinks that, given sufficient time, normal sea water might dolomitize unstable aragonite or high-magnesium calcite. DALY (1909), STEIDTMANN (1911) and MENNIG and VATAN (1959) also favour reaction with ordinary sea water, but most other students of the problem favour the action of connate water in which the Mg/Ca ratio has been increased by precipitation of $CaCO_3$ (ADAMS and RHODES, 1960; BLACKWELDER, 1913; BROWN, 1959; ILLING, 1959; KRYNINE, 1958; NOLAN, 1935; TEODOROVICH, 1959; etc.). NEWELL et al. (1953), while not denying the possibility that some dolomitization can be accomplished by normal sea water, believe that, at least for the Texas–New Mexico Permian reef complex, "shelf-brines" were the most effective agent. This is based on the observation that the lagoonal sediments are mostly dolomitized. The solutions are pictured as undergoing evaporation, which increases the concentration of Mg, and precipitation of $CaCO_3$, which increases the Mg/Ca ratio.

Salinity

In general, it is not possible to determine the salinity of water that produced dolomitization. The consensus appears to be that the solutions are at least as concentrated as normal sea water and in some cases may have been much higher in salts. STRAKHOV (1958), for example, says that dolomite will form up to about 14 per cent salinity, beyond which $CaCO_3$ and $MgCO_3$ form separate phases. He also says that $MgSO_4$, $MgCl_2$ and $CaSO_4$ appear to aid the formation of dolomite. ZELLER et al. (1959), on the basis of experimental work, conclude that the presence of sulphate may be necessary for precipitation of dolomite. TEODOROVICH (1959), on the other hand, says that in a saline lake dolomite forms until $CaSO_4$ begins to precipitate, then ceases. BLACKWELDER (1913) considers the concentration to have been somewhat above normal. ADAMS and RHODES (1960) postulate heavy brines saturated with respect to some of the salts of sea water. TWENHOFEL (1932) believes that "the effects of concentration and gas content of the sea water are uncertain."

Temperature

Agreement is general that elevated temperatures favour the formation of dolomite (BLACKWELDER, 1913; VAN TUYL, 1916; TWENHOFEL, 1932; CLOUD and BARNES, 1948, 1957; KRAMER, 1959; ALDERMAN and SKINNER, 1957; BARON, 1960; TEODOROVICH, 1959). ADAMS and RHODES (1960) say that the temperature at times probably exceeded 35°C.

Carbon dioxide

In general, excess carbon dioxide is thought to favour formation of dolomite (BARON, 1960; TEODOROVICH, 1959; FAIRBRIDGE, 1950; BISSELL and CHILINGAR, 1958). Precipitation would be favoured temporarily, however, by local abstraction of CO_2 with consequent increase of pH and CO_3^{2-} ion concentration. CHILINGAR (1956b) reports synthesis of dolomite at a CO_2 pressure of 4 atm, but SKEATS (1918) found that at this pressure the $MgCO_3$ of dolomite can be removed almost quantitatively. At 1 atm $CaCO_3$ is more soluble, so somewhere between these two pressures the solubilities of $MgCO_3$ and $CaCO_3$ should be equal; he suggests that perhaps this would be the most favourable pressure for formation of dolomite.

Depth

Numerous field studies have led to an impression that much dolomite is formed in shallow water with little or no burial (ADAMS and RHODES, 1960; MENNIG and VATAN, 1959; TWENHOFEL, 1932; NOLAN, 1935; TEODOROVICH, 1942, 1959; CLOUD and BARNES, 1948, 1957; VAN TUYL, 1918; GRAF et al., 1959; KHODAK, 1953). BLACKWELDER (1913) mentions depths of 100 to 200 m. SKEATS (1918), on the basis of $P = 1$ atm +, suggests depths of 15 to 30 ft. FAIRBRIDGE (1950), after reviewing the literature, thinks that it may take place in the 0–150 ft zone. The results of OWEN and BRINKLEY (1941) showing considerable increases in the solubility of carbonates with increase of total pressure are consistent with this picture of dolomitization under near surface conditions. NEWELL et al. (1953) picture dolomitization of a Permian reef as going on while the reef was still active. NEWELL and RIGBY (1957), on the other hand, point out that in the bore hole in the Bahamas there is no dolomite nearer the surface than 530 ft. They do not demonstrate, however, that the dolomitization took place after the sediments had subsided to their present positions. ILLING (1959) suggests that some Paleozoic reefs in western Canada were dolomitized after they were buried to depths of 2000 to 3000 ft, at which time compaction forced connate water out through the porous reefs and dolomitized them.

If dolomitization is dominantly a shallow-water phenomenon and if it takes place rather slowly (see below), then non-dolomitization of certain atolls, reefs and other carbonate sediments may be simply explained by rate of subsidence (SKEATS, 1918).

pH

The consensus is that high pH favours the formation of either primary or secondary dolomite. ADAMS and RHODES (1960) suggest pH values in excess at 9 in some lagoons during dolomite formation.

Time

There is much evidence that dolomitization may begin, at least, soon after precipitation of calcareous sediments, or very early in diagenesis (DALY, 1909; STEIDTMANN, 1911; BLACKWELDER, 1913; TWENHOFEL, 1932; FAIRBRIDGE, 1957; NOLAN, 1935; KRYNINE, 1958; ADAMS and RHODES, 1960, NEWELL *et al.*, 1953). CLOUD and BARNES (1948, 1957) cite lateral persistence of dolomite as evidence of this "penecontemporaneous" formation, especially if the dolomite is fine-grained and interspersed with beds of limestone. Primary dolomite could explain such occurrences, but CLOUD and BARNES believe that the published information on the solubility of calcium and magnesium salts in sea water favours penecontemporaneous origin of dolomite by replacement of $CaCO_3$ rather than primary precipitation and that primary dolomite, if it occurs at all, is of local and exceptional occurrence. HATCH *et al.* (1938) believe most dolomite to be of penecontemporaneous origin and list the following criteria for this interpretation: (1) Limestone beds above and below the dolomite, (2) dolomitization of contained fossils, and (3) dolomite pebbles in the overlying limestone. In contrast, they say that later (subsequent) dolomitization is commonly irregular, incomplete and related to structures such as faults, joints, bedding planes, etc.

On the other hand, dolomitization may proceed very slowly (KRAMER, 1959; FAIRBRIDGE, 1957; BARON, 1960) or may fail to take place at all—witness Paleozoic limestones that are completely undolomitized. Much will depend on local situations —circulation, pH, temperature, Mg/Ca ratio, etc.

Dolomitization undoubtedly proceeds much more rapidly in hypersaline environments whether or not primary dolomite is precipitated under those conditions.

Dolomites in geologic time

No one doubts that dolomites are much more prevalent in Precambrian and Paleozoic rocks than in later ones; it is not nearly as easy to agree on the explanation of this fact. VINOGRADOV's curve of the Ca/Mg ratio for carbonate rocks of the Russian Platform since Precambrian time shows that something drastic geochemically must have happened after the close of the Paleozoic. What was it?

The present Mg/Ca ratio in sea water is about 3 by weight. Do the variations of this ratio in rocks of different ages mean that it has varied in sea water by factors of manyfold during geologic time, or have other factors made these two ratios independent of each other, within certain limits? Several possible explanations of the decrease in the Ca/Mg ratio of rocks with increasing geologic age, which do not mention the ratio in the oceans, are to be found in the literature.

Foremost among these is the simple fact that the older rocks have had more time to pick up Mg. This is almost certainly a part of the explanation; much dolomite that is obviously hydrothermal occurs in older rocks, and there must be considerable areas of epigenetic dolomite that are not obviously related to hydro-thermal activity. It is even possible that reaction with connate waters continues over long spans of time or is renewed by heating when rocks are deeply buried. Once a rock is dolomitized it is unlikely that it will undergo large-scale dedolomiti-zation. The mechanism described for this process by FAUST (1949) can only operate locally, and most of the derived Mg is probably deposited farther out in the lime-stone as hydrothermal dolomite or, exceptionally, magnesite.

On theoretical grounds, therefore, we should expect dolomitization to increase and the Ca/Mg ratio to decrease concomitantly with time. If time were the only factor, however, a plot of Ca/Mg against geologic time would approximate a straight line.

Variation in the amount of carbon dioxide in the atmosphere has been men-tioned as a possible cause. Since we have no way of estimating variations in CO_2 in the atmosphere in the geologic past, and since the effect of changing CO_2 pressures on the formation of dolomite in nature is not known, this factor must remain just hypothetical for the time being.

Times at which particular organisms appeared have been mentioned as a possible factor. No group of $CaCO_3$-secreting organisms in which $MgCO_3$ ap-proaches 50 mol per cent is known (with the possible exception of some bacteria), but there might have been some in the past.

Widespread aridity with consequent high evaporation and increased salinity has been proposed to explain some of the variations. If this were an important factor, then the value of Ca/Mg should be at a minimum during the periods characterized by evaporites and red beds (Devonian, Permian, Triassic). Actually, on VINO-GRADOV'S curve there is a maximum for the Devonian, and the Permian and Triassic are on an ascending segment of the curve.

DALY (1909) and STEIDTMANN (1911) assert that progressively less Mg (with respect to Ca) has been supplied to the oceans throughout a great deal of geologic time, but they do not attempt an explanation.

A recent paper by TIKHOMIROV (1958) contains a suggestion that could explain part of the more or less systematic increase in Ca/Mg from the Precambrian through the Paleozoic. He assumes that at the beginning of geologic time the surface of the earth was covered entirely with ultrabasic rocks and that the first differentiation was by weathering rather than crystallization differentiation. Iron and magnesium were leached out; the iron formed the large-scale Precambrian iron deposits by precipitation of $FeCO_3$ and/or Fe_2O_3. The Mg was available for the wide-spread dolomitization observed in Precambrian and Paleozoic rocks. The residual material, rich in silica, alumina, alkalies, etc., after burial formed granite of the continental areas by melting and/or metasomatism.

This lighter material was elevated farther and covered progressively more of the exposed ultrabasics. Following the Paleozoic, weathering was predominantly of these elevated continental masses that are low in iron and magnesium, and there has not been a sufficient supply of Mg for dolomitization on the same scale as formerly.

No matter what their other merits, it does not appear that any of these factors can explain the tremendous jump in the value of Ca/Mg from 10 to 90 in one geologic period (Jurassic to Cretaceous). This is 10 times the maximum variation before the Jurassic. Neither do they explain the sharp decrease in the value of Ca/Mg since the Cretaceous, which CHILINGAR's curve indicates has continued through the Tertiary to the Recent.

If dolomitization takes place largely by inorganic reaction, perhaps it is catalysed by some trace element, or assemblage of trace elements, that were furnished by the meteorites and meteors that were more abundant in early geologic times and/or by weathering of ultrabasic rocks. If these elements were largely used up from the sea water by the end of the Jurassic, this might explain a sudden decrease in the amount of dolomitization at that time.

The large increase in Ca/Mg in the Cretaceous, however, was probably brought about more largely by increased formation of $CaCO_3$ than by a decrease in the amount of dolomitization that was taking place, Since the formation of limestone in the Cretaceous was almost entirely by organic means (RODGERS, 1957), the availability of an inorganic catalyst to aid the precipitation of $CaCO_3$ at the beginning of the Cretaceous will not explain the enormous increase in limestone. The development of an enzyme that would enable organisms to fix $CaCO_3$ more readily could explain this surge. We should look for such an enzyme in modern organisms. If it is found, its active metal should be determined, and we should find out whether limestones of Cretaceous and more recent ages show this metal more abundantly and/or more consistently than older ones do.

Other problems of dolomitization

TWENHOFEL (1932) makes the following suggestions for further investigations:

1. The extent of dolomitization in present seas, especially where reducing conditions prevail.

2. Experiments on the laboratory production of dolomite in the presence of Fe^{2+} under conditions that simulate, as closely as possible, those in the ocean.

3. Geologic studies to determine environmental conditions under which the sediments (that now exist as dolomites) originated; conditions such as depth of the water, position of the bottom with respect to continuous deposition, concentration of solutes and temperature.

FAIRBRIDGE (1957) mentions the following "unsatisfied" problems regarding dolomite:

1. Hydrologic questions: Were salinity and CO_2 pressure the same in the Proterozoic and Paleozoic as they are now?

2. Is contemporaneous dolomitization taking place at depth? If so, is it due to increased CO_2 pressure and/or greater concentrations of Mg^{2+} (above that in sea water)?

3. What is the nature of Mg enrichment in organic debris after death of the organisms? Does Mg/Ca increase with depth in cores? With depth of water? (Do remnant enzymes or enzymatically derived trace elements that brought about fixation of Mg in the shells catalyse further precipitation of Mg? [E. I.])

4. If aragonite carries much Sr, what becomes of it when the aragonite is dolomitized?

5. More information is needed on the physicochemical factors that control the stability/instability of aragonite and high-magnesium calcite.

CLOUD *et al.* (in press) suggests that, in undeformed basins, the proportion of dolomite to total basin sediments be estimated. He calculates that about 1·2 per cent of the total can be explained as dolomite by squeeze-out and reaction of interstitial waters. Much larger amounts would require another explanation.

In addition to projects mentioned above, I should like to suggest further:

1. That the experiments of bacterial precipitation of dolomite described by LALOU (1957) and by NEHER and ROHRER (1958) be pursued with an eye to possible application to dolomite rocks.

2. That laboratory studies of dolomite precipitation and formation by replacement be continued, but that conditions attainable in nature be employed rather than changing parameters drastically until dolomite is produced in relatively short runs. Perhaps these processes can be followed by making long runs and using ingredients with C^{14} and/or spikes of stable isotopes so that rates of reaction can be determined even if they are very slow.

3. An experimental attempt should be made to evaluate Skeat's "optimum CO_2 pressure" for the precipitation of dolomite.

[*Editors' Note:* Material has been omitted at this point. Only those references cited in the preceding excerpt have been reproduced on the following pages.]

REFERENCES

ADAMS J. E. and RHODES M. L. (1960) Dolomitization by seepage refluxion. *A.A.P.G. Bull.* 44, 1912-1920.

ALDERMAN A. R. (1959) Aspects of carbonate sedimentation. *J. Geol. Soc. Australia* 6, 1-10.

ALDERMAN A. R. and SKINNER H. C. W. (1957) Dolomite sedimentation in the southeast of South Australia. *Amer. J. Sci.* 255, 561-567.

ALDERMAN A. R. and VON DER BORCH C. C. (1960) Occurrence of hydromagnesite in sediments in South Australia. *Nature, Lond.* 188, 931.

BARON G. (1960) Sur la synthèse de la dolomite. Application au phénomène de dolomitization. *Rev. Inst. France Petrol.* No. 1, 3-68.

BISSELL H. J. and CHILINGAR G. V. (1958) Notes on diagenetic dolomitization. *J. Sed. Pet.* 28, 490-497.

BLACKWELDER E. (1913) Origin of the Bighorn dolomite of Wyoming. *Geol. Soc. Amer. Bull.* 24, 607-624.

BROWN C. W. (1959) Diagenesis of late Cambrian oolitic limestone. *J. Sed. Pet.* 29, 260-266.

CHILINGAR G. V. (1956b) Relationship between Ca/Mg ratios and geologic age. *A.A.P.G. Bull.* 40, 2256-2266.

CLOUD P. E., JR. and BARNES V. E. (1948) *The Ellenburger Group of Central Texas.* University of Texas Publication 4621, esp. pp. 90-107.

CLOUD P. E., JR. and BARNES V. E. (1957) Early Ordovician Sea in Central Texas, in *Treatise on Marine Ecology and Paleoecology,* Vol. 2, Paleoecology, pp. 163-214.

CLOUD P. E., JR. *et al.* (in press), Carbonate deposition west of Andros Island, Bahamas. *U.S. Geol. Surv. Prof. Pap.* 350.

DALY R. A. (1909) First calcareous fossils and the evolution of the limestones. *Geol. Soc. Amer. Bull.* 20, 153-170.

FAIRBRIDGE R. W. (1950) Recent and Pleistocene coral reefs of Australia. *J. Geol.* 58, 330-401, esp. 385-389 and 393, the dolomite problem.

FAIRBRIDGE R. W. (1957) The dolomite question, in *Regional Aspects of Carbonate Deposition.* S. E. P. M. Sp. Publ. No. 5, 125-178.

FAUST G. T. (1949) Dedolomitization and its relation to a possible derivation of a magnesium-rich hydrothermal solution. *Amer. Mineral.* 34, 789-823.

GRAF D. L., EARDLEY A. J. and SHIMP N. F. (1959) Dolomite formation in Lake Bonneville, Utah (abstr.). *Geol. Soc. Amer. Bull.* 70, 1610.

HATCH F. H., RASTALL R. H. and BLACK M. (1938) *The Petrology of Sedimentary Rocks,* Allen & Unwin, London, esp. 175, 183-197.

ILLING L. V. (1959) Deposition and diagenesis of some Upper Paleozoic carbonate sediments in Western Canada. *5th World Petrol. Cong.* Section 1, Paper 2, 1-28.

KHODAK YU. A. (1956) The genesis of Lower Cambrian dolomite of the Aldan region of the Yakut, A. S. S. R. *Dokl. Akad. Nauk SSSR.* 106, 328-330.

KRAMER J. R. (1959) Correction of some earlier data on calcite and dolomite in sea water. *J. Sed. Pet.* 29, 465-467. (Study of calcite and dolomite in sea water abstr. *Geol. Soc. Amer. Bull.* 69, 1600, 1958.)

KRYNINE P. D. (1958) Dolomitization and the diagenesis of carbonates (abstr.). *API Geology Domain Symposium on Diagenesis, Brea, California.*

LALOU C. (1957) Studies of bacterial precipitation of carbonates in sea water. *J. Sed. Pet.* 27, 190-195.

MENNIG J. J. and VATAN A. (1959) Répartition des dolomies dans le Dinantien des Ardennes. *Rev. Inst. Franc. Petrol.* 519-534.

NEHER J. and ROHRER E. (1958) Dolomitbildung unter Mitwirkung von Bakterien. *Eclog. Geolog. Helvet.* 51, 213-215.

NEWELL N. D. et al. (1953) *The Permian Reef Complex of the Guadalupe Mountains.* Freeman, San Francisco.

NEWELL N. D. and RIGBY J. K. (1957) Geological studies on the Great Bahama Banks, in *Regional Aspects of Carbonate Deposition.* S.E.P.M. Spec. Publ. No. 5, 15-79.

NOLAN T. B. (1935) The Gold Hill Mining District, Utah. *U.S. Geol. Surv. Prof. Pap. No.* 177.

OWEN B. B. and BRINKLEY S. R., JR. (1941) Calculation of the effect of pressure upon ionic equilibria in pure water and in salt solutions. *Chem. Rev.* 21, 461-474.

RODGERS J. (1957) The distribution of marine carbonate sediments: a review, in *Regional Aspects of Carbonate Deposition.* S.E.P.M. Sp. Publ. 5, 2-14.

SANDER B. (1936) Beiträge zur Kenntniss der Anlagerungsgefüge (Rhythmische Kalke und Dolomite aus der Trias). *Min. Petr. Mitt.* 48, 27-139. (Translated by KNOPF E. B. and published by the A.A.P.G., Tulsa, 1951.)

SKEATS E. W. (1918) The formation of dolomite and its bearing on the coral-reef problem. *Amer. J. Sci.* 195, 185-200.

STEIDTMANN E. (1911) Evolution of limestone and dolomite. *J. Geol.* 19, 323-345, 393-428.

STRAKHOV N. M. (1953) Diagenesis of sediments and its significance for sedimentary ore formation. *Izv. Akad. Nauk S.S.S.R.,* (Ser. Geol.) No. 5, 12-49.

STRAKHOV N. M. (1958) Facts and hypotheses concerning genesis of dolomite rock. *Izv. Akad. Nauk S.S.S.R.,* (Ser. Geol.) No. 6, 1-18 (of transl.).

TEODOROVICH G. I. (1942) Dolomitization of reefogenic formations. *Dokl. Akad. Nauk S.S.S.R.* No. 6, 160-164; Process of dolomitization in the eastern massif of the Ishimbaevo oil-bearing region. *Dokl. Akad. Nauk S.S.S.R.* No. 7, 199-201.

TEODOROVICH G. T. (1959) A contribution on the origin of limestone and dolomite. *I.G.R.* 1, No. 3, 50-73; translated from *Trans. Pet. Inst. Akad. Sci., S.S.S.R.* 5, (1955).

TIKHOMIROV V. V. (1958) The development of the earth's crust and the nature of granite. *Izv. Akad. Nauk S.S.S.R.* (Ser. Geol.) No. 8 (Eng. transl. May, 1960, 1-11).

TWENHOFEL W. H. (1932) *Treatise on Sedimentation* esp. pp. 330-351. Williams and Wilkins, Baltimore.

VAN TUYL F. M. (1916) The origin of dolomite. *Iowa Geol. Surv., Ann. Rep.* 1914, 25, 251-422.

VAN TUYL F. M. (1918) Depth of dolomitization. *Science* 48, 350-352.

VINOGRADOV A. P. (1953) *The Elementary Chemical Composition of Marine Organisms* (translation) Yale University Press, New Haven.

ZELLER E. J., SAUNDERS D. F. and SIEGEL F. R. (1959) Laboratory precipitation of dolomite carbonate (abstr.). *Geol. Soc. Amer. Bull.* 70, 1704.

Part II

HOLOCENE DOLOMITE SEDIMENTOLOGY (MARGINAL MARINE)

Editors' Comments
on Papers 9, 10, and 11

9   ALDERMAN and SKINNER
Dolomite Sedimentation in the South-East of South Australia

10   CURTIS et al.
*Association of Dolomite and Anhydrite in the Recent Sediments
of the Persian Gulf*

11   SHINN, GINSBURG, and LLOYD
Recent Supratidal Dolomite from Andros Island, Bahamas

The following three papers discuss dolomite forming in Recent peritidal environments, essentially at or immediately beneath the sediment-air interface. The significance of these early contributions lies, in part, in the geological evidence they and later workers provided as to the timing, mechanisms, and sources of dolomitizing solutions (Kinsman, 1964, 1966; Butler, 1965, 1969; Illing, Wells, and Taylor, 1965; Hsü, 1967; Liebermann, 1967; Goodell and Garman, 1969; Hsü and Siegenthaler, Paper 13); Mg^{2+} in these examples is derived from concentrated seawater.

In Paper 9, Alderman and Skinner described the occurrence of very finely crystalline (1-20 μm) dolomite in carbonate-evaporite muds from saline coastal lagoons and lakes of the Coorong region of South Australia (first described by Mawson, 1929). They reported the dolomite to be ordered (also see Degens and Epstein, Paper 6), although Skinner (1963) and Skinner et al. (1963) suggested it is nonideal protodolomite. They related its formation as a primary precipitate to annual cycles of Mg/Ca enrichment and relatively higher salinity and pH increase due to evaporation and abstraction of CO_2 by plants, respectively. However, the role of these factors in dolomite formation is still unclear, as is their interpretation of its primary rather than early replacement origin (Bathurst, 1975). Work has continued in this interesting and relatively widespread area of modern dolomite formation largely through the efforts of C. C. von der Borch (von der Borch, 1965; von der Borch and Lock, 1979; von der Borch et al., Paper 24; von der Borch et al., 1964).

Curtis et al. (Paper 10) reported protodolomites forming in

sabkha sediments in the Abu Dhabi region of the Persian Gulf, approximately 200 miles east of where Wells (1962) first described it in similar facies. Whereas Wells was uncertain as to whether the dolomite was a primary precipitate or the product of penecontemporaneous replacement, Curtis et al. concluded it was an early diagenetic replacement of aragonite mud. A more detailed description and interpretation of similar dolomite occurrences in this region were later presented by Illing, Wells, and Taylor (1965; also in Benchmark Volume *Sedimentary Rocks,* ed. A. V. Carozzi, 1975), which in turn was followed by numerous other papers, such as Hsü and Schneider (1973), Patterson and Kinsman (1977), and most recently McKenzie et al. (1980). Based on their study of nonsabkha supratidal flats in the Bahamas, Shinn, Ginsburg, and Lloyd (Paper 11) and others (Shinn, 1964; Shinn and Ginsburg, 1964) reported very finely crystalline (1-2 μm) replacement protodolomites comprising as much as 80 percent of lithified crusts forming in areas of extreme Mg^{2+} concentrations (Mg/Ca ratio as high as 40 to 1) of interstitial brines (disputed by Atwood and Bubb, 1970). In all of these examples, the possibility of a seepage reflux origin (see Part III) is not yet dismissed (Bathurst, 1975).

The further significance of these dolomite occurrences, as well as those on the island of Bonaire (Deffeyes, Lucia, and Weyl, 1965), is also manifest in their analogy to the depositional environments of many ancient dolomites (Part IV, this volume). Of interest is the observation that these dolomites, with the possible exception of those in the Bahamas, are virtually identical to the primary dolomites recognized in earlier studies (see the references in Part I of this book) in terms of their fine crystallinity, association with textureless and unfossiliferous carbonates, and the fact that relatively thin dolomite beds of considerable areal extent commonly are interbedded with unaltered limestones. The current consensus of opinion is that many such dolomites, and their Recent counterparts, are of early replacement origin. However, the reader is referred to the papers in Parts III through VI for further evaluation of such processes and products.

REFERENCES

Atwood, D. K., and J. N. Bubb, 1970, Distribution of Dolomite in a Tidal Flat Environment, Sugarloaf Key, Florida, *Jour. Geology* **78**:499-505.
Bathurst, R. G. C., 1975, *Carbonate Sediments and Their Diagenesis,* 2d ed., Elsevier, Amsterdam, 658p.
Butler, G. P., 1965, *Early Diagenesis in the Recent Sediments of the Trucial Coast of the Persian Gulf,* Ph.D. dissertation, Imperial College, London, 163p.

Butler, G. P., 1969, Modern Evaporite Deposition and Geochemistry of Coexisting Brines, the Sabkha, Trucial Coast, Arabian Gulf, *Jour. Sed. Petrology* **39**:70-89.

Deffeyes, K. S., F. J. Lucia, and P. K. Weyl, 1965, Dolomitization of Recent and Plio-Pleistocene Sediments by Marine Evaporite Waters on Bonaire, Netherlands Antilles, in *Dolomitization and Limestone Diagenesis,* ed. L. C. Pray and R. C. Murray, Soc. Econ. Paleontologists and Mineralogists Spec. Pub. 13, Tulsa, Ok., pp. 71-88.

Goodell, H. G., and R. K. Garman, 1969, Carbonate Geochemistry of Superior Deep Test Well, Andros Island, Bahamas, *Am. Assoc. Petroleum Geologists Bull.* **53**:513-536.

Hsü, K. J., 1967, Chemistry of Dolomite Formation, in *Carbonate Rocks: Origin, Occurrence, and Classification,* ed. G. V. Chilingar, H. J. Bissell, and R. W. Fairbridge, Elsevier, Amsterdam, pp. 169-191.

Hsü, K. J., and J. Schneider, 1973, Progress Report of Dolomitization-Hydrology of Abu Dhabi Sabkhas, Arabian Gulf, in *The Persian Gulf,* ed. B. H. Purser, Springer-Verlag, Berlin, pp. 409-422.

Illing, L. V., A. J. Wells, and J. C. M. Taylor, 1965, Penecontemporary Dolomite in the Pesian Gulf, in *Dolomitization and Limestone Diagenesis,* ed. L. C. Pray and R. C. Murray, Soc. Econ. Paleontologists and Mineralogists Spec. Pub. 13, Tulsa, Ok., pp. 89-111.

Kinsman, D. J. J., 1964, *Recent Carbonate Sedimentation Near Abu Dhabi, Trucial Coast, Persian Gulf,* Ph.D. dissertation, Imperial College, London, 302p.

Kinsman, D. J. J., 1966, Gypsum and Anhydrite of Recent Age, Trucial Coast, Persian Gulf, in *Second Symposium on Salt I,* ed. J. L. Rau, Northern Ohio Geol. Soc., pp. 302-326.

Liebermann, O., 1967, Synthesis of Dolomite, *Nature* **213**:241-245.

McKenzie, J. A., K. J. Hsü, and J. F. Schneider, 1980, Movement of Subsurface Waters Under the Sabkha, Abu Dhabi, U.A.E., and Its Relation to Evaporative Dolomite Genesis, in *Concepts and Models of Dolomitization,* ed. D. H. Zenger, J. B. Dunham, and R. L. Ethington, Soc. Econ. Paleontologists and Mineralogists Spec. Pub. 28, Tulsa, Ok., pp. 11-30.

Mawson, D., 1929, South Australian Algal Limestones in Process of Formation, *Geol. Soc. London Quart. Jour.* **85**:613-621.

Patterson, R. J., and D. J. J. Kinsman, 1977, Marine and Continental Groundwater Sources in a Persian Gulf Sabkha, in *Reefs and Related Carbonates—Ecology and Sedimentology,* ed. S. H. Frost, M. P. Weiss, and J. B. Saunders, Am. Assoc. Petroleum Geologists Studies in Geol. 4, Tulsa, Ok., pp. 381-397.

Shinn, E. A., 1964, Recent Dolomite, Sugarloaf Key, *Geol. Soc. America Guidebook for Field Trip No. 1,* pp. 62-67.

Shinn, E. A., and R. N. Ginsburg, 1964, Formation of Recent Dolomite in Florida and the Bahamas (Abstract), *Am. Assoc. Petroleum Geologists Bull.* **48**:547.

Skinner, H. C. W., 1963, Precipitation of Calcian Dolomites and Magnesian Calcites in the South-East of South Australia, *Am. Jour. Sci.* **261**:449-472.

Skinner, H. C. W., B. J. Skinner, and M. Rubin, 1963, Age and Accumulation Rate of Dolomite-Bearing Carbonate Sediments in South Australia, *Science* **139**:335-336.

von der Borch, C. C., 1965, The Distribution and Preliminary Geochemistry of Modern Carbonate Sediments of the Coorong Area, South Australia, *Geochim. et Cosmochim. Acta* **29**:781-799.

von der Borch, C. C., and D. E. Lock, 1979, Geological Significance of Coorong Dolomites, *Sedimentology* **26**:813-824.

von der Borch, C. C., M. Rubin, and B. J. Skinner, 1964, Modern Dolomite from South Australia, *Am. Jour. Sci.* **263**:1116-1118.

Wells, A. J., 1962, Recent Dolomite in the Persian Gulf, *Nature* **194**:274-275.

9

Reprinted from *Am. Jour. Sci.* **255**:561–567 (1957)

DOLOMITE SEDIMENTATION IN THE SOUTH-EAST OF SOUTH AUSTRALIA

A. R. ALDERMAN and H. CATHERINE W. SKINNER

ABSTRACT. Fine dolomite-calcite sediments of Pleistocene to Recent age occur over a wide area in the South-East province of South Australia. Similar sediments are being formed in a number of saline lakes and in a shallow inlet of the sea. The origin of the dolomitic sediments is ascribed to elevated pH caused by plant growth in shallow saline water.

INTRODUCTION

The physiography of the South-East province of South Australia is dominated by a series of ranges of consolidated sand dunes. They trend NNW parallel to the present coast and are separated by long flat interdunal areas. The ranges represent former coastal dunes left stranded during Quaternary time by the retreating sea. The latest phase of this is to be seen where the Coorong, a long and narrow stretch of sea water, is separated from the open sea by Younghusband Peninsula, a strip of sand dunes. At its southern end a former extension of the Coorong is now represented by a series of lakes and lagoons from which the sea has retreated and which indicate the similar origin of the interdunal areas further inland.

The surface rocks of the region are predominantly carbonates. Limestones of Early to Middle Tertiary age, which are partly dolomitic, outcrop over wide areas in the south and apparently underlie the Pleistocene and Recent marine and freshwater sediments and dune rocks of the central and northern parts of the province. The widespread and prominent dunes consist of varying proportions of quartz sand and shell fragments and are mostly capped by a layer of kunkar limestone. The geology of the region has been discussed by Hossfeld (1950) and Sprigg (1952).

The main drainage of the whole region is underground. Meteoric waters disappear very rapidly in the dunal areas either by penetration or by surface run-off from the kunkar. The flat interdunal areas are however relatively impervious and lead to the development of swamps and shallow lakes after rain. Some lakes are permanent but others are shallow and generally become dry during the summer season.[1] Water plants, the most abundant being *Ruppia maritima* Linn., and a fauna of which copepods and small molluscs are the most obvious members, regenerate in the intermittent lakes during the winter and populate the permanent lakes throughout the year.

The sediment from one of these intermittent lakes, situated 10 miles north of Kingston and referred to in this paper as "Kingston Lake," was found by Sir Douglas Mawson (1929) to consist largely of dolomite. A recent examination of similar material from the same locality shows that it is a very fine aggregate of dolomite and calcite with some halite and small amounts of quartz and clay minerals. Sediments of this nature which are extremely fine-grained and possess marked plastic properties when wet have been locally

[1] The average annual rainfall varies from 15 inches in the north to 34 inches in the south. Kingston has an average of 24.3 inches. Average maximum temperatures are highest, 80°F., in February and lowest, approximately 40°F., in July.

132

known as "pipe-clay". During arid summers this carbonate sediment forms a hard white surface on the dry beds of the lakes, especially those which were formerly the southward continuation of the Coorong. Similar sediment also occurs in the southernmost parts of the Coorong itself and is exposed during dry summers when, at the head of this long and narrow inlet of the sea, water-level is lowered by evaporation.

Our investigations were concentrated on the sedimentation in Kingston Lake. Similar sediments have been found between the dune ranges in an area which extends over many hundreds of square miles. Morover, Rogers, Quirk and Norrish (1956) have described dolomite nodules and dolomite-aluminian-sepiolite marl below a surface soil of aluminian-sepiolite, montmorillonite, il-lite, kaolin, quartz and dolomite from a depression at Tintinara, South Australia, 60 miles NNE of Kingston.

Other examples of plastic dolomitic sediments have been recorded in areas remote from the South-East province. A specimen obtained from East-ern Cove on Kangaroo Island, 140 miles NW of Kingston, has similar com-position and properties and seems to have been formed as a shallow water marine sediment. At the base of an evaporite sequence in the bed of Lake Eyre (nearly 600 miles north of Kingston), C. W. Bonython (1956) has re-corded beds of dolomite which also contain small amounts of palygorskite, kaolinite, illite, quartz and gypsum. It appears that the occurrence of dolomite in the southeast part of South Australia is not of restricted importance. Direct comparison can be made with beds of similar material formed in comparative-ly recent times over a very wide area.

COMPOSITION OF THE SEDIMENTS

A typical sample of the dolomitic sediment is white with a grey tinge when dry. Microscopic examination shows an aggregate of particles most of which are so fine that, apart from the fact that they are birefringent, ordinary optical examination gives very little information. Larger particles of sandy and shelly material as well as organic fragments make up varying proportions of each sample. The carbonates are always very fine-grained and were identi-fied by X-ray powder diffraction. Dolomite, even in the surface samples, has the true dolomite structure (space group R3). Dolomite is always accompanied by calcite but calcite may occur without dolomite. Some X-ray powder diffrac-tion lines are very broad. Intensity profiles of these lines are highly asymmetric and work is proceeding on them. The unit cell of the calcite is generally small-er than that of pure calcite, suggesting solid solution of magnesium in the calcite. The dolomite generally has a slightly larger unit cell than pure dolo-mite. Further X-ray analyses are in progress.

Six samples collected from widely spaced localities and at various depths were selected to illustrate the general mineral and chemical composition of the carbonate-bearing sediments. Compositions are given in table 1 and localities in figure 1. Samples A. 78.1, A. 78.17, D 41 and USE 7-4 can be described as white, extremely fine-grained carbonates cementing quartz grains and organic fragments. D 45 and W 6-4 are dark-gray quartz sands with discrete carbonate

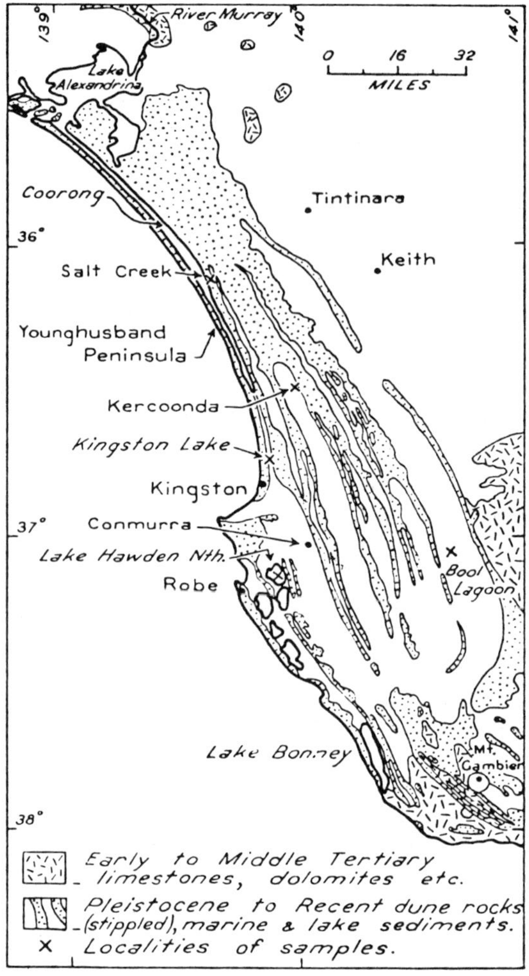

Fig. 1. The South-East of South Australia. Based on S. A. Dept. of Mines geological map, 1953.

grains and many shell fragments. All samples contain minor amounts of clays and halite as well as dolomite and calcite.

Kingston Lake can be taken as typical of many intermittent lakes. Post hole bores put down in this lake and Lake Hawdon North showed 2 to 4 feet of fine carbonate sediment, largely dolomite, above shelly beds and dark sulphur-bearing muds. Fine white sediment occurs on the surface of the Kingston Lake bed at the end of the summer. An annual sedimentation cycle begins when the winter rain accumulates to a depth of 1 to 2 feet in the lake, and plant and animal life begins to reappear in the water. As the weather becomes warmer, water plants growing on the lake floor increase in size and number. It is at this time (November and December) that the lake water may become turbid with fine white sediment which is essentially a mixture of

TABLE 1

Composition of Carbonate Sediments

	A78.1	A78.17	D.41	D.45	USE 7-4	W 6-4
SiO_2	8.24	6.86	19.42	29.00	17.24	54.31
Al_2O_3	1.11	1.21	1.21	2.28	1.19	13.92
Fe_2O_3	0.81	0.29	0.25	0.40	0.43	4.44
CaO	28.61	34.77	28.61	31.78	30.70	3.11
TiO_2	—	—	—	—	—	0.56
MgO	10.64	11.22	9.42	1.34	10.83	3.75
CO_2	33.26	39.39	32.50	26.05	35.90	6.43
SO_3	1.46	0.27	0.43	0.63	0.02	0.19
NaCl	7.88	1.01	4.29	5.18	0.07	0.48
Total Water	5.64	2.67	1.84	2.26	2.96	12.06
Acid Insoluble Organic Matter as Carbon	1.28	0.34	0.17	0.57	0.42	0.59
	98.93	98.03	98.14	99.49	99.76	99.87

SAMPLE A78.1 10 miles south of Salt Creek (Coorong). Surface layer. Calcite and dolomite in approximately equal proportions.

SAMPLE A78.17 Lake Hawdon North. 12-18 in. depth. Dolomite dominant carbonate mineral.

SAMPLE D.41 Kingston Lake. 1½-6 in. depth. Dolomite dominant carbonate mineral.

SAMPLE D.45 Kingston Lake. 33-36 in. depth. Calcite dominant carbonate mineral.

SAMPLE USE 7-4 Kercoonda. 4-7 in. depth. Calcite dominant carbonate mineral.

SAMPLE W 6-4 Bool Lagoon. 24-36 in. depth. Calcite and dolomite in approximately equal proportions.

calcite and dolomite. The fine precipitate settles onto the lake floor. During the summer the salinity of the lake increases due to high evaporation and low rainfall. The last remnant of the water deposits a crust of salt which, as it dries, forms a loose layer above the dolomite. Most of the dolomite-forming lakes are merely slightly lower areas in broad flat plains and from these the salt layer is removed by the prevailing westerly wind. Where, much more rarely, the lake bed is protected from the wind by trees and sandhills the salt accumulates. In lakes where dolomitic sediment has been forming for very many years the proportion of salt in the consolidated sediment is quite minor even though, as we will see, sodium and chlorine are overwhelmingly the most abundant constituents of the water in the present lakes.

A very similar cycle must prevail in the shallow sea waters at the head of the Coorong. Water level in the Coorong must be mainly controlled by two factors—first, the seasonal rise during the wet winter and the corresponding fall due to evaporation in the summer, and secondly, by the amount of water flowing out of the River Murray. A large volume of water flowing from the Murray mouth will bank up water at the open end of the Coorong and keep the level high. A small flow from the Murray will allow a generally lower level in the Coorong. Variations in these two factors cause differences from year to year but in an average year water-level in the Coorong is high in the winter and low in the summer. In a normal summer, lowering of water-level exposes

135

the bed of the Coorong for several miles at its southern end. The sedimentation sequence must be very similar to that in Kingston Lake. Any salt crystallizing on the surface of the dolomite is removed either by wind erosion or is redissolved as the water rises during the winter season.

The interdunal areas, which seem to be of the nature of former "Coorongs", may show more than one sequence of muds, shelly beds and the fine precipitated carbonate sediment. A drain near Conmurra shows two such sequences separated by a layer of kunkar limestone, the whole section having a thickness of 8 feet. Such sedimentary cycles must have been frequent in the South-East province during Quaternary time.

Partial analysis of a sample of suspended sediment shows: CaO, 39%; MgO, 31%; CO_2, 30%. X-ray data show calcite and dolomite in approximately equal proportions. The sediment (SW 15) in the sample studied was present to the extent of about 0.05 gms in 100 ml of lake water or about 2 percent of the total saline content. As will be seen later the amount of suspended sediment varies from place to place in the lake water.

The composition of natural waters from the Coorong and from Kingston Lake is given in table 2. The Kingston Lake water is typical of water from many similar sources in the southeast. For comparison we give analyses of what can be taken as a typical groundwater in the province (Drain L) and two analyses of sea water, one from the South Australian coast near Adelaide, the other being Sverdrup's (1942) world average.

These analyses show that the water in Kingston Lake has essentially the composition of sea water in various degrees of dilution and concentration. The Coorong water has its saline constituents in similar proportions. Drain L and tributary drains were constructed to lower the water table over a very wide area in the southeast province so that its water can be taken as typical of the groundwater of the region. It differs markedly both in salinity and composition from the lake and sea waters. The high proportions of Na and Cl in all the waters of the region are undoubtedly due to cyclic salt blown inland from the sea by the prevailing westerly winds. The water which accumulates to form lakes in the interdunal flats is rainwater which has dissolved the cyclic salt dust either in the air or distributed over the land. It is from such water, which in composition is essentially diluted or concentrated sea-water, that dolomite precipitates in Kingston Lake, e.g. analysis 5, table 2.

ORIGIN OF THE SEDIMENTS

Microscopic examination of the suspended sediment (sample SW15) referred to above does not show the presence of any structures which would indicate that the sediment had been formed by an organism, although a small proportion of faecal aggregates suggest ingestion by small crustacea. Electron micrographs of this sediment have been made by Mr. J. Farrant of the Chemical Physics Section, C. S. I. R. O. Division of Industrial Chemistry. These showed the presence of well formed rhombohedra of up to 20 microns cross section, but there was no morphological evidence of organic origin.

TABLE 2

Saline Content of Water Samples

Month	Coorong 3½m. S.E. Salt Ck.			Kingston Lake			Sea Near Adelaide	Sverdrup	Drain L Near Robe		
	June	Nov.	Jan.	June	Nov.	Jan.			June	Nov.	Jan.
No.	1	2	3	4	5	6	7	8	9	10	11
Cl	55.9	55.8	55.8	56.0	55.9	55.3	55.5	55.04	48.5	49.9	50.1
SO_4	7.5	7.6	7.6	6.3	6.2	7.5	7.9	7.68	9.5	8.7	9.6
CO_3	0.1	0.2	0.1	0.3	0.5	0.1	0.2	0.41	6.1	5.2	4.8
Na	31.8	31.9	31.7	34.3	34.1	34.2	32.2	30.61	27.3	29.0	28.3
Ca	1.0	0.9	1.0	0.8	0.6	0.4	1.1	1.15	4.0	2.8	1.3
Mg	3.7	3.6	3.7	2.3	2.5	2.6	3.8	3.69	4.6	4.4	5.8
% Salinity	7.05	4.7	5.4	3.33	2.44	22.6	3.8	3.5	0.27	0.25	0.28

Our observations have shown that carbonate precipitation can take place in waters which have only half the salinity of average sea water. Precipitation of carbonates in such solutions could be brought about by a rise in pH. The presence of vigorous plant life in water can have a very considerable effect on the pH of the solution. Baas-Becking (1934) has shown that the extraction of carbon dioxide from lake water by plants during photosynthesis can raise the pH to 9.3 in strong sunlight. The pH may fall to below 8 at night as the water absorbs CO_2 from the air.

That there is a close relation between plant growth, rise in pH, and precipitation of dolomitic sediment in Kingston Lake seems certain. Much work still remains to be done before more complete details of the processes are known. More or less continuous records of pH changes and correlation with precipitation and such factors as sunlight, temperature and salinity appear to be necessary. However, our observations to date have shown that precipitation of dolomitic sediment takes place during the spring and early summer when plant growth is in its most vigorous phase. It has also been observed that during sedimentation the water is most turbid, i.e. the precipitate is most plentiful, where plant masses are most plentiful. Such water has shown a pH of 9.2, although it has not yet been possible to determine pH maxima. The normal figure for these waters in equilibrium with air is 8.2-8.4. It has been mentioned earlier that the most common plant in the lakes and obviously the one most responsible for pH changes is *Ruppia maritima* Linn., although sedges and small algae are also present.

The association described here of shallow saline water, plant growth, elevated pH and dolomitic sedimentation has undoubtedly, we believe, been effective in an otherwise insignificant salt lake near Kingston. We see no reason to believe that this association is not of widespread occurrence. We know that dolomitic sediments of this nature have been formed over hundreds of square miles in the southeast province. We know that identical sediments are forming in sea water in the Coorong today. Admittedly the Coorong is an unusually attenuated gulf but similar conditions could presumably develop in many places where areas of shallow sea water are relatively undisturbed and the climate is suitable for vigorous plant growth.

Although the associations described in this paper may have wide implications, the limited nature of the results so far achieved is freely admitted by the authors. Much more detailed continuous observations are necessary at Kingston Lake which however provides an excellent subject as it has some of the characters of a "closed system" with a well-defined annual cycle. A study of the sediments and sedimentary processes in the Coorong is also necessary and could keep many investigators employed for a long time. However, it could easily supply information of the utmost importance. We still do not know the precise conditions under which dolomite forms in such precipitates in preference to one of the forms of calcium carbonate. Perhaps these dolomitic sediments will provide a solution to this problem.

ACKNOWLEDGMENTS
Our thanks are due to Dr. B. J. Skinner, Professor J. G. Wood, Dr. H. B. S. Womersley, Professor E. A. Rudd and Sir Douglas Mawson for their interest and help both in field and laboratory; to the Director and the Chief Analyst of the South Australian Department of Mines (Mr. T. R. Frost) for analyses of water samples and sediments; Mr. John Hutton, C.S.I.R.O. Division of Soils for anlysis of suspended sediment; Mr. J. Farrant, C.S.I.R.O. Division of Industrial Chemistry for making electron micrographs; and to Dr. B. G. Forbes who collaborated in the early stages of this investigation.

REFERENCE
Baas-Becking, L. G. M., 1934, Geobiologie: The Hague, W. P. Van Stockum & Son.

Bonython, C. W., 1956, The salt of Lake Eyre—its occurrence in Madigan Gulf and its possible origin: Roy. Soc. S. Aust. Trans., v. 79, p. 66-90.

Hossfeld, P. S., 1950, The late Cainozoic history of the South-East of South Australia: Roy. Soc. S. Aust. Trans., v. 73, no. 2, p. 232-279.

Mawson, D., 1929, South Australian algal limestones in process of formation: Geol. Soc. Lond. Quart. Journ., v. 85, no. 4, p. 613-621.

Rogers, L. E. R., Quirk, J. P., and Norrish, K., 1956, Occurrence of an aluminum-sepiolite in a soil having unusual water relationships: Jour. Soil Sci., v. 7, no. 1, p. 177-184.

Sprigg, R. C., 1952, The Geology of the South-East Province, South Australia: Geol. Surv. S. Aust., Bull. 29.

Sverdrup, H. U., Johnson, M. W., and Fleming, R. H., 1942, The Oceans: New York, Prentice-Hall.

SCHOOL OF GEOLOGY
  UNIVERSITY OF ADELAIDE
    SOUTH AUSTRALIA

ASSOCIATION OF DOLOMITE AND ANHYDRITE IN THE RECENT SEDIMENTS OF THE PERSIAN GULF

R. Curtis, G. Evans, D. J. J. Kinsman, and D. J. Shearman

*Department of Geology,
Imperial College of Science and Technology,
London, S.W.7*

A. Wells[1] has recorded the presence of dolomite associated with aragonite and gypsum in the Recent sediments of the higher parts of the tidal flats around the Qatar peninsula in the Persian Gulf. A similar environment of coastal lagoons and tidal flats extends eastwards from Qatar along the Trucial Coast towards Oman. This communication refers to a short core collected from the higher part of the coastal flats (the sabkha) in the Sheikdom of Abu Dhabi, some 200 miles east of Qatar. The core, 20 cm in length, was taken from the surface of the sabkha on the mainland approach to the Abu Dhabi causeway.

Wells interpreted the dolomite and gypsum around the Qatar peninsula as having been "derived from sea-water left behind in the sediment on the upper part of the tidal flats and concentrated by evaporation during the time between spring or storm high tides, although it has not yet been established to what extent the dolomite is primary and to what extent it is a penecontemporaneous replacement of pre-existing carbonate". The distribution of dolomite in the core from Abu Dhabi, which is described in this communication, indicates that the dolomite is an early diagenetic replacement mineral.

In the Abu Dhabi core, aragonite, calcite, dolomite and halite have been identified. Gypsum, which occurs in the sample described by Wells[1] from Qatar, has not been found, but anhydrite occurs in considerable amounts. The vertical distribution of these components is shown in Fig. 1. The uppermost 8 cm consists essentially of an accumulation of sand- and silt-sized grains of quartz and

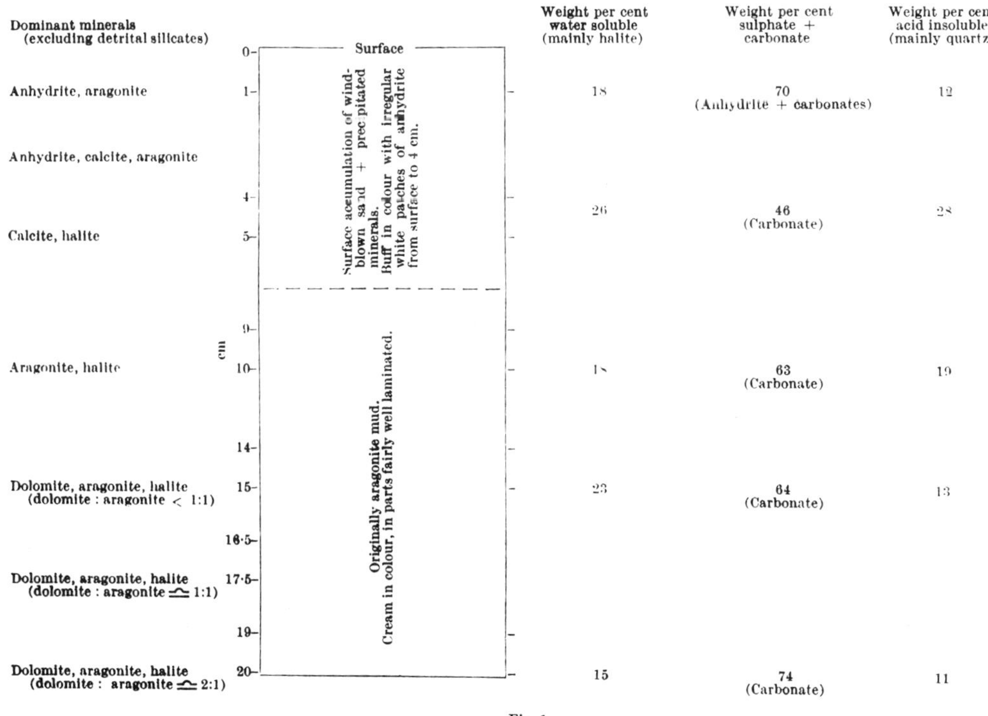

Fig. 1

carbonates, associated with evaporitic halite and anhydrite. This buff-coloured upper part rests on a cream-coloured, very fine-grained carbonate mud, which is well laminated at certain levels. Halite occurs throughout, but anhydrite, which characterizes the superficial parts of the core, has not been identified below 4 cm. At a depth of 9–14 cm from the surface the carbonate is almost exclusively aragonite, but below this level dolomite makes its appearance in detectable amounts. The ratio of dolomite to aragonite increases downwards from less than 1 : 1 at 15 cm to approximately 2 : 1 at 20 cm.

The anhydrite is clearly an evaporitic component, but the vertical distribution and mutual relationships of the dolomite and aragonite suggest that the dolomite is an early diagenetic replacement of aragonite. Although wind-blown grains are scattered throughout the length of the core, there is nothing to suggest that the dolomite is detrital. All the available evidence indicates that it was formed *in situ* and is chemical in origin.

Owing to the extremely low annual rainfall, the pore fluids in the sabkha are supplied by the high salinity sea water of the adjacent lagoons. The waters of the tidal lagoons are alkaline with pH values up to 8·0. In the sabkha, which extends inland for 7 miles near Abu Dhabi, the ground-water is acid with the pH varying between 6·0 and 6·2. Transitional pH values are found in a narrow zone along the seaward margin of the sabkha. During October 1962 the ground water-level in the sabkha stood at approximately 75 cm below the surface, although it varied slightly from place to place. The ground water temperature averaged 35° C. A lateral and vertical migration of these fluids within the sediment is caused by evaporation from the sabkha surface. It is thought that the pore fluids, probably rich in magnesium, bring about the transformation of aragonite to dolomite. However, until a more detailed field and laboratory study has been completed this conclusion can only be a tentative one.

The mineral identifications were made by the X-ray diffraction powder method. The dolomite consistently showed an absence of the superstructure reflexions (221) and ($11\bar{1}$). This feature is characteristic of protodolomites[2]. The dolomite identified by Wells is rich in calcium ($Ca_{54}Mg_{46}$) and has an expanded lattice which suggests that it may also be a protodolomite, although Wells did not record the absence of the superstructure reflexions.

It is of interest at this stage of the investigations, that the dolomite is associated in the Qatar region with gypsum and at Abu Dhabi with anhydrite, in what at first sight appear to be similar environments.

Recent work on a series of short cores taken across the seven mile width of the sabkha near Abu Dhabi has revealed the presence of abundant gypsum. A complex, and as yet not understood, series of relationships exists between the gypsum and the anhydrite. The results of these later investigations will be more fully described elsewhere.

This communication is based on work forming part of a programme of research on the Recent sediments of the Trucial Coast, Persian Gulf, financed by the Department of Scientific and Industrial Research.

[1] Wells, A., *Nature*, **194**, 274 (1962).
[2] Graf, D. L., and Goldsmith, J. R., *J. Geol.*, **64**, 173 (1956).

11

Reprinted from *Dolomitization and Limestone Diagenesis,* ed. L. C. Pray and R. C. Murray, Soc. Econ. Paleontologists and Mineralogists Spec. Pub. 13, 1965, pp. 112–123

RECENT SUPRATIDAL DOLOMITE FROM ANDROS ISLAND, BAHAMAS[1,2]

EUGENE A. SHINN, R. N. GINSBURG, AND R. MICHAEL LLOYD
Shell Development Company, Coral Gables, Florida and Houston, Texas

ABSTRACT

Calcium rich dolomite is in the process of formation on exposed supratidal mud flats in the Bahama Islands. On western Andros Island up to 80 percent dolomite occurs at or near the surface in an area covering hundreds of square miles. In all the Bahama localities the site of formation is inches above mean high tide level.

The dolomite crystals, < 3 microns, occur in pelleted muds that are associated with laminations, stromatolites, and mud cracks. The concentration of dolomite is highest in hardened surface crusts that are cemented with Recent calcite, aragonite, and dolomite. It is believed that much of the dolomite is a penecontemporaneous replacement of calcium carbonate.

The dolomite forms where tidal flooding and storm sedimentation is followed by many days of subaerial exposure. Surface evaporation during these periods of exposure increases the concentration of dissolved salts near the surface but no evaporites are preserved. Magnesium calcium ratios as high as 40 to 1 are present in the concentrated interstitial waters where dolomite is forming.

INTRODUCTION

Contemporary dolomitization[3] has been found to be active in an area covering hundreds of square miles in the Bahamas. The sites of formation are on the tidal flats of western Andros and Abaco islands (fig. 1). Because this dolomite, as well as the other occurrences of Recent dolomite reported by Wells (1962) and Deffeyes, Lucia, and Weyl (1964), and Illing, Wells, and Taylor (1965) (elsewhere in this publication) is an integral part of tidal-flat accumulation, a short summary of the significant tidal flat features of the Bahamas is included here.

The Andros Island tidal flats comprise all the Recent unconsolidated carbonate sediments (mostly pelleted muds) that are within the range of storm tides. Unless otherwise specified, the term "tides" will pertain to normal tides, that is, the range between mean high water and mean low water. Flooding by spring high tides and "storm tides" transports sediment from the offshore marine environment and deposits them on the tidal flats. Although spring and storm tides are the most effective depositors of lime sediments, the periodic normal tidal fluctuation is responsible for the zonation of sedimentary features on the Bahama tidal flats. Based on the normal tidal fluctuations, two zones of deposition are evident:

1. The *intertidal zone* lies entirely within the range of normal tides. These sediments are flooded twice a day. Sediments in this zone contain few primary bedding laminations.
2. The *supratidal zone* (marsh, salting, or sebkha) lies above normal high tide but can be flooded by spring and storm tides. This zone is subaerially exposed for long periods between storms or excessively high spring tides.

There are several kinds of supratidal accumulations but topographic highs, called "palm hammocks," constitute the supratidal zones referred to in this report (see figures 2 and 3). These highs are populated by cabbage palms which grow wherever the sediment surface is one to two feet above mean high water.

OCCURRENCE AND STRATIGRAPHY OF DOLOMITE

In most places the Recent dolomite on Andros Island occurs in a single "crust-like" layer above and below normal high tide level. The crust crops out above normal high tide level on the flanks of palm hammocks. Extending outward from these hammocks the crust forms a more or less continuous layer that is interbedded with Recent sediments. This hard crust can be traced, by probing, to a depth of 3 feet in unconsolidated lime mud at a depth of 4 feet below normal high tide level (fig. 4). Where exposed at the surface the layer is generally cracked and forms polygonal pieces up to 30 inches across. In some areas the polygonal pieces (fig. 5) are raised and tilted by the breathing roots of black mangroves and by burrowing fiddler crabs. Detailed probing and coring and excavating in the lime mud showed that

---

[1] Manuscript received, December 1, 1964.

[2] EPR Publication 392, Shell Development Company, (A Division of Shell Oil Company), Exploration and Production Research Division, Houston, Texas.

[3] The dolomite described here is a poorly ordered high calcium dolomite. X-ray powder camera photographs show that the ordering is similar to that of artificial dolomites called "protodolomites" by Graf and Goldsmith (1956).

141

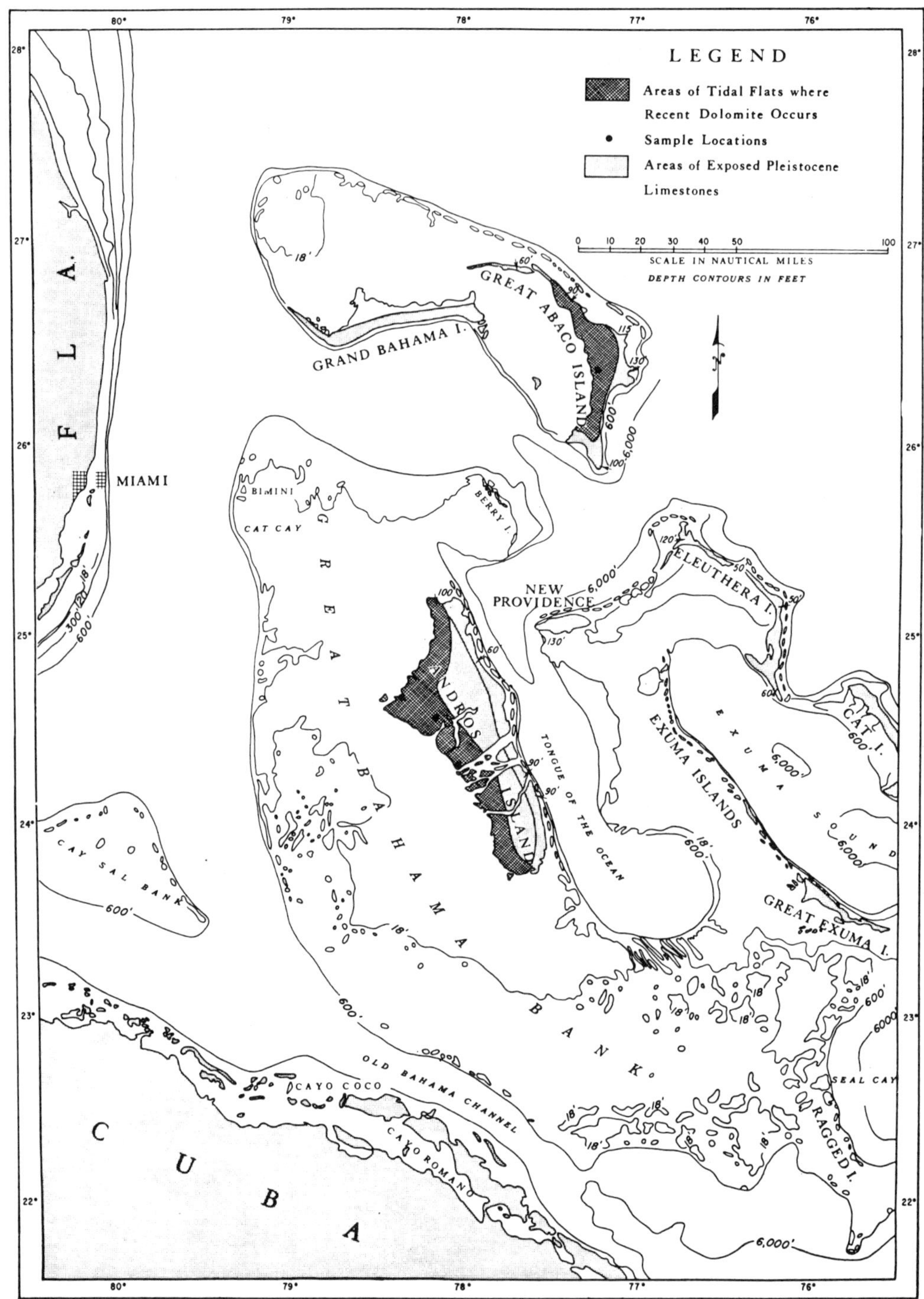

FIG. 1.—Distribution of Recent dolomite off Andros and Abaco Islands, Bahamas.

the buried crust is also broken into polygonal pieces that are tilted at angles. In areas in which the exposed crust has been weathered and broken, small chips of the crust and possibly even disaggregated crystals have become incorporated in the adjacent and underlying unconsolidated sediment. In the intertidal zone, where the crust is buried, small traces of dolomite (less than 5 percent by X-ray analysis) have been detected in sediments immediately above and below the crust horizon. Smaller traces of dolomite have been detected elsewhere in the intertidal sediments, but it is impossible to determine whether the dolomite was formed there or whether it was derived from a nearby crust layer.

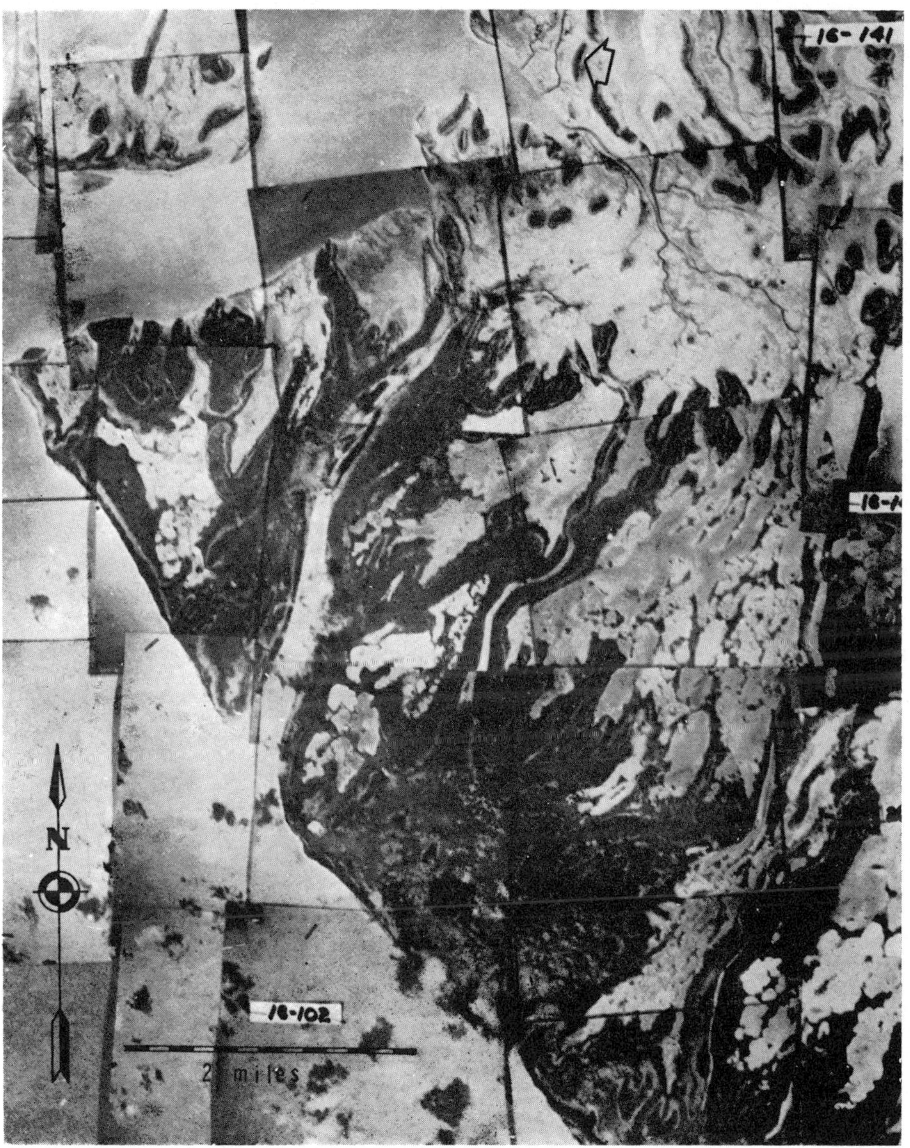

Fig. 2.—Aerial mosaic of the area at the western tip of Andros Island adjacent to Williams Island. The dark linear areas are cabbage palm hammocks. The arrow shows the palm hammock shown in Figure 3.

Fig. 3.—Aerial oblique view of palm hammock shown in Figure 2. Two large arrows point to dark band of dolomitic crust which rings hammock. The layer of dolomitic crust dips under the adjacent intertidal sediments and underlies about three feet of sediment in the intertidal areas shown by the smaller arrow. The distance between the two large arrows is about 500 feet.

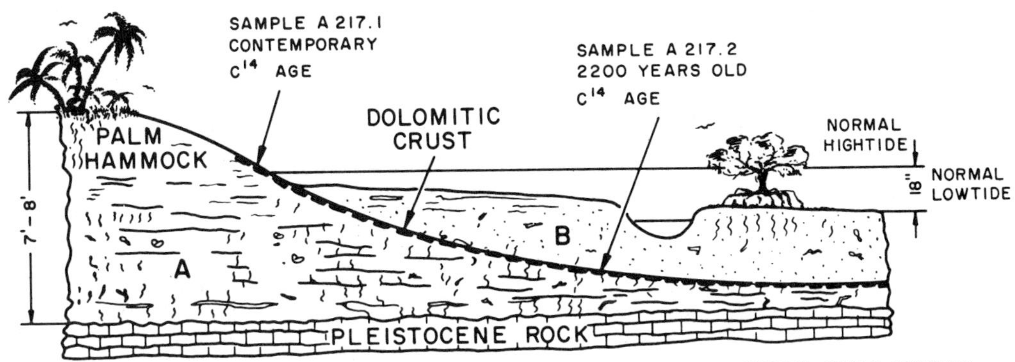

FIG. 4.—Stratigraphic cross section of the flank of a palm hammock. Schematic cross section of a portion of the palm hammock shown in Figure 3. Note that the crust layer is continuous but is contemporary in age where exposed above normal high tide level, and 2200 years old where buried and approximately 4 feet below normal high tide level. Sediments below the crust have a different character than those above the crust.

FIG. 5.—Exposed dolomitic crust. The layer here is broken into polygonal slabs by the breathing roots of black mangroves (vertical shoots to left of observer). Black mangroves grow only in the supratidal zone. Where the crust is buried and below the supratidal zone, it is broken in a similar way. Water in upper right-hand corner shows normal high tide level.

The sediment immediately above the Pleistocene bed rock surface, which in this area is less than 10 feet below sea level, is a dark gray pelletal mud containing land snails and occasional small marine gastropods and foraminifera (fig. 4, sediment A), and it does not contain dolomite. The most characteristic features of this sediment are numerous roots and root hairs and desiccation layers of light tan semiconsolidated sediment. The abundance and kinds of these dead roots and root hairs are dissimilar to those formed by the presently living vegetation. This "sediment type" is several feet thick and is overlain by the crust layer. The sediments immediately above the crust are compositionally and texturally different from the sediment below the crust. This sediment (fig. 4, sediment B) is characterized by larger and more numerous foraminifera and gastropods and also by an almost complete lack of desiccation features. It contains mainly living roots and root hairs which are of a different species than the dead ones in the sediment beneath the crust. The sedimentary structures in the sediment above the crust suggest that it was deposited in the *intertidal zone.*

STRUCTURES AND COMPOSITION OF THE DOLOMITE CRUST

The crust surface is pitted and weathered. It is dark gray to black and can easily be seen from the air (fig. 3). The upper surface can best be described as having the appearance of an old asphalt road coating (fig. 5). The dark color is only a few millimeters thick. Below the dark coating the lithified sediment is white or light tan.

The thickness and hardness of the crust are variable. The average thickness is between 2 and 3 cm thick. In some areas where crust is pitted by weathering the thickness is less than 1 cm. Brittle pieces of crust often ring if struck together. Where the crust is buried or where it is exposed but does not show signs of weathering, it is sometimes greater than 5 cm thick. At Sugarloaf Key in the Florida Keys similar dolomitic crusts up to 10 cm thick occur in the supratidal zone, Shinn (1964).

Often the crust contains no obvious sedimentary structures, but when they are present (fig. 6) they include (1) mud cracked polygons (in place), (2) current-transported mud cracked polygons and chips of dried mud, (3) laminations similar to those associated with algal mats living nearby, and (4) burrows. The most abundant skeletal remains that can be seen with the unaided eye include the marine gastropod *Batillaria* sp., foraminifera of the family Peneroplidae, and the land snail *Cerion*. Often these shells have been leached and are easily crushed between one's fingers.

TEXTURE OF DOLOMITIC CRUST

Approximately 24 cross sectional thin sections were examined petrographically. Inspection of all sections under low power showed that the main constituent is fecal pellets. Clasts composed of foraminifera, mollusk fragments, noncarbonate organic detritus, and vugs filled with clear acicular carbonate make up the remainder of the crust (figs. 7A and 7B).

The samples from which the thin sections were made have been analyzed[4] for dolomite by X-ray diffraction techniques.[5] The percentage of dolomite varies from crust to crust and ranges from 20 to over 80 percent.

Formic acid separation of the dolomite crystals showed that they range in size from 1 to 2 microns. When viewed in thin section under high power on the petrographic microscope, however, these same crystals are extremely hard to distinguish; and because of the small size, the characteristic dolomite crystal shape is not recognizable. The preponderance of these crystals confirms the percentages obtained by X-ray diffraction analyses. The formation of these crystals has almost obscured the boundaries of most of these pellets. Some pellets are quite distinct, whereas others appear only as fuzzy dark areas in a matrix of tiny dolomite crystals. Mollusks and foraminifera have been altered, but because of the extremely small crystal size, it is impossible to determine petrographically whether they have been dolomitized. X-ray analysis of a shell removed from the crust indicated that the shell was very slightly dolomitic. The exposed crusts are extremely porous. Some of the pores are lined with "dogtooth" crystals of calcium carbonate. The acicular crystals grow inward from the vug walls and, in some, completely fill the voids (fig. 7B). This pore filling carbonate appears to be the chief cause of lithification.

RADIOCARBON AGE DETERMINATION[6]

Carbon-14 dates were obtained for eight samples of crust and for the dolomite crystals separated from each of these samples. Results are

[4] X-ray analyses were performed by K. S. Deffeyes and R. Michael Lloyd.
[5] X-ray data were obtained on a Norelco diffractometer using Cu radiation and pulse height discrimination. A scan speed of $\frac{1}{4}$°/minute and a chart speed of 60"/minute were used to give broad peaks at a scale of 4"/minute of arc. Mineral composition was determined by comparing peak height measurements to pure standards. The accuracy is about ±10 percent of the amount present.
[6] Measurements made by E. L. Martin of Shell Development Company.

shown in table 1. For three of the eight samples dated, the total sample proved to be older than the dolomite. One sample was about 1000 years older than the dolomite, and the other two were between 600 and 800 years older than the dolomite. All three were samples of exposed crust. The total carbonate ages of the remaining five samples were the same as those of the dolomite from each of the samples.

Three of these five samples were taken below high tide level, and two were taken above high tide level. The dolomite separated from the three exposed in situ crusts yielded contemporary ages. According to Martin (personal communication), the dates for the dolomite in the exposed crusts are similar to those obtained for the shells of living clams. Measurements of two samples now below low tide level gave ages slightly in excess of 2000 years. One of these subtidal samples (A 217.2) and one of the exposed samples (A 217.1) are from the same locality. They were collected about 500 feet apart, and the layer between them is unbroken (fig. 4). Samples A 207 and A 210 shown in table 1 represent a similar situation at another locality.

CHEMISTRY

The dolomite described in this report is a non-ideal dolomite. The term "non-ideal" was coined by Goldsmith and Graf (1958) to identify certain dolomites of sedimentary origin which do not possess the ideal dolomite structure. Ideally the mineral dolomite is a rhombohedral carbonate containing equal molar proportions of calcium carbonate and magnesium carbonate. The crystal structure consists of layers of carbonate groups separated alternately by layers of all calcium ions and all magnesium ions. It is the ordering of the cations in alternating layers which makes dolomite a separate mineral phase rather than a solid solution midpoint between the rhombohedral $CaCO_3$ calcite and the rhombohedral $MgCO_3$ magnesite.

Most non-ideal dolomites differ from the ideal mineral in that they (1) contain an excess amount of $CaCO_3$ and (2) show a departure

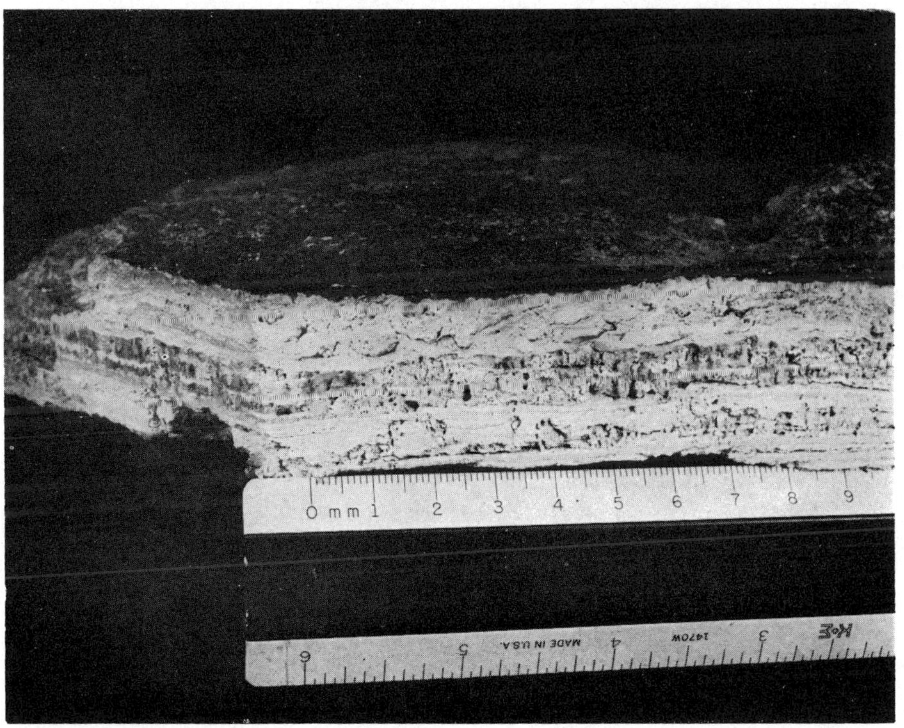

Fig. 6.—Sawed surface of a piece of dolomitic crust containing sedimentary structures. Molds of mat-forming filamentous algae (band containing vertical voids in center of crust) are visible as well as sedimentary laminations and voids. Note dark upper surface. Sample contains about 25% dolomite.

Fig. 7a.—Cross-sectional thin section of dolomitic crust photographed under low magnification in plain light. Distance from top to bottom is 2 cm. Note irregular laminations, pellets, clasts, and vugs. Most vugs are lined with acicular carbonate as shown in Figure 7b.

from perfect ordering of calcium and magnesium atoms in the crystal structure. Both of these characteristics are seen in X-ray diffraction patterns. Because calcium is a larger atom than magnesium it follows that the substitution of calcium for magnesium will cause an expansion of the structure. For instance, the characteristic spacing of atomic planes parallel to the [211] crystallographic plane is 2.88Å in ideal dolomite. When 5 percent of the magnesium positions are occupied by calcium the d-spacing is about 2.895Å. The ordered array of calcium and magnesium atoms in ideal dolomite gives rise to a set of X-ray reflections which have no counterpart in calcite or magnesite. The presence or absence of these superstructure reflections, as they are called, their sharpness and relative intensities give a qualitative indication of perfection of ordering in the dolomite.

In figure 8 a sample of Andros dolomite is compared to two other dolomites by means of X-ray diffraction patterns. The Permian dolomite is very near ideal. The [211] d-spacing is 2.88Å and all three ordering peaks are sharply defined. The Cretaceous dolomite shows a slight shift of the [211] peak to the right in the direction of higher d-spacing indicating the presence of excess calcium carbonate—in this case the composition is about 55 mole percent $CaCO_3$. All three ordering peaks are present but they tend to be somewhat diffuse and of lower intensity relative to nearby peaks when compared to the Permian sample. The Andros dolomite shows a peak shift of about the same magnitude as the Cretaceous sample indicating 5 percent excess $CaCO_3$ but the ordering peaks are absent in the tracing. A powder camera picture of the same Andros sample (fig. 9) after long exposure shows the [211] and [111] order reflections to be present but extremely weak, while the [100] order reflection cannot be seen.

The Andros sample used in figure 8 is from a piece of very hard crust which contained approximately 80 percent dolomite. The dolomite was easily purified by slow leaching of the calcite and aragonite with dilute acid. Many of the Andros samples contain less dolomite and display excess calcium carbonate up to 62 mole percent. For the higher calcium dolomites the differential acid leaching technique does not work. After an initial relative increase in the dolomite concentration, further leaching reduces calcite, aragonite, and dolomite at approximately the same rate. It is not clear whether this is due to a smaller dolomite crystal size or to greater solubility of very high calcium dolomites.

The amount of excess calcium carbonate present in the Andros material is unusually high compared to the values reported by Goldsmith and Graf (1958). The highest value they list is 56.2 mole percent from a core in the Funafuti Atoll. Seventeen out of the 20 Andros samples are greater than this value. An expansion of the dolomite structure by the incorporation of other large ions such as iron or manganese could give a false impression of the amount of excess calcium carbonate present. A check of two Andros samples by spectrochemical methods, however, shows the concentration of such ions to be so low as to probably have a negligible effect. Most of the samples described by Goldsmith and Graf (1958 show some weaken-

ing of the ordering reflections probably similar to that shown by the Cretaceous sample in figure 9, but none as weak as the Andros material.

The Andros dolomite, then, represents a greater departure from ideal crystal structure than other natural non-ideal dolomites. In this respect the Andros material most clearly resembles the protodolomite produced in the early stages of dolomite synthesis in experimental bombs (Graf and Goldsmith, 1956). In time synthetic protodolomites convert into well-ordered dolomite with an equilibrium $CaCO_3$ composition appropriate for the temperature of the experiment. According to the experiments of Goldsmith and Heard (1961) the equilibrium $CaCO_3$ composition is 55 mole percent at 900°C and decreases with decreasing temperature to 50 mole percent at room temperature. The sedimentary non-ideal dolomites must, therefore, be metastable. Realizing this possibility, Goldsmith and Graf (1958) attempted to produce an equilibrium composition dolomite by heat treating of natural non-ideal dolomites. They were able to accomplish this equilibration only after

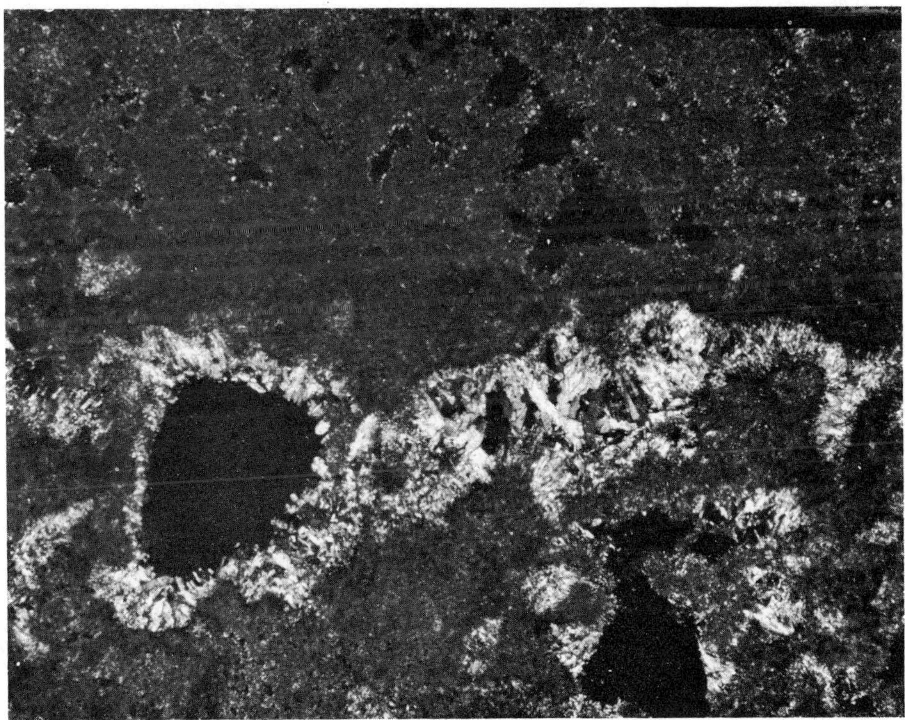

Fig. 7b.—Vugs lined with acicular carbonate. Vague outlines of pellets are visible in lower right-hand corner of photograph. Tiny 2μ sized dolomite crystals form matrix. Magnification ×90.

adding a flux to promote recrystallization.

In terms of its crystal chemistry the Andros dolomite appears to represent a very early stage of dolomite formation. The large range of $CaCO_3$ compositions suggests that some changes do occur toward a higher quality dolomite within the early diagenetic environment. However, from studies of other non-ideal dolomites, it appears that recrystallization is necessary to reorganize the structure into truly ideal dolomite.

DISCUSSION

The carbon-14 age determinations, the sedimentary structures, and the stratigraphic relationships indicate that the formation of dolomite has been taking place slightly above normal high tide level during a gradual transgression of the sea (fig. 4). The sedimentary evidence is as follows:

1. Mud cracks similar to those within the crusts in the supratidal zone were found in several samples from the *intertidal zone* and also from some samples from below the *intertidal zone*. One of the mud cracked samples from below the intertidal zone was found to be over 2000 years old by carbon-14 analysis.

2. Laminations like those associated with algal mats that grow in conjunction with the supratidal crusts are often present in the crusts collected from below the intertidal zone. These algal mats never grow on the sediments below the intertidal zone and when present on the intertidal sediments they populate only the upper few inches of the intertidal zone.

3. Where the crust is buried beneath intertidal sediments it is generally broken into polygonal shaped pieces that are tilted at various angles, just as it is where it is exposed above normal high tide level. The same agencies that crack and tilt the exposed crust in the supratidal zone (roots and sand burrowers) are assumed to be responsible for the cracking and tilting of the buried crusts.

There is no evidence that the dolomite is of detrital origin. We have been unable to find any source for detrital dolomite other than the Recent dolomite itself. Offshore marine sediments contain no dolomite. Even if these sediments did contain small amounts of dolomite it would require an extremely efficient and unknown sorting mechanism to separate and deposit dolomite only in the supratidal zone.

If we can depend upon the dating method and the sedimentary structures, we must conclude that dolomite is forming and has been forming during the past 2000 years in a narrow zone slightly above normal high tide level. The lateral continuity of the crust layer shown in figure 4 indicates that the relative rise of sea level in this area has been gradual and continuous. The crust layer has drawn a natural graph of relative sea level rise.

Some of the dolomite may be a direct precipitate from the interstitial water but the authors have reason to believe that much of the dolomite has formed by replacement. It is thought that much of the supratidal dolomite is a product of dolomitization because the dolomitic crusts are largely composed of fecal pellets. The most common pellet producer, a small gas-

TABLE 1.—*Radiocarbon Ages of Crusts*[1]

Sample No.	Relationship to Tide Level	Dolomite (%)	Calcium Carbonate (%)	Radiocarbon Ages in Years	
				Dolomite	Dolomite + $CaCO_3$
Western tip of Andros Island (Williams Island area)					
*A217.1	Exposed and above normal high tide level	—	—	0–160	0–160
*A217.2	Exposed in gully 4′ below high tide level	—	—	2200 ± 150	2150 ± 160
A204	Exposed above high tide level	50	50	0–160	1270 ± 140
West Coast of Andros Island near Loggerhead Creek					
†A207	Buried 1′ below sediment and at low tide level	20	80	1090 ± 170	1000 ± 140
†A210	Buried 6″ below sediment at high tide level	60	40	410 ± 150	390 ± 120
A211	Broken slab on beach, not in place	70	30	0–160	820 ± 140
Interior portion of Andros tidal flats near Turner Sound					
A213	Exposed and above high tide level	80	20	0–160	790 ± 130
A215.2	Not buried and 2′ below low tide level	30	70	2120 ± 150	2270 ± 130

Aragonite from land snails (*Cerion* sp.) from underside of crust Sample A217.1 dated 1690 years.

* Sample shown in Figure 4.
† Samples from a situation similar to that shown in Figure 4.
[1] Carbon from living coral (*Porites*) used as standard for measuring Carbon-14 activity.

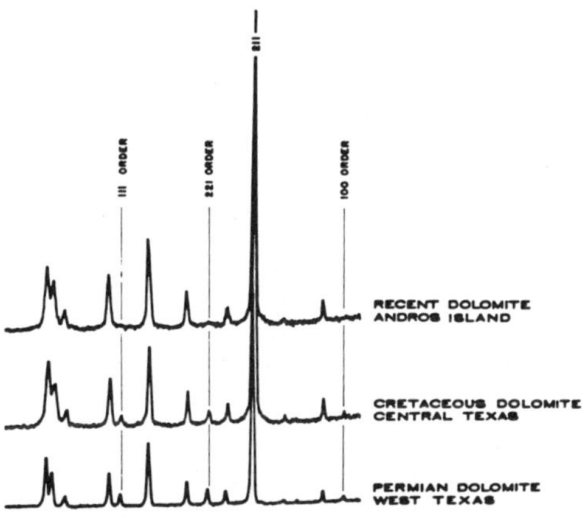

FIG. 8.—X-ray diffractometer comparison of Recent dolomite and Cretaceous and Permian dolomites from Texas.

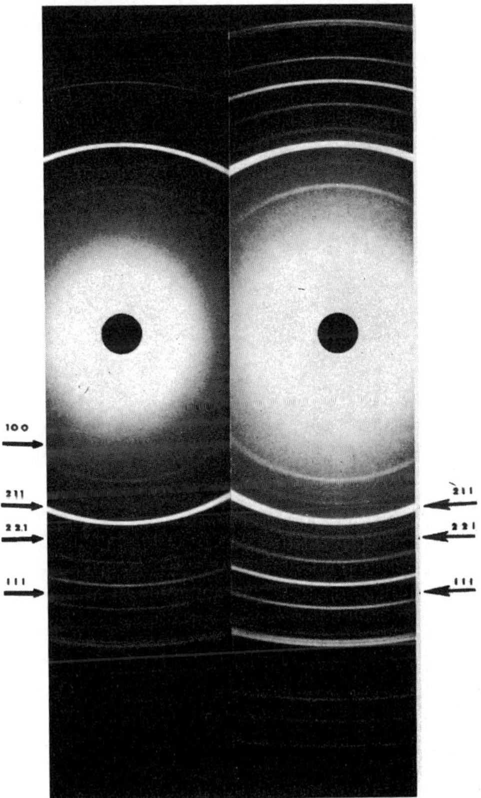

FIG. 9.—X-ray powder camera photograph showing comparison of ordering reflections between a Cretaceous sample, left (same as sample in Figure 8), and Recent Andros dolomite on right. Ordering lines [221] and [111] are only barely visible and the [100] reflection is absent.

tropod *Batillaria* sp. is rarely found in the supratidal zone but lives in great numbers in the nearby intertidal zone. In order for these pellets to get to the supratidal zone they must be transported from the intertidal zone. Since pellets in the intertidal zone contain no appreciable amount of dolomite, then they must become dolomitized after they have arrived in the supratidal zone. Dolomitic mollusk shells within the crust also indicate replacement, although the authors recognize that the dolomite measured in these shells may represent dolomite that infilled the numerous tiny algal borings that penetrate the shells. Thin sections of such shells show what looks like dolomite within the shell wall, but it is always associated with numerous microscopic borings.

Field observations show that the supratidal sediments on Andros Island are derived from the nearby marine and intertidal sediments. During a four-day stay at the location shown in figure 3, a mild storm occurred. Wind waves raised water level and stirred up the offshore sediments causing the flooding tide to be milky with suspended sediment. When the tide ebbed, waters leaving the flats were relatively clear. The suspended sediment had been deposited. This observation was made during a relatively mild storm that lasted only a few hours, and confirms the observations of Ball, Shinn, and Stockman (1963) who reported that several inches of lime mud was deposited on the supratidal surfaces of islands throughout Florida Bay during Hurricane Donna.

Unconsolidated sediments from the nearby marine zone and from the nondolomitic intertidal zone contain 1 to 3 percent magnesium carbonate. This small amount of indigenous magnesium could not account for the observed amount of dolomite. An outside source must exist. The most logical source is the sea, but how is magnesium supplied to the supratidal zone which is generally above sea level?

On Andros and Abaco islands spring and storm tides flood the supratidal zone at least once a month and sometimes more often. Still it is unlikely that dolomitization occurs in spurts during the short periods when these surfaces are inundated. Dolomite must form when these surfaces are above sea level!

Analyses of capillary water squeezed from dolomitic supratidal sediments at Sugarloaf Key in the Florida Keys show a very high Mg to Ca ratio, (Shinn, 1964). Normal sea water has a molar ratio of Mg to Ca of about 5 to 1. The supratidal sediments at Sugarloaf Key, which are similar to those on the Andros tidal flats, have molar ratios of Mg to Ca in excess of 40 to 1 and salinities 5 to 6 times that of sea water. There is little doubt that such a water could supply the Mg necessary for dolomite formation. There is some problem, however, in explaining how the water becomes Mg enriched.

In all the supratidal areas that have a dolomitic crust there is a saline ground water table a few inches below the surface. The level of this water fluctuates slightly in response to the normal tidal fluctuations and its salinity is generally near that of normal sea water. This vertical fluctuation acts like a pump and causes air to be alternately drawn in and out of the overlying sediments. At the same time the saline ground water, which lies at approximately the level of normal high tide, supplies a continual source of water which is carried by capillary movement to the surface where it is evaporated. Salts are precipitated and an interstitial brine is produced in the upper few cm of sediment. In the upper few cm of sediment the Mg content relative to Ca in the interstitial water is somehow increased. It is likely that the relative increase of Mg is due to the precipitation of Ca in the form of gypsum or as carbonate cement, as suggested by Deffeyes et al. (1964). In this humid climate we do not see gypsum preserved, but it is possible that it is continually being washed away by .spring and storm tides and rain. The high salinities indicate that gypsum should precipitate.

These observations are not intended to explain the chemistry of dolomite formation but rather to show the environmental contitions in the supratidal zone that apparently are capable of producing the chemical conditions necessary for dolomitization. Other factors such as H_2S, CO_2, desiccation, rain water, temperature fluctuations, humidity, and the catalytic effects of organisms, which are beyond the scope of this report, may also be essential to the formation of dolomite.

GEOLOGIC IMPLICATIONS

The discovery of Recent dolomite on carbonate tidal flats has immediate geologic meaning. Contemporary formation of dolomite on the surface of unconsolidated Recent sediment confirms the view of many geologists that dolomite can be penecontemporary or early diagenetic. This paper shows that in the supratidal zone of the Bahama tidal flats

1. sea water does provide $Mg++$, although other unknown factors may be necessary to complete dolomitization;
2. high pressure and high temperatures are not necessarily needed to form dolomite;
3. dolomitization is an integral part of sedimentation; and
4. dolomite does form in the supratidal zone.

ACKNOWLEDGMENTS

We wish to express our appreciation to K. S. Deffeyes, F. J. Lucia, and P. K. Weyl whose initial discovery of Recent dolomite on the Island of Bonaire stimulated our efforts and led to its recognition in the Bahamas. Special thanks go to Deffeyes, now at the University of Minnesota, for working out the mineralogy of our samples and for perfecting a technique for estimating percent dolomite with the X-ray machine, and to E. L. Martin of Shell Development Company for separation of dolomite and for performing the carbon-14 analyses, and to Heinz Studer for making the electron micrographs. We are much indebted to D. V. Higgs of Shell Development Company for his encouragement and criticism of the manuscript. Also of invaluable service was M. M. Ball of Shell Oil Company who stimulated discussions and advice throughout the study. Special thanks go to W. E. Bloxom, Philip Fickman, and E. S. Earle of Shell Oil Company who while enduring mosquitoes, horseflies, muck and mire, aided in the collection of material and stimulated several ideas in the present text.

REFERENCES

BALL, M. M., SHINN, E. A., AND STOCKMAN, K. W., 1963, Geologic effects of hurricane Donna (abs.) Am. Assoc. Petroleum Geologists Bull., v. 47, no. 2, p. 349.

DEFFEYES, K. S., LUCIA, F. J., AND WEYL, P. K., 1964, Dolomitization; observations on the island of Bonaire, Netherlands Antilles: Science, Feb. 14, v. 143, No. 3607, p. 678-79.

GOLDSMITH, J. R., AND GRAF, D. L., 1958, Structural and compositional variations in some natural dolomites: Jour. Geology, v. 66, p. 678-693.

———, AND HEARD, H. C., 1961, Subsolidus phase relations in the system $CaCO_3$-$MgCO_3$: Jour. Geology, v. 69, p. 45-74.

GRAF, D. L., AND GOLDSMITH, J. R., 1956, Some hydrothermal synthesis of dolomite and protodolomite: Jour. Geology, v. 64, p. 173-186.

ILLING, L. V., AND WELLS, A. J., 1964, Dolomitization in the Persian Gulf (abs.) : Am. Assoc. Petroleum Geologists Bull., v. 48, No. 4, p. 532.

ILLING, L. V., WELLS, A. J., AND TAYLOR, J. C. M., 1965. Penecontemporary dolomite in the Persian Gulf, *in* Dolomitization and Limestone Diagenesis, A Symposium: (PRAY, L. C., and MURRAY, R. C., editors), Soc. Econ. Paleontologists and Mineralogists Spec. Publ. No. 13, p. 89-111.

SHINN, E. A., 1964, Recent dolomite, Sugarloaf Key, *in* Guidebook for GSA Field Trip No. 1, South Florida Carbonate Sediments (compiled by R. N. GINSBURG) : p. 62-67.

WELLS, A., 1962, Primary dolomitization in Persian Gulf: Nature, v. 194 (4825), p. 274-275.

Part III

MODELS FOR DOLOMITIZATION BY SURFACE-GENERATED HYPERSALINE SOLUTIONS

Editors' Comments
on Papers 12 and 13

12 ADAMS and RHODES
Dolomitization by Seepage Refluxion

13 HSÜ and SIEGENTHALER
Preliminary Experiments on Hydrodynamic Movement Induced by Evaporation and Their Bearing on the Dolomite Problem

Dolomitization resulting from the reflux of surface-generated, hypersaline solutions in evaporative, penemarine environments is discussed in this part. Based on their research on Permian carbonates of West Texas and southeastern New Mexico, Adams and Rhodes (Paper 12) concluded that restrictions of normal marine circulation are necessary for the formation of most dolomites. Following the earlier comments of R. H. King (1947) on the incompleteness in the Permian Basin of an idealized salt section precipitated from a desiccating basin, they envisioned ancient, hypersaline epicontinental shelves as the source of the brines that dolomitized subjacent lagoonal sediments as the dense "potent" solutions percolated through these metastable carbonates. They termed this process *seepage refluxion,* although a similar model had been proposed previously by Newell et al. (1953) and by Teodorovich (1955). Adams and Rhodes noted the presence in the Permian rocks examined of four "footprints" in support of their theory: anhydrite inclusions, halite inclusions, relict brines trapped in the Permian section, and thick dolomite sections beneath the implied evaporitic lagoon. Although their model appears convincing, it suffers from a number of difficulties: (1) Is the finely crystalline, sublagoonal dolomite secondary or primary (Chilingar and Bissell, 1961)? (2) Why did replacement rigidly respect lithologic boundaries in not dolomitizing the Capitan reef and forereef beds? (3) There is not a sufficient hydraulic head to enable such brines to flow through relatively impermeable carbonate muds, especially when inherently low permeabilities were further reduced by evaporite plugging (Hsü and Siegenthaler, Paper 13). (4) Although reflux dolomitization has been proposed in the Canary Islands (Müller and Tietz, 1966), Cretaceous of Texas (Fisher and Rodda, 1969), and on Bonaire (Deffeyes, Lucia, and Weyl, 1965), its reality and effectiveness in the latter areas have been questioned (Lucia, 1968; Murray, 1969).

In a set of elegantly simple experiments, Hsü and Siegenthaler (Paper 13) tested the theory of evaporative pumping, which suggests that dolomitization may occur on arid supratidal flats by the upward flow of seawater-derived, interstitial fluids. They demonstrated that evaporation provides the energy necessary for vertical fluid movement and that flow rates (and Mg^{2+} supply rates) are governed solely by evaporation and are not dependent on permeability. They contrasted their model with that of capillary concentration (see Friedman and Sanders, 1967, p. 281), which relates to capillary movement of water in a vadose zone. Their model is seemingly verified by field observations on Bonaire (Lucia, 1968; Murray, 1969) and to some extent in the Bahamas (Shinn, Ginsburg, and Lloyd in Paper 11), and is most plausible because it explains why dolomitization is commonly, but not always, fabric selective in favor of fine-grained sediments; it provides for an unlimited quantity of Mg^{2+} ions necessary for dolomitization, regardless of whether the Mg/Ca ratio of metasomatic solutions needs to be particularly high (Langmuir, 1965; Berner, 1967; Hsü, 1967; Usdowski, 1967; Folk and Land, Paper 19); and based on a process effectiveness of 15 percent or less, it approximates presumed rates of dolomite formation (10 km/3.5 my) that are in agreement with inferences on certain ancient marine dolomite sequences. Recent work by McKenzie et al. (1980) offers much hydrological data that seem to support the validity of an evaporative pumping cycle (initiated by flooding) under the Abu Dhabi sabkha in the Persian Gulf.

The preeminent conclusion of these two papers is that dolomitization is an eogenetic process related to the interstitial movement of saline waters in arid coastal regions. Ancient analogues of peritidal, evaporitic dolomitization (for example, see Friedman and Sanders, 1967) are discussed in Part IV of this book and alternatives to these models in Parts V through VII. Hsü (1966) and Hsü and Siegenthaler (Paper 13) further maintained that dolomitization by relatively fresh groundwaters is an ineffective process in nature. To the extent this conclusion is now regarded as correct is discussed in Papers 17 through 20 of this book.

REFERENCES

Berner, R. A., 1967, Comparative Dissolution Characteristics of Carbonate Minerals in the Presence and Absence of Aqueous Magnesium Ion, *Am. Jour. Sci.* **265**:45-70.

Chilingar, G. V., and H. J. Bissell, 1961, Discussion:Dolomitization by Seepage Refluxion, *Am. Assoc. Petroleum Geologists Bull.* **45**:679-681.

Deffeyes, K. S., F. J. Lucia, and P. K. Weyl, 1965, Dolomitization of Recent and Plio-Pleistocene Rocks by Marine Evaporite Waters on Bonaire, Netherlands Antilles, in *Dolomitization and Limestone Diagenesis,* ed. L. C. Pray and R. C. Murray, Soc. Econ. Paleontologists and Mineralogists Spec. Pub. 13, Tulsa, Ok., pp. 71-88.

Fisher, W. L., and P. U. Rodda, 1969, Edwards Formation (Lower Cretaceous), Texas:Dolomitization in a Carbonate Platform System, *Am. Assoc. Petroleum Geologists Bull.* **53**:55-72.

Friedman, G. M., and J. E. Sanders, 1967, Origin and Occurrence of Dolostones, in *Carbonate Rocks: Origin, Occurrence, and Classification,* ed. G. V. Chilingar, H. J. Bissell, and R. W. Fairbridge, Elsevier, Amsterdam, pp. 267-348.

Hsü, K. J., 1966, Origin of Dolomite in Sedimentary Sequences:A Critical Analysis, *Mineralium Deposita* **2**:133-138.

Hsü, K. J., 1967, Chemistry of Dolomite Formation, in *Carbonate Rocks: Origin, Occurrence, and Classification,* ed. G. V. Chilingar, H. J. Bissell, and R. W. Fairbridge, Elsevier, Amsterdam, pp. 169-191.

King, R. H., 1947, Sedimentation in Permian Castile Sea, *Am. Assoc. Petroleum Geologists Bull.* **31**:470-477.

Langmuir, D., 1965, Stability of Carbonates in the System $MgO\text{-}CO_2\text{-}H_2O$, *Jour. Geology* **73**:755-780.

Lucia, F. J., 1968, Recent Sediments and Diagenesis of South Bonaire, Netherlands Antilles, *Jour. Sed. Petrology* **38**:845-858.

McKenzie, J. A., K. J. Hsü, and J. F. Schneider, 1980, Movement of Subsurface Waters under the Sabkha, Abu Dhabi, U.A.E., and Its Relation to Evaporative Dolomite Genesis, in *Concepts and Models of Dolomitization,* ed. D. H. Zenger, J. B. Dunham, and R. L. Ethington, Soc. Econ. Paleontologists and Mineralogists Spec. Pub. 28, Tulsa, Ok., pp. 11-30.

Müller, G., and G. Tietz, 1966, Recent Dolomitization of Quaternary Biocalcarenites from Fuerteventura (Canary Islands), *Contr. Mineralogy and Petrology* **13**:89-96.

Murray, R. C., 1969, Hydrology of South Bonaire, N.A.—A Rock Selective Dolomitization Model, *Jour. Sed. Petrology* **39**:1007-1013.

Newell, N. D., J. K. Rigby, A. G. Fischer, A. J. Whiteman, J. E. Hickox, and J. S. Bradley, 1953, *The Permian Reef Complex of the Guadalupe Mountains Region, Texas and New Mexico,* W. H. Freeman, San Francisco, Ca., 236p.

Teodorovich, G. I., 1955, K Voprosu o Proiskhozhdenii Osadochnykh Izvest Kovodolomitovykh Porod, *Akad. Nauk SSSR Izv. Ser. Geol.* **5**:75-107. (A Contribution on the Origin of Limestones and Dolomites, trans. M. Burgunker, *Internat. Geology Rev.* **1**:50-73, 1959.)

Usdowski, H.-E., 1967, Die Genese von Dolomit in Sedimenten, in *Mineralogie und Petrographie in Einzeldarstellungen 4,* Springer-Verlag, Berlin, 95p.

DOLOMITIZATION BY SEEPAGE REFLUXION[1]

JOHN EMERY ADAMS[2] AND MARY LOUISE RHODES[2]

ABSTRACT

Bedded dolomites in the Permian Basin were formed by the alteration of metastable limestones by hypersaline brines refluxing from evaporite lagoons. The hot, heavy, highly alkaline, carbon dioxide-free, magnesium-supercharged brines displaced connate waters to provide both a chemically favorable environment for magnesium-calcium exchange and a vehicle for removing displaced calcium. Fossil lagoonal brines and fillings of halite and anhydrite in the dolomite pores offer proof of the brine invasion. Sedimentary dolomites in other areas are commonly associated with evaporites, and for these dolomites a similar origin is postulated.

INTRODUCTION

The stratigraphic column from the mid-Precambrian to the Recent contains many bedded dolomites. In spite of their abundance, stratigraphic span, and geographic range, the origin of these dolomites is a controversial subject. Recent regional mapping of depositional and diagenetic relationships supplies the data for interpreting the origin of some major Permian dolomites. The explanation here advanced may aid in determining the origin of other deposits of sedimentary dolomites.

The abundance and widespread distribution of major sedimentary dolomites indicate that these rocks are products of common, relatively simple natural processes. The environments in which the now dolomitized rocks were deposited, as recorded in the dolomites themselves and in associated sediments, support this conclusion. The environmental range represented is largely identical with that responsible for normal marine limestones, but in addition, includes a restricted lagoon phase. Major dolomite deposits of both normal marine and restricted environments are limited to, or closely associated with, shallow epicontinental shelves or oceanic banks. Thus, all adequately explored major sequences of sedimentary dolomite are known to include or interfinger with calcitic limestones, characterized by abundant shallow-water fossils. Surprisingly high percentages of bedded dolomites are associated with evaporites. Close associations of evaporites, dolomites, and marine limestones are common

enough to suggest that restricted lagoon-open ocean couples are essential for the formation of extensive dolomites. This idea is not new, but a workable, theoretical basis for applying it has been lacking. Indeed, Fairbridge's (1957) excellent review and appraisal of the dolomite problem ranks evaporitic control with the least plausible of many discredited dolomitization hypotheses.

Geologists generally agree that most bedded dolomites were produced by post-depositional alteration of limestones. Disagreements arise over the time of alteration and the nature of the physiogeological and physiochemical processes involved. The main problems are: (1) supplying enough chemically active magnesium to complete the dolomitizing process; (2) transporting this magnesium into and distributing it through the host limestones; (3) providing a chemically favorable environment in which approximately half the calcium can be displaced by magnesium; (4) completing the replacement while active introduction of magnesium is in progress; and (5) removing the displaced calcium. A discussion of the chemical processes involved in displacing the calcium by magnesium is beyond the scope of this report. The other problems are essentially geological and yield to geological analyses.

In the more extensive dolomite formations, hundreds or even thousands of cubic miles of bedded limestone were dolomitized. Processes capable of dolomitizing these large masses are more critical than those effective on bodies of limited size. The method of dolomitizing an extensive limestone body after its consolidation and calcitization is difficult to explain. Theories of secondary replacement requiring diffusion of magnesium through solid rock sound plausible for small metasomatic deposits, but not for whole diagenetic environments.

---

[1] Manuscript received, May 19, 1960.
Published by permission of the Standard Oil Company of Texas.

[2] Geologists, Standard Oil Company of Texas.
The manuscript was read and corrections made by W. T. Holzer.

Comparing depositional environments characterized by dolomites with those dominated by non-dolomitic limestones provides an interesting approach. The most obvious difference is the presence or absence of associated lagoonal evaporites. It is readily apparent, where complete depositional sequences are preserved, that limestones closely associated with contemporaneous evaporite lagoons seldom escaped extensive dolomitization. Limestones lacking evaporitic associations at the time of deposition normally remain undolomitized even after hundreds of millions of years. The chemical, geometric, and hydrodynamic reasons for this relationship become apparent on analyzing the two enviromnents.

Modern Limestone Shelves

The intensively studied modern Florida shelf-Bahama Bank area typifies the environments where undolomitized limestones form. These shallow limestone shelves covering 100,000 square miles perch on steep-walled carbonate pedestals dating back to the mid-Mesozoic. Water depths on the limestone banks range to perhaps 1,000 feet, with 95 per cent of the bank floor above the 25-foot level. Discontinuous island chains and marginal reefs line the windward shelf edges. Tides, currents, and storms push great quantities of sea water onto and across the shelves. Waves and currents keep the water on the shelf mixed and almost unstratified and at low tide the excess escapes over the margins of the shallow bank. The shelf limestones are largely of organic origin (Epstein and Lowenstam, 1957). Living plants and animals carpet much of the shelf; other areas are covered by submerged limestone dunes and mud flats (Illing, 1954). Bottom sediments are relatively permeable, and tides flow through as well as over the floor.

The loosely cemented Florida-Bahama limestones possibly resulted from rapid upgrowth of banks and shelf surfaces, or from solution of exposed areas during the Pleistocene lowering of sea-level (Newell, 1955). Limestones in South Florida are only slightly less permeable than a sponge. Some of these porous and permeable limestones have been in contact with sea water for thousands of years. They show no evidences of dolomitization, although composed of metastable aragonite and organically deposited calcite. The only magnesium present is preserved as scattered molecules of $MgCO_3$ in the crystal lattice of the organic calcite. Apparently concentrations of magnesium in modern sea water are too low for direct precipitation of dolomite or active replacement of calcium. Restricted lagoons are lacking and losses due to evaporation are quickly replaced by sea-water inflow and by rainfall. The Recent Florida-Bahama limestones have never been saturated with hypersaline brines. Wells penetrating the pre-Pleistocene bank sections, however, encounter bedded evaporites and dolomites alternating with calcitic limestones. The current cycle of free circulation may eventually be restricted, and evaporites with dolomites may again be deposited.

Ancient Limestone Shelves
open marine environment

Depositional environments on wide-spreading ancient limestone shelves were somewhat more diversiform than those characterizing the Florida-Bahama banks. Many early limestone shelves were hundreds of miles wide and their floors were relatively flat (Fig. 1). Seaward slopes on these shelves commonly averaged less than an inch per mile. Permian deposits in the southwestern United States classically illustrate the wide sedimentary sequential range possible on shallow shelves. Exploration for Permian and older oil fields furnishes opportunities and incentives for studying environmental and facies variations in these Permian rocks on a three-dimensional front.

The early Permian seaway resembled many other limestone shelf-seas recorded in the stratigraphic column. It occupied a partly landlocked embayment extending into the continental interior more than 1,000 miles. An irregular trident of steep-walled bathyal troughs carried ocean waters to the interior of this intra-continental shelf-sea. Exuma Sound, Tongue of the Ocean, and similar channels perform like service for the Bahama Banks. Flat shallow shelves surrounded and separated the three deep Permian troughs. Steep slopes bordering these shelves plunged down to the nearby flat floors of the troughs. The shelves, slopes, and basin floors have been named undaform, clinoform, and fondoform segments, respectively (Rich, 1951). Shallow early Permian shelves were open, and fresh sea waters circulated freely across them. Irregularities on the bottoms, adverse winds, and fluid inertia undoubtedly combined to limit or check cross-shelf circulation. Wherever this happened, evaporitic controls

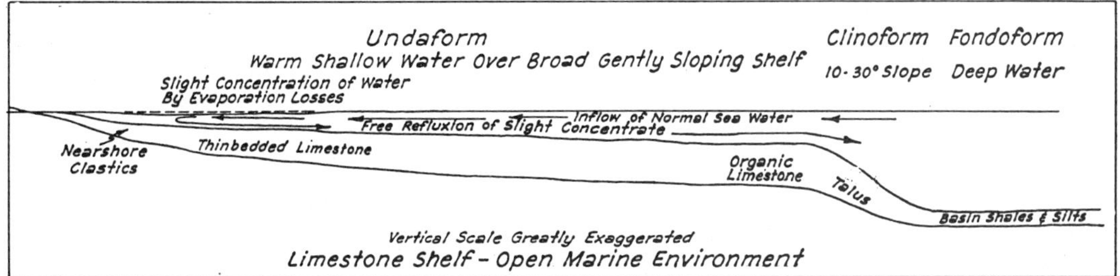

Fig. 1.—Diagram of open marine environment on limestone shelf.

directed movement of the waters.

Loss of water by evaporation lowered the water level on the shelf and increased the concentration and specific gravity of the remaining brine. This heavier brine sank and drained seaward down the sloping shelf. Seaward escape of dense brine is called reflux action (King, 1947). Water flowing in from the open ocean, replaced that lost by refluxion and by evaporation, and maintained circulation across the shallow shelves. With free refluxion, concentrations on even the broadest shelves seldom became great enough to inhibit organic activity. Increased temperatures, lack of food, and the accumulation of waste products were probably more important than increased salinity in limiting carbonate-secreting animal growth. Limestones on early epicontinental flats accumulated more slowly and were denser than those forming on the rapidly growing modern oceanic banks. Most ancient shelf carbonates were torn up and reworked time after time. During this reworking they were almost continuously bathed in sea water. The fact that they are limestones probably indicates that normal sea water has never been an effective dolomitizing agent. The thick Early Permian marine section indicates that a balance between rate of subsidence and rate of deposition prevented restriction and maintained free circulation throughout the Wolfcamp epoch.

RESTRICTED SHELF ENVIRONMENT

The onset of post-Wolfcamp tectonic stabilization slowed subsidence of the shelves to perhaps 0.007 foot per year. Cool, food- and mineral-charged sea water, pushed onto the shallow sunlit shelves, produced lush plant and animal growths. Aggressive shelf-edge organisms combed a disproportionate share of food and minerals from the inflowing currents. Shell- and skeleton-forming inhabitants in this favored marginal zone built

the shelf upward much faster than normal subsidence lowered it. Sediment-binding, agitation-resisting organisms congregated here and built reef mounds. When the shelf-edge buildups reached wave base the lateral extension of these mounds into continuous ridges resulted in shelf-edge barrier reefs. Once established, these barriers were largely self-perpetuating. Storm waves breaking against the reefwalls swept detritus into the protected backreef zones. The waves also piled rubble into the forereef deeps. This rubble built shallow foundations of talus across which the reefs forestepped seaward.

Eventually the barrier reef zones became wide and high enough to check free surface refluxion. Development of circulation-restricting barriers and elimination of free refluxion led to the formation of shelf-wide evaporite lagoons (Fig. 2). The appearance of dolomites in the Permian section practically coincided with the formation of these lagoons.

Several times during the Permian, sporadic increases in rate of subsidence allowed fresh sea water to flood the lagoons. During these open-shelf intermissions normal marine limestones spread shoreward across the lagoonal evaporites. In intervening periods of quiescence the reef barriers re-established themselves and lagoonal deposits buried the widespread limestone ledges. Alternating subsidences, stillstands, and even minor upwarps produced, overwhelmed, and recreated a series of shifting environmental provinces. The surprisingly uniform sedimentary profiles resulting from each recurrent fluctuation show that the environmental patterns developed were not fortuitous.

EVAPORITE LAGOON EVIRONMENT

In the Permian evaporite lagoons stream inflow and rainfall were insufficient to offset evaporation losses. Resulting deficiencies were balanced by

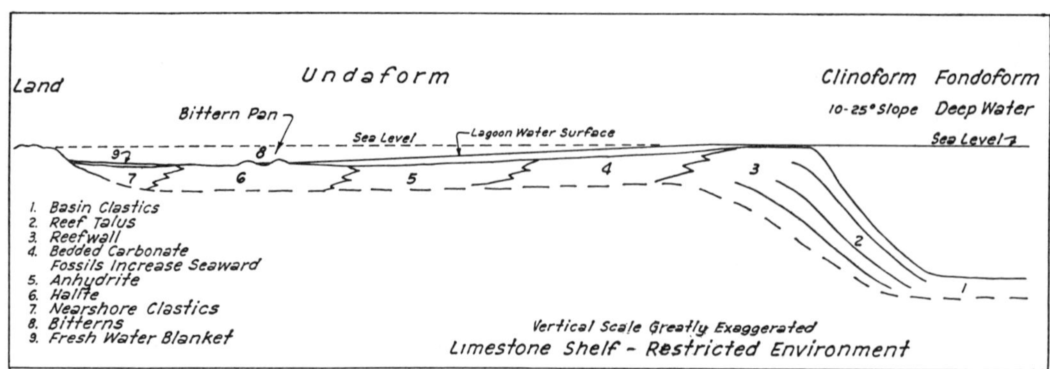

FIG. 2.—Diagram of restricted environment on limestone shelf.

ocean waters drawn inward across the barrier reefs. Influxing sea water flowed toward the depressions caused by evaporation. During this extended journey across the wide Permian lagoons the inflowing waters became more and more concentrated. Permian lagoonal geometry prevented this dense brine from refluxing prematurely. Lagoons 200 miles wide and 10 feet deep had floor slopes averaging about 1 foot in 20 miles. Hydrodynamic currents pushing down these almost imperceptible slopes were persistent rather than precipitous. With free reflux checked, concentrations continued increasing until the brines became dense enough to force a passage through the lagoon floor. Effective dolomitization apparently occurred as these hypersaline brines seeped downward through the underlying carbonates. For this reason a discussion of the origin and composition of the brines is essential.

Most sediments deposited in the shelf-wide Permian lagoons behind the circulation-restricting barriers were derived from sea water (Fig. 3). Modern sea waters are characterized by salinities of approximately 35,000 parts per million. For simplification this value is commonly expressed in parts per thousand and written 35°/oo. Having already undergone several billion years of pre-Permian development, the Permian ocean waters probably approached this same 35°/oo. Relative proportions of solutes making up the 35°/oo also probably resembled those present in modern oceans.

BRINE STRATIGRAPHY
REEF AND VITASALINE ENVIRONMENTS

Waters crossing the lagoonal barriers came from the upper oceanic layers where salinities, temperatures, and other factors favored organic activity. Rocks of the reef and near-backreef areas reflect this vitasalinity (Lang, 1937). Reef and vitasaline limestones include skeletons of framework-builders, shells and tests of mobile or attached forms, and bioclastic debris. Primary textures were relatively coarse. As temperatures and concentrations increased farther within the lagoons, habitability of the inflowing waters declined. Inner limits of the vitasaline zones are poorly defined biologically because different groups of organisms have widely varying toler-

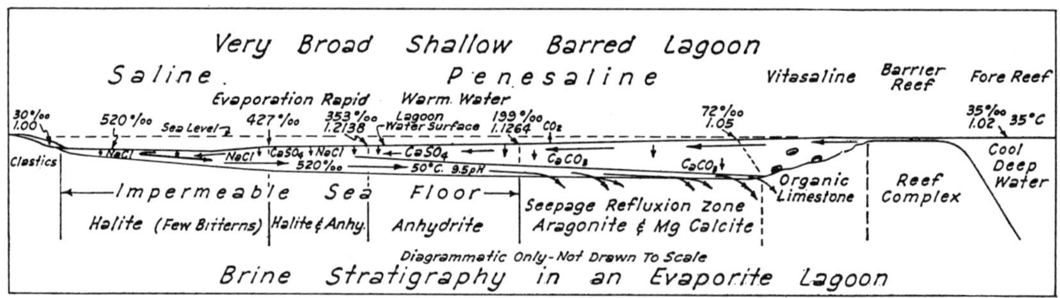

FIG. 3.—Idealized section of brine stratigraphy in evaporite lagoon.

ances to salt. Few forms, however, thrive and reproduce in salinities above 72°/oo, at which concentration chemical precipitation of $CaCO_3$ begins (Scruton, 1953, after Clarke, 1924). The intersection of the 72°/oo isosalinity contour and the lagoon surface thus supplies a practical boundary between the vitasaline and penesaline parts of the inflowing current. Hypersaline brines refluxing across the lagoon floor had concentrations far greater than those of the inflowing surface currents. Where the influxing water rested on hypersaline brine at the base of the lower backreef slope, bottom salinities increased from less than 72°/oo to more than 199°/oo within a few inches.

PENESALINE ENVIRONMENT

Chemically extracted, penesaline limestones precipitated from the inflowing waters until salinities of 199°/oo and densities of 1.126 were reached. A rise in temperature and increase in the concentration from 72°/oo to 199°/oo caused expulsion of carbon dioxide from the influxing sea water. This loss of CO_2 from solution led to the cessation of limestone deposition. With expulsion of CO_2, the pH of the brine probably rose to 9.0 or more. Continued increases in temperature and concentration in later evaporation stages prevented any resorption of CO_2 from the atmosphere.

Limestones of the outer penesaline zone contain tests and skeletons of swimming and floating organisms killed by increasing concentrations of the inflowing currents. These shells increased the textural range and permeability of the normally, fine-grained penesaline carbonates. Because of the high temperature, aragonite was the commonly precipitated carbonate mineral. Calcite, other than that in the shells and skeletons, was apparently magnesium-free. Presence of calcite laminae in the Castile anhydrite (Adams, 1944) indicates that some, and probably all, chemically precipitated carbonates were originally non-dolomitic.

After the practical disappearance of CO_2 and the concentration of brine to 199°/oo, precipitation of calcium sulphate began. Gypsum, rather than anhydrite, was deposited at temperatures below 34°C. (McDonald, 1953) but Permian brines were seldom that cool. The gypsum crystals that did form later dehydrated after settling into the dense brines refluxing across the lagoon floors (Conley and Bundy, 1958). Anhydrite precipitated until salinities of 427°/oo and den-

sities of 1.257 were reached. Deposition of anhydrite reduced the quantity of sulphate ions in solution and this decrease further raised the pH of the remaining brine. Salinity increases expelled free oxygen and rendered the brine essentially Eh neutral. In some Permian lagoons anhydrite was the highest-rank evaporite precipitated. In such lagoons brine densities never rose above 1.213, but in most lagoons more soluble salts were also deposited. Small amounts of magnesite were frequently precipitated with the anhydrites but most of the magnesium originally in the influxing water remained in solution. $CaCO_3$ introduced into brines with concentrations above 200°/oo might have combined with magnesium to form dolomite, but natural methods for introducing the $CaCO_3$ were lacking.

SALINE ENVIRONMENT

Halite, or rock salt, characterized evaporite deposits in the 353°/oo to 523°/oo salinity range. Brine density in the halite environment varied between 1.213 and 1.306, or roughly 18 to 28 per cent greater than that of normal sea water. Halite and anhydrite both precipitate through the range from 353°/oo to 427°/oo, but crystallize only at the brine-atmosphere interface. The anhydrite zone was thus limited by the intersections of the 199 °/oo to 427°/oo isosalinity contours with the surface of the lagoon. Halite was deposited beyond the 353°/oo isosalinity contour.

Halite deposition, however, was not limited to submerged areas in the Permian evaporite lagoons. Highly concentrated brines yielded large amounts of salt with only slight fluid losses. Excessive precipitation beneath the centers of evaporation raised the lagoon floors to sea-level. Brine films spreading over the exposed surfaces then built atoll-like salt ramparts rising well above the water level of the lagoon. In the locally arid climates hypersaline brines migrated over these ramparts and deposited their bittern loads of hygroscopic potash and magnesium within the desiccating pans. Bitterns were preserved only where deposited far enough from shore to escape solution during infrequent stream water flooding. This salt rampart explanation for bittern pans is separate from, and entirely unrelated to, the dolomitization hypothesis here advocated.

Bitterns, being the final products of complete evaporation, should include all the sea water solutes not previously deposited. Magnesium and potassium salts, deposited from a unit volume of sea water evaporated to dryness, should be several times as abundant as the anhydrite. Instead, the Permian bittern deposits are very minor compared with other evaporites. Bitterns are only one of many shorted evaporite clan members in the Permian lagoonal sequences. Sea water evaporated to dryness should yield 19 times as much halite as anhydrite and 14 times as much anhydrite as limestone (Delwig, 1955, p. 104). Instead of this ideal 1:14:266 ratio, the Permian average is closer to 1:8:40. It was this disproportionate lack of high rank evaporites that led King (1947) to develop the refluxion theory.

DOLOMITIZATION BY SEEPAGE REFLUXION

The scarcity of high-rank salts in the Permian evaporite deposits indicated that highly concentrated brines escaped before complete desiccation. Refluxion was a simple and generally effective process. The heaviest brines migrated to the lowest closed depressions on the lagoon floor. These heavy brines were covered by a blanket of lighter water which protected them from further desiccation. As a result they accumulated, filled the depressions, and flowed on, seeking a natural spillway. Those parts of the floor of the lagoon which were covered by tightly recrystallized anhydrite or halite were effectively impermeable. The slightly greater permeability of the penesaline carbonates provided the spillway for the heavy brines. Pores in these chemically precipitated limestones were of capillary dimensions. Connate water in these pores prevented the entrance of light fluids. Even heavy brines seeped so slowly through them that the lagoonal circulation was uninterruptedly maintained. Slight variations in rates of seepage refluxion determined the maximum concentrations attainable in the lagoons.

The highly concentrated brines were chemically potent solutions. They were hot, with temperatures of 35°C. or above; highly alkaline, with a pH of 9.0 or higher; and had nearly neutral redox potentials. The carbon dioxide, calcium, oxygen, and sulphide contents were abnormally low. Concentrations of magnesium, potassium, sodium, bromine, fluorine, and some other elements were excessively high. The Mg-Ca ratio was many times greater than that in normal sea water. While the evaporite lagoon persisted, the supply of this chemically unbalanced brine was assured.

The refluxing hypersaline brine seeping through the lagoon floor was much denser than the connate water it displaced. The brine, therefore, tended to seep directly downward and only followed bedding planes when vertical migration paths were unavailable. Rocks through which the brine seeped were composed of metastable aragonite and high-magnesian calcite. These chemically and physically unstable carbonates were readily dolomitized by the potent hypersaline brines. In rocks with varying permeabilities, the seeping brines migrated mainly through the more porous zones and bypassed the denser limestone lenses. Permeability, though low in the penesaline carbonates, was uniformly effective; therefore, these sections are almost completely dolomitized. In vitasaline carbonate sections, however, relatively permeable dolomites are commonly interbedded with unaltered lithographic limestones. Coarse-grained porous Permian dolomites are limited to beds previously composed of coarse-grained porous limestones. Fine-grained dense dolomites occupy those zones in which lithographic limestones normally form on open shelves. Dolomite textures, thus, are caused by primary permeability and crystallinity rather than by dolomitization.

Constantly renewed supplies of brine continued to pour through the Permian carbonates for appreciable spans of geologic time. Spillways through which the brines entered the rock column shifted seaward as the shelves regressed. The lagoonal sedimentary profile followed the progressively forestepping reefwall. Old escape zones, overlapped and sealed off by the advancing evaporites, were replaced by similar outlets farther seaward. With each forestep, previously uninvaded vitasaline and reefwall limestones were exposed to the dolomitizing brines. Regressions were normally so slow that all limestones susceptible to alteration were dolomitized before the supply of brine was cut off.

Changes in the composition and character of the refluxing brine and of the invaded rocks limited the downward extent of dolomitization. Magnesium losses and calcium additions, during dolomitization, reduced the overwhelming Mg-Ca ratio of the primary brine. Loss of heat to the cooler limestones and to the displaced connate waters lowered the temperature of the refluxing brine. Carbon dioxide from the displaced connate waters tended to neutralize the alkalinity of the brine. Calcitization of the thick foreset talus was largely completed before brine invasion. Aragonites in these heavily loaded zones which escaped early calcitization were stabilized by increasing overburden pressures (Jamieson, 1953). Dolomitization of calcite and stabilized aragonite has not been recognized in Permian rocks.

REFLUXION FOOTPRINTS

If the hypersaline brines followed the paths through the Permian limestones here suggested, they should have left footprints behind them. Even a cursory examination of the Permian carbonates associated with evaporite lagoons shows many of these footprints. Among these are inclusions of anhydrite that are hundreds or thousands of feet vertically and miles horizontally from deposits of bedded anhydrite (Adams, 1932). These inclusions fill worm borings, fractures, fossil casts, solution cavities, and primary interstitial openings. Many shells are completely replaced. Others, especially the fusulinids, retain parts of the original wall structures suspended in the anhydrite. Benthonic animals whose shells are replaced by anhydrite, and worms that dug the anhydrite-filled borings did not live in an anhydrite-depositing environment. The anhydrite is clearly secondary. The increasing content of anhydrite, as the evaporite boundary is approached, points to the source. The decrease in pressure resulting from the escape of hypersaline brines through capillary passages into open cavities probably explains the deposition of the inclusions. Loss of heat may have aided the process. The inclusions of relatively insoluble anhydrite are preserved in cuttings and cores from wells, but have been dissolved from outcrops. The total volume of the inclusions within the Permian dolomites may well run into cubic miles.

The second set of footprints is similar to the first. Some dolomite cores cut with oil-base muds or highly saturated brines, instead of fresh water, show inclusions of halite with the anhydrite. The halite is washed out of most cuttings and its presence was unsuspected in the early drilling and in examination of outcrops. Halite inclusions normally take the shape of the vugs, fossil casts, and other openings that they occupy. Increased oil production, obtained when porous reservoir dolomites are washed with fresh water, suggests an abundance of crystalline halite in these rocks. Inclusions of anhydrite could form from brines with concentrations of $200°/oo$, but crystalline halite presumably required concentrations above $353°/oo$. The escape of concentrated saline brines demonstrated by these inclusions explains the common lack of high rank evaporites in many Permian lagoons. Anhydrite and halite inclusions within many Permian dolomite reservoirs are regarded as proof that emplacement occurred before entrapment of oil. The abundance of hypersaline brines trapped in pre-evaporite and evaporite-equivalent carbonates furnishes a third set of footprints. Salinities of these brines range up to $360°/oo$ or slightly above the lower limits for precipitation of halite. Still higher concentrations probably exist in favorably located traps. Salinities between $199°/oo$ and $300°/oo$ are common in off-structure wells. These brines trapped high in the Permian section thus have concentrations similar to those in evaporite lagoons.

The thick sections of sub-lagoon dolomites supply a fourth group of seepage refluxion footprints. Adequate magnesium for dolomitizing these 100- or 1,000-cubic-mile masses was available only in the evaporite lagoons. Heavy brines seeping downward through unstable carbonates were readily capable of emplacing the amounts of magnesium required within the time available. In their downward passage, they carried the displaced calcium with them. Part of this calcium was redeposited as vein fillings and other secondary growths. The remainder, still in solution, gives the trapped hypersaline brines an otherwise unexplainable non-lagoonal character. Dolomitization, as evidence of hypersaline brine refluxion, gains significance when supported by inclusions of halite and anhydrite, and by trapped supersaline brines. These four footprints and associated data furnish the basis for this approach to dolomitization.

STRATAL GEOMETRY OF DOLOMITIZATION

The Permian dolomites (Fig. 4) were formed on and around shallow inundated shelves whose interiors were occupied by restricted evaporite lagoons. These deposits developed during the regressive phase of a major sedimentary cycle. Aggressive organic activity along the edges of the shelves maintained the crests of the barrier reefs at or near sea-level. Talus, broken from the reef crests by wave action, built shallow forereef platforms across which the reefwalls progressively forestepped during periods of slow subsidence of

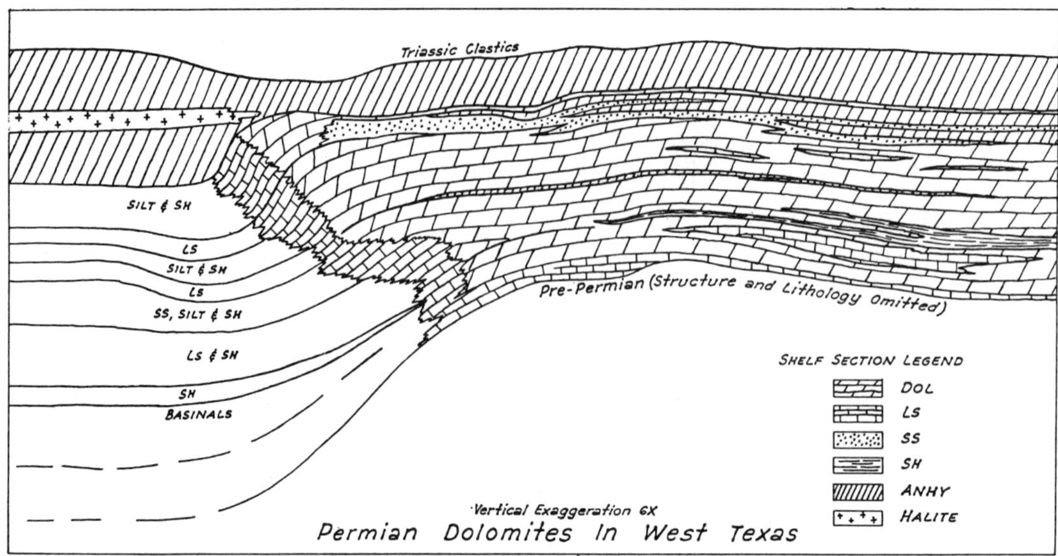

FIG. 4.—Basin-shelf cross section showing generalized lithologic and structural relations of Permian dolomites and associated sediments.

the basin. Concurrent subsidence, upbuilding, and forestepping allowed thick evaporite sections to accumulate in the continuously shallow lagoons. During the migration of environmental provinces seaward, evaporites overlapped penesaline carbonates, which in turn submerged reefwall deposits while new reefwalls formed over the freshly deposited talus. The refluxing brines recrystallized large parts of these topographically lower sections to produce the extensive dolomite complex. Carbonates deposited above refluxing brine levels remain undolomitized unless subsequently invaded by brines from higher lagoons.

The geological settings for other dolomites are more diversified than those associated with the reef-bordered shelf-wide Permian Basin lagoons. Dolomites apparently form in both the transgressive and regressive phases of sedimentary cycles. Environmental variations, however, are largely differences in proportion rather than in kind. Restrictions of circulation are indicated for most dolomites. Dolomite-linked evaporites may form behind longshore bars, within atoll rings, on alternately flooded and emergent mud flats, and in many other environments. Truncation in many places removed the bedded evaporites, leaving inclusions of anhydrite as the only evidence of their former existence.

Ordovician, Devonian, and Cretaceous dolomites of Texas which have varying amounts of evaporites associated with them show evidence of deposition on partially restricted shelves. Similar evaporite-dolomite-limestone sequential relationships are known or suspected for deposits of dolomite the world over.

SUMMARY

Dolomitization was a normal process in carbonate rocks on barred shelves. The reef barrier prevented free refluxion of waters from the lagoon. Following this restriction of circulation the salinity of the brines increased and evaporites formed. The hypersaline brines eventually became heavy enough to displace the connate waters and seep slowly downward through the slightly permeable carbonates of the lagoon floor. During this seepage, the magnesium replaced part of the calcium and the metastable aragonites and high magnesian calcites recrystallized as dolomites. Chemical and physical changes, in the brines and in the rocks through which they flowed, limited the extent of dolomitization.

REFERENCES

ADAMS, J. E., 1932, "Anhydrite and Associated Inclusions in the Permian Limestones of West Texas," *Jour. Geol.*, Vol. 40, pp. 30–45.
————, 1944, "Upper Permian Ochoa Series of Delaware Basin, West Texas and Southeastern New

Mexico," *Bull. Amer. Assoc. Petrol. Geol.*, Vol. 29, pp. 1596–1625.

CLARKE, F. W., 1924, "Data of Geochemistry," 5 ed., *U. S. Geol. Survey Bull. 770.*

CONLEY, R. F., AND BUNDY, W. M., 1958, "Mechanism of Gypsification," *Geoch. Cosmoch. Acta.*, Vol. 15, pp. 57–72.

DELLWIG, L. F., 1955, "Origin of the Salina Salt of Michigan," *Jour. Sed. Petrology*, Vol. 25, pp. 83–110.

EPSTEIN, S., AND LOWENSTAM, H. A., 1957, "On the Origin of Sedimentary Aragonite Needles of the Great Bahama Banks," *Jour. Geol.*, Vol. 65, pp. 364–75.

FAIRBRIDGE, R. W., 1957, "The Dolomite Problem," in "Regional Aspects of Carbonate Deposition," *Soc. Econ. Paleon. and Mineral. Spec. Pub. 5*, pp. 125–78.

ILLING, L. V., 1954, "Bahaman Calcareous Sands," *Bull. Amer. Assoc. Petrol. Geol.*, Vol. 38, pp. 1–95.

JAMIESON, J. C., 1953, "Phase Equilibrium in the System Calcite-Aragonite," *Jour. Chem. Phys.*, Vol. 21, pp. 1385–90.

KING, R. H., 1947, "Sedimentation in Permian Castile Sea," *Bull. Amer. Assoc. Petrol. Geol.*, Vol. 31, pp. 470–77.

LANG, W. B., 1937, "The Permian Formations of the Pecos Valley of New Mexico and Texas," *ibid.*, Vol. 21, pp. 833–98.

McDONALD, G., 1953, "Anhydrite and Gypsum Relations," *Amer. Jour. Sci.*, Vol. 252, p. 884.

NEWELL, N. D., 1955, "Bahamian Platforms," in "Crust of the Earth," *Geol. Soc. America Spec. Paper 62*, pp. 303–19.

RICH, J. L., 1951, "Three Critical Environments of Deposition and Criteria for Recognition of Rocks Deposited in Each of Them," *Bull. Geol. Soc. America*, Vol. 62, pp. 1–20.

SCRUTON, P. S., 1953, "Deposition of Evaporites," *Bull. Amer. Assoc. Petrol. Geol.*, Vol. 37, pp. 2498–2512.

13

Reprinted from *Sedimentology* **12**:11–25 (1969)

PRELIMINARY EXPERIMENTS ON HYDRODYNAMIC MOVEMENT INDUCED BY EVAPORATION AND THEIR BEARING ON THE DOLOMITE PROBLEM

K. JINGHWA HSÜ AND CHRISTOPH SIEGENTHALER

Swiss Federal Institute of Technology, Zürich (Switzerland)

(Received March 6, 1969)

SUMMARY

Experiments were performed to demonstrate that saline interstitial waters under an arid coastal plain could be set in hydrodynamic movement to replace the evaporative loss near the surface.

In contrast to movement in a vadose zone, which is related to capillary forces caused by surface tension, the type of movement we observed was induced by a vertical hydraulic gradient under the evaporated area. Fluid flow through porous media was induced by an upward decrease in hydrodynamic potential during evaporation. We propose to call this type of movement *evaporative pumping*.

Experiments further verified that flow rate induced by evaporative pumping was directly governed by evaporation alone; linear flow rate through coarse sand and that through very fine silt was approximately the same under the same evaporative condition. Yet the movement apparently obeyed Darcy's law. Since the permeability of the medium was fixed, and the flow rate was dictated by evaporation, only the hydraulic gradient could be the dependent variable in the Darcy equation. We observed that the gradient was indeed much greater in relative impermeable silt than in permeable sand, when the water to replace evaporative loss must flow with the same rate through those media.

Our experiments suggested that hydrodynamic movement induced by evaporation could be an effective mechanism to transport magnesium-bearing solutions through relatively impermeable sediments of an arid coastal plain. Computations showed that dolomitization by evaporative pumping could proceed at a rate to account for the origin of Recent dolomite crusts, and for the great thickness of ancient supratidal dolomites. In contrast, we believed that dolomitization by seepage reflux or by groundwaters cannot be extensive, because of the inadequate magnesium-supply rates by such waters.

THE DOLOMITE PROBLEM

Recent carbonate sediments consist almost entirely of calcite and aragonite, whereas ancient carbonate rocks include abundant dolomites. This striking contrast represents a great puzzle to geologists. Different ideas have been advocated to explain the origin of dolomites, resulting in a controversy, also known as the dolomite problem. Dolomites have been formed in the past in different environments at different times, and different theories have been invoked to explain their various origins. Yet certain chemical conditions must have been fulfilled and a certain hydrological framework must have prevailed when large bodies of sedimentary dolomites were formed.

In a previous article the senior author analyzed the chemistry of dolomite formation (HsÜ, 1967). The conclusions that the solubility of dolomite under atmospheric conditions is of the order of 10^{-17} and that the Mg/Ca activity ratio of dolomitizing solution need not be particularly high are being repeatedly confirmed by recent investigations (e.g., LANGMUIR, 1965; BERNER, 1967; USDOWSKI, 1967). That dolomite is not being precipitated from normal marine conditions may be explained in terms of chemical kinetics (BERNER, 1966). Yet the same lack of equilibration does not seem to exist between a subsurface water and its enclosing carbonate phases. The geochemistry of Florida groundwaters indicates that they are more or less equilibrated (HsÜ, 1963; HANSHAW et al., 1965; BECK et al., 1966). Pore fluids of Recent dolomites are apparently also equilibrated so that their Mg/Ca-ratio is about the same as that of normal sea water (ILLING et al., 1965, p.101). Chemical considerations suggest that an interstitial flow of natural waters with Mg/Ca-ratio > 1 would probably induce dolomitization (HsÜ, 1967). The problem of subsurface dolomitization is thus less a chemical paradox, but more a question of magnesium-supply rate.

A treatment of the hydrology of possible dolomitization in humid temperate regions was attempted by the senior author. He found that deeper groundwaters do not flow fast enough, and shallower groundwaters do not contain enough magnesium for extensive dolomitization (HsÜ, 1966, p.137).

ADAMS and RHODES (1960) postulated a mechanism of dolomitization by seepage reflux. The idea was attractive because the concentration of a supersaline brine in a lagoon seemingly fulfills both the chemical and the hydrological requirements of dolomitization: the Mg/Ca-ratio of the brines is increased through the precipitation of a calcium sulphate phase, and the density of the Mg-rich brine may provide the hydrodynamic potential for its return to the oceans. The theory received considerable support after DEFFEYES et al. (1964; 1965) found Recent dolomites on the side of the Pekelmeer, a supersaline lagoon on Bonaire, Netherlands Antilles. They further performed a demonstrative experiment to illustrate the possibility of reflux under a density head.

The simplicity of the reflux idea is persuasive, the circumstantial field evidence seemed convincing, and the demonstrative experiments seductive. The seepage reflux theory of dolomitization is thus becoming very popular. Various indications that Recent and ancient dolomites were formed under arid conditions were cited or implied as supporting evidence in favor of the theory (BERNER, 1965; COTTER, 1966; MÜLLER and TIETZ, 1966; ROEHL, 1967; KINSMAN, 1968).

Actually, the same difficulty of having to pump sufficient magnesium-rich solution through the sediments also plagues the reflux theory: the slight difference in density between a supersaline brine and normal seawater is not an effective hydraulic head to drive the brine through fine-grained lime sediments back to the oceans. For effective refluxing on the Pekelmeer, DEFFEYES et al. (1964; 1965) estimated that the permeability along the path of refluxing must be 0.24 darcies, which is considerably higher than that of an average lime mud. Deposition of anhydrite or gypsum cement by brines would further reduce the permeability of the refluxing path. To test the reflux theory, LUCIA (1967, p.852) made a boring through the sediments under the Pekelmeer: *he found no indication of seepage reflux!*

Our dilemma consisted in our ignorance of the hydrodynamics of an arid coastal plain. What is the driving force for the movement of the interstitial water and what are the factors controlling its rate? After puzzling over the problem for ten years, it finally occurred to us that the supratidal flat is not necessarily the *source*, but may actually be a *sink* of a hydrodynamic system!

Our experiments have been designed to simulate the conditions of an arid coastal plain, where the relief on land is negligible, and where evaporation results in a significant loss of pore fluids. Our results led us out of the magnesium-supply dilemma, and led us to the idea that supratidal dolomites could be formed consequent to movements of saline waters induced by evaporative pumping.

DEMONSTRATIVE EXPERIMENT OF EVAPORATIVE PUMPING

Evaporative pumping experiment

Our initial experiment was designed to demonstrate that seawater has been drawn to replace the loss of pore fluids in tidal-flat sediments due to evaporation.

A tank with glass walls, and a $100 \times 60 \times 25$ cm^3 dimension was constructed. The tank was half filled with sand, which has a median diameter of 0.3 mm and a permeability of 70 darcies. The sand surface was nearly flat on the right side in the middle, but sloped 10° leftward on the left side of the tank. The depression was covered by a 3% NaCl solution, simulating a shelf sea on the side of a tidal flat (Fig.1). The solution was colored violet by KMnO$_4$. The simulated "sea-level" was kept constant by a regulator, which was linked to a salt-solution reservoir. The regulator consisted of a float and a valve, which would open automatically when the "sea-level" should drop 1 mm (consequent to seepage loss), so that salt solution could flow from the reservoir through the regulator into the "shelf sea" to re-establish the "sea-level" to a pre-set height. The reservoir was graduated so that the amount of daily flow from the reservoir could be determined. The right side of the sand tank was connected to a second regulator, which was linked to a fresh-water reservoir. A 250-W lamp was positioned 10 cm above the middle of the sand flat to provide evaporation energy. The lamp was enclosed by a box with perforated walls, 35 cm long, so that evaporation could be concentrated within the fenced-in area.

At the start of the experiment the pores of the sand were filled with the fresh water, the model "shelf sea" was covered by the salt solution, and the water levels at both regulators were set equal, slightly below that of the sand flat surface. The lamp was turned on for the duration of the experiment.

This initial condition was unstable. Salt water began to seep beneath the simulated "shelf" because of the density head (Fig.1, top). The movement was further enhanced by the retreat of the fresh-water front caused by the evaporative loss of the fresh water under the lamp. On the fifth day more than half of the sand body was filled by salt water (Fig.1, middle). Salt water continued then to advance in a broad front upward. On the ninth day, salt water occupied the pore space of the sand in the area under evaporation. Only a small fresh-water lense was present at the upper right corner of the tank. Meanwhile, the height level at the fresh-water regulator was raised to about 1 cm above that at the salt-water regulator, so that the height head of the fresh water balanced the density head of the salt water. At this stage, hydrodynamic movement of the salt water would cease when the lamp was turned off. However, salt-water movement continued when the lamp was on, so that the evaporative loss could be supplied by salt water from the reservoir through the regulator. On the thirteenth day, a slight amount of blue ink was injected into the moving salt water zone in order to produce flow lines (Fig.1, bottom). Flow of salt water continued during the next ten days at a steady rate of 2,000 cm³/day from the reservoir to replace the evaporative loss. However, the hydrodynamic movement of the saline water ceased after the lamp was turned off at the end of the experiment.

The experiment indicates that *evaporation provides the energy for the movement of subsurface waters* in an arid coastal region underlain by sediments with saline interstitial waters. Water loss due to evaporation from the sediments near the surface draws replacement from deeper zones, which in turn would be recharged by saline water from other sources, which, in our experiment, consisted of seepage

---

Fig.1. Hydrodynamic movement induced by evaporative pumping. The experiment began first on May 3, when the basal part of the sand body was invaded by salt water directly introduced from the reservoir. A new start was made on May 6, when a simulated shelf sea, the source of salt-water invasion, was used. At the beginning, the salt water seeped down under a gravity head (top). The wedge spread laterally until May 10 (middle). Afterwards, the salt–fresh water interface moved mainly upward. The bottom picture shows flow lines after ink was injected into the moving salt-water body. The rate of movement was governed solely by the evaporative loss, which remained constant at 2,000 cm³/day.

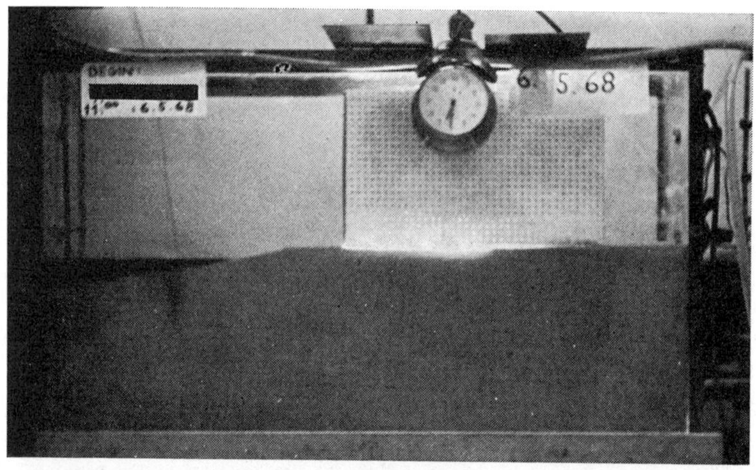

from a simulated "shelf sea". The role of evaporation might thus be comparable to that of a pump, inducing hydrodynamic movement of an otherwise hydrostatic salt-water body!

Gravity drive experiment

During a second experiment, the water level at the fresh-water regulator was raised about 2–4 cm above that at the salt-water regulator, in order to simulate the condition under a tidal flat where, or when, a dynamic fresh-water exists. With the increased fresh-water head at right, the fresh–salt water contact moved leftward until a dynamic equilibrium was re-established. Already on the second day, the area under evaporation was underlain in part by the enlarged fresh-water lense, and the evaporation loss had begun to be replaced by flow from the fresh-water reservoir. The steady state was established on the third day, thereafter the replacement for evaporative loss of 2,000 cm³/day was drawn completely from the fresh-water reservoir.

The energy for the fresh-water movement during this experiment was provided also mainly by evaporation, as long as the lamp was on. After the lamp was turned off, the maintenance of a slight height-head was sufficient to induce a gravity-driven movement to drive the salt water back to the "simulated sea". This phenomenon is, by the way, well known to hydrologists working against salt-water invasion under a humid coastal region.

Geological interpretations

Our experiments demonstrate that *an arid tidal flat under evaporation could be the sink, rather than the source of subsurface water movement*, and that Mg^{2+}-bearing solution for supratidal dolomitization may be drawn from the sea or from deeper ground waters to replace the evaporated pore fluids in tidal-flat sediments. Since evaporation acts like a pump to induce a hydrodynamic subsurface movement from a sea-water source, we suggest the term *dolomitization by evaporative pumping* as an alternative mechanism for dolomitization by seepage reflux. The postulated directions of movement are exactly opposite to each other.

We recognize that the end result of dynamic salt-water flow due to evaporative pumping is the deposition of an evaporite crust on the tidal flat, as we have observed on our model. In fact, the origin of some Recent and ancient supratidal evaporites may be related to this mechanism, rather than to deposition from a body of standing lagoonal water. NaCl and other soluble salts should also be present as residues of evaporation. ILLING et al. (1965) reported salt crust on Persian Gulf sabkhas although the amount present is not commensurate to account for the evaporative losses. This fact may be explained by the combination of several mechanisms: (*1*) high spring tides, which bring in normal sea water, undersaturated with respect to gypsum and other salts, and which can remove part or whole of the evaporative residue from a supratidal flat; (*2*) strong winds, experienced by workers on a sabkha, that effectively erode and remove the salt crust (ILLING et al., 1965, p.95); (*3*) dissolution by dynamic fresh water after a rainfall.

The second series of experiments demonstrates that dynamic fresh water may be present under a tidal flat in a humid region, or in an arid region after a heavy rainfall. Crusts of dominantly calcitic composition, e.g., those on Andros Island in part (SHINN et al., 1965, p.121), may have been formed during the times of fresh-water flows. We envision, therefore, that the interlaminated supratidal dolomite and limestone may signify alternating dry and wet conditions, as those rocks may have resulted from alternating hydrodynamic movements by sea and fresh waters to replace the evaporative loss on a tidal flat.

That a dynamic fresh-water wedge may exist temporarily after a heavy rain under an arid coastal plain may be a cause of the wide-spread submarine cementation reported by various authors in this issue (SHINN, 1969; TAYLOR and ILLING, 1969; DE GROOT, 1969; ALEXANDERSSON, 1969). Under such circumstances and when evaporation was reduced, the interstitial saline waters, originally underlying a coastal plain, must be driven partly back to the sea, while the boundary between the fresh and salt waters moved seaward (as dictated by a simple consideration of the conservation of matter). Interstitial saline waters saturated with $CaCO_3$ at pH $\leq$ 7 (ILLING et al., 1965, p.101) should become supersaturated under normal marine conditions (pH $>$ 8), leading to submarine cementation at a sediment–water interface where such waters seeped out.

CRITICAL EXPERIMENTS ON RATES OF EVAPORATIVE PUMPING

Experiments on flow rates

During our demonstrative experiments, we already noted that the flow rate by evaporative pumping seemed to depend upon the evaporation rate: the salt water remained hydrostatic when there was negligible evaporation, but moved actively to replace evaporative loss when the lamp was turned on. We proceeded, therefore, to settle by experiments the question, if the rate of movement by evaporative pumping depends also upon the permeability of sediments.

Eleven glass cylinders, with a cross-sectional area of about 40 cm², were used: ten were filled with sediments, ranging in size from very coarse sand to very fine silt, and an eleventh (*H*) was filled with water (Table I). Each cylinder had an inlet near its base which was fed from a level regulator. The height of the water level within the various regulators is given in Table I. Minus values denote levels beneath the sand surface and plus values denote those above. The water level within the sediment column in all cases was the same as that of the level in the regulator before evaporation started, and when the pore water was static. After evaporation was started, the levels remained at about the same height in cylinders filled with coarse or medium sand, but apparently dropped in cylinders

TABLE I

RESULTS OF CRITICAL EXPERIMENTS ON RATE OF EVAPORATIVE PUMPING

Cylinders:	A	B	C	D	E	F	G	H	I	J	K
Lamp power (W)	60	60	60	60	60	60	60	60	60	15	15
Lamp height (cm)	2	2	2	2	2	2	2	2	1	3.5	3.5
Sediment type	very coarse sand	very coarse sand	medium sand	medium sand	very fine silt	very fine silt	medium sand	water only	medium sand	medium sand	medium sand
Sediment size (mm)	1.2	1.2	0.3	0.3	0.01	0.01	0.3	—	0.3	0.3	0.3
Sediment height (cm)	20	20	20	19	20	20	20	—	20	19	19
Water level at regulator (cm)	—10	—2	—10	—2	—2	—10	+1	0	+1	—10	—2
Permeability (millidarcies)	$8.1 \cdot 10^5$	$8.1 \cdot 10^5$	$7.3 \cdot 10^4$	$6.9 \cdot 10^4$	15.4	15.4	$7.3 \cdot 10^4$		$7.3 \cdot 10^4$	$7.3 \cdot 10^4$	$7.3 \cdot 10^4$
Evaporation rate (cm^3/cm^2/day)	1.21	2.48	2.54	2.24	1.94	1.94	2.03	3.48	4.11	0.236	0.236
Linear flow rate (cm/day)	3.0	6.2	6.3	5.6	4.8	4.8	5.1	3.5	10.3	0.59	0.59
Evaporation condition (relative)	strong	strong	strong	strong	strong	strong	strong	strong	very strong	weak	weak

E and *F*, which were filled with silt. The levels in *E* and *F* could not be exactly determined in these series of experiments; colored tracers suggested that the pores were at least partially filled with water up to the original water level. The levels of water within the cylinders *G* and *I* remained at 1 cm above the sand surface, and were apparently the same as those in the regulator.

The regulators were kept at constant level during the experiments. Each was linked to a fresh-water reservoir, so that the evaporation rate could be measured. Lamps with a 60- or a 15-W bulb were used, positioned at various heights to investigate the dependence of the flow rates upon the conditions of evaporation. A 60-W lamp was positioned 2 cm above the sediment (or water) surface in the cylinders *A–H*, to induce the relatively strong evaporation. The 60-W lamp in the cylinder *I*, however, was positioned only 1 cm above the surface to induce a very strong evaporation. A lamp of 15 W was placed 3.5 cm above the surface in the cylinders *J* and *K* to induce the relatively weak evaporation. The arrangement of a cylinder under evaporation, connected to a level regulator and a water reservoir, is shown by Fig.2. Considering the experimental data of Table I, the following conclusions can be drawn:

(*1*) Except for the cylinder *A*, the flow rates through sediments of widely different sizes and permeabilities under the same evaporation condition (relatively strong) were approximately the same, ranging from 1.9 to 2.54 cm³/day/cm²

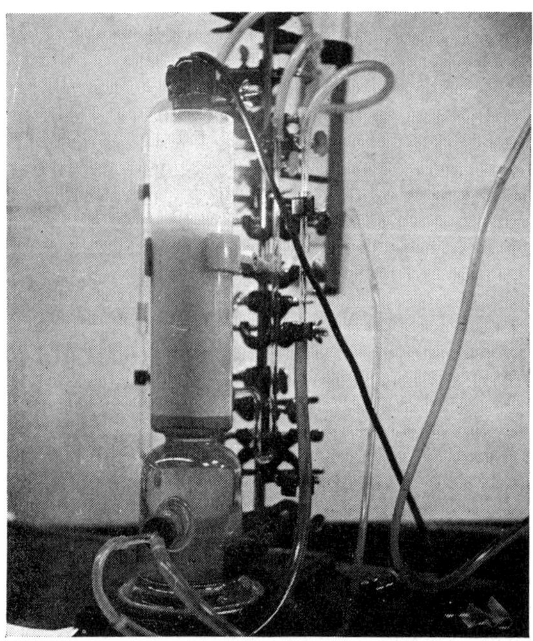

Fig.2. Apparatus used in measuring flow rate under evaporative pumping.

sediment-surface. The rate through the very coarse sand in the cylinder B was only about 20% faster than that through the very fine silt, although the permeability of the sand was 50,000 times greater! The relatively slow evaporative rate of the pore water in the coarse sand of the cylinder A is not known; perhaps the relatively thick (10 cm) dry sand above the "water table" served as an insulation to retard evaporation.

(2) Except for the cylinder A, the flow rate through sediments with different "water tables" under the same evaporation condition was approximately the same. The evaporative rate for the cylinder G, in which a medium sand was covered by 1 cm water, was just about the same as the average evaporative loss for cylinders with "water tables" beneath the sand. The volumetric evaporation rate of pure water in cylinder H was 75% higher than the average rate of water loss through the sediments. The difference between that of cylinders G and H was surprising, until we noticed that the surface of the water layer in cylinder G was covered by a thin film of sediment particles, floating as a result of the surface tension of the water layer. Consequently, the surface area of fluid under evaporation in G was comparable to that of other sand cylinders, in contrast to that of cylinder H.

Table I also shows that the linear flow rates through the porous sediments were apparently greater than the linear evaporative rate measured over a pure water body. This only reflects the fact that the loose sediments had a porosity of about 40%, so that a 2 $cm^3/cm^2/day$ water-loss must be replaced by a linear flow rate of 5 cm/day.

(3) The flow rate through sediments of the same permeability under varying evaporation conditions was greatly different, ranging from 4.11 to 0.236 $cm^3/day/cm^2$ sediment-surface.

Our experiments revealed the surprising fact that the flow rate of pore water through sediments by the evaporative pumping mechanism was apparently almost independent of permeability. The single decisive factor seemed to have been the conditions of evaporation, which in nature would be related to temperature, wind velocity, air humidity, etc.

Questions were raised as to the physical nature of the evaporative pumping movement. Is it to be compared to capillary movement in a vadose zone? Or is it a form of Darcy flow? Is Darcy's law violated as the flow rate was apparently independent of permeability?

The movement of water through cylinder G is certainly not to be identified as capillary movement through a vadose zone: the sand body was completely saturated and was covered under 1 cm of water. Using the Darcy equation (HUBBERT, 1940):

$$q = \frac{k}{\mu}\rho g(dh/dl)$$

where q, the volume of water flowing through the sediment column per cross-sectional area per sec, $= 2.2 \cdot 10^{-5}$ cm/sec; k, permeability in c.g.s.-unit $= 7.3 \cdot 10^{-7}$ cm²; ρ, density of water, $= 1$ g/cm³; g, gravitational acceleration $= 9.8 \cdot 10^{-2}$ cm²/sec; μ, viscosity of water, is about 1 centipoise, or 10^{-2} g/cm/sec; and dh/dl is the hydraulic gradient necessary to induce the darcy flow at the observed rate, we could compute, since the height of the sand column is 20 cm, the difference in manometer height, Δh, between the water level at the regulator and that within the cylinder G. We found Δh to be $6.5 \cdot 10^{-3}$ cm. In other words, the measured flow rate through the sand could be induced when the level within the cylinder E was kept by evaporative loss at a height of 0.0065 cm below that in the regulator. This difference was too small to be visible.

Similarly, our computations show that Δh for cylinders with water-filled coarse or medium sands would also be too small to be observed. Only for cylinders E and F could there be a significant difference. We are, therefore, carrying out another series of experiments to measure this difference and to test if the flow rate by evaporative pumping is consistent with that predicted by the Darcy equation. Our results so far gave semi-quantitative confirmation, but we could not complete our experiments before the publication of this special issue. Nevertheless we gained an insight into the evaporative pumping movement by the following experiment.

We used a cylinder set up like cylinder E, except three manometers were connected to the silt column in this new cylinder E', one at the bottom, one in the middle, and one near the top. Also the water level at the regulator was exactly even with the sediment-surface within the cylinder E'. Before the lamp was turned on, the silt column was filled with water till its very top. There was no vadose zone, nor were there any hydrodynamics. Then the lamp was turned on. The manometer connected to the bottom of the column remained stationary. Meanwhile, the level in the manometer connected to the upper part of the sediment column started to descend, until a steady state was achieved. This height difference between the two manometers was responsible for water movement, and Δh would be maintained as long as the steady-state evaporative conditions existed. As soon as the lamp was turned out, the water level within the sediment column returned to the top, equal to that at the regulator. Then hydrostatic condition again prevailed. *We have thus proven that the upward flow was related by a height difference in the two manometers.* Since the movement was not caused directly by the action of surface tension in a capillary tube to draw water *above* the water level within the regulator, the movement cannot be considered a capillary movement in a vadose zone.

This mechanism also explains why the flow rate within the porous medium by evaporative pumping was apparently not governed by permeability. Using the pumping of a well as an example, the flow rate through the porous medium around the borehole must be accelerated when the pump is operating. This increased rate in a medium of fixed permeability can only be induced by a steepening of the hydraulic gradient perpendicular to the borehole. Similarly the rate of flow induced by an evaporative pump is governed by the need to replace evaporative loss. A vertical hydraulic gradient would be set up under a horizontal surface of evaporation to induce darcy flow within the porous medium to replace evaporative loss. This vertical gradient is much larger in a fine sediment than that in a coarse sediment, if the same amount of evaporative loss is to be replaced.

Geological interpretations

Our experiments permit us the following interpretations of the observed dolomite distribution pattern in nature:

(*1*) Evaporative pumping is an effective mechanism to induce hydrodynamic movement within an otherwise hydrostatic system. The commonly observed association of dolomites with shelf-evaporites can be explained in terms of the rapid magnesium-supply rate by water movements under evaporative pumping in an arid climate.

(*2*) The lack of permeability-control in dolomitization can be explained by the fact that the rate of magnesium-supply by evaporative pumping is practically independent of sediment permeability. Our experiment also provides an explanation of the paradox reported by MURRAY (1960, p.74), that "the preferentially dolomitized limestone was the lime mud, ... rather than the associated higher-permeability sorted carbonate sands" (see also MURRAY and LUCIA, 1967, p.28). Evaporative pumping caused effective dynamic salt-water movement in lime mud to induce dolomitization, whereas porous lime sand served as aquifer for waters moving under a height head, leading to cementation by sparry calcite. The fact that the Recent submarine sedimentation in the Persian Gulf region is confined to sand layers (TAYLOR and ILLING, 1969) may also be explained if the calcite cementation has been caused by the flow of interstitial waters under a gravity head, which is governed by the sediment permeability.

(*3*) In arid regions where evaporative pumping is probably effective during much of the year, annual evaporation rate permits an estimate of the upward flow rate of pore water to replace evaporative loss.

Assume, for the sake of quantitative estimates, an evaporative pump for dolomitization is 100% effective, if (*a*) evaporation loss was entirely replaced by interstitial flow through the sediment, and (*b*) all Mg^{2+} carried by the passing water movement was used up in dolomite formation. The rate of optimal dolomite formation under an arid climate could thus be estimated.

Using the figure that annual evaporation is 100 cm/year or 100 cm^3/cm^2 water surface per year in an arid coastal region (DEFFEYES et al., 1965; SCHLANGER, 1965), the annual evaporative loss of interstitial water, according to our experiment, would be about half as much for sediments of 40% porosity, or about 50 $cm^3/$

cm² sediment-surface per year. Assume the source to replace evaporative loss is the normal sea water, which contains $1.3 \cdot 10^{-3}$ g/cm³ Mg, the annual magnesium-supply rate would be $6.5 \cdot 10^{-2}$ g/year/cm² sediment-surface. To form a 1-cm thick dolomite crust from lime sediment requires 0.23 g Mg^{2+}/cm² sediment-surface, which could be supplied by the optimal evaporative pumping in 3.5 years. If the pump operates at 100% efficiency, the rate of dolomite formation is thus 10 m/3.5 thousand years, or 10 km/3.5 million years.

Recent dolomite crust apparently started to form when the sea level reached its present stand about 5,000 years ago, and the crust thickness ranges from a few cm on Andros Island (SHINN et al., 1965) to about 2 m on Persian Gulf sabkhas (ILLING et al., 1965). The rates of formation suggest that dolomitization by evaporative pumping operated at a 0.2% efficiency under the relatively humid climate of Andros, but at almost 15% efficiency on some parts of the arid Persian Gulf sabkhas. The Hauptdolomit of the Alps formed during the Norian Epoch, approximately 1/6 of the Triassic time span, or about 7 million years, and has a maximum thickness of about 1,000 m (FISCHFR, 1964). The efficiency of dolomitization by evaporative pumping to produce this thickness was about 5%, not much different from the average now prevailing on the sabkhas.

(4) The relation of hydrodynamic flow to evaporation rate does not apply if the movement was induced by a gravity head. Under coastal plains in a humid climate, ground waters move under a height head, and cannot induce extensive dolomitization (HSÜ, 1966).

CONCLUSIONS

(1) Dynamic movement of salt waters under an arid coastal plain could be induced by evaporation; evaporative loss of interstitial waters could be replaced by seepage flow from the sea. This type of hydrodynamic movement will be referred to as evaporative pumping. The path of water movement is opposite to that of seepage reflux.

(2) The rate of flow induced by evaporative pumping is independent of the permeability of the porous media, but is mainly governed by the evaporation conditions.

(3) Magnesium-supply considerations suggest that neither ground waters moving under a height head nor supersaline waters refluxing under a density head could adequately account for extensive dolomitization of not very permeable lime sediments. In contrast, magnesium could be supplied by the evaporative pumping at such a rate so that ancient and Recent dolomites may have been formed by this mechanism, even if the efficiency of dolomitization by evaporative pumping operated at a 15% or less efficiency.

REFERENCES

ADAMS, J. E. and RHODES, M. L., 1960. Dolomitization by seepage refluxion. *Bull. Am. Assoc. Petrol. Geologists*, 44:1912–1920.

ALEXANDERSSON, T., 1969. Recent littoral and sublittoral high-Mg calcite lithification in the Mediterranean. *Sedimentology*, 12:47–61.

BACK, W., CHERRY, R. N. and HANSHAW, B. B., 1966. Chemical equilibrium between the water and minerals of a carbonate aquifer. *Natl. Speleology Soc., Bull.*, 28:119–126.

BERNER, R. A., 1965. Dolomitization of mid-Pacific atolls. *Science*, 144:1297–1299.

BERNER, R. A., 1966. Chemical diagenesis of some modern carbonate sediments. *Am. J. Sci.*, 264:1–36.

BERNER, R. A., 1967. Comparative dissolution characteristics of carbonate minerals in the presence and absence of aqueous magnesium ion. *Am. J. Sci.*, 265:45–70.

COTTER, E., 1966. Limestone diagenesis and dolomitization in Mississippian carbonate banks in Montana. *J. Sediment. Petrol.*, 36:764–774.

DEFFEYES, K. S., LUCIA, F. J. and WEYL, P. K., 1964. Dolomitization; observations on the island of Bonaire, Netherlands Antilles. *Science*, 143:678–679.

DEFFEYES, K. S., LUCIA, F. J. and WEYL, P. K., 1965. Dolomitization of Recent and Plio-Pleistocene sediments by marine evaporite waters on Bonaire, Netherlands Antilles. *Soc. Econ. Paleontologists Mineralogists, Spec. Publ.*, 13:71–88.

DE GROOT, K., 1969. The chemistry of submarine cement formation at Dohat Hussain in the Persian Gulf. *Sedimentology*, 12:63–68.

FISCHER, A. G., 1964. The Lofer cyclothems of the Alpine Triassic. *Kansas Geol. Surv., Bull.*, 169:107–149.

HANSHAW, B. B., BACK, W. and MEYER, R., 1965. Carbonate equilibria and radiocarbon distribution related to groundwater flow in the Floridan limestone aquifer. *Intern. Assoc. Hydrol., Symp. Dubrovnik, 1965*:601–614.

HSÜ, K. J., 1963. Solubility of dolomite and composition of Florida ground waters. *J. Hydrol.*, 1:288–310.

HSÜ, K. J., 1966. Origin of dolomite in sedimentary sequences: a critical analysis. *Mineralium Deposita*, 2:133–138

HSÜ, K. J., 1967. Chemistry of dolomite formation. In: G. V. CHILINGAR, H. J. BISSELL and R. W. FAIRBRIDGE (Editors), *Developments in Sedimentology, 9B, Carbonate Rocks*. Elsevier, Amsterdam, pp.169–191.

HUBBERT, M. K., 1940. The theory of groundwater motion. *J. Geol.*, 48:785–944.

ILLING, L. V., WELLS, A. J. and TAYLOR, J. C. M., 1965. Penecontemporary dolomite in the Persian Gulf. *Soc. Econ. Paleontologists Mineralogists, Spec. Publ.*, 13:89–111.

KINSMAN, D. J. J., 1968. AAPG sponsored research project: early diagenesis of carbonate sediments in a supratidal evaporitic setting. *Bull. Am. Assoc. Petrol. Geologists*, 52: 2071–2072.

LANGMUIR, D., 1965. Stability of carbonates in the system $MgO-CO_2-H_2O$. *J. Geol.*, 73: 755–780.

LUCIA, F. J., 1967. Recent sediments and diagenesis of south Bonaire, Netherlands Antilles. *J. Sediment. Petrol.*, 38:845–858.

MÜLLER, G. and TIETZ, G., 1966. Recent dolomitization of Quaternary biocalcarenites from Fuerteventura (Canary Islands). *Contrib. Mineral. Petrol.*, 13:89–96.

MURRAY, R. C., 1960. Origin of porosity in carbonate rocks. *J. Sediment. Petrol.*, 32:59–84.

MURRAY, R. C. and LUCIA, F. J., 1967. Cause and control of dolomite distribution by rock selectivity. *Bull. Geol. Soc. Am.*, 78:21–36.

ROEHL, R. O., 1967. Stony mountain (Ordovician) and interlake (Silurian) facies analogues of Recent low-energy marine and subaerial carbonates, Bahamas. *Bull. Am. Assoc. Petrol. Geologists*, 51:1979–2032.

SCHLANGER, S. O., 1965. Dolomite-evaporite relations on Pacific islands. *Tohoku Univ., Sci. Rept.*, 37:15–29.

SHINN, E. A., 1969. Submarine lithification of Holocene carbonate sediments in the Persian Gulf. *Sedimentology*, 12:109–144

SHINN, E. A., GINSBURG, R. N. and LLOYD, R. M., 1965. Recent supratidal dolomite from Andros Island, Bahamas. *Soc. Econ. Paleontologists Mineralogists, Spec. Publ.*, 13: 112–123.

TAYLOR, J. C. M. and ILLING, L. V., 1969. Holocene intertidal calcium carbonate cementation Qatar, Persian Gulf. *Sedimentology*, 12:69–107.

USDOWSKI, H. E., 1967. *Die Genese von Dolomit in Sedimenten — Mineralogie und Petrographie in Einzeldarstellungen*, 4. Springer, Berlin, 95 pp.

Part IV

ANCIENT ANALOGS OF PERITIDAL, EVAPORITIC DOLOMITIZATION

Editors' Comments
on Papers 14, 15, and 16

14 LAPORTE
Carbonate Deposition Near Mean Sea-Level and Resultant Facies Mosaic: Manlius Formation (Lower Devonian) of New York State (Abstract)

15 MATTER
Tidal Flat Deposits in the Ordovician of Western Maryland

16 ZENGER
Significance of Supratidal Dolomitization in the Geologic Record

The discoveries in the 1950s and 1960s of dolomite forming in Recent marine supratidal and intertidal deposits caused a surge of studies dealing with ancient analogues of peritidal dolomitization (Ginsburg et al., 1954; Fisher, 1964; Roehl, 1967; Schenk, 1967; West et al., 1968, Braun and Friedman, 1969; Fisher and Rodda, 1969; Textoris, 1969), although Reuling had postulated the intertidal dolomitization of some Funafuti carbonates as early as 1934.

In this part, Papers 14 and 15 by Laporte and Matter, respectively, are typical of these many references that describe the textures and depositional environments of ancient carbonate rocks presumed to have been syndepositionally dolomitized in the peritidal environment. In a well-received paper dealing with supratidal, intertidal, and subtidal carbonate facies in the Manlius Formation (Lower Devonian, New York), Laporte presented evidence for supratidal deposition: algal laminae, mudcracks, birdseye structures, rare fossils, and dolomite. Appealing to the Florida-Bahama model, he proposed that the dolomitization was penecontemporaneous. His publication was the first in that vintage year (1967) for papers reporting ancient examples of penecontemporaneous peritidal dolomitization. Because Laporte was only partly concerned with dolomitization in this relatively long paper, we include only his abstract here.

Matter appealed to various bedding types in Ordovocian carbonates of western Maryland to distinguish tidal flat subenvironments. Thin-bedded, mud-cracked, lumpy carbonates were deposited in the marsh; laminated dolomites formed in the supratidal zone; and ribboned and stromatolitic limestones were deposited in the more regularly flooded intertidal zone. His comparison with modern analogs in the Persian Gulf and the Caribbean clearly implied that he assumed the dolomite was penecontemporaneous. Although Matter did not specifically infer coexisting conditions of hypersalinity, this has been a prevailing theme in such studies of ancient shallow-marine dolomites.

In a thought-provoking article, Zenger (Paper 16) questioned the dogmatic analogy between modern and ancient dolomite occurrences and asked us to consider whether most ancient dolomites represent a peritidal site of accumulation of the original carbonates, and if so, whether most of the dolomite itself formed there penecontemporaneously. Referring to his examinations of Cambrian through Devonian carbonates from New York and California and other published studies, he stressed that there are numerous examples in which dolomitization is believed to have occurred in environments other than peritidal. Indeed, a number of recent studies demonstrate a primary (Behrens and Land, 1972) or replacement origin (Scholle, 1971; Mossler, 1971; Harris, Paper 23; see also Parts V-VII) of dolomites in modern and ancient subtidal environments. Also, dolomite replacing peritidal carbonates could be later diagenetic. He further questioned what he considered to be the overgeneralized concept of evaporitic dolomitization, referring to Holocene and ancient hypersaline environments in which dolomite is not present. Some of his examples, in particular from the early Paleozoic of New York, may be misleading insofar as hypersaline depositional conditions are becoming more apparent there (Mazzullo and Friedman, 1977) and elsewhere in Cambro-Ordovician epicontinental carbonates (Barnes, 1959; Saxby and Lamar, 1957). Zenger referred to the less than subtle crystallographic and chemical differences between modern and ancient dolomites, some of which he suggested may be related to our limited Holocene time-reference vantage point. In any event, he illustrated that the dolomite problem remains unsolved.

REFERENCES

Barnes, V. E., 1959, *Stratigraphy of the Pre-Simpson Paleozoic Subsurface Rocks of Texas and Southeast New Mexico*, vols. 1-2, Univ. Texas Pub. 5924, Austin, Tx., 836p.

Behrens, E. W., and L. S. Land, 1972, Subtidal Holocene Dolomite, Baffin Bay, Texas, *Jour. Sed. Petrology* **42**:155-161.

Braun, M., and G. M. Friedman, 1969, Carbonate Lithofacies and Environments of the Tribes Hill Formation (Lower Ordovician) of the Mohawk Valley, New York, *Jour. Sed. Petrology* **39**:113-135.

Fischer, A. G., 1964, The Lofer Cyclothems of the Alpine Triassic, in *Symposium on Cyclic Sedimentation,* ed. D. F. Merriam, Kansas Geol. Survey Bull. 169, pp. 107-149.

Fisher, W. L., and P. U. Rodda, 1969, Edwards Formation (Lower Cretaceous), Texas:Dolomitization in a Carbonate Platform System, *Am. Assoc. Petroleum Geologists Bull.* **53**:55-72.

Ginsburg, R. N., L. B. Isham, S. J. Bein, and J. Kuperberg, 1954, Laminated Algal Sediments of South Florida and Their Recognition in the Fossil Record, *Univ. Florida Marine Lab Report 54-21,* Coral Gables, Fl., 33p.

Mazzullo, S. J., and G. M. Friedman, 1977, Competitive Algal Colonization of Peritidal Flats in a Schizohaline Environment:The Lower Ordovician of New York, *Jour. Sed. Petrology* **47**:398-410.

Mossler, J. H., 1971, Diagenesis and Dolomitization of Swope Formation (Upper Pennsylvanian), Southeast Kansas, *Jour. Sed. Petrology* **41**:962-970.

Reuling, H. T., 1934, Der Sitz der Dolomitisiering. Versuch einer Neuen Auswertung der Bohr-Ergebnisse von Funafuti, *Senckenberg. Naturforsch. Ges. Abh.* **428**:1-44.

Roehl, P. O., 1967, Stony Mountain (Ordovician) and Interlake (Silurian) Facies Analogs of Recent Low-Energy Marine and Subaerial Carbonates, Bahamas, *Am. Assoc. Petroleum Geologists Bull.* **51**:1979-2032.

Saxby, D. B., and J. E. Lamar, 1957, Gypsum and Anhydrite in Illinois, *Illinois Geol. Survey Circ. 226,* Urbana, Il., 26p.

Schenk, P. E., 1967, The Macumber Formation of the Maritime Provinces, Canada—A Mississippian Analogue to Recent Strand-Line Carbonates of the Persian Gulf, *Jour. Sed. Petrology* **37**:365-376.

Scholle, P. A., 1971, Diagenesis of Deep-Water Carbonate Turbidites, Upper Cretaceous Monte Antola Flysch, Northern Apennines, Italy, *Jour. Sed. Petrology* **41**:233-250.

Textoris, D. A., 1969, Supratidal Origin of Appalachian Basin Dolostone (Abstract), *Geol. Soc. America Spec. Paper 121,* pp. 470-471.

West, I. M., A. Brandon, and M. Smith, 1968, A Tidal Flat Evaporitic Facies in the Viséan of Ireland, *Jour. Sed. Petrology* **38**:1079-1093.

14

CARBONATE DEPOSITION NEAR MEAN SEA-LEVEL AND RESULTANT FACIES MOSAIC: MANLIUS FORMATION (LOWER DEVONIAN) OF NEW YORK STATE

LEO F. LAPORTE
Providence, Rhode Island

ABSTRACT

Study of the Lower Devonian Manlius Formation along 250 miles of outcrop in New York State indicates that, by analogy with Recent carbonate environments, it was deposited in an environment very near mean sea-level, and having great lateral extent with very low relief. Three facies can be recognized within the Manlius: *supratidal, intertidal,* and *subtidal,* representing carbonate environments slightly above, at, and just below, mean sea-level, respectively. Small fluctuations in water level, caused by lunar tides, storms, or seasonal climatic variations, as well as local differences in sediment accumulation and erosion, resulted in sharp lateral and vertical facies changes with the result that the Manlius today has a complex internal stratigraphy and fossil distribution.

The *supratidal facies* is characterized by non-fossiliferous, laminated, mudcracked, dolomitic, pelletal carbonate mudstone. Mudcracks and spar-filled vugs ("bird's-eye" structure) indicate frequent subaerial exposure; thin bituminous films separating individual carbonate laminae suggest the presence of algal mats. Dolomitization was penecontemporaneous and is inferred to have been similar to present-day supratidal dolomite formation in Florida and the Bahamas.

The *intertidal facies* is composed of alternating thin beds of sparsely fossiliferous, pelletal, carbonate mudstone and skeletal calcarenite; individual beds commonly lie uncomformably on those below. Many primary structures are evident, especially scour-and-fill, cross-stratification, and limestone-pebble conglomerate. Many fossil occurrences have abundant individuals but only a few taxa are represented. Algal stromatolites and oncolites also are common. A few mudcracks and minor erosional relief on carbonate mudstone beds indicate intermittent subaerial exposure.

The *subtidal facies* is a skeletal, pelletal, carbonate mudstone commonly with well-developed, tabular masses of stromatoporoids. The relatively diverse biota of this facies required continuous marine submergence; its close association and juxtaposition with rocks that clearly indicate periodic subaerial emergence further suggest that water depths were very shallow. The stromatoporoids are interpreted as having grown either as encrusting masses within tidal creeks or channels, or else as closely crowded, individual heads in front of the tidal flats of the intertidal facies.

The three facies of the Manlius existed contemporaneously, retreating and advancing continuously with the result that, today, they form a complex facies mosaic. Manlius deposition as a whole took place within and marginal to a broad and shallow, though somewhat restricted, lagoon that developed on the landward side of a wide belt of crinoid meadows, represented by the laterally equivalent Coeymans Formation, which is a brachiopod-rich crinoidal calcarenite. During the Early Devonian marine transgression in the northern part of the Appalachian basin, these environments represented by the Manlius and Coeymans migrated westward with time across New York State.

15

TIDAL FLAT DEPOSITS IN THE ORDOVICIAN OF WESTERN MARYLAND[1]

ALBERT MATTER

Department of Geology, University of Berne, Switzerland

ABSTRACT

The lower part of the New Market limestone (Lower Middle Ordovician) was studied in the Wilson Quarry east of Clear Spring, Maryland. It was subdivided according to bedding into six types: laminated dolomite, ribboned, lumpy, stromatolitic, intraclastic and bioclastic dolomitic limestones which are closely interbedded. Sediment filled and smaller spar filled mud cracks were found throughout the section, indicating that most of it accumulated above mean low water level. The six rock types are believed to represent sub-environments within the gross inter-supratidal environment: 1) marsh: lumpy limestones, 2) supratidal: laminated dolomite, 3) intertidal: ribboned and stromatolitic limestones.

INTRODUCTION

During Cambro-Ordovician times about 3000 m of carbonates accumulated in the Central Appalachian area. This thick sequence is built up of a few basic lithologic types which frequently recur in cycles (Donaldson, 1960, Sarin, 1962). Contrasting with the fairly uniform lithology is a great variety of sedimentary structures. On the basis of the cyclic arrangement of the rock types, their lithology, particle size, fossil assemblages and sedimentary structures, various authors (Wilson, 1952, Neuman, 1951, 1960, Long, 1953, Sando, 1957, Donaldson, 1960, and others) have pointed out that most of the rock types had been deposited in warm and shallow seas with occasional exposure.

To pinpoint as closely as possible how shallow these seas were at a given time, the Middle Ordovician New Market limestone of the St. Paul group (Neuman, 1951) was studied in detail as a representative small portion of the Cambro-Ordovician sequence of Western Maryland. The New Market limestone is lithologically uniform, most of the rocks are pelmicrites and contain only a sparse fauna. The only observable parameters which show appreciable variation are the sedimentary structures. In recent years the value of sedimentary structures for the interpretation of the hydrodynamic (Middleton, 1965, ed.) as well as the geologic environment (Potter and Pettijohn, 1963) has been clearly recognized. In this study, therefore, emphasis is laid on the bedding and other sedimentary structures, their aspect, variation and origin from which hopefully also the depositional environment may be inferred more closely.

SEDIMENTARY STRUCTURES

The section of the New Market limestone in the Wilson Quarry (section 19 of Neuman, 1951)

[1] Manuscript received November 9, 1966.

4 miles east of Clear Spring, Maryland, where about 50 m of the formation are extremely well exposed, was measured and sampled in detail to evaluate the vertical succession and distribution of the different structures (fig. 1a). The New Market formation, the thickness of which, according to Neuman (1951) seems to vary from 30 m in Virginia to 240 m in Southern Pennsylvania, measures about 125 m in the Wilson area.

In fresh outcrops hardly any sedimentary structures show up in the New Market limestone because of its dark grey color and aphanitic texture. The well weathered faces and the acetate peels made from many specimens allowed the section in the Wilson Quarry to be . subdivided into six structurally different rock types (fig. 1b): laminated, lumpy, · ribboned, stromatolitic, intraclastic and bioclastic dolomitic limestones. Their characteristic properties are compiled in table 1. These six rocks types differ from each other mainly in thickness and form of the beds. The laminated and ribboned rocks both have in common a regular alternation of parallel dolomite and limestone layers (figs. 2A and 2B). They differ mainly in the thickness of these layers which are 3–20 times thinner in the laminated dolomite. Laminae of similar thickness but different form are found in the stromatolitic limestones which occur as head shaped masses and as undulous or parallel sided beds. They consist of an alternation of crenulated light and dark laminae (fig. 2C). Intercalated dolomite laminae are rather rare in the stromatolites of the New Market formation, though they are common in other Cambro-Ordovician formations of the area. The rocks termed lumpy limestones display the widest variation ranging from something similar to a ribboned rock but with more discontinuous limestone beds to a rock with coarse limestone clasts in a dolomite matrix (fig. 2D). Stratifica-

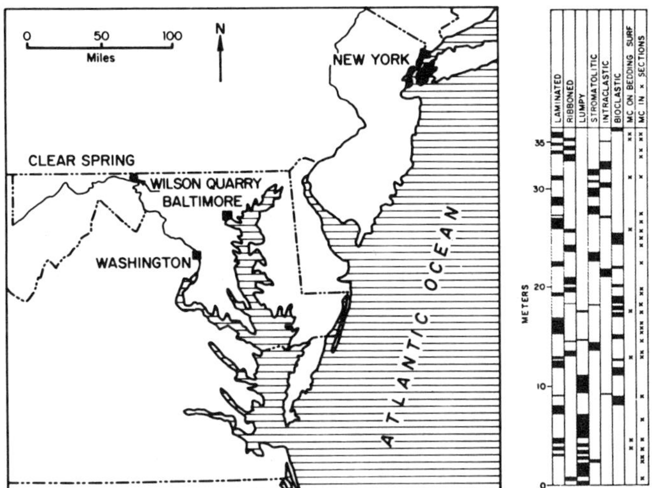

FIG. 1.—This figure shows the location of the Wilson Quarry locality (1a) and the measured section of the lower part of the New Market limestone (1b). This section shows the vertical distribution of rock types and mud cracks (MC).

tion is generally absent in the massive intraclastic limestones, yet sometimes a slight stratified aspect is produced by alternation of coarser and finer grained layers, where the clasts are embedded in a dolomite matrix (fig. 2E). Rocks similar to ribboned limestones but with more irregular beds and abundant fossil debris and those similar to lumpy rocks with irregular pieces of bioclastic limestone in a dolomite matrix were lumped together as bioclastic limestones (fig. 2F).

Dolomite is important compositionally as well as for the structural aspect of the different rock types. The emphasis in this paper, however, is laid on the environment of deposition and the environmental significance of the various rock types as defined by different bedding. Because the origin of dolomite has no direct bearing on this problem it will be dealt with in a separate paper.

Desiccation Features

Although the six rock types show different kinds of stratification and also minor variations in texture and composition, we do not understand stratification nor do we know precisely where the different types formed, except for the stromatolites. However, different types of shrinkage cracks were observed in many specimens throughout the section. If it can be shown that these cracks were formed while the sediment was subaerially exposed and not by shrinkage resulting from dehydration following burial of the muds, the cracks would indicate that the depositional surface was at the air-water interface.

Mud cracks.—Mud cracks are most obvious if they have been weathered out on bedding surfaces. In the Wilson Quarry several such bedding planes showed beautiful first and second order shrinkage polygons (figs. 3 and 4). The polygons were observed in various sizes from 30 cm to 1 cm in their largest diameter. In sectional views, however, the cracks are more difficult to see, especially in an aphanitic limestone like the New Market. Therefore planar and sectional views of the same mud cracked beds were compared in order to recognize and relate the typical mud crack features seen in section to the polygonal pattern on the bedding plane surfaces. This proved to be very useful, because by detailed sampling and careful study of acetate peels made from every specimen a large number of mud cracks was found in horizons, which in the field did not seem to contain any shrinkage features (fig. 16). Basically, two different types of mud cracks are distinguished 1) sediment filled, and 2) spar filled.

The first type of mud crack is most commonly present in the ribboned limestones and the laminated dolomites. Occasionally the cracks can be traced to the bedding plane surfaces where they form a polygonal pattern (fig. 3). In cross sections the cracks show more or less vertical walls (figs. 2A and 3) or very irregular form (fig. 2B), but are rarely V-shaped. Some of them were probably deformed by soft sediment deformation. In the ribboned limestones they

TABLE 1.—*Major properties of the rock types of the New Market limestone*

Rock type	Stratification	Other sedimentary structures	Composition	Average grain size of particles	Fossil content
Laminated	*Thickness of units:* 10–150 cm *Character:* Laterally discontinuous laminae. Lamination due to regular alternation of limestone and dolomite or due to alternation of coarser and finer dolomite crystals. *Thickness of laminae:* 0.15–3 mm	Mud cracks	Pellets and euhedral dolomite. Widely variable in proportions from lamina to lamina, ranging from 20% dolomite in pelmicrite laminae to 100% in pure dolomite laminae.	Pellets: 40–60μ Dolomite: 10–50μ	Very scarce
Ribboned	*Thickness of units:* 20–80 cm *Character:* Regular alternation of limestone and dolomite beds. Bed surfaces undulous or planar-smooth. Dolomite beds laminated. *Thickness of beds:* 0.5–3 cm	Mud cracks Ripples (a <0.5 cm) Small scale slump structures Cut and fill	Lime beds: Pelmicrite with <5% dolomite. Insoluble residue 1.2–3.5%. Dolomite beds: Dolomite rhombs with interstitial pelletal micrite. Insoluble residue 5.0–10.7%.	Pellets: 40–80μ Dolomite: 10–20μ in pelmicrite beds and 20–40μ in dolomite beds.	Scarce
Lumpy	*Thickness of units:* 10–170 cm *Character:* Highly variable. Discontinuous beds or isolated fragments of limestone interbedded or surrounded by dolomite. Some specimens show transition from discontinuous beds to a completely fragmented rock. *Thickness of beds* (if present): 0.5–3 cm	Mud cracks Ripples (a <0.5 cm)	Lime beds and clasts: Pelmicrite with less than 5% dolomite. Dolomite beds: Dolomite rhombs with interstitial pelleted micrite.	Clasts: 0.2–50 mm Pellets: 40–50μ Dolomite: 30–40μ	Scarce, fragments of tetracorals, brachiopods, trilobites, ostracods.
Stromatolitic	*Thickness of units:* 0.5–80 cm *Character:* Regular alternation of crenulated light and dark laminae. Either flat lying or undulous or head shaped. *Thickness of laminae:* 0.10–2 mm	Shrinkage cracks "Birdseyes" Edgewise conglomerates	Dark laminae: Pelmicrite. Light laminae: Meshwork of spar filled molds of algal filaments surrounded by pellets. In other specimens also dolomite with little interstitial micrite	Pellets: 30–70μ Dolomite: 10–40μ	Very scarce
Intraclastic	*Thickness of units:* 14–80 cm *Character:* Massive to poorly bedded due to different size of intraclasts.		Intraclasts floating in a slightly calcitic dolomite matrix. Intraclasts are large pellets and true pelmicrite clasts.	Intraclasts: 3 cm Pellets: 70–1200μ Dolomite: 30–50μ	Scarce, fragments of brachiopods, trilobites, ostracods.
Bioclastic	*Thickness of units:* 10–80 cm *Character:* Either alternation of irregular bioclastic limestone beds with dolomite laminae or diffuse irregular biogenic limestone lumps in a dolomite matrix. *Thickness of beds:* Dolomite 1 cm Limestone 2–4 cm	Oncolites Burrows	Limestone beds: Biopelmicrites to biogenic intramicrites with scattered dolomite rhombs. Dolomite laminae: Dolomite rhombs with interstitial pelletal calcimicrite.	Pellets: 30–70μ Dolomite: 10–30μ	Abundant, fragments of tetracorals, brachiopods, pelecypods, gastropods, trilobites, bryozoans.

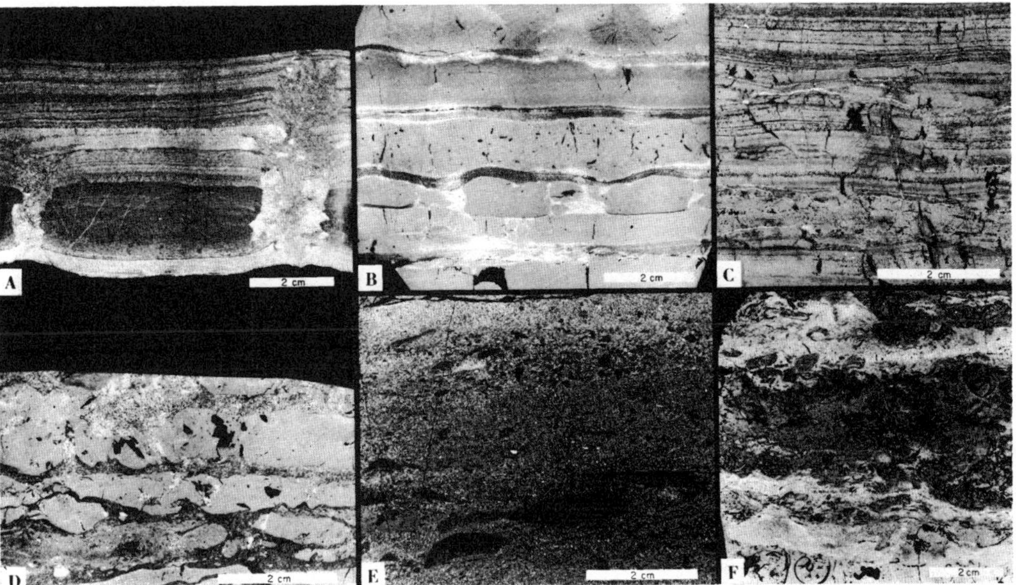

FIG. 2.—Rock types of the New Market limestone (negative prints of peels).

A.—Laminated dolomite; dark layers are more dolomitic than light layers. Mud cracks are filled with dolomite rhombs, limestone chips and pellets.

B.—Ribboned limestone with alternating thicker smooth or rippled permicrite and thinner dolomite beds; one bed shows irregular mud cracks and another distinct burrows.

C.—Stromatolitic limestone consisting of alternating dense and "spongy" pelleted limestone laminae with small spar filled mud cracks. Spongy aspect is caused by spar filled molds of algal filaments.

D.—Lumpy limestone; discontinuous and partly burrowed limestone beds and intraclasts in a dolomite "matrix".

E.—Poorly bedded intraclastic limestone; intraclasts and pellets with dolomite rhombs in interstices.

F.—Bioclastic limestone; irregular dolomitic limestone beds with fragments of brachiopods, gastropods, ostracods and other unidentified shells alternating with dolomite beds (light). Note oncolite in upper left.

are usually confined to the finer grained limestone beds, which ultimately consist of calcimicrite in the form of silt sized pellets. The cracks are fairly easy to recognize in acetate peels because they are filled with silt-sized dolomite rhombs similar to those of the adjacent dolomite bed and with limestone chips and pellets. Generally the thicker the layers the deeper and wider the cracks and the larger the polygons. This relationshlp between thickness of the beds and depth of the cracks is exemplified by the laminated dolomites where the cracks are generally of microscopic size.

The second type of mud cracks are those filled with sparry calcite. Such spar filled cracks are frequently present in stromatolites. They are generally V-shaped and either open towards the top or towards the bottom of the beds. The open-upward spar filled cracks could be traced to a bedding plane surface where they formed a polygonal pattern (fig. 4). These cracks are small, a few tenths of a millimeter to several millimeters deep and a fraction of a millimeter

wide. They are occasionally found in the same bed with sediment filled cracks.

Mud cracks almost invariably form when water-saturated muds become exposed to the air, but they also have been observed in a tidal channel under several meters of water (Van Straaten, 1954). These subaqueous cracks, however, do not look like those commonly produced subaerially and would, if present in ancient rocks hardly be mistaken as subaerial mud cracks. Burst (1965) investigated experimentally with different mixtures of clay minerals the subaqueous formation of shrinkage cracks. He was able to produce small scale cracks which displayed a polygonal surface pattern very similar to that of the subaerial crack pattern of modern mud flats. However, the cracks only formed if more than 2 percent of swelling clay minerals were present in the total sample. The clay mineral content of the New Market limestone is distinctly less than 2 percent of the total sample as estimated from X-ray diffraction patterns of the insoluble residue. Therefore, it is almost certain, that the cracks described in this paper are of subaerial origin.

Mud cracks may form in various environments such as flood plains, abandoned river channels, tidal flats, etc. (Shrock, 1948, p. 195), but the presence of pelleted layers and marine fossils rule out the former two. The formation and the aspect of the mud cracks of the New Market limestone described above are best explained by the mechanism outlined by Shinn (1964) for mud cracks in recent sediments. Shinn (1964) noticed that on modern supratidal flats the polygons sometimes remain exposed for years. During such long exposures they weather considerably by spalling off small particles of various sizes. These chips will not only partly fill the cracks but they will also cover the polygons and protect them from further weathering. The next reflooding eventually may tear loose larger pieces of the polygons themselves which are then transported along with the smaller chips and deposited with the fresh sediment carried ashore by the storm.

If it is believed that the cracks were filled this way it is on first sight somewhat surprising to find cracks which are filled with sparry calcite instead of with lime chips and dolomite rhombs. This might indicate that these cracks formed below the surface or as has been proposed by Fischer (1964) even at the surface where they were quickly overgrown by an algal mat which

Fig. 4.—Upper bedding surface of a stromatolitic limestone showing a pattern of spar filled mud cracks; the cracks are narrow as shown on the negative print of a peel below (arrow points upward).

protected them from getting filled with detritus. In some cases the uppermost part of these cracks is filled with a few dolomite rhombs and small pellets. This might indicate that the narrowness of the crack alone might suffice to protect it from complete infilling with sediment.

Lumpy limestones.—The lumpy rocks consist of irregular rounded or angular limestone fragments which are surrounded by a dolomite matrix (fig. 2D) or which are "welded" together in the absence of dolomite. In some specimens the cross-sectional view shows flat fragments strung out in discrete but discontinuous layers parallel to continuous layers (fig. 5A). Clearly the discontinuous layers were once continuous too for the fragments can be pieced together easily (fig. 5A). The bedding surfaces of these layers show a pattern of irregular flat polygons, which indicates that the fragmentation was a desiccation phenomenon.

In other specimens none of the layers are continuous and only a few fragmented layers can be pieced together. Between these layers the fragments have a sub-parallel arrangement not

Fig. 3.—Lower bedding surface of a laminated dolomite with polygonal pattern of mud cracks and cross-sectional view. Arrow points upward.

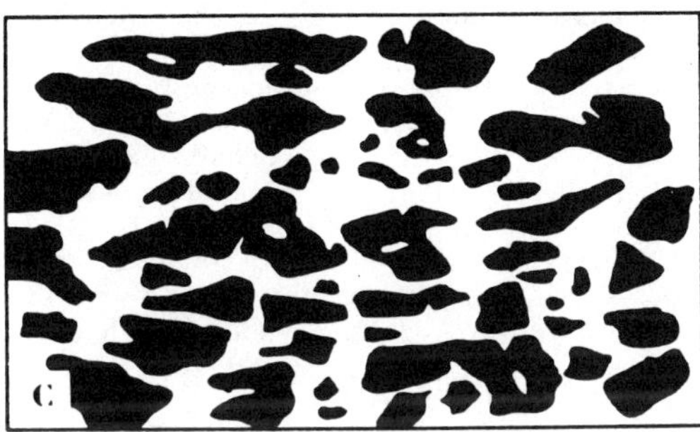

FIG. 5.—Tracings of peels showing the evolution of lumpy limestones caused by disruption following desiccation (black = limestone, white = dolomite).
A.—Beds discontinuous (except for topmost layer), but polygons still in place.
B.—All beds disrupted, clasts have been transported, only the fragments of one bed are in place.
C.—Bedding completely disrupted, all clasts have been transported and their outlines considerably modified by burrowing organisms.

unlike that of a flat pebble conglomerate (fig. 5B). Presumably these are fragments of polygons torn loose and transported during storm flooding (Shinn, 1964).

Fig. 5C shows a lumpy limestone in which the shape of the fragments has been modified considerably by burrowing organisms. These fragments show a completely random distribution and are interpreted as the result of complete reworking of mud cracked limestone polygons.

Some rocks which have many deep, wide shrinkage cracks filled with lime mud, show a lumpy structure in cross sections oblique to the cracks. In this, a fourth variety of lumpy limestone it is the crack filling which appear as lumps. This type of rock might be confused with the ribboned rock, but the latter is less frequently mud cracked and has smooth instead of irregular bedding surfaces.

Although it is claimed that desiccation is the process which produced the lumpy limestones, it is by no means the only one possible. Specimens from the New Market limestone show that burrowing organisms can disturb the original alternation of beds by reworking the lime mud beds to irregularly shaped "clasts." Complete churning by the organisms leads to a nearly structureless bioturbite. In argillaceous limestones, on the other hand, purely diagenetic concretionary accumulation of lime in nodules also results in a lumpy rock (Hadding, 1958) though its aspect differs somewhat from the lumpy structure described above.

Stromatolites

Stromatolites of various growth patterns, both head-shaped and digitate masses as well as flat lying parallel or undulous beds, are common throughout the Cambro-Ordovician of Maryland (Long, 1953, Sando, 1957). Within the New Market limestone laterally continuous undulous stromatolites and Collenia like heads 10–20 cm high are most frequent. They are sometimes associated with edgewise conglomerates. The distribution of stromatolites in the section of the Wilson Quarry is plotted in fig. 1b.

It is not in the scope of this paper to describe the stromatolites in detail or to discuss the problems of stromatolitic growth. For this study the stromatolites are of interest only in so far as they represent an indicator of the depositional environment. Several lines of evidence lead to the conclusion that the stromatolites are typical of intertidal and supratidal environments. First, studies of recent stromatolites show that today Collenia like heads and more or less planar algal mats grow most frequently above mean low water level on mud flats which are periodically flooded and exposed (Ginsburg and others, 1954, Ginsburg, 1960, Logan, Rezak, and Ginsburg, 1964). Secondly, the recent as well as the ancient stromatolites are generallly interbedded with laminated, mud cracked sediments and associated with flat pebble conglomerates, which in some beds in the New Market limestone were clearly derived by erosion from algal heads, possibly after prior desiccation of the algal masses. Thirdly, the algal layers themselves contain microscopic desiccation cracks filled with sparry calcite (fig. 4). Moreover, the ribboned limestones with distinct mud cracks frequently drape over or underlie the stromatolites. In addition spar filled molds of erect algal filaments are clearly preserved in some planar stromatolitic beds and irregularly swelling and thinning horizontal spar filled voids impose a birdseye aspect to the rock. Unfortunately, voids of as diverse an origin as spar filled fossils, algal threads, burrows or intergranular pores have been included under the "hodge podge" term birdseye structure. The birdseyes dealt with here are identical with the shrinkage pores of Fischer (1964). Although, the mechanism of their formation is as yet uncertain a subaerial origin of the birdseye structure is also suggested by a few surface samples from modern supratidal flats which showed the typical birdseye arrangement of the voids. Thus, the association of stromatolites with algal threads and spar filled mud cracks and birdseyes confirms that the ancient stromatolites like their recent analogue grew at or above the air-water interface.

ENVIRONMENT OF DEPOSITION
Gross Environment

Taking into consideration the aspects of the various rock types and their mutual relationships in space as well as the information gained from the sedimentary structures it is possible to infer the depositional environment more precisely than previously has been reported (Neuman, 1960). Fig. 1b shows the frequency within the measured section of mud cracks, both those which are exposed on bedding planes and those which were observed in acetate peels taken from sections perpendicular to bedding. These mud cracks are so densely distributed over the section that they would suggest a supratidal to intertidal origin for most of the section even if no algal stromatolites and lumpy limestones were present. If we take into account the presence of lumpy limestones, which have been interpreted as a product of intense desiccation and the presence of stromatolites this becomes even more likely. Moreover, the general paucity of marine fossils is another, though on first hand less convincing, indicator for intertidal environments. However, if we consider that the fossils present are disarticulated and broken shells and compare this with the present day situation on mud flats, the sparsity of marine fossils fits rather nicely. One of course might object that a sparse marine fauna could also be found in subtidal environments. But if all the evidence (mud cracks, birdseyes, stromatolites, paucity of fossils) is taken collectively it seems highly probable that the laminated dolomites, ribboned and lumpy limestones and the stromatolites are of inter- to supratidal origin.

Taken together, these rocks represent about nine tenths of the whole section (fig. 1b). The remaining small fraction is made up of intraclastic and bioclastic limestones. It is interesting to note that the lumpy limestones occur mainly in the lower part, the bioclastic limestones in the middle and the stromatolites, ribboned and intraclastic limestones in the upper part of the section, whereas the laminated dolomites are found throughout the section (fig. 1b). Future work will have to show whether this distribution has any significance. Although the bioclastic and intraclastic limestones do not show any direct evidence of intermittent exposure they are so closely interbedded with the intertidal limestones, which represent the bulk of the section that they must have formed within or close to the same environment. The whole section therefore is believed to have been deposited at the air-water interface.

Sub-environments

If the different rock types were formed in the same environment why do they show such a variation of stratification which ranges from laminated and ribbon to lumpy limestones?

It is common experience that recent tidal flats are subjected to a wide range of conditions. They are periodically flooded and exposed to desiccation and weathering and thus represent the ideal site for the formation of a wide range of sediment types. The wide mud flats covered with innumerable mud cracks and spotted with calm lagoons are dissected by many tidal creeks, which rework the semi-consolidated muds; and from the low cliffs which border the creeks and shores flat fragments are eroded (Ginsburg, 1960) and mixed with the fresh particles. Such local sub-environments within a tidal flat account for the different limestone types of the New Market formation. The two most important differences between the various types are the different thickness of the beds and the frequency of mud cracks. Similar differences have also been reported from modern tidal flats. However, the precise site where a distinct rock type may form is largely controlled by the geographic and local settings.

The laminated, mud cracked dolomites of the New Market limestone (fig. 2A) show all the features of modern laminated muds. These are found today either in the intertidal zone as the Persian Gulf (Illing, Wells, and Taylor, 1965) or in the supratidal zone as in the Carribbean, Florida Bay or the Bahamas (Ginsburg and others, 1954, Purdy and Imbrie, 1964, and pers. experience). In the former case the laminations are strongly controlled by the algae. The ancient equivalents of these intertidal laminated sediments would be stromatolitic limestones. However, in the supratidal zone in the Carribbean, the Bahamas and Florida Bay, where the algae play a less dominant role, strong winds and storms are capable of transporting new sediments onto the flat. The surface of the flat is very uneven because it cracks and weathers during long subaerial exposure. In addition the surface is scoured by the flooding currents as well as the runoff. The sediment laminae are therefore irregular and discontinuous. Repetitive flooding of the supratidal flats leads to a laminated sediment. The mud cracked laminated dolomite of the New Market formation is similar to these mud cracked supratidal sediments, but lacks the algal features of the intertidal laminated sediments of the Persian Gulf. Therefore, the laminated dolomites of the New Market limestone are interpreted as supratidal deposits.

If the amount of sediment deposited per time interval increases, or if the conditions during sedimentation of a particular bed remain constant over a longer period of time, a bed instead of a lamina will form. Moreover, if sedimentation is rhythmic, a ribboned rock instead of a laminite will result. The ribboned limestones of the New Market formation frequently drape over stromatolites and over small coral bioherms. Their individual limestone layers always have smooth surfaces, sometimes with small ripples and occasionally with spar and sediment filled mud cracks. Beds 1 to 3 cm thick are most frequently observed in recent marsh sediments but similar beds, consisting of alternating smooth muddy and sandy layers have also been reported from the intertidal zone of well zoned modern mud flats (Reineck, 1962, Evans, 1965). The thick layers of the marsh are deposited by unusually heavy storms and remain exposed over long periods of time. Such layers are usually heavily mud cracked and show very uneven surfaces (Shinn, 1964). Because the ribboned limestones show smooth bedding surfaces, rare desiccation features and drape-over stromatolites and coral bioherms, they were most likely deposited in the intertidal zone of a mud flat.

The lumpy limestones share with the supratidal laminated dolomites the frequency of desiccation phenomena and the scarcity of fossils. However, the individual beds of lumpy limestones are usually thicker and more irregular and uneven; in addition the desiccation features are more pronounced. The lumpy rocks thus represent a deposit of a different sub-environment. Depending on the time gap between deposition of successive beds the intensity of desiccation will vary and give rise to the observed variety of lumpy limestones ranging from less cracked ribbon-like limestones to heavily cracked ones. On the basis of the matching attributes of the lumpy limestones and modern marsh sediments (Van Straaten, 1954), it is concluded that the lumpy limestones represent a high marsh deposit.

CONCLUSIONS

The considerable variation in stratification features among laminated, ribboned and lumpy limestones might suggest that each represents a very different depositional environment, but the evidence presented in this paper indicates that this variation reflects only minor differences within the gross tidal flat environment.

The thin bedded, mud cracked lumpy rocks were deposited on the marsh where they were subjected to heavy shrinkage; the laminated dolomites accumulated in the supratidal zone and the ribboned limestones together with the stromatolitic limestones were deposited in the the regularly flooded intertidal zone.

Each of these rock types, correlated with a specific sub-environment of a tidal flat, is not unique to the New Market limestone. The same rock types are found in most Paleozoic limestone formations of the Central Appalachians and other areas e.g. in the Conococheaque (Long, 1953) and Beekmantown (Sando, 1957) groups of Western Maryland, in the Ellenburger group of Texas (Cloud and Barnes, 1948).

While on one hand the recognition of the gross environment will suffice for certain geologic considerations e.g. distribution of open marine and restricted intertidal facies, a further subdivision of the kind outlined in the present paper on the other hand may prove useful for other purposes. Because the coastal areas are most sensitive to changes of sea level or rate of subsidence, minor changes will be recorded in the intertidal sediments. The distribution of the gross environments in time and space will reflect major changes only, whereas from the distribution of the sub-environments much can be learnt about the development of the coastal topography which in turn had controlled sedimentation.

ACKNOWLEDGEMENTS

The author gratefully acknowledges the financial support received from the Swiss National Fund for Scientific Research and from the Government of the Canton of Berne. Facilities were kindly provided by Dr. F.J. Pettijohn, Dept. of Geology, and Dr. D.W. Pritschard, Dept. of Oceanography of The Johns Hopkins University.

The writer wishes to thank Dr. Robert N. Ginsburg for many inspiring discussions, critical reading of the manuscript and many suggestions. He also wants to express his appreciation to Dr. Lawrence A. Hardie for his valuable suggestions and for his help in improving the style of the paper, and to Mr. R. Linfield for preparing the figures.

REFERENCES

BURST, J. F., 1965, Subaqueously formed shrinkage cracks in clay: Jour. Sedimentary Petrology, v. 35, no. 2, p. 348–353.

CLOUD, P. E., JR., AND BARNES, V. E., 1948, The Ellenburger group of Texas: Univ. Texas Bull. 4261.

DONALDSON, A. C., 1960, Interpretation of depositional environments of Lower Ordovician carbonates in central Appalachians: West Virginia Academy Science Proc., 1959 and 1960, v. 31 and 32, p. 153–161.

EVANS, GRAHAM, 1965, Intertidal flat sediments and their environments of deposition in the Wash: Quart. Jour. Geol. Soc. London, v. 121, p. 209–245.

FISCHER, A. G., 1964, The Lofer cyclothems of the alpine Triassic: Kansas Geol. Survey Bull., v. 169, p. 107–149.

GINSBURG, R. N., 1960, Ancient analogues of recent stromatolites: Int. Geol. Cong., pt. 22, p. 26–35.

———, ISHAM, L. B., BEIN, S. J., AND KUPERBURG, J., 1954, Laminated algal sediments of South Florida and their recognition in the fossil record: Unpub. rept. no. 54-21, Coral Gables, Florida, Marine Lab., Univ. Miami, p. 1–33.

HADDING, ASSAR, 1958, The Pre-Quaternary sedimentary rocks of Sweden: Lunds Universitets Årsskrift, N. F., Avd. 2, v. 54, no. 5, p. 1–262.

ILLING, L. V., WELLS, A. J., AND TAYLOR, J. C. M., Penecontemporary dolomite in the Persian Gulf: in Pray, L. C., and Murray, R. C. (eds.), 1965, Dolomitization and limestone diagenesis—a symposium: Soc. Econ. Paleontologists and Mineralogists Spec. Pub. 13, p. 89–111.

LOGAN, B. W., REZAK, RICHARD, AND GINSBURG, R. N., 1964, Classification and environmental significance of algal stromatolites: Jour. Geology, v. 72, no. 1, p. 68–83.

LONG, M. B., 1953, Origin of the Conococheague limestone: The Johns Hopkins Univ., Baltimore, unpub. Ph.D. thesis, 213 p.

MIDDLETON, G. V. (ed.), 1965, Primary sedimentary structures and their hydrodynamic interpretation: Soc. Econ. Paleontologists and Mineralogists Spec. Pub. 12, 265 p.

NEUMAN, R. B., 1951, St. Paul group: A revision of the "Stones River" group of Maryland and adjacent states: Geol. Soc. America Bull., v. 66, p. 267–324.

———, 1960, The Saint Paul group of Maryland: in Gates, O. (ed.), Lower Paleozoic carbonate rocks in Maryland and Pennsylvania: Guidebook 3, Am. Assoc. Petroleum Geologists, Atlantic City Conf., p. 16–17.

POTTER, P. E., AND PETTIJOHN, F. J., 1963, Paleocurrents and basin analysis. Academic Press, New York, 296 p.

PURDY, E. G., AND IMBRIE, JOHN, 1964, Carbonate sediments, Great Bahama Bank: Guidebook for Geol. Soc. America field trip no. 2, Miami, Florida, 66 p.

REINECK, H.-E., 1962, Schichtungsarten in Wattenböden: Zeitschr. für Pflanzenernährung, Düngung, Bodenkunde, v. 99, p. 154–159.

SANDO, W. J., 1957, Beekmantown group (Lower Ordovician) of Maryland: Geol. Soc. America Mem. 68, p. 1–161.

SARIN, D. D., 1962, Cyclic sedimentation of primary dolomite and limestone: Jour. Sedimentary Petrology, v. 32, no. 2, p. 451–471.

SHINN, E. A., 1964, Recent dolomite, Sugarloaf Key: in Guidebook for Geol. Soc. America field trip no. 1, South Florida carbonate sediments (compiled by R. N. Ginsburg), p. 62–67.

SHROCK, R. R., 1948, Sequence in layered rocks. McGraw-Hill Book Co., New York, 507 p.

VAN STRAATEN, L. M. J. U., 1954, Composition and structure of recent marine sediments in the Netherlands: Leidse Geol. Medel., v. 19, p. 1–110.

WILSON, J. L., 1952, Upper Cambrian stratigraphy in the central Appalachians: Geol. Soc. America Bull., v. 63, p. 275–322.

Significance of Supratidal Dolomitization in the Geologic Record

DONALD H. ZENGER   *Geology Department, Pomona College, Claremont, California 91711*

ABSTRACT

Numerous discoveries of Holocene penecontemporaneous dolomite forming mainly in the supratidal zone have prompted environmental analogies with ancient dolomites. Although most of these may be justified, overgeneralization has resulted. Many dolomites lack definitive evidence of either supratidal accumulation or supratidal dolomitization.

Detailed field and laboratory studies of three dolomitic units in New York (Little Falls, Upper Cambrian; Herkimer, Middle Silurian; and Lockport, Middle Silurian) and one in California (Lost Burro, Middle and Upper Devonian) show that the dolomite, apparently metasomatic, represents original calcium carbonate sediments of variable environmental deposition. Portions of the Little Falls and Upper Lockport Formations contain evidence (algal stromatolites, desiccation cracks, intraclasts) of tidal flat (peritidal) deposition, whereas the other occurrences (Herkimer, Lost Burro, and lower Lockport) contain evidence (bioherms, abundant open-marine fossils) of subtidal accumulation. Dolomitization is primarily diagenetic. Even in those cases which possibly represent supratidal dolomitization, the dolomite is ordered and of much coarser grain size than that in the modern analogues.

There is considerable evidence that much Holocene dolomite, as well as a substantial part of the dolomite in the geologic record, formed under evaporitic conditions, especially in contact with brines of high Mg^{++}/Ca^{++} ratios. However, findings in Holocene sediments as well as the lack of indication of the initial presence of evaporites in many ancient dolomites suggest that dolomite need not be an evaporitic mineral.

INTRODUCTION

Excellent reviews of work on dolomites and dolomitization have been presented by Fairbridge (1957), Ingerson (1962), Sonnenfeld (1964), Friedman and Sanders (1967), and Lovering (1969). Prior to 1957, modern marine dolomite was reported as occurring primarily as isolated grains in shallow- to deep-water sediments (Fairbridge, 1957, p. 164). When compared with the impressive abundance of dolomite in the late Precambrian and early Paleozoic, this paucity of Holocene dolomite prompted the formulation of the "dolomite problem"; there was essentially no opportunity to invoke the principle of uniformitarianism in attempting to unravel the mystery of the origin of ancient dolomite. However, during the past dozen years or so, numerous occurrences of Holocene dolomite have been discovered. It is the widespread and problematic use of analogy between these and ancient dolomites that I would like to examine. A problem still exists.

Other questions, such as the significance of primary dolomite in Holocene and ancient sediments and the perplexing difficulty of synthesizing true dolomite at earth surface conditions, are important in the over-all treatment of dolomitization but will not be considered here. In the present paper, the focus is on the widespread dolomites of the past which were deposited in or immediately adjacent to the marine environment; I will not consider more "local" varieties that have been controlled by tectonic, metamorphic, or weathering processes.

DISCUSSION

Supratidal Dolomitization

The great bulk of Holocene dolomite occurs in the landward portion of tidal flats, the supratidal zone, or that area above mean high tide. Some examples, with one key reference for each, include: the Persian Gulf (Illing and others, 1965); and Bahamas (Shinn and others, 1965; *see* Fig. 1); the Florida Keys (Shinn, 1968); the island of Bonaire in the Netherlands Antilles (Deffeyes and others, 1965); and South Australia (Alderman and Skinner, 1957).

Figure 1.　Holocene mud cracked dolomitic crust in supratidal zone, west side of Andros Island, Bahamas.

Largely subsequent to these discoveries, geologists have recognized an abundance of ancient analogues—supratidal dolomites from the geologic record (for example, Laporte, 1967; Matter, 1967; Schenk, 1967; Textoris, 1969; Thompson, 1970; West and others, 1968). Whereas the evidence appears to support most of these cases, it does appear that over-generalization has resulted (Textoris, 1969; Zenger, 1970). The interpretation of a supra-tidal setting for ancient dolomites may be based as much on the fact that most Holocene dolomite is supratidal as on the evidence in the rocks themselves! In fact, some of the evidence itself may be equivocal. The main purpose of this paper is to direct attention to the prob-ability that a considerable portion of late Precambrian and early Paleozoic dolomite did not form in the supratidal environment.

Textoris (1969) has summarized evidence that he considered indicative of a penecon-temporaneous, supratidal origin for Paleozoic dolomites in the Appalachians, such as the following: dolomite occurs closer to terrigenous sources and limestone farther away; intercalated limestones show characteristics of intertidal and shallow subtidal deposits; desiccation cracks; birdseye structures; intraclasts; evap-orite crystal molds; some types of burrows [vertical]; algal stromatolites forming low, laterally linked hemispherical (LLH; Logan and others, 1964) structures; ostracodes; and dolomite rhombs in micrite. He noted that the paucity of fossils is due *not* to obliteration by dolomitization but rather to original scarcity. To his list could be added thinly laminated dolomite that is suggestive of algal mat activity (Freeman, 1966; Shinn and others, 1969).

As suggested above, many of these pieces of evidence are in themselves questionable, and it is the collective nature of the characteristics in certain units that is convincing. For example, there is some question concerning the conten-tion (Hatfield and Rohrbacher, 1966) that dolomite generally has a greater terrigenous content than associated limestones, even dolo-mites showing supratidal characteristics (Zen-ger, 1965a). Whereas algal mats are character-istic of the supratidal zone in south Florida and the Bahamas, stromatolites in the Persian Gulf and Shark Bay, Australia (Davies, 1970; Logan, 1961), are primarily intertidal. Although modern subtidal algal stromatolites (excluding oncolites) are uncommon, they do occur (Gebelein, 1969) and possibly they may have been much more impressive in the late Pre-cambrian and early Paleozoic (Garrett, 1970). Recently, Playford and Cockbain (1969) discovered columnar algal stromatolites in Devonian rocks of the Canning Basin, western Australia, which they interpreted as being the products of *deep-water* nonskeletal algae. Laminated sediments and evaporite crystals, although suggestive of supratidal conditions, are not infallible criteria. Gutstadt (1968, p. 1288) felt that birdseye structures cannot be considered diagnostic of the supratidal environ-ment. Also, it might be very difficult to distin-guish between the intertidal and supratidal paleoenvironments in certain instances; in such circumstances, the use of the term peritidal, as used by Folk (1970), to include both tidal flat zones seems more appropriate.

Over the past decade, I have examined in detail four dolomitic units as follows (Fig. 2): Lockport Formation, Middle Silurian, western and central New York (Zenger, 1965b); Herkimer Formation, Middle Silurian, central

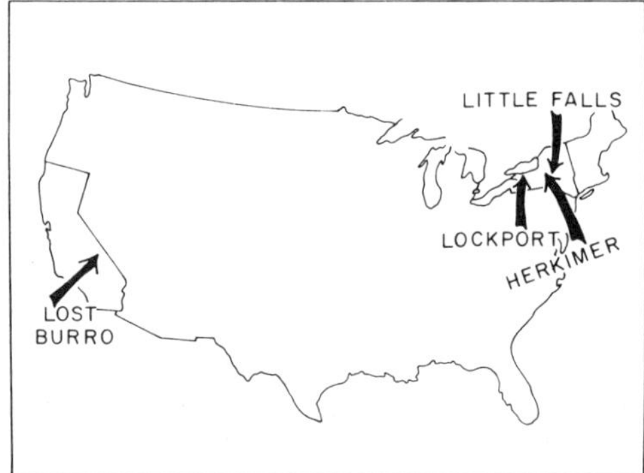

Figure 2.　Location of type sections of dolomitic units in New York and California that were studied by writer.

New York (Zenger, 1966, 1971); Little Falls Formation, Upper Cambrian, east-central New York (Zenger, 1969); Lost Burro Formation, Middle and Upper Devonian, east-central California (Zenger and Pearson, 1969).

The Lockport Formation (0 to 200 ft thick), including the fossiliferous Gasport Member (20 to 40 ft thick) in its lower part, consists primarily of associated limestone and dolomite. Underlying the Lockport in central New York is the Herkimer Formation which in its western portion (Joslin Hill Member, 80 to 100 ft thick) includes intercalated fossiliferous, sandy dolomite and dolomitic sandstone. The Little Falls Formation (400 ft thick) is mainly dolomite with considerable associated dolomitic sandstone. Lost Burro carbonates, about 2,000 ft thick in the western Basin and Range province, are divisible into a lower dolomite and an upper limestone. Details of the stratigraphy and petrology of these units may be found in the references given above. Table 1 summarizes their characteristics and suggested general environments of deposition as gauged by criteria which permit a comparison with modern and reported ancient supratidal dolomites.

There is ample, unequivocal evidence that the great bulk of dolomite in these units resulted from metasomatism of calcium carbonate sediments which represented various environments of deposition. With the exception of dolomitized bioherms, biostromes, and "non-reef" megafossils, most of the sound evidence is observed microscopically and consists of partially or wholly replaced skeletal debris, oöids, peloids, calcispheres, and grapestone, as well as calcitic poikilotopic dolomite.

Hemispherical, laterally linked stromatolites (Fig. 3), desiccation cracks, peloids, and oöids in parts of the upper Lockport and throughout much of the Little Falls suggest very shallow water to primarily tidal flat, or peritidal, conditions of deposition of the original sediments. Vertically stacked hemispheroids (SH; Logan and others, 1964) and flat laminated horizons are relatively common in the Little Falls. Despite the questionable nature of some of the individual criteria, the collective evidence appears quite reliable in suggesting a peritidal, possibly supratidal, depositional environment for a considerable portion of the sediments of these units. With regard to even these probable examples of supratidal occurrences, there are two immediately apparent differences from

the Holocene penecontemporaneous "analogues." The grain size in these particular Paleozoic dolomites ranges from micron-sized to coarse centimicron-sized but is generally between 40 and 200 μ (Fig. 4), considerably coarser than the modern dolomite, which is commonly in the 1 to 5 μ range. Micrite is relatively rare in the dolomites I have studied, existing mainly in cavities in fossils or associated with stromatolitic structures. Ranges of grain size of some reported ancient supratidal dolomites are: 5 to 10 μ and 30 to 70 μ (Laporte, 1967, p. 79); 10 to 25 μ (Schenk, 1967, p. 368, 372); and 10 to 50 μ (Matter, 1967, p. 603). Furthermore, the dolomite I have studied appears to be well-ordered, whereas modern dolomite is almost exclusively calcium-rich and

Figure 3. Algal stromatolite (LLH) in Little Falls Formation at type locality, Little Falls, New York (note pen knife in upper right corner for scale). Environment of deposition is interpreted as peritidal.

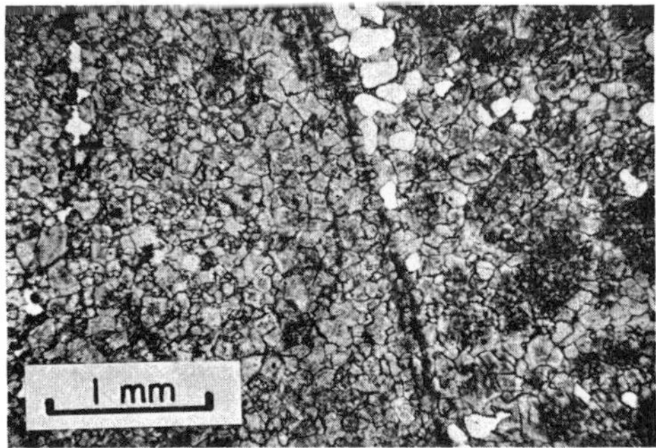

Figure 4. Photomicrograph of thin section (ordinary light) of margin of completely dolomitized algal stromatolite (left half of figure—note relic laminations trending steeply near center) and detrital filling (quartz grains, relict peloids) on right side. Note particularly the coarseness of dolomite crystals. Little Falls Formation, Middleville, New York.

TABLE 1. SUMMARY OF CHARACTERISTICS AND SUGGESTED ENVIRONMENTS OF DEPOSITION OF STUDIED DOLOMITE UNITS

Selected Points for Comparison with Modern and Reported Ancient Supra-tidal Dolomites	Units and Descriptions					
	Lockport Formation		Herkimer Formation	Little Falls	Lost Burro Formation	
	Gasport Member	"Upper" Lockport	Joslin Hill Member	Formation	"Lower"	"Upper"
Algal stromatolites	None observed	Common at certain horizons; mainly LLH	None observed	Relatively common; LLH, SH, possible flat algal laminations	Possible minor flat algal laminations	None observed
Evaporites	Relatively common; mainly sulfates in vugs	Common, mainly sulfates in vugs; pseudomorphs may represent syngenetic origin	None observed	None observed	None observed	None observed
Sedimentary structures	Minor cross laminations	Intraclasts; vuggy and porous zones; birdseyes rare; ripple marks; oöids; peloids; minor desiccation cracks	Cross laminations; ripple marks (para-ripples common); intraclasts minor	Intraclasts; cross laminations; desiccation cracks; oöids; peloids; grapestone; vuggy, birdseyes rare; vertical burrows?	Desiccation cracks, cross laminations in lower quartzitic portion; peloids	Peloids
Fossils	Bioherms; abundant marine fauna: brachiopods, corals, pelmatozoans, stromatoporoids, etc.	Fossiliferous zones, some biostromal; corals and stromatoporoids most common; brachiopods, ostracodes, trilobites	Abundant marine fauna: brachiopods, mollusks, pelmatozoans, ostracodes, trilobites, etc.	Extremely rare; sporadic zones with lingulid brachiopods	Rare; few zones with brachiopods, corals, and stromatoporoids	Stromatoporoid biostromes; less common brachiopods and corals; calcispheres
Insoluble content versus dolomite/calcite ratio	Untested; apparently no significant correlation	No significant correlation	Not applicable; essentially no calcite present	Not applicable; calcite very minor and late diagenetic	Not applicable; calcite very minor	No correlation
Nature of dolomite	Crystal size mainly 20 to 100μ; well ordered; apparently stoichiometric	Crystal size mostly 40 to 200μ; well ordered; apparently stoichiometric	Crystal size mainly 40 to 250μ; well ordered; dolomite is ferroan	Crystal size mainly 40 to 180μ; well ordered; apparently stoichiometric	Crystal size mainly 25 to 300μ; well ordered; apparently stoichiometric	Crystal size mainly 40 to 100μ; well ordered; apparently stoichiometric
Nature of dolomitization	Mainly diagenetic replacement; no conclusive evidence of any penecontemporaneous dolomitization	Mainly diagenetic replacement; local occurrences of rhombs in micrite may indicate some initial penecontemporaneous dolomitization	Mainly early diagenetic replacement; no conclusive evidence for any penecontemporaneous dolomitization	Mainly diagenetic replacement; complete dolomitization precludes determination of any penecontemporaneous dolomitization	Mainly diagenetic replacement; complete dolomitization precludes determination of any initial penecontemporaneous dolomitization	Mainly diagenetic replacement
General environment of deposition	Shallow subtidal shelf; normal salinity	Shallow subtidal, at times partly restricted to peritidal and in part evaporitic	Shallow subtidal shelf; normal salinity	Shallow subtidal to mainly peritidal (chiefly intertidal); no evidence of hypersaline conditions	Shallow subtidal to perhaps partly peritidal; normal or near-normal salinity	Shallow subtidal; normal or near-normal salinity

disordered. Figure 5 shows the lack of certain ordering reflections as well as displacement and broadening of the [211] dolomite line in a sample of a dolomitic crust from the Bahamas and compares this with an ordered, stromatolitic, supratidal (?) dolomite from the Cambrian Little Falls Formation. I do not wish to make a strong case for either of these differences (textural and structural) being used as evidence supporting my contention regarding the possible overemphasis on the importance of penecontemporaneous supratidal dolomite in the past. Time and the attendant subjection of penecontemporaneous protodolomite to the diagenetic environment could account for the differences mentioned. Graf and Goldsmith (1956, p. 184–185) suggested that ancient ordered dolomite may have gone through a protodolomite stage. They concluded that at earth-surface temperatures, the crystallochemical differences between Ca^{++} and Mg^{++} are insufficient to permit the energy available there to drive the ordering beyond the protodolomite stage with any degree of speed. Interestingly, in a later paper, Goldsmith and Graf (1958) reported some calcium-rich dolomite with weakened ordered reflections from the geologic record. P. W. Choquette (1971, written commun.) also has observed that dolomites of presumed early diagenetic origin from several Paleozoic units contain an excess of calcium. On the basis of the peak [211] position, the particular ordered Paleozoic dolomites I have studied do *not* appear to be calcium-rich.

In certain of my samples, there is some questionable evidence suggestive of dolomitization having begun penecontemporaneously. Obviously much of the process occurred later during diagenesis; the dolomites concluded their evolution in the decimicron- to centimicron-sized range. We may be too close to pass judgment on the textural appearance of Holocene dolomite beyond the penecontemporaneous or earliest diagenetic stage. Perhaps more thorough study of earlier Holocene and late Pleistocene dolomitization will prove helpful in this regard. Deffeyes and others (1965, p. 86) stated that Pliocene-Pleistocene dolomite rhombs on Bonaire averaged 75 μ.

Despite the compelling evidence for a supratidal site of deposition for much of the Little Falls and upper Lockport Formations, dolomitization need not have occurred there. As intimated above, it is difficult to ascertain whether the inception of dolomitization in a proposed supratidal environment was penecontemporaneous. In many reports of ancient supratidal dolomite (Matter, 1967; Thompson, 1970), the main reason for accepting a penecontemporaneous origin is the association of dolomite with sediments exhibiting supratidal characteristics. Assuming that Thompson (1970) has correctly identified subtidal and supratidal facies in Upper Ordovician rocks of the southern Appalachians, his conclusion of a supratidal origin for the dolomite appears reasonably justified; he (p. 1282) reported a high correlation between dolomite content and beds interpreted as supratidal. Fabric evidence, such as dolomite rhombs in micrite (Textoris, 1969) and preferential dolomitization of mud crack fillings (Laporte, 1967, p. 79–80), as criteria for penecontemporaneous dolomitization is commonly suggestive but not definitive. Freeman (1966) interpreted the dolomitization of two Ordovician formations in Arkansas as being late diagenetic (postlithification) despite their supratidal characteristics such as stromatolites, laminations, mud cracks, burrows, molds of halite crystals, and dolomite. The dolomite does *not* seem to be related to desiccation and the late origin is indicated by dolomite rhombs floating within sparry cement and the concentration of dolomite along stylolites, not as insoluble residue (*see also* Fig. 6). Freeman's observations indicate that the presence of dolomite should not be considered a solid piece of evidence for a supratidal occurrence!

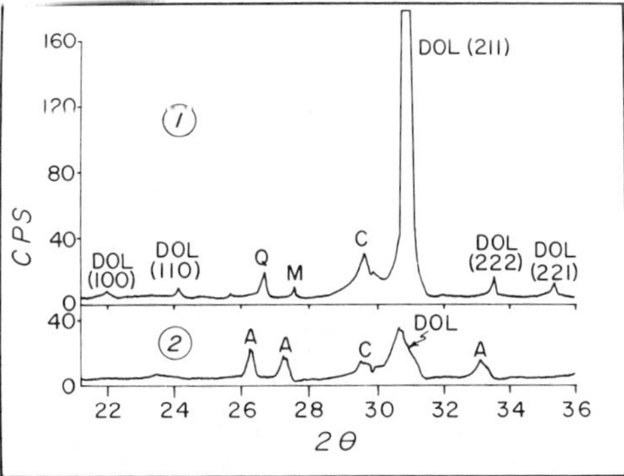

Figure 5. Comparison of x-ray diffractograms of sample of algal stromatolitic dolomite from Cambrian Little Falls Formation (1) and one from recent dolomitic crust from supratidal zone in Bahamas (2). Note line broadening and displacement of main dolomite reflection, [211], in the protodolomite (2). Also note apparent absence or insignificance of ordering reflections [100] and [221]. Although not shown here, a longer scan showed the absence of the [111] in the recent sample.

Figure 6. Late diagenetic dolomitization in Lower Ordovician Ritchie Limestone, Saratoga Springs, New York. Dolomite rhombs (dark) centered in fracture-filling spar (light). Micritic matrix (dark) shows no diagnostic evidence of supratidal deposition. Photomicrograph of acetate peel (ordinary light).

near-shore environment. The presence of stromatoporoid biostromes, rare brachiopods, corals, conodonts, and calcispheres suggest subtidal conditions. Dolomite forming most of the lower part of the formation at the type section (Panamint Mountains, California; Fig. 2) contains few fossils, but the apparent lack of burrows and evaporite crystal molds, and the rarity of possible flat algal laminations, make tenuous an interpretation of a supratidal environment for much of the section. However, sandstone-filled cracks in the lowest member of the unit probably represent desiccation structures and may reflect peritidal deposition.

Obviously, interpretations regarding the timing of dolomitization become much more difficult for a completely dolomitized sequence.

The upper three-fifths of the Lost Burro, consisting primarily of limestone with some intercalated dolomitic limestone and dolomite, accumulated in a shallow, mainly quiet-water,

The upper Lockport, part of which may represent initial supratidal accumulation, as discussed above, contains zones of *in situ* tetracorals, tabulates (Fig. 7), and stromatoporoids. Further, microscopic study reveals that many apparently unfossiliferous samples of Lockport dolomite contain ostracode and brachiopod shells essentially as abundant as in the associated intercalated limestones but in varying degrees of obliteration (Fig. 8). These are taken to represent subtidal accumulations.

Figure 7. Close-up of favositid biostrome, with vugs, Lockport Formation, just east of Niagara Falls, New York. Note abundance of *in situ Favosites*. Biostrome formed in subtidal environment and is completely dolomitized.

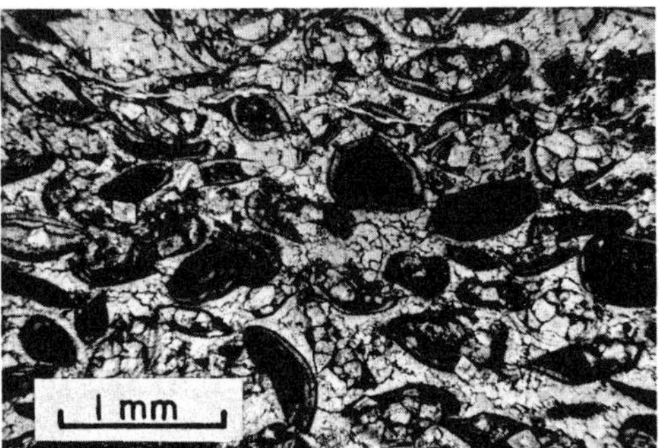

Figure 8. Photomicrograph of thin section (ordinary light) of tests of ostracode *Eukloedenella* partially replaced by dolomite crystals (large, light crystals within and transecting tests). Deposition interpreted as subtidal. Lockport Formation, Sherrill, New York.

Figure 9. Photomicrograph of thin section (crossed nicols) of sample of dolomitized bioclastic Joslin Hill Member (Herkimer Formation, Utica, New York). Note relatively well-preserved bryozoan zoaria and pelmatozoan fragments in quartzose (most of lightest grains are quartz) dolomite matrix. Deposition interpreted as subtidal and dolomitization as early diagenetic.

Fossils and ripple marks, commonly pararipples, are very common in the Joslin Hill Member of the Herkimer Formation which apparently formed in a neritic environment of normal salinity. Similar conditions probably prevailed during deposition of the lower fossiliferous and biohermal Lockport (Gasport Member). Although much of the abundant fossil material (brachiopods, bivalves, trilobites, ostracodes, pelmatozoans, and so on; Fig. 9) is fragmental, intact shells are common—I think they are too common to represent subtidal skeletons washed into the supratidal zone. The Gasport and Herkimer lack evidence of burrowing and desiccation. Some laminated dolomitic siltstone and dolarenite exist in the Herkimer; however, it seems quite clearly *not*

to be due to algal mat activity because of the extreme evenness of the laminae, the excellent sorting within most laminae, and the pronounced orientation of elongate grains parallel to the bedding. One could argue that, in a manner similar to the Canary Island occurrence of Müller and Tietz (1966), emergence of a shoreline area could transfer a subtidal accumulation to the supratidal zone where diagenetic dolomitization could occur. (Interestingly, in the Persian Gulf area, much of the "sabkha sediments" that have been dolomitized recently are algal-mat, intertidal sediments which now occur beneath supratidal deposits and landward of the present intertidal zone as a result of regression (Illing and others, 1965, p. 97; Kinsman, 1971).) However, the lack of evidence of penecontemporaneous evaporites having been associated with Gasport and Herkimer sediments does not suggest supratidal dolomitization.

In addition to these specific dolomite units, there are many others containing abundant neritic fossils or biohermal structures (or both) that must have accumulated subtidally. General examples of which most readers are aware are the widespread Middle Silurian reef dolomites of the Great Lakes region and the Middle Devonian reefs of Alberta. The Ely Springs Dolomite (Upper Ordovician), as I have observed it in east-central California, is richly fossiliferous, and except for some barren zones, almost certainly was not deposited in a tidal flat environment. In fact, after cursory examination of other dolomitic units in the Great Basin, I contend that there may be several unlikely candidates for a supratidal origin. Other reports of dolomites which represent original subtidal calcium carbonate sediments are numerous (Asquith, 1967; Fadeyev, 1962; Schmidt, 1965; and many others). Recently, Mossler (1972), in studying the Swope Formation (Pennsylvanian) of southeastern Kansas, concluded that the dolomite is not of supratidal origin; he noted a normal marine biota and lack of supratidal structures and textures in a dolomitic biomicrite subfacies. Behrens and Land (1970) reported a pure Holocene dolomite bed, dated 2,400 yrs B.P., in Baffin Bay, Texas; they concluded that the bed was deposited subtidally, in probably 5 or 6 m of water.

Lindholm (1969) proposed a detrital origin for some dolomites that contain evidence of accumulation below the strandline. Diagenetic overgrowths on the detrital nuclei could produce the typical crystalline fabric observed in most dolomite. He suggested that the source could be ancient dolomite or, appealing to the Holocene dolomite occurrences, supratidal dolomitic sediments forming nearby. Such supratidal dolomite is absent in the Onondaga Limestone of New York (Lindholm, 1969, p. 1037), but he considered that it may have been eroded. Parts of all the units I have studied show clastic textures and structures, but these seem to have been imposed on the grains by marine processes and do not necessitate a terrigenous source. Two problems immediately recognized in attempting to apply Lindholm's hypothesis to these units are the general lack of suitable source areas of dolomite and the lack of significant correlation between the amount of insoluble residue and dolomite content (Table 1; Zenger, 1965a; Zenger and Pearson, 1969, p. 52).

Seepage refluxion of Mg-rich brines (Adams and Rhodes, 1960) and "capillary concentration" or evaporative pumping of fluids moving through sea marginal sediments toward the sediment surface subjected to evaporation (Shinn and Ginsburg, 1964; Friedman and Sanders, 1967, p. 267, 280–281; Hsü and Siegenthaler, 1969) are two commonly invoked mechanisms for dolomitization under evaporative or hypersaline conditions. R. C. Murray and S. R. Bereskin (1970, oral commun.) have independently discussed with me the problem of sequences consisting of alternations of fossiliferous, subtidal dolostone and barren dolomite that might represent the supratidal environment. They envision seepage refluxion of brines from the supratidal zone represented by barren zones as the mechanism of dolomitization of the subtidal sequences below. This hypothesis is attractive, providing there are earmarks of supratidal deposition in the barren beds. Also, one should seek evidence of brines having moved through the subtidal succession, particularly evaporite crystals or pseudomorphs of such.

Evaporitic Dolomitization

There is considerable evidence that much Holocene dolomite, as well as a substantial part of the dolomite in the geologic record, accumulated under evaporitic conditions, especially in contact with brines with high Mg^{++}/Ca^{++} ratios. In fact, Friedman and Sanders (1967, p. 334, 338) go so far as to maintain that dolomite

is an evaporative mineral. Regardless of the type of a particular dolomite, they claimed it resulted from the action or reaction of hypersaline brines. It is not the intent of this paper to consider at great length whether dolomite should be universally treated as evaporitic; many modern occurrences and laboratory investigations favor the hypothesis. However, some doubts remain. In the Coorong area of Australia, maximum dolomite seems to be forming when the ionic proportions of the lakes and lagoon are those of normal sea water and pH is high (as much as 10.3) probably owing to abundant plant growth (Skinner, 1963, p. 459, 471). As carbonate deposition, including protodolomite, occurs under a wide range of salinites overlapping that of normal sea water, the controlling factor appears to be pH. A. M. Gaines (1970, written commun.), who has synthesized ordered dolomite at 100°C, favors dolomitizing waters of high Mg^{++}/Ca^{++} ratios owing to the effect of this ratio on the kinetics of dolomite formation. He does question the necessity of high solution concentration. Although Shinn (1968, p. 615) recorded very high concentrations and Mg^{++}/Ca^{++} ratios of interstitial waters of the supratidal dolomite crust on Sugarloaf Key, Atwood and Bubb (1970) have recently disputed these findings, reporting chlorinities and Mg^{++}/Ca^{++} ratios of the waters from the same crust as being very near that of normal Florida Bay water. They suggested that normal sea water might be the dolomitizing agent. Their view was supported by O. P. Bricker (1970, written commun.) who has measured near-normal chlorinities and Mg^{++}/Ca^{++} ratios of interstitial waters of supratidal dolomite crusts on northern Andros Island, Bahamas. He noted that despite the slight change in chlorinity from below normal to slightly above, the Mg^{++}/Ca^{++} ratios remained essentially constant. He considers a process whereby large quantities of water move through the sediment with small amounts of Mg^{++} (too little to be detected analytically against the background of Mg^{++} in sea water) being removed. Recently, Kinsman (1971) claimed that the variable Mg^{++}/Ca^{++} ratios in the Persian Gulf sabkha show little obvious relation to dolomite occurrence. It should be mentioned that whereas the radiocarbon age for the dolomite in the Bahaman crusts is contemporary (Shinn and others, 1965, p. 118), elsewhere it appears to be hundreds of years old. There is the possibility, then, of a change in the chemistry of interstitial waters between the time of dolomitization and the present. From a different experimental approach, Lippmann (1968) has presented a possible analogy between the formation of norsethite [$BaMg(CO_3)_2$] and dolomite. He suggested that high $CO_3^=$ content, hence alkaline conditions, may be critical for dolomite formation and that perhaps hypersalinity and high Mg^{++}/Ca^{++} ratios are not required. It is of interest and possibly of significance in this regard to note that the presence of authigenic feldspars in the Little Falls Formation and in overlying Ordovician dolomites in New York (Braun and Friedman, 1969, p. 116–117) may signify alkaline conditions of deposition.

On the other side of the coin, Davies (1970, p. 179, 197) reported evaporitic conditions and gypsum in supratidal sediments of Shark Bay, Australia, but most of the associated crusts are composed of aragonite with no trace of dolomite. Furthermore, there are exceptions to the common association of dolomite and evaporites in the geologic record. For example, there is essentially no dolomite in the highly evaporitic Purbeckian limestones of the Dorset Coast, England (I. N. West, 1969, oral commun.). Hudson (1970) has reported calcite pseudomorphs after gypsum from the Great Estuarine Series (Middle Jurassic) of Scotland, but no dolomite was observed. Sonnenfeld (1964, p. 126) earlier remarked that ". . . the occurrence of evaporites in juxtaposition with limestones as well as dolomites indicates that evaporation with the resulting magnesium concentration is not the decisive factor."

Some Lockport dolomite contains gypsum and anhydrite which undoubtedly have their source in the evaporitic Salina Group above. However, the great bulk of the sulfates occurs in vugs and fractures which postdate the main metasomatism; it is difficult to determine whether seepage refluxion was important in the dolomitization process. Farther east, the Herkimer (Joslin Hill Member) is separated from the evaporites by overlying impervious dark shales. This relation, coupled with the lack of evidence of evaporites in the unit, suggests the improbability of dolomitization by seepage refluxion. It is conceivable that some dolomitization of the Little Falls may have occurred through capillary concentration or by seepage refluxion from overlying lagoonal waters, but there is no evidence of laterally or vertically adjacent evaporite deposits nor have any evaporite molds been recognized under the microscope. Certainly there is no direct evidence that hypersaline brines moved through the sediments constituting the Lost Burro Formation.

CONCLUSIONS

It is indeed easier to pose questions than to resolve them. In another paper, I hope to consider more fully various questions concerning dolomitization including those mentioned here and the main dolomite problem itself. For example, one interesting aspect is the large discrepancy between the extent of Holocene dolomite and that along any horizon in the early Paleozoic. In the present paper, I wish to point out the possible overextrapolation between Holocene, penecontemporaneous supratidal dolomite and ancient dolomites. I want to reinforce the fact that many ancient carbonates dolomitized partially or wholly, penecontemporaneously or diagenetically, either did not accumulate in a supratidal environment or were not replaced there. Collective evidence, with little reliance on the presence of dolomite itself, should be sought and considered carefully before assuming a supratidal setting not only for the site of deposition of the original sediments but for initial dolomitization. Higher than normal concentrations and Mg^{++}/Ca^{++} ratios may be associated in "environments" conducive to dolomite formation, but are they necessary? Despite the common association of dolomite and evaporites, the evidence presented here suggests that it may also be an overgeneralization to categorize all dolomites as being evaporitic.

ACKNOWLEDGMENTS

Acknowledgment is made to the donors of the Petroleum Research Fund, administered by the American Chemical Society, for partial support of my research. Field work undertaken during the past 10 years was supported primarily by the New York State Geological Survey and partly by Geological Society of America Grant 928–63. I am very indebted to A. M. Gaines and O. P. Bricker for providing unpublished information. The manuscript has been critically read by S. R. Bereskin, O. P. Bricker, P. W. Choquette, J. F. Davis, D. W. Fisher, A. M. Gutstadt, W. B. Rogers, D. A. Textoris, A. O. Woodford, and my wife Ann, to whom I am grateful.

REFERENCES CITED

Adams, J. E., and Rhodes, M. L., 1960, Dolomitization by seepage refluxion: Am. Assoc. Petroleum Geologists Bull., v. 44, p. 1912–1921.

Alderman, A. R., and Skinner, H.C.W., 1957, Dolomite sedimentation in the southeast of South Australia: Am. Jour. Sci., v. 255, p. 561–567.

Asquith, G. B., 1967, The marine dolomitization of the Mifflin Member Platteville Limestone in southwest Wisconsin: Jour. Sed. Petrology, v. 37, p. 311–326.

Atwood, D. K., and Bubb, J. N., 1970, Distribution of dolomite in a tidal flat environment, Sugarloaf Key, Florida: Jour. Geology, v. 78, p. 499–505.

Behrens, E. W., and Land, L. S., 1970, Subtidal, Holocene dolomite in Baffin Bay, Texas: Geol. Soc. America, Abs. with Programs (Ann. Mtg.), v. 2, no. 7, p. 491–492.

Braun, M., and Friedman, G. M., 1969, Carbonate lithofacies and environments of the Tribes Hill Formation (Lower Ordovician) of the Mohawk Valley, New York: Jour. Sed. Petrology, v. 39, p. 113–135.

Davies, G. R., 1970, Algal-laminated sediments, Gladstone Embayment, Shark Bay, Western Australia, *in* Logan, B. W., Davies, G. R., Read, J. F., and Cebulski, D. E., eds., Carbonate sedimentation and environments, Shark Bay, Western Australia: Tulsa, Am. Assoc. Petroleum Geologists Mem. 13, p. 169–205.

Deffeyes, K. S., Lucia, F. J., and Weyl, P. K., 1965, Dolomitization of Recent and Plio-Pleistocene sediments by marine evaporite waters on Bonaire, Netherlands Antilles, *in* Pray, L. C., and Murray, R. C., eds., Dolomitization and limestone diagenesis: Soc. Econ. Paleontologists and Mineralogists Spec. Pub. 13, p. 71–87.

Fadeyev, F. I., 1962, Upper Carboniferous dolomite in the Kuybyshev area of the Volga: Akad. Nauk SSSR Doklady, v. 147, p. 180–182.

Fairbridge, R. W., 1957, The dolomite question, *in* LeBlanc, R. J., and Breeding, J. G., eds., Regional aspects of carbonate deposition: Soc. Econ. Paleontologists and Mineralogists Spec. Pub. 5, p. 125–178.

Folk, R. L., 1970, Evidence of peritidal origin for part of the Caballos Novaculite (Devonian, radiolarian chert), Marathon Basin, Texas: Geol. Soc. America, Abs. with Programs (Ann. Mtg.), p. 552–553.

Freeman, T., 1966, Post-lithification dolomite in the Joachim and Plattin Formations (Ordovician), northern Arkansas: Geol. Soc. America, Abs. for 1965, Spec. Paper 87, p. 59.

Friedman, G. M., and Sanders, J. E., 1967, Origin and occurrence of dolostones, *in* Chilingar, G. V., Bissell, H. J., and Fairbridge, R. W., eds., Carbonate rocks, 9A: Amsterdam, Elsevier, p. 267–348.

Garrett, P., 1970, Phanerozoic stromatolites: Noncompetitive ecologic restriction by grazing and burrowing animals: Science, v. 169, p. 171–173.

Gebelein, C. D., 1969, Distribution, morphology, and accretion rate of Recent subtidal algal stromatolites, Bermuda: Jour. Sed. Petrology, v. 39, p. 46–69.

Goldsmith, J. R., and Graf, D. L., 1958, Structural and compositional variations in some natural dolomites: Jour. Geology, v. 66, p. 678–693.

Graf, D. L., and Goldsmith, J. R., 1956, Some hydrothermal syntheses of dolomite and protodolomite: Jour. Geology, v. 64, p. 173–186.

Gutstadt, A. M., 1968, Petrology and depositional environments of the Beck Spring Dolomite (Precambrian), Kingston Range, California: Jour. Sed. Petrology, v. 38, p. 1280-1289.

Hatfield, C. B., and Rohrbacher, T. J., 1966, Dolomite-insoluble residue relationships in the Ten Mile Creek Dolomite (Middle Devonian) near Toledo, Ohio: Jour. Sed. Petrology, v. 36, p. 828–831.

Hsü, K. J., and Siegenthaler, C., 1969, Preliminary experiments on hydrodynamic movement induced by evaporation and their bearing on the dolomite problem: Sedimentology, v. 12, p. 11–25.

Hudson, J. D., 1970, Algal limestones with pseudomorphs after gypsum from the Middle Jurassic of Scotland: Lethaia, v. 3, p. 11–40.

Illing, L. V., Wells, A. J., and Taylor, J. C. M., 1965, Penecontemporaneous dolomite in the Persian Gulf, *in* Pray, L. C., and Murray, R. C., eds., Dolomitization and limestone diagenesis: Soc. Econ. Paleontologists and Mineralogists Spec. Pub. 13, p. 89–111.

Ingerson, E., 1962, Problems of the geochemistry of sedimentary carbonate rocks: Geochim. et Cosmochim. Acta, v. 26, p. 815–847.

Kinsman, D. J. J., 1971, Early diagenesis of carbonate sediments in a supratidal evaporitic setting: Am. Assoc. Petroleum Geologists Bull., v. 55, p. 167–168.

Laporte, L. F., 1967, Carbonate deposition near mean sea-level and resultant facies mosaic: Manlius Formation (Lower Devonian) of New York State: Am. Assoc. Petroleum Geologists Bull., v. 51, p. 73–101.

Lindholm, R. C., 1969, Detrital dolomite in Onondaga Limestone (Middle Devonian) of New York: Its implications to the "dolomite question": Am. Assoc. Petroleum Geologists Bull., v. 53, p. 1035-1042.

Lippmann, F., 1968, Synthesis of $BaMg(CO_3)_2$ (norsethite) at 20°C and the formation of dolomite in sediments, *in* Müller, G., and Friedman, G. M., eds., Recent developments in carbonate sedimentology in central Europe: Heidelberg, Springer-Verlag, p. 33–37.

Logan, B. W., 1961, *Cryptozoon* and associated stromatolites from the Recent, Shark Bay, Western Australia: Jour. Geology, v. 69, p. 517–533.

Logan, B. W., Rezak, R., and Ginsburg, R. N., 1964, Classification and environmental significance of algal stromatolites: Jour. Geology, v. 72, p. 68–83.

Lovering, T. S., 1969, The origin of hydrothermal and low temperature dolomite: Econ. Geology, v. 64, p. 743–754.

Matter, A., 1967, Tidal flat deposits in the Ordovician of western Maryland: Jour. Sed. Petrology, v. 37, p. 601–609.

Mossler, J. H., 1972, Diagenesis and dolomitization of Swope Limestone (Upper Pennsylvanian) southeast Kansas: Jour. Sed. Petrology (in press).

Müller, G., and Tietz, G., 1966, Recent dolomitization of Quaternary biocalcarenites from Fuerteventura (Canary Islands): Contr. Mineralogy and Petrology, v. 13, p. 89–96.

Playford, P. E., and Cockbain, A. E., 1969, Algal stromatolites: Deepwater forms in the Devonian of Western Australia: Science, v. 165, p. 1008–1010.

Schenk, P. E., 1967, The Macumber Formation of the Maritime Provinces, Canada—a Mississippian analogue to Recent strand-line carbonates of the Persian Gulf: Jour. Sed. Petrology, v. 37, p. 365–376.

Schmidt, V., 1965, Facies, diagenesis, and related reservoir properties in the Gigas Beds (Upper Jurassic), northwestern Germany, *in* Pray, L. C., and Murray, R. C., eds., Dolomitization and limestone diagenesis: Soc. Econ. Paleontologists and Mineralogists Spec. Pub. 13, p. 125–178.

Shinn, E. A., 1968, Selective dolomitization of some Recent sedimentary structures: Jour. Sed. Petrology, v. 38, p. 612–616.

Shinn, E. A., and Ginsburg, R. N., 1964, Formation of Recent dolomite in Florida and the Bahamas [abs.]: Am. Assoc. Petroleum Geologists Bull., v. 48, p. 547.

Shinn, E. A., Ginsburg, R. N., and Lloyd, R. M., 1965, Recent supratidal dolomite from Andros Island, Bahamas, *in* Pray, L. C., and Murray, R. C., eds., Dolomitization and limestone diagenesis: Soc. Econ. Paleontologists and Mineralogists Spec. Pub. 13, p. 112–123.

Shinn, E. A., Lloyd, R. M., and Ginsburg, R. N., 1969, Anatomy of a modern carbonate tidal flat, Andros Island, Bahamas: Jour. Sed. Petrology, v. 39, p. 1202–1228.

Skinner, H. C. W., 1963, Precipitation of calcian dolomites and magnesian calcites in the southeast of South Australia: Am. Jour. Sci., v. 261, p. 449–472.

Sonnenfeld, P., 1964, Dolomites and dolomitization: A review: Canadian Petroleum Geology Bull., v. 12, p. 101–132.

Textoris, D. A., 1969, Supratidal origin of Appalachian Basin dolostone: Geol. Soc. America, Abs. with Programs for 1968, Spec. Paper 121, p. 470–471.

Thompson, A. M., 1970, Tidal-flat deposition and early dolomitization in Upper Ordovician rocks of southern Appalachian Valley and Ridge: Jour. Sed. Petrology, v. 40, p. 1271–1286.

West, I. M., Brandon, A., and Smith, M., 1968, A tidal flat evaporitic facies in the Visean of Ireland: Jour Sed. Petrology, v. 38, p. 1079–1093.

Zenger, D. H., 1965a, Calcite-dolomite ratios vs. insoluble content in the Lockport Formation (Niagaran) in New York State: Jour. Sed. Petrology, v. 35, p. 262–265.

—— 1965b, Stratigraphy of the Lockport Formation (Middle Silurian) in New York State: New York State Mus. and Sci. Service Bull. 404, 210 p.

—— 1966, Revision of the Herkimer Sandstone (Middle Silurian), New York: Geol. Soc. America Bull., v. 77, p. 1159–1165.

—— 1969, Stratigraphy of the Little Falls Formation (Upper Cambrian) in the type area, east-central New York: Geol. Soc. America, Abs. with Programs for 1969, pt. 1 (Northeastern Sec.), p. 67–68.

—— 1970, Supratidal dolostones: An overemphasis on their significance in geologic record? [abs.]: Am. Assoc. Petroleum Geologists Bull., v. 54, p. 561.

—— 1971, Uppermost Clinton (Middle Silurian) stratigraphy and petrology, east-central New York: New York State Mus. and Sci. Service Bull. 417, 58 p.

Zenger, D. H., and Pearson, E. F., 1969, Stratigraphy and petrology of Lost Burro Formation, Panamint Range, California, *in* Short contributions to California geology: California Div. Mines and Geology Spec. Rept. 100, p. 45–65.

MANUSCRIPT RECEIVED BY THE SOCIETY APRIL 2, 1971

REVISED MANUSCRIPT RECEIVED JULY 6, 1971

PUBLISHED BY PERMISSION OF THE ASSISTANT COMMISSIONER, NEW YORK STATE MUSEUM AND SCIENCE SERVICE, JOURNAL SERIES NO. 128

Part V

DOLOMITIZATION IN A MIXING ZONE OF MARINE-METEORIC WATER

Editors' Comments
on Papers 17 Through 20

17   HANSHAW, BACK, and DEIKE
A Geochemical Hypothesis for Dolomitization by Ground Water

18   BADIOZAMANI
The Dorag Dolomitization Model—Application to the Middle Ordovician of Wisconsin

19   FOLK and LAND
Mg/Ca Ratio and Salinity: Two Contols over Crystallization of Dolomite

20   DUNHAM and OLSON
Diagenetic Dolomite Formation Related to Paleozoic Paleogeography of the Cordilleran Miogeocline in Nevada

The papers in this part discuss alternatives to the seepage refluxion and evaporative pumping models for dolomitization. The authors agree that in many ancient (and modern—Zenger, Paper 16) carbonates, the widely accepted association between dolomitization and hypersalinity is not applicable, and each discussed examples in which subtidal carbonates have been diagenetically dolomitized by mixed meteoric-marine groundwaters. Indeed, the concept of dolomitization by groundwater is not a new idea; it was introduced as early as 1835 by de Collegno. Runnells's important study (1969) on diagenetic effects, such as dolomitization, promoted by mixed subsurface waters of various kinds was important in the later development of the concept of dolomitization by mixed meteoric and marine groundwaters. Many studies dealing with mixing zone dolomites have appeared in the recent literature, including those of Hsü (1963, 1966), Land (1973), Land and Epstein (1970), Land et al. (1975), and others. It is noteworthy that the process of mixing zone dolomitization was appreciated through two different approaches: carbonate petrology (Land, 1973) and water geochemistry (Hanshaw, Back, and Deike, Paper 17).

In their important study of Tertiary aquifers of Florida and Yucatán, Hanshaw, Back, and Deike (Paper 17) evaluated the role of groundwaters in the formation of regional dolomites. They concluded that significant amounts of replacement dolomite generally cannot be formed in deep phreatic aquifers (Steidtmann, 1911; Hsü, 1966), despite the fact that waters may be saturated with respect to dolomite and may approach critical Mg/Ca ratios of one-to-one unless sufficient quantities of Mg^{2+} are supplied by "solution-cannibalism" (Goodell and Garman, 1969). Instead they suggested that regional dolomitization may occur by an increase in the Mg/Ca ratio through a mixing of meteoric and subjacent marine waters along, for example, a freshwater-saltwater interface.

Following this theme, Badiozamani (Paper 18) developed a geochemical model ("Dorag") that he used to demonstrate that a mixing of meteoric waters with from 5 to 30 percent seawater results in a hybrid solution that becomes increasingly more saturated with respect to dolomite. His model, tested in studies of Middle Ordovician carbonates in Wisconsin and adjacent states, presumed to explain the origin of dolomite along positive structural elements without requiring Mg/Ca ratios greater than unity.

No one deserves more credit than L. S. Land for popularizing the concept of mixing zone dolomitization. In another paper, Land and his colleague, R. L. Folk (Paper 19), maintained that ordered dolomite can form most easily from fluids of Mg/Ca ratios of one-to-one if crystallization rates are very slow or if solutions are dilute, with few interfering ionic species; protodolomite instead forms under higher Mg^{2+} concentrations (Mg/Ca five-to-one to ten-to-one) when extreme ionic concentrations are attained and Mg^{2+} is forced to precipitate. However, these suggestions must be qualified considering the apparent field evidence from Florida and the Bahamas (Shinn et al., Paper 11; Atwood and Bubb, 1970) in which higher than unity values of Mg/Ca ratios may be common during dolomitization. Folk and Land concluded that although dolomite can and probably does form in hypersaline and normal-marine environments, dolomitization can be affected by freshwater dilution and consequently may also occur in "schizohaline" subsurface environments, mixed subsurface waters (fresh and hypersaline or normal marine), or, less commonly, phreatic waters enriched in cannibalized Mg^{2+}.

Dunham and Olson (Paper 20) provide another example of the groundwater dolomitization of Paleozoic subtidal carbonates in which freshwater was derived from coeval, subaerially exposed land masses. Citing Folk and Land (Paper 19) and Land, Salem, and Morrow (1975), they offered as supportive petrographic evidence the lack of evaporites and the presence of relatively large (>5 μm) replacement dolomite crystals (see also Behrens and Land, 1972). The question of relating dolomite crystal fabrics and geochemistry to various replacement modes, however, is a subject deserving further study.

REFERENCES

Atwood, D. K., and J. N. Bubb, 1970, Distribution of Dolomite in a Tidal Flat Environment, Sugarloaf Key, Florida, *Jour. Geology* **78**:499-505.

Behrens, E. W., and L. S. Land, 1972, Subtidal Holocene Dolomite, Baffin Bay, Texas, *Jour. Sed. Petrology* **42**:155-161.

Collegno, P. de, 1835, Notes sur quelques Points des Alpes Suisses, *Soc. Géol. France Bull.* **6**:106-114.

Goodell, H. G., and R. K. Garman, 1969, Carbonate Geochemistry of Superior Deep Test Well, Andros Island, Bahamas, *Am. Assoc. Petroleum Geologists Bull.* **53**:513-536.

Hsü, K. J., 1963, Solubility of Dolomite and Composition of Florida Ground Waters, *Jour. Hydrology* **1**:288-310.

Hsü, K. J., 1966, Origin of Dolomite in Sedimentary Sequences:A Critical Analysis, *Mineralium Deposita* **2**:133-138.

Land, L. S., 1973, Contemporaneous Dolomitization of Middle Pleistocene Reefs by Meteoric Water, North Jamaica, *Bull. Marine Sci.* **23**:64-92.

Land, L. S., and S. Epstein, 1970, Late Pleistocene Diagenesis and Dolomitization, North Jamaica, *Sedimentology* **14**:187-200.

Land, L. S., M. R. Salem, and D. W. Morrow, 1975, Paleohydrology of Ancient Dolomites: Geochemical Evidence, *Am. Assoc. Petroleum Geologists Bull.* **59**:1602-1625.

Runnells, D. D., 1969, Diagenesis, Chemical Sediments, and the Mixing of Natural Waters, *Jour. Sed. Petrology* **39**:1188-1201.

Steidtmann, E., 1911, The Evolution of Limestone and Dolomite, *Jour. Geology* **19**:323-345, 392-428.

17

A Geochemical Hypothesis for Dolomitization by Ground Water[1]

BRUCE B. HANSHAW, WILLIAM BACK, AND RUTH G. DEIKE

Abstract

Most modern disordered dolomite has been found in dynamic environments. However, solutions associated with modern dolomite formation do not have a common Mg/Ca ratio; the ratio ranges from about 3 to 100. Ground-water circulation may have a significant role in formation of regional dolomites; one of the primary requirements for regional dolomite formation is a large supply of magnesium ions.

An X-ray study of well cuttings from the Tertiary limestone aquifer of central Florida indicates that it is composed primarily of calcite and dolomite with minor amounts of quartz and apatite. The magnesium content of the calcite is slightly lower ($0–2$ percent $MgCO_3$) in the recharge areas than in the deeper confined parts of the aquifer system ($2–4$ percent $MgCO_3$).

Our data support other recent work and indicate an equilibrium constant for dolomite of 2×10^{-17}. Inasmuch as this value is exactly the square of the calcite equilibrium constant ($10^{-8.35}$), the Mg/Ca ratio must be unity for the three-phase equilibrium, calcite-dolomite-water. The Mg/Ca ratio in water from the aquifer is as low as 0.05 in the recharge area where the water is also undersaturated with respect to both calcite and dolomite. With time and length of travel path in the system, the water increases systematically in Mg/Ca ratio, which approaches unity; saturation with respect to the two carbonates also increases downgradient until the solution apparently becomes oversaturated with respect to both carbonates. In Tertiary limestones of the Yucatan Peninsula, the Mg/Ca range in water is similar to that for Florida.

The small amount of magnesium available from the solution of magnesium calcites and dolomite in the potable zone of active circulation is insufficient to provide the amount required for extensive dolomitization unless enormous quantities of rock are available for dissolution. However, dolomite may be forming in the zones of brackish water that underlie the Florida and Yucatan Peninsulas. The required magnesium may be derived from the readily available ocean water or reflux brines as the hydrologic regimen is changed because of relative fluctuations of sea level.

Introduction

ONE of the enigmas of carbonate geochemistry is that the oceans are apparently at or beyond saturation with respect to both pure calcite and dolomite, whereas modern marine sediments are composed of the metastable phases, aragonite and high-magnesium calcite plus low-magnesium calcite. Aragonite and high-magnesium calcite are generally, though not always, associated with shell fragments. Ancient carbonate rocks, on the other hand, are chiefly low-magnesium calcite and dolomite. Study of the geochemistry and hydrology of the Tertiary carbonate aquifer systems of the Florida and Yucatan Peninsulas bear on this perplexing problem. The purpose of this paper is to evaluate the role of ground water in the formation of regional dolomites.

Carbonate Geochemistry

General Background.—The discovery of modern disordered dolomite (similar to the "protodolomite"

synthesized by Graf and Goldsmith in 1955) by Alderman and Skinner (1957) provided new data on the mode and environment of dolomite formation. Since the initial work in the Coorong lagoon of southern Australia, the area has been studied extensively (Alderman and von der Borch, 1961, 1963; Alderman, 1965; von der Borch, 1965a, 1965b; Skinner, 1963; Skinner, Skinner, and Rubin, 1963; and Clayton et al., 1968). The Coorong dolomite occurs in a restricted lagoonal environment and is associated with an assemblage of evaporite minerals. The pH of the associated water can be high ($\sim$10) and the molal Mg/Ca ratio in solution ranges from a low of 7 to as high as 100. These significant findings in the Coorong lagoon have spurred other workers to look for modern dolomites elsewhere and in other environments. The search has been rewarding.

Wells (1962) and Curtis et al. (1963) reported modern dolomite and its association with evaporite minerals from coastal lagoons and tidal flats in the Persian Gulf. A comprehensive report on modern

[1] Publication authorized by the Director, U. S. Geological Survey.

211

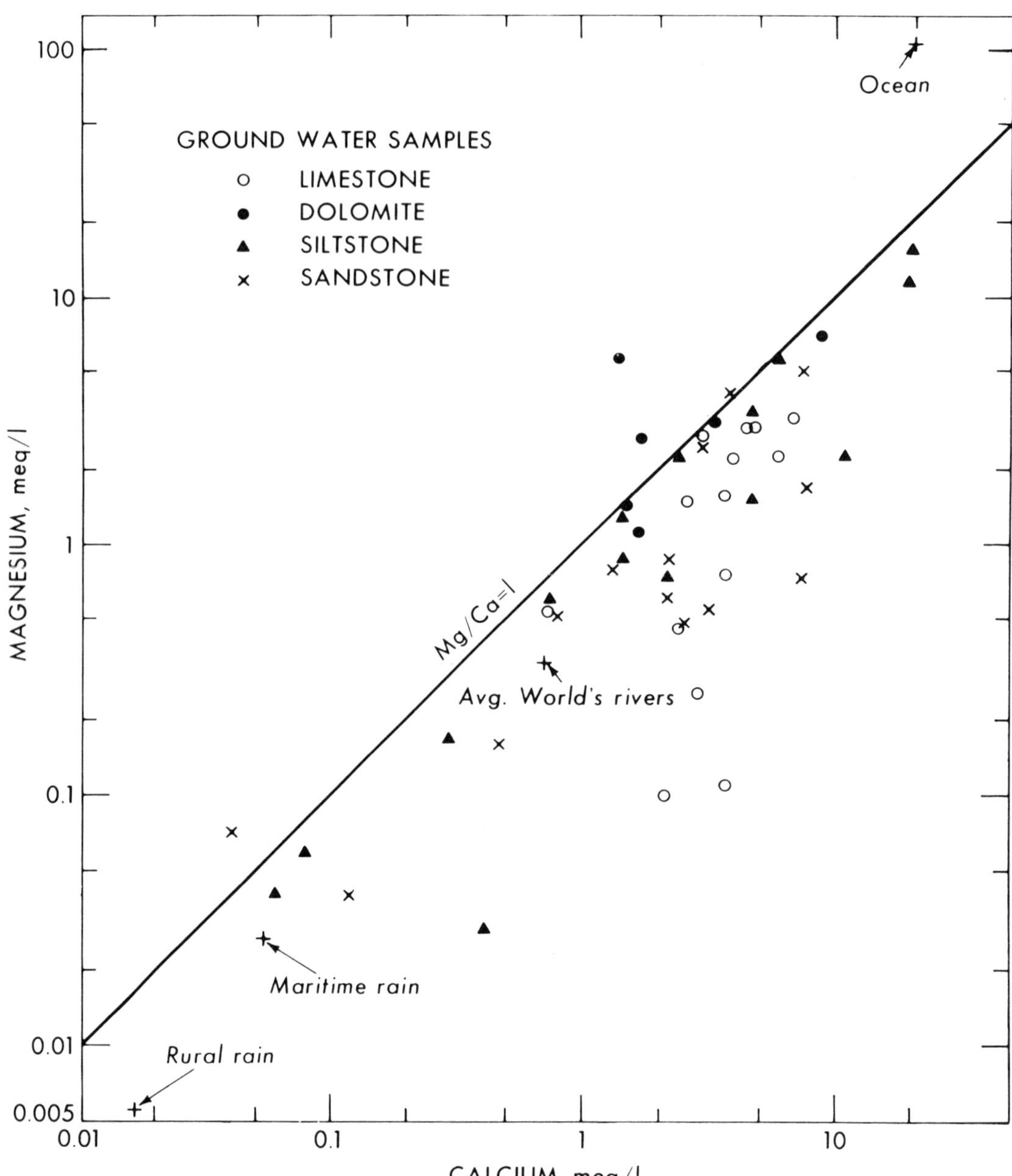

Fig. 1. Plot of Mg versus Ca for water samples from various aquifers. Ground water data from White, Hem, and Waring (1963) and average composition of world's rivers from Holland (written communication, 1969).

dolomites from these salt flats is given by Illing, Wells, and Taylor (1965). The Mg/Ca ratio (all Mg/Ca ratios used in this paper are molal ratios) in interstitial water on the flats is greater than that of normal ocean water (5.2) and apparently increases to a value roughly twice that of ocean water owing to precipitation of calcium carbonates and sulfates in sediments on the flats. Farther inland, the Mg/Ca decreases to as low as 3, presumably owing to dolomitization.

Modern dolomite from supratidal flats in the Florida Keys was reported by Shinn (1964, 1968). Modern dolomite from a similar environment in the Bahamas was discussed in detail by Shinn, Ginsburg, and Lloyd (1965), who reported Mg/Ca ratios as high as 40 in the interstitial solution. However, Atwood and Bubb (1970) found that the Mg/Ca ratio is close to ocean water in interstitial water from tidal flat sediments on Sugarloaf Key, Florida, which contains modern dolomite (Shinn, 1964). Also, Bricker (oral communication, 1969) reports a Mg/Ca ratio of approximately 3 in interstitial water from supratidal flats on Andros Island, Bahamas. These sediments also contain penecontemporaneous dolomite.

The formation of dolomite by the diagenetic action of magnesium brines on limestone sediments was suggested by Newell et al. (1953) and discussed at length by Adams and Rhodes (1960), who termed the mechanism "seepage refluxion." According to Deffeyes, Lucia, and Weyl (1965), the mechanism is operative today within permeable carbonate sediments on Bonaire Island in the Netherlands Antilles. These authors suggested that a brine produced by evaporation and more concentrated and dense than sea water flows downward into the carbonate sediments owing to density differences. This brine, with a Mg/Ca ratio of about 30, presumably dolomitizes the permeated sediments. An interesting example of this type of dolomitization was recently described by Fisher and Rodda (1969) from the Edwards Formation of Texas. Here an extensive evaporite facies occupies the center of a former lagoonal area. A dolomitic facies encircles the former lagoon. The MgCO₃ content of the dolomitic facies decreases away from the former lagoon and grades into a normal limestone facies. The origin of the dolomite was attributed to metasomatic action of magnesium-enriched brines on calcium carbonate sediments.

Modern dolomites have also been reported from continental closed basins. Jones (1961) described dolomite in saline sediments from Deep Springs Lake, a small playa in California. By means of radiocarbon dating, Peterson, Bien, and Berner (1963) subsequently proved the dolomite to be modern. According to Clayton, Jones, and Berner (1968), the Mg/Ca ratio of coexisting solutions averages about 6. Modern dolomites from saline lake environments have also been reported in Texas by Reeves and Parry (1965) and in the Great Salt Lake Desert by Graf, Eardley, and Shimp (1961). Unfortunately, these reports provided no information on the Mg/Ca ratio of interstitial water.

All of the environments of modern dolomite formation cited above should provide diagnostic sedimentary features if the material were preserved in the rock record. These would include association with evaporites and/or shallow water or supratidal phenomena, such as mud cracks and broken shell debris. None of these mechanisms can account for many of the dolomites of regional extent which are common in the geologic record. Where one of these mechanisms may have been operative on a large scale, as described by Fisher and Rodda (1969), we find the typical association with a large lagoonal environment and attendant evaporite minerals. More detailed study of ancient rocks may reveal other regional dolomite-evaporite associations (Sloss, 1953).

From the foregoing review, we see that the Mg/Ca ratio reported for interstitial water in dolomite sediments varies from a low of about 3 to as high as 100. Land (1967) points out that this ratio is important in the *kinetics* of dolomitization; the higher values may not be necessary for dolomitization to occur, but they do speed up the process. Therefore, the formation of dolomite is at least a twofold problem: (1) kinetics, i.e., sufficient time, and (2) source of sufficient magnesium ion. In ground-water systems, sufficient time is generally available; the problem then reduces to determining whether there is sufficient magnesium available. In order to do this, we must know the specific Mg/Ca ratio in solution for the three-phase equilibrium, dolomite-calcite-solution. If this ratio is exceeded in any limestone terrane, then dolomitization should proceed spontaneously, given enough time.

At the three-phase equilibrium, a unique species ratio in solution must exist, provided that the fluctuations of temperature and pressure are kept reasonably small, as in most ground water environments. Consider the reaction:

$$CaMg(CO_3)_2 = Ca^{+2} + Mg^{+2} + 2CO_3^{-2}.$$

The equilibrium constant, K_d, is equal to the activity, α, of the ions as follows:

$$K_d = \alpha_{Ca^{+2}} \cdot \alpha_{Mg^{+2}} \cdot \alpha_{(CO_3^{-2})}^2.$$

Multiplying the equation by $\dfrac{\alpha_{Ca^{+2}}}{\alpha_{Ca^{+2}}}$ gives:

$$K_d = \alpha_{Mg^{+2}} \cdot \frac{\alpha_{(Ca^{+2})^2}}{\alpha_{Ca^{+2}}} \cdot \alpha_{(CO_3^{-2})^2}$$

and because:

$$\alpha_{(Ca^{+2})^2} \cdot \alpha_{(CO_3^{+2})^2} = (K_c)^2$$

we arrive at:

$$\frac{K_d}{K_c^2} = \frac{\alpha_{Mg^{+2}}}{\alpha_{Ca^{+2}}} = \text{a constant}$$

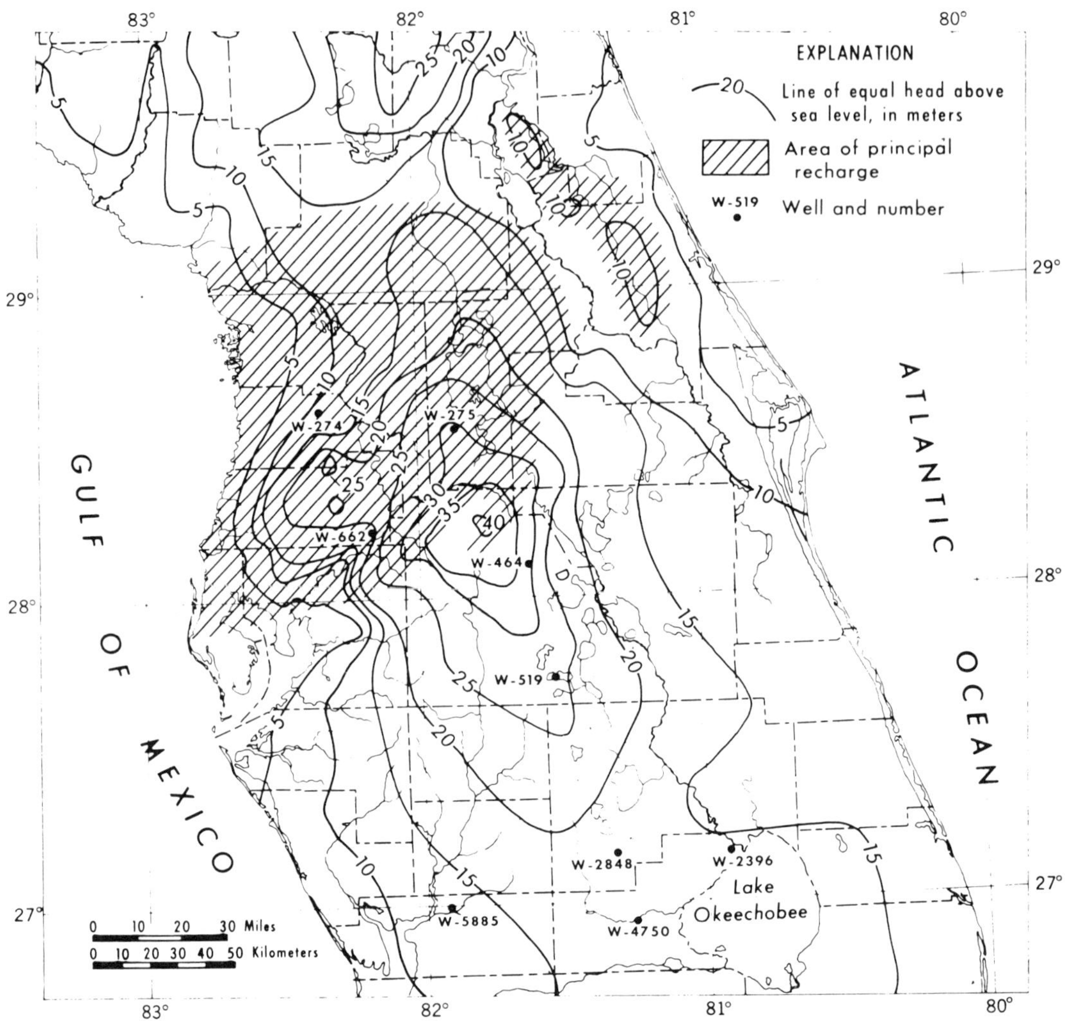

Fig. 2. Potentiometric map of central Florida showing areas of principal recharge (after Stringfield, 1936 and Healy, 1962) and location of wells from which cuttings were studied.

which means that for isothermal three-phase equilibrium, dolomite-calcite-solution, the $\alpha_{Mg^{+2}}/\alpha_{Ca^{+2}}$ is fixed at some constant value. This constant may vary slightly if we are not dealing with pure end members; that is, if the calcite contains a minor amount of magnesium ion or if the dolomite contains extra calcium or magnesium instead of a 50–50 stoichiometric amount. The equilibrium constant for calcite is close to $10^{-8.35}$ at 25° C. The equilibrium constant for dolomite is, however, somewhat in doubt; K_d determined by several laboratories differs by as much as three orders of magnitude. If

the three-phase equilibrium occurs in nature, then a study of water chemistry from various geologic terranes, and especially carbonates, should reveal a unique value for the Mg/Ca ratio. Figure 1 is a plot of Mg versus Ca in water samples from various types of aquifers throughout the world; data are mostly from White, Hem, and Waring (1963). With the exception of three analyses, all of the ratios of magnesium to calcium lie at or below 1, and two of these three exceptions are from dolomitic rocks. If calcite is not present as a discreet phase, then there is no constraint on the Mg/Ca ratio (other

[*Editors' Note:* Table 1, which appears here in the original, has been reproduced on the following page.]

conditions, such as temperature, being equal) which may have any value *greater* than that for the three-phase equilibrium. Further, water from limestones all have a calcium content ranging between 2 and 6 meq/l (milliequivalents per liter), whereas the magnesium content covers a much broader range. This figure suggests a major control on the Mg/Ca ratio in potable ground-water aquifers which limits the ratio to 1 or less.

It would be more correct to use activities of dissolved ionic species rather than molal ratios in Figure 1, but it is not possible to do so from some of White, Hem, and Waring's (1963) data. However, the activity coefficient of magnesium is approximately equal to the activity coefficient of calcium over a wide range of low ionic strengths so that molal ratios of these two species are about the same as activity ratios. Our Mg/Ca data will be reported as molal concentrations so that they may be immediately compared with Figure 1 and the preceding Mg/Ca data.

Tertiary Carbonate Aquifers of Florida and Yucatan Peninsulas.—Our study of the principal artesian carbonate aquifer of Tertiary age in Florida and rocks of equivalent age and mineralogic composition on the Yucatan Peninsula includes analyses of water chemistry and isotopic compositions of the dissolved species and minerals, coupled with a detailed X-ray mineralogic study of well cuttings from selected wells in Florida. Results of some of this work have been published by Back (1963), Hanshaw, Back, and Rubin (1965, 1966), and Back and Hanshaw (1970).

Table 1.--Mineral composition of selected well cuttings from the Floridan aquifer

Recharge area--Downgradient area

Mineralogy	Corresponding d-spacing Å	Texture	No. 274		No. 275		No. 464		No. 519		No. 662		No. 2396		No. 2848	No. 4750		No. 5885	
			A-81-94 m	B-193-200 m	A-113-116 m	B-268-271 m	A-187-193 m	B-276-279 m	A-111-114 m	B-240-244 m	A-37-39 m	B-64-66 m	A-245-250 m	B-327-330 m	A-362-364 m	A-81-84 m	B-397-406 m	A-89-97 m	B-127-137 m
Calcite (mol % $MgCO_3$)																			
0-2%	3.036-3.0295	Incrustation on well casing	---	---	---	---	---	---	---	---	---	---	---	---	2%	---	---	---	---
		Chalky, fine-grained	---	---	---	<1%	2%	---	---	---	---	---	---	---	---	---	---	pnm[2]	---
		Fossiliferous	60%	5%	---	---	---	---	---	3%	<1%	80%	60%	<1%	2%	---	---	pnm	---
		Fossil cement	15%	---	---	---	---	---	---	---	---	---	---	---	1%	---	---	---	---
		Micrite	2%	---	2%	---	---	---	---	5%	---	---	---	---	---	pnm	pnm	---	---
		Microcoquina	---	---	45%	1%	2%	---	---	30%	---	---	---	---	---	---	---	---	---
		Microcoquina cement	---	---	1%	---	---	---	---	---	---	---	---	---	---	---	5%	---	---
		Travertine	---	---	---	---	<1%	---	---	2%	---	---	---	---	1%	pnm	pnm	pnm	---
		Microsparite	---	---	---	---	---	---	---	---	60%	10%	---	---	1%	pnm	5%	---	---
>2-4%	>3.0295-3.0235	Travertine	---	---	---	---	---	---	3%	---	---	---	---	3%	1%	pnm	---	---	---
		Microsparite	20%	---	5%	<1%	---	---	65%	60%	---	10%	3%	50%	20%	60%	3%	---	---
		Microcoquina	---	---	---	---	---	---	---	---	---	---	---	---	2%	---	30%	---	---
		Fossiliferous	---	---	10%	1%	---	---	10%	---	---	---	<1%	10%	5%	---	---	---	---
		Incrustation on well casing	---	---	---	---	1%	---	---	---	---	---	---	---	8%	---	---	---	---
		Cement around quartz grains	---	---	---	---	3%	---	---	---	---	---	---	---	1%	---	---	---	---
		Coquina conglomerate	---	---	---	---	---	---	7%	---	---	---	---	---	---	pnm	---	---	---
		Fine-grained tightly cemented	---	---	---	---	---	---	2%	---	---	---	---	---	---	---	---	---	---
		Fossil cement	---	---	---	---	---	---	---	---	---	---	5%	25%	1%	---	---	---	---
>4-6%	>3.0235-3.0175	Fossiliferous	---	---	---	---	---	---	2%	---	---	---	---	---	---	---	---	---	---
Dolomite (mol % excess $CaCO_3$)																			
<0%	<2.8882	Coarse grained	---	---	---	35%	---	2%	---	---	---	---	---	---	---	---	---	---	---
		Fine grained	---	---	---	30%	---	95%	2%	---	---	---	---	---	---	---	---	---	---
		Crystalline	---	---	---	1%	---	2%	---	---	---	---	---	---	---	---	---	---	---
>0-2%	>2.8882-2.8952	Very fine grained	---	50%	---	---	<1%	---	---	---	---	---	---	---	1%	---	---	2%	---
		Incrustation on well casing	---	---	---	---	1%	---	---	---	---	---	---	---	---	---	---	---	---
		Sugary, medium grained	---	---	---	---	---	---	---	---	---	---	---	---	---	---	---	3%	---
		Fossiliferous	---	---	---	---	---	---	---	---	---	---	---	---	---	---	---	pnm	---
>2-4%	>2.8952-2.9020	Sugary, very fine grained	1%	---	5%	---	5%	---	---	---	---	---	---	---	1%	pnm	35%	<1	---
		Very fine grained	---	---	2%	---	---	---	---	---	---	---	---	---	---	---	---	3%	---
		Conglomerate	---	---	---	---	---	---	---	---	---	---	---	---	---	pnm	---	---	---
>4-6%	>2.9020-2.9085	Sugary, medium grained	---	40%	15%	---	---	---	2%	---	---	---	---	---	20%	pnm	---	25%	---
		Incrustation on well casing	---	---	---	---	---	---	---	---	---	---	---	---	8%	---	---	---	---
		Microcoquina	---	---	---	---	---	---	---	---	---	---	---	---	2%	---	---	---	---
		Crystal layer	---	---	---	---	---	---	---	---	---	---	---	---	---	---	---	pnm	---
		Shell?	---	---	---	---	---	---	---	---	---	---	---	---	---	---	---	pnm	---
		Coral?	---	---	---	---	---	---	---	---	---	---	---	---	---	---	---	pnm	---
6-8%	>2.9085	Ostracods	---	---	5%	---	---	---	---	---	---	---	---	---	---	---	---	---	---
Apatite	-------		---	---	---	---	---	---	1%	---	---	---	---	---	---	---	---	6%	---
Quartz grains	-------		---	---	10%	---	85%	---	---	---	---	---	---	---	20%	---	---	---	95%

[1] d-spacings which correspond to mol % excess $MgCO_3$ in calcite or $CaCO_3$ in dolomite were taken from best fit curve of $2\theta_{104}$ vs. d_{104} vs. mol % $MgCO_3$. Internal measurement error in °2θ at 95% confidence = ±1.4 mol %.

[2] pnm = present, but not measured.

Results of the X-ray mineralogic investigation by one of us (R. G. D.) are reported in Table 1. Locations of the wells from which cuttings were examined are shown on Figure 2. Silt to pebble-sized well cuttings were taken at two depths (labeled A and B on Table 1) in each of nine wells. To separate the mineral compositions shown in Table 1, the cuttings were handpicked under a binocular microscope. Homogeneous fragments were separated by grain size and texture into microcrystalline or micritic (Folk, 1959), chalky, and sugary. Nonhomogeneous fragments were separated by type, size, cementing material, packing density, and color. Considerable portions of the aquifer which are described as "microcoquina" in Table 1 are cemented medium- to fine-grained, tiny fossil (usually ostracod) fragments. These were carefully crushed and both cement and fossils individually examined for possible differences in composition. Fossils from different phyla were also analyzed separately for magnesium content. Secondary crystalline coatings were removed from the underlying host rock and analyzed separately to compare conditions of overgrowth precipitation with conditions of replacement.

Representative pieces of separated material were powdered and Duco-cemented onto glass slides for analysis by X-ray diffraction. Initial mineralogic identifications were made from 2° 2θ per minute scans with $CuK\alpha$, nickel-filtered radiation utilizing pulse-height analysis. The $MgCO_3$ content in calcite and also the excess $CaCO_3$ in dolomite were estimated from measurements of the d_{104} reflections scanned at 1/8° 2θ per minute, using a plot of d-spacing versus composition derived from the cell dimension data of Goldsmith, Graf, and Heard (1961). Because the principal d-spacings for disordered dolomite apparently vary from those for an ordered structure with the same composition, the plot is not fully reliable at values near Ca_{50}-Mg_{50}. The plot of d-spacing versus composition is a best-fit curve between measurements from ordered dolomite of known compositions and dolomite of similar compositions disordered by heating. The values for excess $CaCO_3$ in dolomite shown in Table 1 are therefore subject to revision as more data are obtained on nonstoichiometric dolomites. For this reason, the actual measurements from which our compositions were derived are also shown on Table 1. Quartz which occurs in nearly all of the samples studied was used as a correction for the internal error which, at the 95 percent confidence level, is ±1.4 mole percent $MgCO_3$.

Over half of the total material analyzed is calcite divided about equally into 0–2 and >2–4 mole percent $MgCO_3$. The purest calcites are composed of foraminifera and ostracod microcoquina, with minor microsparite, and occur in well samples 274A, 275A, 519B, 662A and B, and 2396A. Most of these wells are in the recharge area. Calcites with the highest $MgCO_3$ contents are mostly microsparite with some fossil material. These calcites occur in well samples 519A and B, 2396B, 2848A, and 4750A, which are mainly in the downgradient area.

There are other calcites present in relatively small amounts which are of interest for their $MgCO_3$ content. For example, the cement between grains of quartz and fossils is low in magnesium, ranging from <1 to >3 mole percent. An incrustation on the casing in a downgradient well, composing roughly 8 percent of the sample, is a mixture of calcite containing >2–4 mole percent $MgCO_3$ and dolomite with >4–6 percent excess $CaCO_3$. Travertine-like coatings of calcite which contain higher $MgCO_3$ contents occur in downgradient-area well samples. In recharge area wells, calcite coexisting with 20 percent or more dolomite contains 0–2 mole percent $MgCO_3$, but in the downgradient area, calcites coexisting with dolomite contain >2–4 mole percent $MgCO_3$.

Approximately one-third of the well samples are ordered dolomite apparently containing from <0 to as much as 8 mole percent excess $CaCO_3$. X-ray data indicate quartz and apatite make up most of the remainder of the samples. The carbonate peak intensities were such that the small amount of clays and other noncarbonate material could not be identified. All the dolomites with well-defined principal peaks also show order reflections.

The dolomite texture from both recharge and downgradient areas is generally massive, homogeneous, sugary, and varies in grain size. Purest dolomites (apparently <0–2 mole percent excess $CaCO_3$) are found in the deeper portions of recharge area wells 274, 275, and 464. Dolomites with higher (>2–6 mole percent) excess $CaCO_3$ are more abundant in the downgradient well samples; much of this dolomite is fossil replacement. The highest percentage of excess $CaCO_3$ in any of the dolomitic samples was found in dolomitized ostracods in the recharge area.

Fossils are generally calcitic in the recharge area but in the downgradient area shells, coral(?), bryozoans, etc. become dolomitic. Dolomitized forams were not found. In well 5885, cuttings from 89–97 meters contain phosphatic internal molds and skeletal material of the cheilostome bryozoan *Metrarabdotos*(?) cf. *M. tenue* (Busk) altered to dolomite.

X-ray analysis indicates that these bryozoans are composed of dolomite containing only 1.6 mole percent excess $CaCO_3$, minor apatite, and no calcite. In a study of *M. tenue*, Cheetham, Rucker, and Carver (1969) found that modern material is wholly

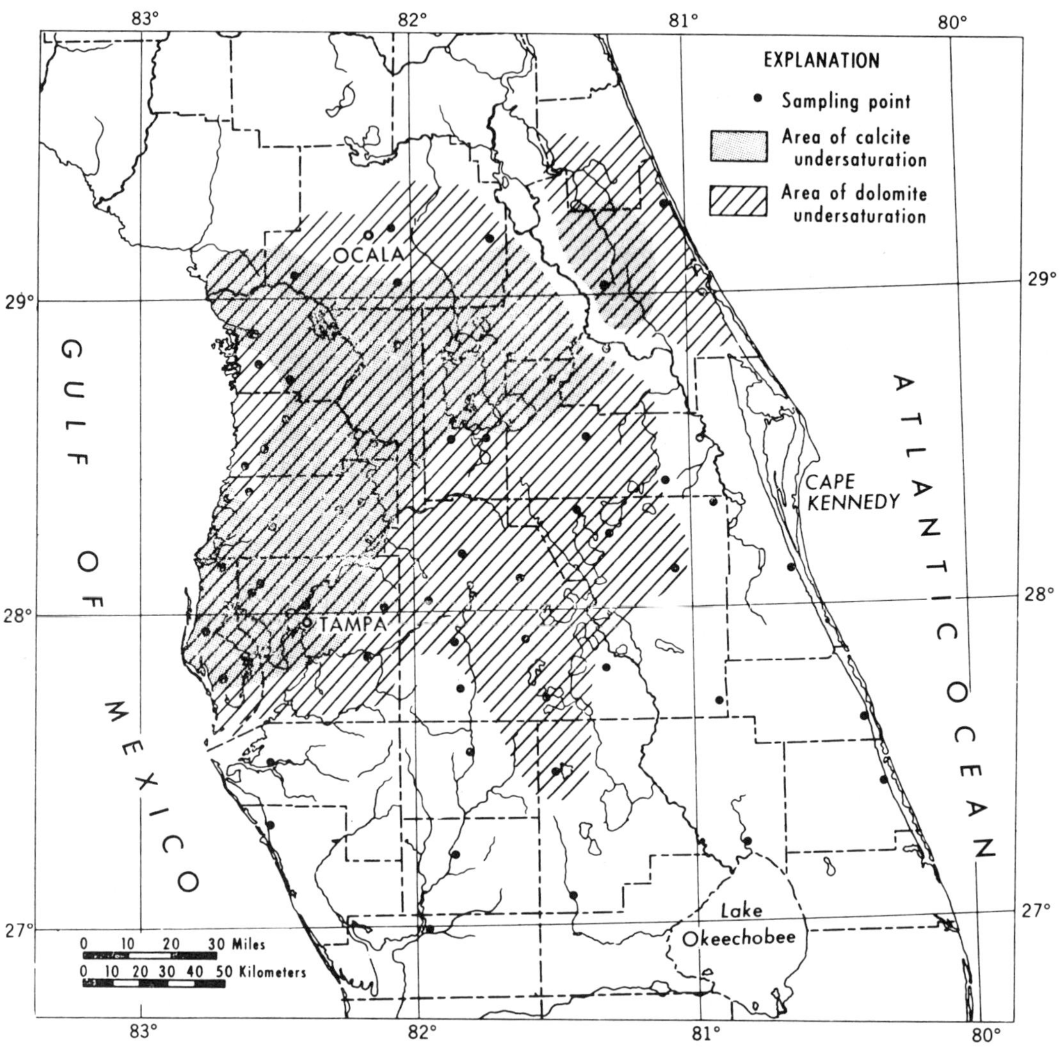

FIG. 3. Map of central Florida showing areas in which water is undersaturated with respect to calcite and dolomite.

calcitic with 4 to 5 mole percent $MgCO_3$; this species has a calcitic rather than an aragonite structure. Thus the bryozoans from the Floridan aquifer have been altered to dolomite. Other fossils from the same well are composed of dolomite with 5.2 mole percent excess $CaCO_3$.

Mass balance calculations by Hsu (1964) indicate that simply moving magnesium ion (obtained from high-magnesium calcites) from the recharge area to a downgradient area cannot dolomitize a significant portion of the Floridan aquifer. Our mineralogic studies concur with this; there is obviously insuffi-

cient magnesium from magnesium calcites available today to accomplish widespread dolomitization. However, Goodell and Garman (1969) described a mechanism for dolomitizing a substantial part of the Bahama platform based on careful study of a deep test well on Andros Island. They describe four major cycles of deposition, uplift and erosion, and dolomitization. The source of magnesium was presumably from solution of thick sequences of high-magnesium calcites and downward percolation of high-magnesium ground water during times of uplift; they term the process "solution-cannibaliza-

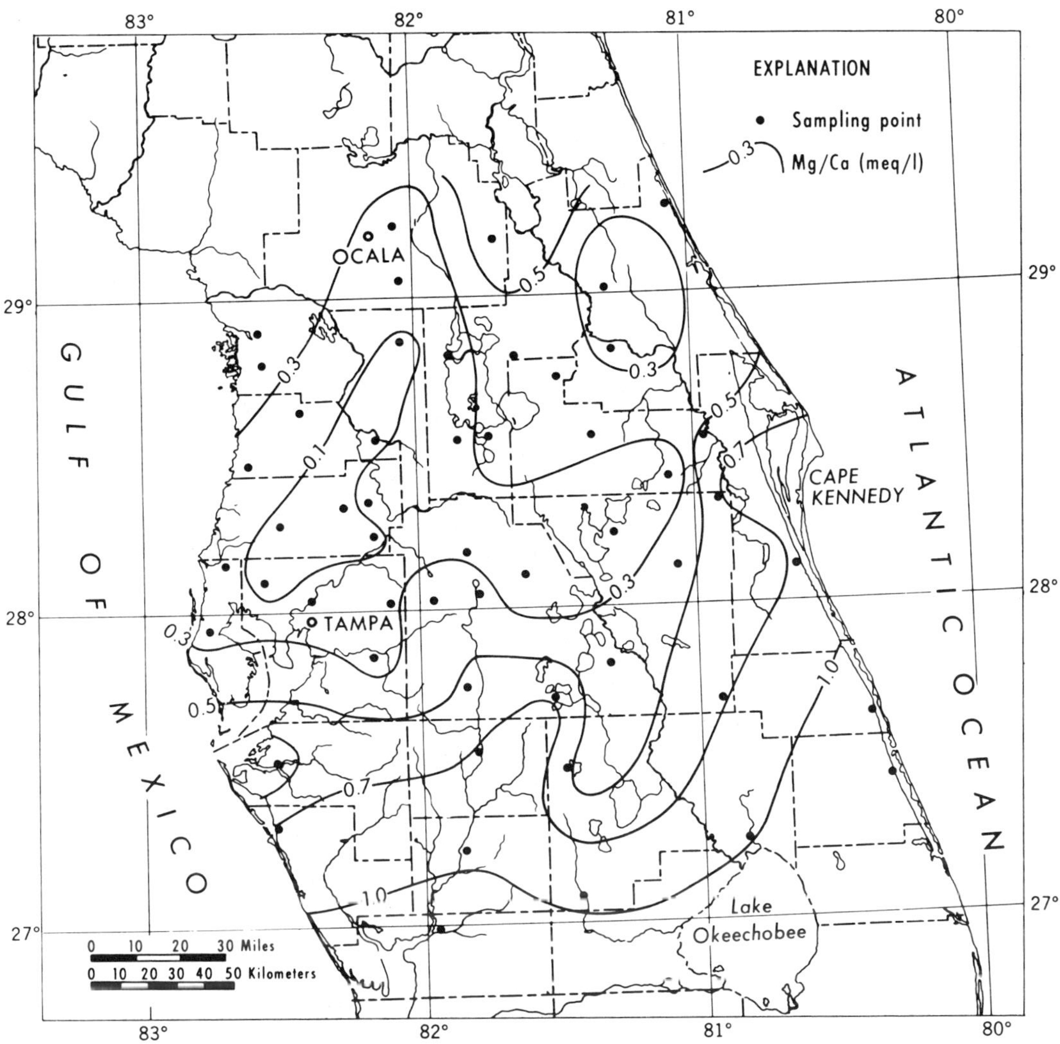

FIG. 4. Distribution of Mg/Ca in ground water from the principal artesian aquifer in central Florida.

tion." It seems plausible that similar events occurred during the deposition of carbonates on the Florida platform during Tertiary time. Chen (1965) showed that the two regions, the Bahamas and Florida, have markedly similar Tertiary geologic histories. Thus, "solution-cannibalization" may account for much of the dolomite found in the potable part of the Floridan aquifer. However, although much of the dolomite in the potable water zone apparently was produced during the Tertiary, and the magnesium source in the potable zone is not sufficient to dolomitize a significant part of the Florida platform,

we will attempt to show that dolomitization is occurring today in the nonpotable part of the aquifer system.

The potentiometric surface for the artesian carbonate aquifer of central Florida is shown in Figure 2. Areas of significant recharge are indicated by shading; the extent of these areas is based on our studies (referenced above) of chemical equilibrium between the aquifer minerals and ground water and on radiocarbon dating coupled with tritium content of ground water. Location of recharge areas are further corroborated by ground-water temperature

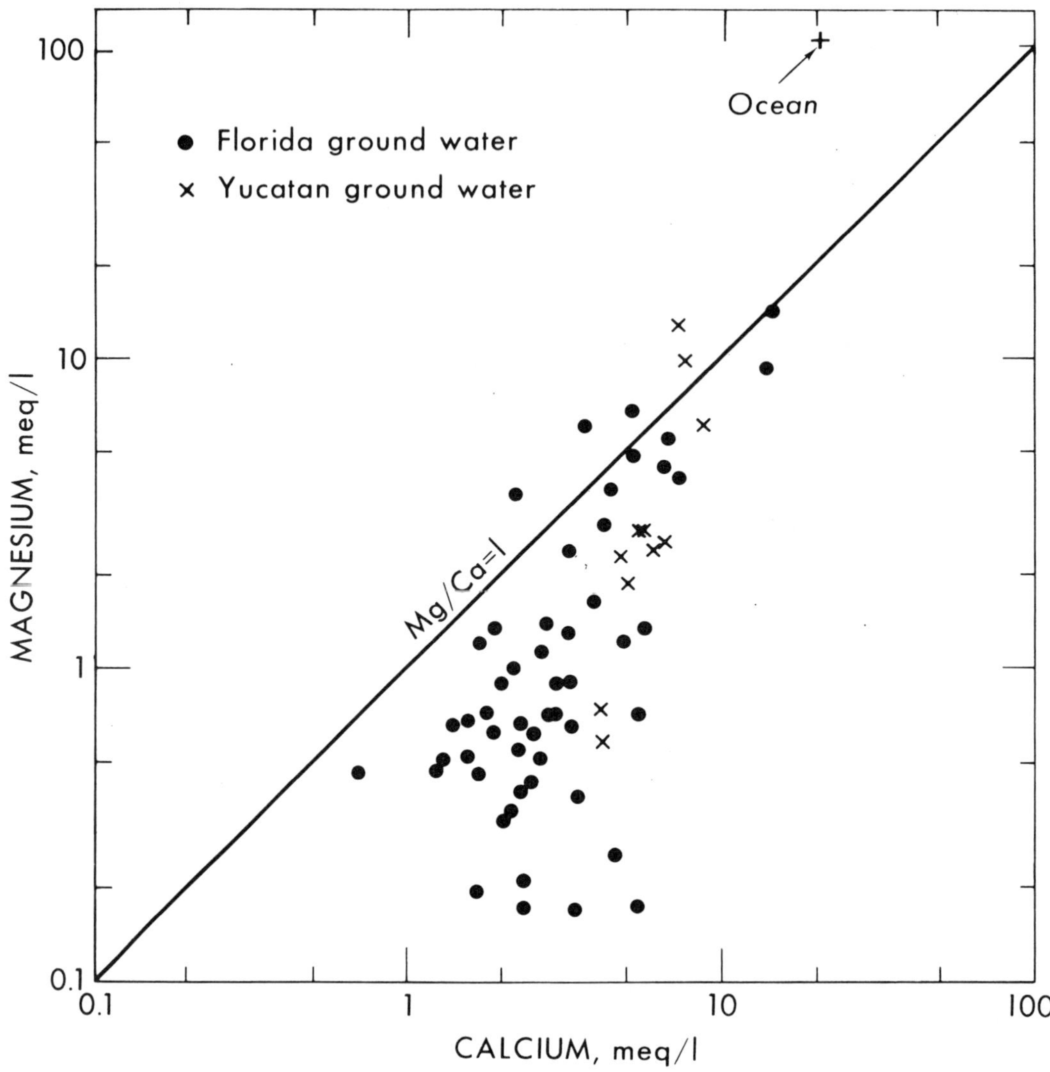

Fig. 5. Plot showing the approach of Mg/Ca to unity in ground water from Tertiary carbonate aquifers of Florida and Yucatan.

studies (Schneider, 1966) and by water-balance calculations (Rodney N. Cherry, oral communication, 1969).

From published chemical analyses of Florida ground water, Hsu (1963) calculated that the equilibrium constant of dolomite lies between 1 and 3×10^{-17}. Barnes and Back (1964) independently arrived at identical conclusions based on their geochemical studies of several carbonate aquifer systems. However, neither of these studies considered the flow system. Additionally, owing to lack of sufficient data, Hsu was forced to make the implicit assumption of dolomite-calcite-solution equilibrium throughout the system. However, our work (Back, 1963; Hanshaw, Back, and Rubin, 1966; and unpublished data) has shown that ground water is undersaturated with respect to both calcite and dolomite over large areas and especially in the principal recharge areas (Fig. 3). This means that three-phase equilibrium is not reached in most parts of the system and, in

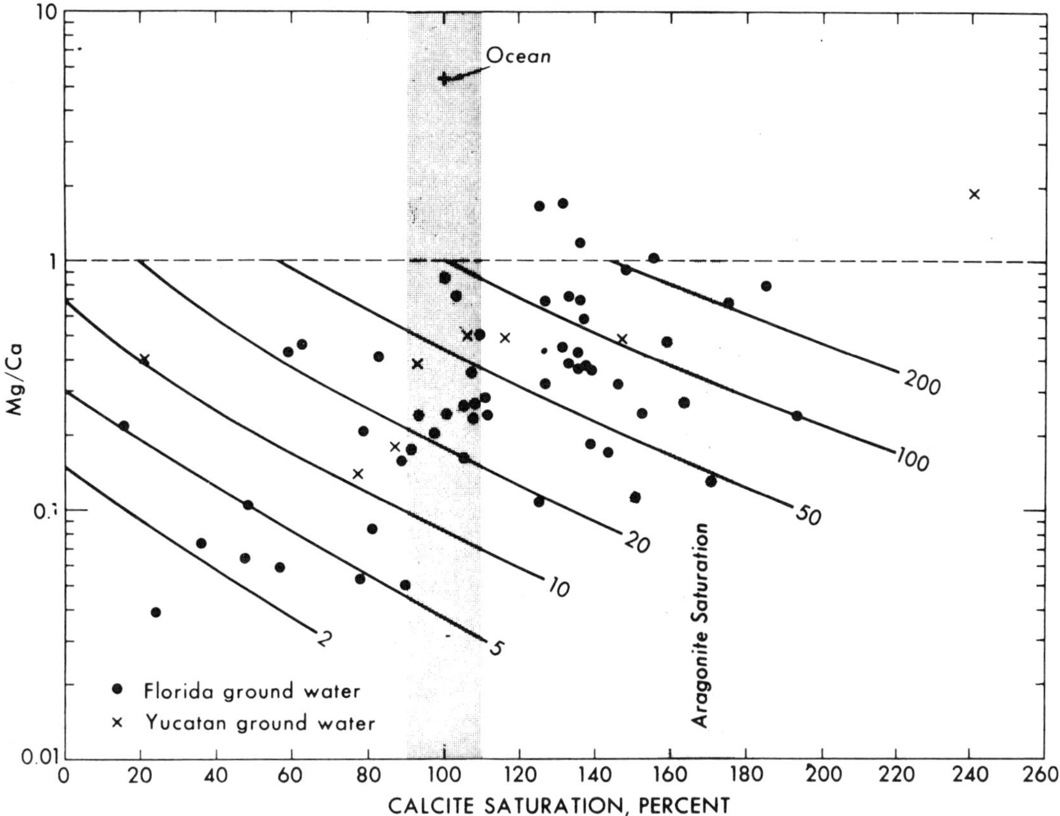

FIG. 6. Diagram showing relationship between the Mg/Ca and equilibrium conditions for aragonite and calcite (shaded areas), and percent dolomite saturation (solid lines). Three Yucatan data points shown on Figure 5 are missing; no chemical saturation data for one and the other two plot above and to the right of the limits on this diagram.

fact, only occurs in the deeper part. Although Hsu did not have these data available, his value for the equilibrium constant of dolomite agrees well with ours.

Because ground water is undersaturated with carbonate minerals in the recharge areas and increases in apparent saturation downgradient (Figs. 2 and 3), we might expect that the Mg/Ca ratio would change systematically along any flow line; Figure 4 shows that this is the case. The lowest Mg/Ca ratio (less than 0.1) closely corresponds to the area of most rapid recharge based on temperature, saturation studies, and radiocarbon and tritium results (unpublished data). By taking into account the degree of saturation with respect to carbonate minerals plus the physical flow system, we can explain the very low Mg/Ca ratios through lack of three-phase equilibrium.

Figure 5 is a plot of the magnesium versus calcium molal concentrations of ground-water samples from Florida and the Yucatan. Again, as with data from Figure 1, most of the points fall near or below a Mg/Ca ratio of 1. Three points from Florida have a Mg/Ca ratio grater than 1; these wells all occur along the coast and produce a mixture of oceanic water and potable water of the aquifer which has a lower Mg/Ca ratio. Apparently these waters are supersaturated with respect to dolomite. On Figure 5, analyses from the Yucatan fall along a nearly straight line which passes through the Mg/Ca point for ocean water; thus, water samples from the Yucatan represent a mixture of a shallow layer of potable water with a large body of saline water which is near oceanic composition and underlies it. Both of the points above the line are samples from close to the coast and represent saline wells similar to those in Florida.

Data presented in Figures 1 and 5 suggest that the Mg/Ca ratio in many aquifer systems is close to unity. Therefore, it appears from the field data that:

$$\frac{\alpha_{Mg^{+2}}}{\alpha_{Ca^{+2}}} = \frac{K_d}{K_c^2} \approx 1$$

and because the value for K_c, at 25° C, is $10^{-8.35}$:

$$K_d = (10^{-8.35})^2 = 10^{-16.7} = 2 \times 10^{-17}.$$

Other recent work has estimated the equilibrium constant for dolomite from the composition of various subsurface waters; these are all in the range of 1.5 to 3.0×10^{-17} (Barnes and Back, 1964; Holland et al., 1964; Hsu, 1963; Langmuir, 1964; and Lerman, 1965).

The relationship between calcite, dolomite, and solution is summarized in Figure 6. The areas of calcite and aragonite equilibrium are shaded to indicate a ± 10 percent uncertainty in the calculations. Note that only three samples appear to be supersaturated with respect to aragonite, and most samples are either in equilibrium or possibly supersaturated with respect to calcite. Lines of equal percent dolomite saturation show that fewer than a dozen samples may be supersaturated. The amount of supersaturation may be more apparent than real owing to the following: (1) Ion activity products are compared with equilibrium constants for a variety of *pure* phases (Back, 1961), but the change in free energy caused by introduction of trace elements into a crystal lattice (e.g., strontium and magnesium in calcite or aragonite) has not been considered. (2) The formation of ion pairs, such as $CaSO_4^0$ and $MgSO_4^0$, in solution effectively lowers the activity of single ion species below the values used in our calculations. No doubt, much of our apparent supersaturation is caused by these two effects. Such problems will be considered in future work, but the maximum effect would be in the downgradient areas and probably would be <20 percent.

Environments of Dolomite Formation

The foregoing discussion indicates that the three-phase equilibrium in ground-water systems occurs at Mg/Ca ratio of approximately unity. This implies that if calcite is immersed in a solution with a higher ratio, then dolomitization should proceed spontaneously, provided sufficient time and magnesium ion are available.

Zen (1960) has shown that the chemical composition of open ocean water is externally controlled and is therefore an independent system variable. In that case, the number of phases present at equilibrium will be reduced by one (Korzhinskii, 1950; Thompson, 1955) so that *either* calcite *or* dolomite should be stable in ocean water. Therefore, a Mg/Ca ratio

greater than unity should cause dolomitization, and with the ratio of Mg/Ca in ocean water at 5.2, why aren't marine carbonates all dolomites? As noted previously, most modern carbonates consist of the metastable phases, aragonite and high-magnesium calcite, as well as the relatively more stable low-magnesium calcite. Furthermore, Berner (1966a) has pointed out that there is no evidence for recrystallization of metastable carbonates to more stable forms in normal sea water. This is in agreement with the findings of many other workers, including Chave (1954), Ginsburg (1957), Gross (1964), and Pilkey (1964). However, in dynamic environments (i.e., active circulation of solutions, including ground water), diagenetic recrystallization does occur (Berner, 1966a; Friedman, 1964; Harris and Mathews, 1968). In fact, it seems extremely significant that the one thing that all environments have in common where dolomite is reported to be forming today is active circulation of solutions. Certainly these environments do not have an Mg/Ca ratio in common; this varies from 3 to 100. In order to explore the static versus dynamic environment further, let us consider a simple two box model for the ocean water-carbonate sediment system, box A and box B, respectively. Let us limit our consideration further to a one centimeter cube within the sediment, box B, which consists of calcite and has a porosity of 20 percent. From the equation:

$$\text{Ca positions} = \frac{\rho_c N (1 - \eta)}{M_c}$$

where

ρ_c = density of calcite
N = Avogadro's number
η = porosity
M_c = molecular weight of calcite

we calculate the total number of calcium positions in the cube to be 1.3×10^{22}. Simple metasomatic replacement of one half of the calcium by magnesium to form dolomite requires 6.5×10^{21} magnesium ions. Assuming that the pore water is essentially ocean water composition, we can calculate the total number of magnesium ions available in the pore solution from

$$\text{Mg ions} = \frac{C \cdot N \cdot \eta}{M_{Mg} \times 10^6}$$

where

C = 1,272 ppm Mg^{+2} in ocean water
M_{Mg} = molecular weight of magnesium.

The number of magnesium ions available is only 6.3×10^{18} in the pore water of the cube. Obviously this solution cannot dolomitize a significant amount of calcium carbonate. If magnesium ion did interact with the mineral grains to form dolomite, then there would be a diffusion gradient between box A and the pore water as the Mg/Ca ratio in the pore water decreased toward the equilibrium value of unity. Magnesium would diffuse from A to B, and calcium would diffuse from B to A. However, much recent work (Berner, 1966a, 1966b, 1967; Bischoff, 1968; Bischoff and Fyfe, 1968) indicates that magnesium ions may interact with mineral surfaces and effectively inhibit the further reaction between grains and interstitial solution. The fact that Siever, Beck, and Berner (1965) found only minor increases or decreases in Mg/Ca ratios in interstitial water from deep sea cores is indicative of nonreactivity and is not surprising. Harriss and Pilkey (1966) also reported no change in Mg/Ca ratio of interstitial waters in two 4-meter cores. Friedman (1964) and Zen (1960) each reported dolomite from the open ocean. Both occurrences were at or near the sediment-water interface and might be interpreted as special cases of local dolomitization. Thus, carbonates in the open ocean should be dolomitized, but the evidence for this is nearly absent; perhaps this is because of the armoring effect. Minor dolomitization seems to occur only at the sediment-water interface, where currents refresh the solution constantly.

From the forefoing, the following emerges: (1) modern dolomites are associated with a wide range of Mg/Ca ratios, from 3 to 100; (2) the equilibrium value is close to unity at $25° \pm 10°$ C; (3) all modern dolomites are found in a dynamic environment; and (4) given enough time and sufficient magnesium ion supply (i.e., an external reservoir and a hydrodynamic system), virtually any calcium carbonate terrane may be dolomitized.

One environment which can fit all these requirements and which can also account for regional dolomitization is a ground-water system. From Figures 1 and 5, we observe that most potable groundwater systems simply approach three-phase equilibrium and are therefore incapable of extensive dolomitizing without an additional source of magnesium. Goodell and Garman's (1969) mechanism of autocannibalization of high-magnesium calcite on the Bahama banks during periods of emergence could account for some dolomites of regional extent; we believe this may have occurred on the Florida platform during Tertiary time. This could account for much of the dolomite found today (Table 1) in cuttings from the potable part of the system.

Underlying the potable water lens is a zone of brackish water, underlain in turn by an extensive zone of hypersaline water. The volume of Tertiary rocks in contact with nonpotable water is much greater than that in contact with fresh water. A diagrammatic cross-section (Fig. 7) of the central Florida Peninsula illustrates the distribution of hydrochemical facies and Mg/Ca ratios within the ground-water system. Table 2 gives the Mg/Ca ratios for representative samples from the brackish zone as well as a few selected examples from the saline and hypersaline zones. Samples from the brackish zone generally contain less than 150 mg/l (milligrams per liter) magnesium ion; magnesium concentrations in the saline zone range from 300 to less than 1,000 mg/l and, in the hypersaline zone, up to 3,500 mg/l. Samples from both the brackish and saline zones generally have Mg/Ca ratios greater than unity and could therefore dolomitize limestone. Mg/Ca ratios from samples in the hypersaline zone vary greatly. However, according to Holland (1967), the Mg/Ca ratio at three-phase equilibrium

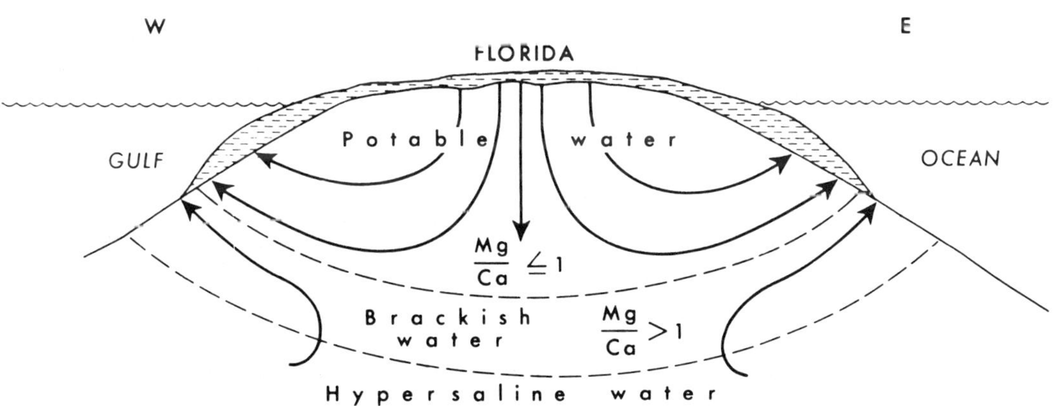

FIG. 7.   Diagrammatic cross section of central Florida with arrows showing generalized flow paths (modified from Vernon, 1969).

TABLE 2.--Mg/Ca molal ratios for water samples from selected non-potable wells in Florida[1]

Well location	Producing depth (meters)	Total dissolved solids (ppm)[2]	Mg/Ca molal ratio
BRACKISH ZONE			
1. Belle Glade, U.S. Sugar Corp. #1	480-520	3,300	2.8
2. Everglades #2	120-150	1,040	2.5
3. LaBelle	195 (total depth)	1,420	1.3
4. W. Palm Beach Coast Guard	320 (total depth)	3,290	2.7
5. Ocean Reef Hotel	340-350	4,640	2.0
6. Pine Island	Not reported	1,660	1.7
7. 40-mile Bend	305 ,750	4,000	2.4
8. Royal Palms Ranger Station	200-430	5,760	1.9
9. Grossman's Hammock	140-400	2,900	2.4
10. Pennecamp Park	Not reported	4,660	2.2
SALINE ZONE			
11. Hot Springs	500-530	33,900	2.8
12. Oil well HE-341, sec. 32, T45S, R29E	475-1,450	10,100	3.2
13. Gulf Coast drilling, 61 km W of Miami	310-1,230	17,900	2.8
14. Groveland sec. 17, T24S, R25E	785-800	43,400	1.1
15. Jacksonville, test well at water tank	795 (total depth)	15,900	.8
16. Normal ocean water		35,000	5.2
HYPERSALINE ZONE			
17. S. of Anna Maria	130 (total depth)	62,000	6.7
18. Burus, sec. 18, T3S, R1E	640 (total depth)	93,400	1.4
19. Oil well HE-341	3,700-3,705	216,000	.3
20. J. A. Curry well, Humble Oil Co.	3,750-3,805	207,000	.2

[1] Samples from brackish zone collected by Bruce B. Hanshaw. Analyses from saline and hypersaline zones were provided by B. Joyner and M. Kaufman, Florida district, U.S. Geological Survey, or were taken from Black and Brown (1951).

[2] Values are in units of parts per million because a correction for density of solution was not made.

is less than one at temperatures above 25° C. But care must be used in interpreting these ratios because at very high ionic strength, activity ratios may deviate greater from molal ratios. Based on mean salt calculations, σ_{Mg}/σ_{Ca} is about 1.4 at an ionic strength of 0.5 molal (close to sea water strength) and approximately 0.7 at an ionic strength of 3. The hydraulic connection between these very deep zones and upper parts of the system is poorly known.

Kohout (1967) has described cyclic flow of salt water in the Biscayne aquifer of southeastern Florida related to thermal convection within the aquifer. He suggests that a similar mechanism may cause slow convective movements in salty water underlying all of peninsular Florida. If this is so, then a mechanism is available for brackish and saline waters to dolomitize much of the carbonate material in the aquifer. Such dolomitization was also suggested by the report (Vernon, 1969) of dolomite crystals from a highly cavernous zone, the Boulder zone, in the Florida aquifer. According to Vernon, the Boulder zone exists over much of central Florida. Parts of the Boulder zone correspond to the zone of brackish water; Vernon postulated that the cavernous Boulder zone may be caused by solution of limestone brought about by mixing potable and saline water. Holland

(1964) and Runnells (1969) point out that when two waters saturated with the same mineral phases are mixed, undersaturation may occur. Thus, mixing of saline with potable water to produce the brackish water zone (Fig. 7) may cause solution of limestone. In this process both Ca^{+2} and CO_3^{-2} would be increased in the solution phase which could cause supersaturation with respect to dolomite. Based on the Mg/Ca ratios in the brackish zone (Table 2), all these water samples should dolomitize limestone; they are, in fact, supersaturated with respect to dolomite. Owing to sea-level changes, climatic changes, and/or occasional uplift or downwarp of the Floridan platform, the brackish and saline zones would change position within the system and dolomitize considerable parts of the rock below the potable water lens.

Concluding Remarks

Field and laboratory data indicate that: (1) An upper limit on the Mg/Ca ratio exists in ground water from various sedimentary aquifers. This ratio results from the water attaining equilibrium with calcite and dolomite. The three-phase equilibrium is not achieved in all areas because of slow reaction rates. (2) Calcite in most carbonate aquifers contains insufficient magnesium to cause extensive dolomitization unless many volumes of magnesium-calcite are dissolved for every volume of dolomite formed. (3) Dolomitization, caused by the circulation of ground water with Mg/Ca ratio >1, occurs in the brackish water zones. Brackish water comes from mixing of potable water with subsurface brines or ocean water and provides a dynamic source of magnesium.

Future work will include interpretation of radiocarbon dating of Floridan ground water and of departure from equilibrium of water with respect to carbonate minerals in order to determine reaction rates. Investigation of the C^{13}/C^{12} ratio of coexisting dolomite and calcite and of the dissolved carbonate species of the water may provide additional insight into dolomitization in the Florida aquifer.

Acknowledgments

During the course of this study, we have benefited greatly from the continual encouragement and help given us by our colleague Blair F. Jones. We have profited from many discussions with him and also with Robert A. Berner, Rodney N. Cherry, and F. Joseph Pearson. We also wish to thank Boyd Joyner and Matthew I. Kaufman for providing us with unpublished brine analyses from Florida. Especial thanks go to Robert O. Vernon, Director of the Florida Geological Survey, for providing access to the well cuttings used in this investigation. Alan

Cheetham and Thomas Gibson kindly provided us with fossil identifications for which we are grateful. Finally, we offer particular expressions of thanks to Blair F. Jones, Matthew Kaufman, and E-an Zen for their critical and thoughtful reviews of the manuscript. These acknowledgments do not imply that the individuals mentioned above necessarily agree with all of our ideas.

U. S. GEOLOGICAL SURVEY,
WASHINGTON, D. C.,
April 20, 1971

REFERENCES

Adams, J. E., and Rhodes, M. L., 1960, Dolomitization by seepage refluxion: Am. Assoc. Petroleum Geologists Bull., v. 44, p. 1912–1920.
Alderman, A. R., 1965, Dolomitic sediments and their environment in the southeast of south Australia: Geochim. et Cosmochim. Acta, v. 29, p. 1355–1365.
——, and Skinner, H. C. W., 1957, Dolomite sedimentation in the south-east of south Australia: Am. Jour. Sci., v. 255, p. 561–567.
——, and von der Borch, C. C., 1961, Occurrence of magnesite-dolomite sediments in south Australia: Nature, v. 192, p. 861.
——, 1963, A dolomite reaction series: Nature, v. 198, p. 465–466.
Atwood, D. K., and Bubb, J. N., 1970, Distribution of dolomite in a tidal flat environment, Sugarloaf Key, Florida: Jour. Geology, v. 78, p. 499–505.
Back, William, 1961, Calcium carbonate saturation in ground water, from routine analyses: U. S. Geol. Survey Water-Supply Paper 1535-D, 14 p.
——, 1963, Preliminary results of a study of calcium carbonate saturation of ground water in central Florida: Internat. Assoc. Sci. Hydrology Bull., v. 8, p. 43–51.
——, and Hanshaw, B. B., 1970, Comparison of chemical hydrogeology of the carbonate peninsulas of Florida and Yucatan: Jour. Hydrology, v. 10, p. 330–368.
Barnes, Ivan, and Back, William, 1964, Dolomite solubility in ground water: U. S. Geol. Survey Prof. Paper 475-D, p. D179–D180.
Berner, R. A., 1966a, Chemical diagenesis of some modern carbonate sediments: Am. Jour. Sci., v. 264, p. 1–36.
——, 1966b, Diagenesis of carbonate sediments: Interaction of magnesium in sea water with mineral grains: Science, v. 153, p. 188–191.
——, 1967, Comparative dissolution characteristics of carbonate minerals in the presence and absence of aqueous magnesium ions: Am. Jour. Sci., v. 265, p. 45–70.
Bischoff, J. L., 1968, Kinetics of calcite nucleation: Magnesium ion inhibition and ionic strength catalysis: Jour. Geophys. Research, v. 73, p. 3315–3322.
——, and Fyfe, W. S., 1968, Catalysis, inhibition, and the calcite-aragonite problem, 1, The aragonite-calcite transformation: Am. Jour. Sci., v. 266, p. 65–79.
Black, A. P., and Brown, Eugene, 1951, Chemical character of Florida's water: Fla. Water Survey and Res. Paper No. 6, 119 p.
Chave, K. E., 1954, Aspects of the biochemistry of magnesium; [Pt.] 2, Calcareous sediments and rocks: Jour. Geology, v. 62, p. 587–599.
Cheetham, A. H., Rucker, J. B., and Carver, R. E., 1969, Wall structure and mineralogy of the cheilostome bryozoan *Metrarabdotos:* Jour. Paleontology, v. 43, p. 129–135.
Chen, C. S., 1965, The regional lithostratigraphic analysis of Paleocene and Eocene rocks of Florida: Fla. Geol. Survey Bull. No. 45, 105 p.
Clayton, R. N., Jones, B. F., and Berner, R. A., 1968, Isotope studies of dolomite formation under sedimentary conditions: Geochim. et Cosmochim. Acta, v. 32, p. 415–432.

——, Skinner, H. C. W., Berner, R. A., and Rubinson, M., 1968, Isotopic compositions of Recent south Australian lagoonal carbonates: Geochim. et Cosmochim. Acta, v. 32, p. 983–988.
Curtis, R., Evans, G., Kinsman, D. J. J., and Shearman, D. J., 1963, Association of dolomite and anhydrite in the Recent sediments of the Persian Gulf: Nature, v. 197, p. 679–680.
Deffeyes, K. S., Lucia, F. J., and Weyl, P. K., 1965, Dolomitization of Recent and Plio-Pleistocene sediments by marine evaporite waters on Bonaire, Netherlands Antilles, *in* L. C. Pray and R. C. Murray, eds., Dolomitization and limestone diagenesis: Soc. Econ. Paleontologists and Mineralogists Spec. Pub. 13, p. 71–88.
Fisher, W. L., and Rodda, P. U., 1969, Edwards Formation (Lower Cretaceous), Texas: Dolomitization in a carbonate platform system: Am. Assoc. Petroleum Geologists Bull., v. 53, p. 55–72.
Folk, R. L., 1959, Petrology of sedimentary rocks: Univ. of Texas, Hemphill's, Austin, Texas, 154 p.
Friedman, G. M., 1964, Early diagenesis and lithification in carbonate sediments: Jour. Sed. Petrology, v. 34, p. 777–813.
Ginsburg, R. N., 1957, Early diagenesis and lithification of shallow-water carbonate sediments in south Florida, *in* R. J. LeBlanc and J. G. Breeding, eds., Regional aspects of carbonate deposition: Soc. Econ. Paleontologists and Mineralogists Spec. Pub. 5, p. 80–100.
Goldsmith, J. R., Graf, D. L., and Heard, H. C., 1961, Lattice constants of the calcium-magnesium carbonates: Am. Mineralogist, v. 46, p. 453–457.
Goodell, H. G., and Garman, R. K., 1969, Carbonate geochemistry of Superior deep test well, Andros Island, Bahamas: Am. Assoc. Petroleum Geologists Bull., v. 53, p. 513–536.
Graf, D. L., Eardley, A. J., and Shimp, N. F., 1961, A preliminary report on magnesium carbonate formation in glacial Lake Bonneville: Jour. Geology, v. 69, p. 219–223.
——, and Goldsmith, J. R., 1955, Dolomite-magnesian calcite relations at elevated temperatures and CO_2 pressures: Geochim. et Cosmochim. Acta, v. 7, p. 109–128.
Gross, M. G., 1964, Variations in the O^{18}/O^{16} and C^{13}/C^{12} ratios of diagenetically altered limestones in the Bermuda Islands: Jour. Geology, v. 72, p. 170–194.
Hanshaw, B. B., Back, William, and Rubin, Meyer, 1965, Radiocarbon determinations for estimating ground-water flow velocities in central Florida: Science, v. 148, p. 494–495.
——, 1966, Carbonate equilibria and radiocarbon distribution related to groundwater flow in the Floridan limestone aquifer, U.S.A.: Internat. Assoc. Sci. Hydrology Pub. 74, Symposium of Dubrovnik, 1965, v. 2, p. 601–614.
Harris, W. H., and Matthews, R. K., 1968, Subaerial diagenesis of carbonate sediments: Efficiency of the solution-reprecipitation process: Science, v. 160, p. 77–79.
Harriss, R. C., and Pilkey, O. H., 1966, Interstitial waters of some deep marine carbonate sediments: Deep-Sea Research, v. 13, p. 967–969.
Healy, H. G., 1962, Piezometric surface of the Florida aquifer in Florida, July 6–17, 1961: Fla. Geol. Survey, Map Ser. No. 1.
Holland, H. D., 1964, Solubility of calcite in NaCl solutions between 50° and 200° C [abs.]: Geol. Soc. America Spec. Paper 82, p. 94–95.
——, 1967, Gangue minerals in hydrothermal deposits, *in* Geochemistry of Hydrothermal Ore Deposits, H. L. Barnes, ed.: Holt, Rinehart and Winston, Inc., New York, p. 382–436.
——, Kirsipu, T. V., Huebner, J. S., and Oxburgh, U. M., 1964, On some aspects of the chemical evolution of cave waters: Jour. Geology v. 72, p. 36–67.
Hsu, K. J., 1963, Solubility of dolomite and composition of Florida ground waters: Jour. Hydrology, v. 1, p. 288–310.
——, 1964, Origin of dolomite in sedimentary sequences: A critical analysis: Mineralium Deposita, v. 2, p. 133–138.
Illing, L. V., Wells, A. J., and Taylor, J. C. M., 1965, Penecontemporary dolomite in the Persian Gulf, *in* L. C. Pray

and R. C. Murray, eds., Dolomitization and limestone diagenesis: Soc. Econ. Paleontologists and Mineralogists Spec. Pub. 13, p. 89–111.

Jones, B. F., 1961, Zoning of saline minerals at Deep Springs Lake: U. S. Geol. Survey Prof. Paper 424-B, p. B199–B202.

Kohout, F. A., 1967, Ground-water flow and the geothermal regime of the Floridan Plateau: Trans. Gulf Coast Assoc. Geological Socs., v. 17, p. 339–354.

Korzhinskii, D. S., 1950, Phase rule and geochemical mobility of elements: XVIII Internat. Geol. Congress Proc., Sec. A., p. 50–57.

Land, L. S., 1967, Diagenesis of skeletal carbonates: Jour. Sed. Petrology, v. 37, p. 914–930.

Langmuir, D., 1964, Stability of carbonates in the system $CaO-MgO-CO_2-H_2O$: Ph.D. thesis, Harvard Univ.

Lerman, Abraham, 1965, Paleoecological problems of Mg and Sr in biogenic calcites in light of recent thermodynamic data: Geochim. et Cosmochim. Acta, v. 29, p. 977–1002.

Newell, N. D., and others, 1953, The Permian reef complex of the Guadalupe Mountains region, Texas and New Mexico—A study in paleoecology: W. H. Freeman, San Francisco, 236 p.

Peterson, M. N. A., Bien, G. S., and Berner, R. A., 1963, Radiocarbon studies of Recent dolomite from Deep Springs Lake, California: Jour. Geophys. Research, v. 68, p. 6493–6505.

Pilkey, O. H., 1964, Mineralogy of the fine fraction of some cores: Marine Sci. Gulf and Caribbean Bull., v. 14, p. 126–139.

Reeves, C. C., Jr., and Parry, W. T., 1965, Geology of west Texas pluvial lake carbonates: Am. Jour. Sci., v. 263, p. 606–615.

Runnells, D. D., 1969, Diagenesis, chemical sediments and the mixing of natural waters: Jour. Sed. Petrology, v. 39, p. 1188–1201.

Schneider, Robert, 1966, An interpretation of the geothermal field associated with the carbonate-rock aquifer system of Florida [abs.]: Geol. Soc. America Spec. Paper 101, p. 191–192.

Shinn, E. A., 1964, Recent dolomite, Sugarloaf Key, Florida, *in* R. N. Ginsburg, Composition, south Florida carbonate sediments: Guidebook for field trip #1, Geol. Soc. Am. Ann. Mtg., Miami Beach, Nov. 1964, p. 26–33.

——, 1968, Selective dolomitization of Recent sedimentary structures: Jour. Sed. Petrology, v. 38, p. 612–666.

——, Ginsburg, R. N., and Lloyd, R. M., 1965, Recent supratidal dolomite from Andros Island, Bahamas, *in* L. C. Pray and R. C. Murray, eds., Dolomitization and limestone diagenesis: Soc. Econ. Paleontologists and Mineralogists Spec. Pub. 13, p. 112–123.

Siever, Raymond, Beck, K. C., and Berner, R. A., 1965, Composition of interstitial waters of modern sediments: Jour. Geology, v. 73, p. 39–73.

Skinner, H. C. W., 1963, Precipitation of calcian dolomites and magnesian calcites in the south-east of south Australia: Am. Jour. Sci., v. 261, p. 449–472.

——, Skinner, B. J., and Rubin, Meyer, 1963, Age and accumulation rates of dolomite-bearing carbonate sediments in south Australia: Science, v. 139, p. 335–336.

Sloss, L. L., 1953, The significance of evaporites: Jour. Sed. Petrology, v. 23, p. 143–161.

Stringfield, V. T., 1936, Artesian water in the Florida peninsula: U. S. Geol. Survey Water-Supply Paper 773-C, p. C115–C195.

Thompson, J. B., Jr., 1955, The thermodynamic basis for the mineral facies concept: Am. Jour. Sci., v. 253, p. 65–103.

Vernon, R. O., 1969, The geology and hydrology associated with a zone of high permeability (Boulder zone) in Florida: Soc. Mining Engineers, Preprint 69-AG-12, 24 p.; [abs.]: Mining Engineers., v. 20, no. 12, p. 58, 1968.

von der Borch, C. C., 1965a, The distribution and preliminary geochemistry of modern carbonate sediments of the Coorong area, south Australia: Geochim. et Cosmochim. Acta, v. 29, p. 781–799.

——, 1965b, Source of ions for Coorong dolomite formation: Am. Jour. Sci., v. 263, p. 684–688.

Wells, A. J., 1962, Recent dolomite in the Persian Gulf: Nature, v. 194, p. 274–275.

White, D. E., Hem, J. D., and Waring, G. A., 1963, Chemical composition of subsurface waters, chap. F *in* Data of geochemistry, 6th edition: U. S. Geol. Survey Prof. Paper 440-F, p. F1–F67.

Zen, E-an, 1960, Carbonate equilibria in the open ocean and their bearing on the interpretation of ancient carbonate rocks: Geochim. et Cosmochim. Acta, v. 18, p. 57–71.

18

Reprinted from *Jour. Sed. Petrology* **43**:965–984 (1973)

THE DORAG DOLOMITIZATION MODEL—APPLICATION TO THE MIDDLE ORDOVICIAN OF WISCONSIN[1,2]

KHOSROW BADIOZAMANI[3]

Department of Geological Sciences, Northwestern University, Evanston, Illinois 60201

ABSTRACT: Most current models of dolomitization are incapable of explaining the genesis of dolostones that lack evidence of supratidal origin and are not associated with evaporites. These models require sea water evaporation and a high Mg^{2+}/Ca^{2+} ratio in solution as essential factors for dolomitization. Calculations show that mixing meteoric ground waters with up to 30% sea water causes undersaturation with respect to calcite, whereas dolomite saturation increases continuously. Therefore, in the range of approximately 5 to 30% sea water, calcite can be replaced by dolomite. A new term "Dorag" (Persian for mixed blood or hybrid) is used for this model of dolomitization.

The Dorag model satisfactorily explains the origin of dolostones along positive elements or epicontinental shelves; it is based on the effect of ionic strength on the solubility of carbonate minerals, and does not require a Mg^{2+}/Ca^{2+} ratio greater than one, a ratio obtained at calcite-dolomite equilibrium.

Dolomitic facies of the Middle Ordovician Series in southwestern Wisconsin and adjacent states is an example of dolostones associated with positive elements. Lithologic and paleontologic evidence, as well as primary structures, within the Mifflin Member (Platteville Formation) suggest the existence of a shallow and open-marine environment over the Wisconsin Arch at the time of deposition of the Member, whereas deeper water conditions prevailed farther from the Arch. Oxygen and carbon isotopic analyses of the Mifflin Member and chemical analyses for Sr and Na in the limestones and dolostones indicate exchange of the Mifflin carbonates with meteoric water. Because (a) evaporites, algal mats, mud cracks and other evidences of restricted environments are not found in the Mifflin Member, (b) isotopic and chemical data suggest exchange of the Mifflin carbonates with meteoric water, and (c) the dolomite facies of this member is only associated with structural highs, the dolomitization of the Mifflin Member is best explained by mixing of sea water and ground water in the phreatic zone. Assuming Dorag dolomitization has been operative, three major lateral shifts in the dolostone-limestone boundary in the Champlainian Series can be shown to represent episodes of transgression-regression during the Middle Ordovician.

Conclusions can be drawn concerning (a) paleo-sea level fluctuation, (b) paleobathymetry, and (c) configuration and position of the paleo-ground water lens.

INTRODUCTION

The genesis of dolomite and processes responsible for dolomitization have attracted the attention of scientists for many years. The observation of paucity of Recent dolomites in comparison with the massive volume of older dolomites and the failure to produce this mineral experimentally at near earth-surface temperature and pressure initiated the so-called "dolomite problem." The complexity of the problem was enhanced by kinetic factors that apparently inhibit dolomite precipitation from sea water. Dolomite does not precipitate from sea water although the water is many times supersaturated with respect to the mineral. Also, dolomitization of carbonate sediments in direct contact with sea water is rare (Sonnenfeld, 1964, p. 101).

Apart from scientific thirst for solving the problem, an understanding of the genesis of dolomite and time of dolomitization has great economic value. Positive correlation between percent dolomitization and mineralization, where mineralizing solutions have been responsible for dolomitization, has been used in mining districts as a guide to discovery of additional ore bodies. Similarly, recognition of secondary dolostone with enhanced porosity has been used in the oil industry as a guide to oil production.

---

[1] Manuscript received September 28, 1972; revised May 30, 1973.

[2] Modified from a dissertation presented in partial fulfillment of the Ph.D. degree, Northwestern University.

[3] Present address, American Metal Climax, Inc., 105 S. Meridian St., Indianapolis, Indiana 46225.

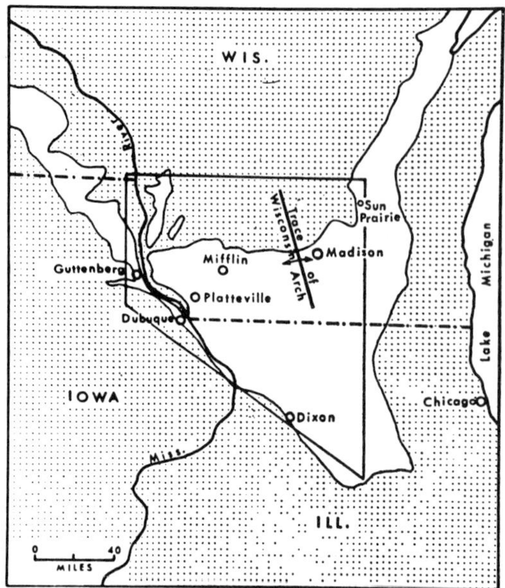

FIG. 1.—Index map, showing the outcrop area of the Champlainian Series (blank area) in Wisconsin and adjacent states. Trace of Wisconsin Arch and the area under study are also shown.

Higher porosity in secondary dolostone is thought to be the result of 13% volume reduction owing to dolomitization (Landes, 1946; Murray, 1960; Weyl, 1960).

Although the mechanisms and physiochemical conditions suitable for dolomitization are still ambiguous and not fully understood, knowledge of the environment of deposition can result in distinction between primary and secondary dolostone. Discovery of Recent dolostones in Southern Australia (Alderman and Skinner, 1957), Andros Island (Shinn *et al.*, 1965), Persian Gulf (Illing *et al.*, 1965) and other localities provided valuable information regarding the environment of deposition of this mineral; results of these studies were quickly applied to the geologic record. Using the present as a key to the past, close association of Recent dolomite with evaporite minerals led to overgeneralizations, such as ". . . most dolostone deposits in the geologic record owe their origin to hypersaline brines. They must, therefore, be considered as evaporite deposits" (Friedman and Sanders, 1967, p. 267).

Widespread dolomitic facies associated with shallow epicontinental shelves or structural highs (Prouty, 1948; Cohee, 1948; Cloud and Barnes, 1957) constitute substantial amounts of dolostones that are devoid of evaporites and lack evidence of supratidal origin. Hypersaline dolomitization models, therefore, fail to provide a suitable mechanism of dolomitization for these sediments. It is the purpose of this study to present a dolomitization model that is independent of evaporation or high Mg^{2+}/Ca^{2+} ratio in the solution and to suggest an explanation of the genesis of dolomites associated with epicontinental shelves or positive elements.

The Champlainian Series (Middle Ordovician) in southern Wisconsin and adjacent states (Fig. 1) provides a classic example of dolomitization along a positive element and was selected as the field area of investigation. Two different kinds of information obtained from detailed study of 69 quarries, roadcuts, and natural exposures were used to establish the environment of deposition and diagenetic history of the Champlainian Series.

The study of primary structures, paleontology, stratigraphy and facies distribution in conjunction with petrographic analysis provided information necessary to interpretation of the variation and relative depth of the depositional environment as well as to the vertical and horizontal configuration of the dolostone-limestone boundary (Badiozamani, 1972). Oxygen and carbon isotopic analyses, chemical analyses of sodium and strontium, and X-ray diffractograms furnished the necessary information to interpret the diagenetic history of the Series. The results of this work show that the dolomitization of the Champlainian Series was not penecontemporaneous, rather it was the result of a relatively early diagenetic process following the subaerial exposure of the positive elements and establishment of fresh water lenses. Dolomitization occurred in the brackish zone where sea and fresh waters were mixed.

DOLOMITIZATION

Introduction

Extensive studies regarding the mechanisms of dolomitization in Recent environments have been carried out in the past decade. A summary of recent thoughts on dolomitization mechanisms is given in Figure 2. Although dolostones have a variety of origins, it is likely that the majority have a secondary origin (e.g., Garrels and Mackenzie, 1971) and only a small fraction of the dolostones which are associated with evaporites are considered to be primary. Secondary dolomitization may occur during early stages of diagenesis, i.e., penecontemporaneous dolomitization, or it may occur later in the history of the rock. A condition of all the models, except Dorag and cannibalization, is

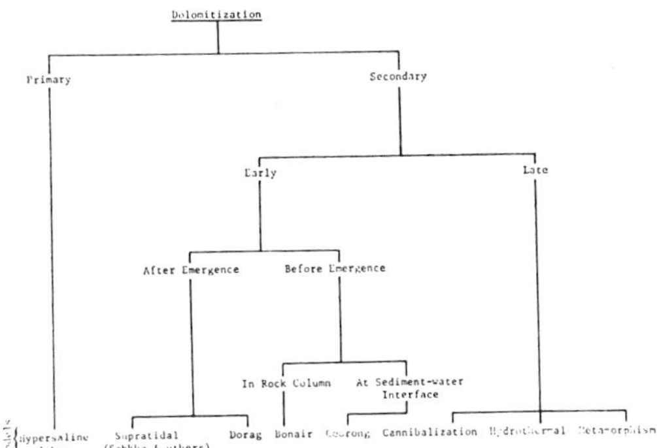

FIG. 2.—Generic classification of dolostones.

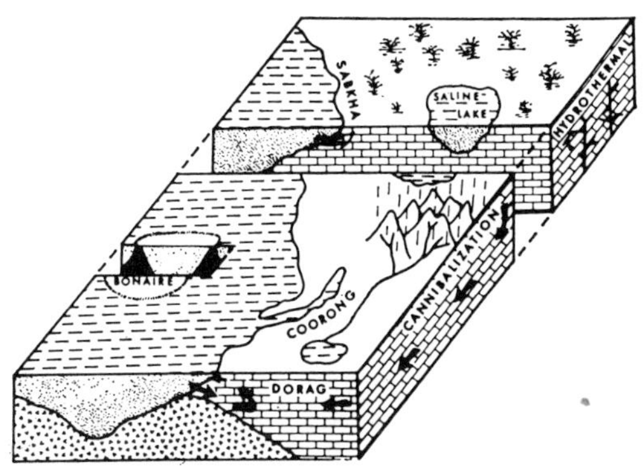

FIG. 3.—Diagrammatic representation of the source of Mg^{2+}, and the circulation path of various dolomitization models. The diagram shows mechanisms currently favored for obtaining circulation such as evaporation (Sabkha; Illing *et al.*, 1965), density gradient (Bonaire, Deffeyes *et al.*, 1965), and hydrostatic head (Dorag, Hanshaw *et al.*, 1971, cannibalization, Goodell and Garman, 1969).

that the solution should be many times supersaturated with respect to calcite and dolomite. Also, common and necessary requirements of all the models for dolomitization are an almost unlimited source of Mg^{2+} and a mechanism for circulation of reactive fluids through the pore spaces. Circulation is required to supply the mass of Mg necessary for the dolomitization reaction. Mechanisms currently favored for obtaining circulation are evaporation (Sabkha model, Illing *et al.*, 1965); density gradients (Bonaire model, Deffeyes *et al.*, 1965); and hydrostatic head models (the ground water model, Hanshaw *et al.*, 1971, and the cannibalization model, Goodell and Garman, 1969). These models, the source of Mg^{2+}, and their circulation path are represented diagrammatically in Figure 3.

Mechanism of Ground Water Dolomitization

In 1916 Van Tuyl published his views regarding the mechanisms of dolomitization and noted that ground water had been suggested to be an agent in dolomitization by Collegno as early as 1834. However, after reviewing the concept, Van Tuyl terminated the discussion by acceptance of Steidtmann's (1911) opinion that ground water dolomitization can be operative only on a local scale. This notion was further supported by Hsu (1966) who used mass balance calculations to show that even if ground water dolomitization was 100 percent effective, the rate would still be too slow to account for any massive dolostones. Hsu's calculations were based on Florida's ground water flow rate of seven meters per year (Hanshaw *et al.*, 1965). The source of Mg was considered to be from alteration of Mg-calcite or other Mg-bearing rock forming minerals.

Dolomitization by ground water was revived by Hanshaw *et al.* (1969, 1971) with some modification to account for the amount of Mg necessary for massive dolomitization. In their modified model, dolomitization is postulated to occur in the brakish water zone that results from mixing of potable and sea waters (also see Vernon, 1969). In this zone, Mg^{2+} ions are supplied by sea water and dissolution of $CaCO_3$ occurs owing to mixing of the two waters. Recently Land (1972) has also invoked the same mechanism of dolomitization for dolomites of the Jamaican reefs.

Holland (1964), and Mathews (1971) point out that when two waters saturated with respect to calcite are mixed, the mixture may become undersaturated with respect to calcite. Runnells (1969) showed that due to the nonlinearity of the $CaCO_3$ solubility curve, a mixture of the two waters can result in undersaturation. Garrels *et al.* (1961) demonstrated that solubility of calcite in NaCl and $MgCl_2$ solution increases with increasing ionic strength. The solubility is maximum at intermediate ionic strength and rapidly decreases at higher ionic strengths.

Analyses of seven Yucatan's ground waters (Back and Hanshaw, 1970, Table 3, p. 351) were used by the author to determine the effect of the mixing of two waters on the degree of saturation of calcite and dolomite. Different proportions of Yucatan's ground waters were mixed theoretically with average sea water (Goldberg, 1957). Distribution of species, their activities, activity coefficients, and ion activity products (IAP) were calculated using a com-

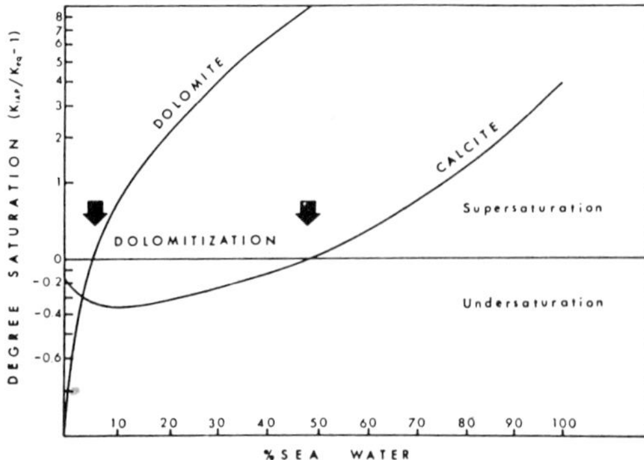

FIG. 4.—The effect of mixing of Yucatan ground water with sea water on saturation degrees of calcite and dolomite. Note that the mixture of Yucatan ground water with up to 30 percent sea water is less saturated with respect to calcite than the initial ground water, whereas dolomite saturation increases continuously. Dolomitization, therefore, should occur in the zone between 5 to 30 percent sea water which is undersaturated with respect to calcite and supersaturated with respect to dolomite (between the arrows).

puter program SOLSAT. In calculation of IAP, effect of complexes on solubility of carbonates (Garrels *et al.*, 1961) was taken into account. The system was maintained closed with respect to CO_2.

A sample calculation for 95% ground water mixed with 5% sea water should suffice to show the procedure used in constructing Figure 4. Compositions of the sea and ground water and mixture produced by mixing the two waters are given in Table 1.

TABLE 1.—*Composition of the waters used in sample calculation*

Constituents	Ground Water* ppm	Sea Water** ppm	Mixture† ppm
K^+	7.6	380.0	26.22
Na^+	25.0	1,050.0	76.25
Ca^{2+}	82.0	400.0	97.90
Mg^{2+}	9.0	1,300.0	73.55
SiO_2	6.8	6.0	6.76
SO_4^{-2}	17.0	2,650.0	148.65
HCO_3^-	252.0	140.0	246.40
Cl^-	40.0	19,000.0	988.00
pH	7.15	8.15	7.17

* Yucatan ground water (Hoctun) (Back and Hanshaw, 1970).
** Average sea water (Goldberg, 1957).
† Mixture of 95% ground water and 5% sea water.

Ionic strength of the mixture was calculated from

$$I = \tfrac{1}{2} \Sigma_i Z_i^2 m_i, \qquad (1)$$

where I is ionic strength of the solution, Z_i the charge on the ith species, and m_i is concentration of the ith species in mole per liter. Activity coefficient of each species was calculated from extended Debye-Hückel equation (Helgeson, 1969)

$$\log \gamma_i = \frac{-A Z_i^2 I^{\frac{1}{2}}}{1 + a^0_i B I^{\frac{1}{2}}} + \dot{B}I, \qquad (2)$$

where γ_i is the activity coefficient of the ith ion in the solution, A and B are constants characteristic of the water, a^0_i refers to the "distance of closest approach" of the ith ion and $\dot{B}$ is the deviation function. After activity coefficients were obtained, concentration of each species was calculated using Garrels and Thompson's (1962) model. For example, the distribution of the CO_3^{2-} bearing species was calculated as follows:

$$m_{CO_3^{2-}\text{total}} = m_{CaCO_3^0} + m_{MgCO_3^0} + m_{NaCO_3^-}$$
$$+ m_{HCO_3^-} + m_{CO_3^{2-}}\text{uncomplexed.} \quad (3)$$

The equilibrium dissociation relations for the various species are:

$$\frac{\gamma_{Ca^{2+}} m_{Ca^{2+}} \gamma_{CO_3^{2-}} m_{CO_3^{2-}}}{\gamma_{CaCO_3^0} m_{CaCO_3^0}} = K_{CaCO_3^0} \quad (4)$$

$$\frac{\gamma_{Mg^{2+}} m_{Mg^{2+}} \gamma_{CO_3^{2-}} m_{CO_3^{2-}}}{\gamma_{MgCO_3^0} m_{MgCO_3^0}} = K_{MgCO_3^0} \quad (5)$$

$$\frac{\gamma_{Na^+} m_{Na^+} \gamma_{CO_3^{2-}} m_{CO_3^{2-}}}{\gamma_{NaCO_3^{2-}} m_{NaCO_3^-}} = K_{NaCO_3^-} \quad (6)$$

$$\frac{\gamma_{H^+} m_{H^+} \gamma_{CO_3^-} m_{CO_3^{2-}}}{\gamma_{HCO_3^-} m_{HCO_3^-}} = K_{HCO_3^-}. \quad (7)$$

Thus, five mass action equations are available to describe the various carbonate species. Similar mass balance and mass action equations for HCO_3^- and SO_4^{2-} bearing species will provide the desired number of equations that can be solved simultaneously with the aid of a high-speed computer to obtain concentrations of each species. Calculated distribution of species for the mixture of 95% ground water and 5% sea water is presented in Tables 2 and 3.

Saturation degree of the mixed solution with regard to calcite and dolomite is calculated and compared with that of the initial ground water in Table 4. The data indicate that the mixture is less saturated with respect to calcite and more saturated with respect to dolomite than the initial ground water.

TABLE 2.—*Summary of ion pairing for sample calculation (in percent)*

Ion	Molality (Total)	Free ion	Me-SO_4 pair	Me-HCO_3 pair	$MeCO_3$ pair
Ca^{2+}	.00249	91.333	5.055	3.436	.177
Mg^{2+}	.00303	93.282	4.769	1.643	.303
Na^+	.02387	99.384	.611	0.000	.004
K^+	.00067	99.632	.368	0.000	0.000

Ion	Molality (Total)	Free ion	Ca-Anion pair	Mg-Anion pair	Na-Anion pair	K-Anion pair
SO_4^{2-}	.00155	73.123	7.975	9.319	9.424	.159
HCO_3^-	.00408	96.831	1.990	1.179	0.000	0.000
CO_3^{2-}	0.000	25.572	22.069	46.922	5.438	0.000

The calculation (Fig. 4) shows that up to a 30% sea water-ground water mixture, the solution becomes less saturated with respect to calcite than the initial ground water. The results also show that dolomite saturation increases continuously with increasing sea water mixture. Therefore, in the range of 5 to 30% sea water where the solution is undersaturated with respect to calcite, but many times super-

TABLE 3.—*Distribution of species in the mixture*

Species	ppm	Molality	Activity
K^+	2.61×10	6.68×10^{-4}	5.50×10^{-4}
Na^+	5.44×10^2	2.37×10^{-2}	1.97×10^{-2}
Ca^{2+}	8.92×10	2.23×10^{-3}	1.13×10^{-3}
Mg^{2+}	6.85×10	2.82×10^{-3}	1.52×10^{-3}
SO_4^{2-}	1.09×10^2	1.13×10^{-3}	5.37×10^{-4}
CO_3^{2-}	2.99×10^{-1}	4.99×10^{-6}	2.42×10^{-6}
Cl^-	9.95×10^2	2.81×10^{-2}	2.32×10^{-2}
OH^-	3.03×10^{-3}	1.78×10^{-7}	1.48×10^{-7}
H^+	7.86×10^{-5}	7.81×10^{-8}	6.76×10^{-8}
H_2O	9.97×10^5	5.55×10	1.00
KSO_4^-	3.33×10^{-1}	2.47×10^{-6}	2.04×10^{-6}
$NaCO_3^-$	8.80×10^{-2}	1.06×10^{-6}	8.89×10^{-7}
$NaSO_4^-$	1.73×10	1.45×10^{-4}	1.22×10^{-4}
$CaCO_3$	4.31×10^{-1}	4.31×10^{-6}	4.35×10^{-6}
$CaHCO_3^+$	8.47	8.39×10^{-5}	7.06×10^{-5}
$CaSO_4$	1.68×10	1.23×10^{-4}	1.25×10^{-4}
$MgCO_3$	7.71×10^{-1}	9.17×10^{-6}	9.26×10^{-6}
$MgHCO_3^+$	4.23	4.97×10^{-5}	4.13×10^{-5}
$MgSO_4$	1.73×10	1.44×10^{-4}	1.47×10^{-4}
HSO_4^-	4.13×10^{-4}	4.26×10^{-9}	3.55×10^{-9}
HCO_3^-	2.48×10^2	4.08×10^{-3}	3.41×10^{-3}
H_2CO_3	3.17×10	5.12×10^{-4}	5.17×10^{-4}
HCl	4.48×10^{-11}	1.23×10^{-15}	1.24×10^{-15}
$Mg(OH)^+$	4.27×10^{-3}	1.03×10^{-7}	8.61×10^{-8}
$Ca(OH)^+$	1.97×10^{-4}	3.46×10^{-9}	2.88×10^{-9}

TABLE 4.—*Saturation degree of the initial ground water and the mixture*

Solution	log IAP		% Saturation degree*	
	Calcite	Dolomite	Calcite	Dolomite
Ground water	–8.47	–17.66	85	22
Mixture	–8.56	–16.99	69	102

* Saturation degree $= K_{IAP}/K_{eq} \times 100$.

saturated with respect to dolomite, dolomitization by replacement could occur. The effect of removal by precipitation has not been considered in this calculation. In general, dolomitization process can be expressed by the chemical equation

$$2CaCO_3 + Mg^{2+} = CaMg(CO_3)_2 + Ca^{2+} \quad (8)$$

Such a process would result in 13% volume reduction on a mole-for-mole basis, if the effect of compaction (Weyl, 1960) or addition of material by the percolating solution are not considered. Dolomitization may not be accompanied by 13% volume reduction if replacement of calcite by dolomite is on a volume-for-volume basis. This type of replacement is apparently taking place in Jamaica where all the structural details of the coralline algae are being preserved after dolomitization (Land, 1970).

Vertical changes in the position of the brackish zone owing to fluctuations in sea level or changes in the supply of fresh water to the aquifer could result in considerable volumes of rocks being dolomitized below the potable water lens.

A new term, *"Dorag" dolomitization*, is used for dolomitization by mixture of sea and ground water. This term differentiates such a mechanism of dolomitization from that presumed to occur through movements of ground water alone. Dorag is a Farsi (Persian language) word meaning mixed blood or hybrid and is used to imply the importance of mixing of fresh and sea water which is the essence of this mechanism of widespread dolomitization.

APPLICATION OF DORAG DOLOMITIZATION TO THE CHAMPLAINIAN SERIES OF WISCONSIN

Stratigraphy

The name Champlainian was proposed in 1914 by Schuchert and Barrell for the Middle Ordovician strata cropping out in the region of Lake Champlain, New York. In Wisconsin and adjacent states time equivalent strata are underlain unconformably by dolomites of the Canadian Series and overlain by Cincinnatian

TABLE 5.—*Summary of the Stratigraphy of the Champlainian Series in Wisconsin*

Series	Formation	Member	Lithologic Description	Stratigraphic Relations	Author	Thickness (ft) Min	Thickness (ft) Max
MIDDLE ORDOVICIAN (CHAMPLAINIAN)	GALENA	Noncherty unit	Thin to medium or thick-bedded, yellowish-buff, fine to medium grain, porous and vuggy, sparsely fossiliferous, dolostone.	Unconformable with unit above, but conformable with strata below.	Agnew, et al. (1956)	110	130
		Cherty unit	As above but cherty and abundance of Receptaculites.	Conformable above and below.	Agnew, et al. (1956)	---	105
	DECORAH	Ion	Medium to thick-bedded, buff and gray, medium to coarse grain with grayish-green calcareous fossiliferous shale partings.	Conformable contacts.	Kay (1929)	20.0	22
		Gutten-berg	Thin to medium-bedded, pink and buff, crystalline, fossiliferous limestone in the west and dolostone in the east, dark-brown shale partings.	Conformable contacts.	Kay (1928)	6.5	16
		Spechts Ferry	Grayish-green fossiliferous shale with buff lenses of limestone toward east, metabentonite near base, and limestone nodules containing phosphatic pellets near top.	Upper contact conformable, unconformable with lower unit.	Kay (1928)	0.0	8
	PLATTEVILLE	Quimbys Mill	Medium-bedded, dark-brown, fine grain, conchoidal fracturing limestone and dolostone with dark brown shale partings.	Unconformable with Spechts Ferry and contains solution pits at the unconformity surface; lower contact controversial.	Agnew and Heyl (1946)	0.0	18
		Magnolia	Thin to medium-bedded, bluished-gray on fresh surface, weathers buff; fine to medium grain, fossiliferous dolostone, less wavy-bedded than Mifflin; a coquina bed is present near the top in many localities; shale partings are greenish-gray.	Lower contact conformable and not clearly defined.	Bays and Raasch (1935)	13.0	18
		Mifflin	Thin to medium, wavy-bedded, fine grain, bluish-gray fossiliferous limestone in the west, and buff dolostone toward the east; valves and tests are not preserved in dolomitic facies; shale partings are bluish-gray and are less pronounced in the dolostone; thickness of the shale partings increases toward west and south; basal beds are sandy in some places.	Disconformable with Pecatonica; disconformity surface marked by a corrosion zone stained with pyrite and iron oxide.	Bays (1937)	12.0	17
		Pecatonica	Medium-bedded, fine to medium grain, saccharoidal dense to vuggy, bluish-gray and buff, sparsely fossiliferous (molds) dolostone with shale parting, lower beds are sandy and contain phosphatic nodules.	Disconformable with both Mifflin and Glenwood; Pre-Mifflin subaerial exposure is indicated by solution pits containing Pecatonica granules and the presence of occasional sandy unit at the contact in some places.	Hershey (1894) Kay (1935)	20	24
	GLENWOOD		Greenish-gray or pale-green argillaceous sandstone, shale, and buff sandy dolostone.	Small magnitude unconformity between Glenwood and St. Peter (Templeton and Willman, 1963).	Calvin (1906) Bevan (1926)	0	3
	ST. PETER		Fine to coarse-grain, moderately well-sorted, poorly cemented, cross-bedded, quartz sandstone.	Lower contact unconformable on Canadian and Tremealeauan units.	Owen (1847)	40	

shales and limestones. Here the Champlainian Series comprise in ascending order the St. Peter, Glenwood, Platteville, Decorah and Galena formations. Except for the St. Peter Formation which is a fine to coarse-grained sandstone marking the base of the Tippecanoe Sequence (Sloss, 1963) and the Glenwood Formation, a green-gray sandy shale, the remaining formations are dominated by dolostone in the east and limestone toward the west. Summary of the stratigraphy of the Champlainian Series in Wisconsin is presented in Table 5.

The Platteville Formation of the Champlainian Series was selected for detailed study of
dolomitization associated with positive elements.
Because the number of outcrops exposing all
members of the Platteville Formation is limited,
a complete analysis of the whole Formation
was impossible. However, the abundant outcrop
of the Mifflin Member and its well-exposed
transition zone between the dolomitic and calcitic facies made this stratigraphic unit ideal
for detailed study. The writer believes that the
conclusions regarding the dolomitization processes involved, based on study of the Mifflin
Member, are not only applicable to the Platteville, Decorah and the Galena formations, but
perhaps to many regionally dolomitized dolostones associated with positive elements or epicontinental shelves.

The problem of widespread dolomitization of
the Champlainian Series has attracted the interest of scientists for many years. Bays (1937)
strongly advocated a secondary origin of dolomitization for the Platteville dolostone but
provided no further explanation regarding
mechanisms involved. Deininger (1964) studied
the transition zone between limestone and dolostone and preferred dolomitization by hydrothermal solution to other processes, but mentioned that there was lack of sufficient evidence
to support one process over the other. A semi-
restricted lagoon, an environment of higher
salinity than normal sea water, was suggested
by Asquith (1967) to be the controlling factor
in dolomitization of the Mifflin Member of
the Platteville Formation. As will be shown
later, none of the proposed models can adequately explain the extent of existing dolomitization in the area.

The faunal distribution, homogeneity in characteristics of the carbonates over very broad
areas, thickness of the unit, the light color of
the sediments and diagnostic primary structures
suggest a broad, shallow and open environment
of deposition for the Mifflin Member (Badiozamani, 1972).

Isotopic Analysis

Introduction.—Application of isotopic analysis to the study of carbonate rocks has opened
a new field in geology. The first application of
oxygen isotopic fractionation to geological problems was by McCrea (1950) followed by that
of Urey *et al.* (1951), and Epstein *et al.* (1951).
Later, the work of Clayton and Epstein (1958),
and Engel *et al.* (1958) on the determination
of the fractionation factor between coexisting
quartz-calcite pairs and quartz-dolomite pairs

in naturally occurring hydrothermal deposits,
led Degens and Epstein (1964) to investigate
the origin of dolomites using oxygen isotopic
ratios. In spite of a prediction made from high
temperature analyses that primary dolomite at
25° C should contain oxygen of about six to
ten per mil (parts per thousand, ‰) heavier
than coexisting calcite (Clayton *et al.*, 1958),
Degens and Epstein (1964) did not find such
a fractionation in Recent dolomites or in those
from older rocks. Degens and Epstein (1964)
concluded that either the analyzed dolomites
had not formed under equilibrium conditions or
that the dolomites were formed from a precursor calcite in which the oxygen isotopic
ratio remained unchanged during dolomitization. The latter interpretation implies that
dolomitization is a replacement phenomenon in
which Mg substitutes for the Ca of precursor
calcite without exchange of oxygen in the CO_3
radical. Currently there is no general agreement among the students of oxygen isotopic
chemistry on the problem of 5–7‰ oxygen isotopic fractionation between primary dolomite
and coexisting calcite at 25° C.

Representative samples of the Mifflin Member in southwestern Wisconsin were analyzed
for oxygen and carbon isotopes with expecta-
tion that such data would provide information
leading to reconstruction of the environment
of deposition and processes of dolomitization.

Sample Preparation.—Samples of rock for
isotopic analysis were ground manually for
short periods of time and the 44–74 μ fraction
size was separated by sieving. Short-period
manual grinding was used to prevent isotopic
exchange with CO_2 of the atmosphere. Jamieson and Goldsmith (1960) found that during
prolonged grinding, calcite can be partially
converted to aragonite. It is possible that during such conversion, atmospheric CO_2 may be
substituted for the CO_2 of the calcite according
to the following equation

$$CaCO_{3\,(calcite)} + CO_{2\,(atm.)} = CaOCO_{2\,(atm.)\,(aragonite)} + CO_2.$$

X-ray diffractograms were employed to obtain a rough estimate of the percentage of calcite and dolomite present in each sample. One
hundred to 200 mg of each sample were reacted
with 5 ml of 100% phosphoric acid according
to the standard procedure of McCrea (1950).
The CO_2 evolved from samples with less than
five percent of either calcite or dolomite was
collected only once after the reaction was completed. The CO_2 from samples with a higher
percentage of calcite or dolomite was collected

233

three times, after reaction time of one-half, three, and 24 hours. This method is a modified version of the procedure used by Degens and Epstein (1964) for mixture of calcite and dolomite which is based on reaction rates and grain size effect. Clayton *et al.* (1968) showed that to obtain better separation of oxygen isotopes for calcite and dolomite the grain size should have a narrow range (see also Fritz and Fontes, 1966). Recently Walter *et al.* (1972) studied the reaction rates and δO^{18} variation of carbonates owing to grain size and time of reaction. Their findings corroborate the effects of grain size on the reaction rate and reduction of isotopic cross contamination between calcite and dolomite in a reaction tube.

Results of Analyses.—CO_2 extracted from each sample was analyzed on a Nier (1947) six inch, 60° sector type mass spectrometer of the Enrico Fermi Institute for Nuclear Studies, University of Chicago, modified for high precision by McKinney *et al.* (1950).

Isotopic values are reported in terms of δ defined as

$$\delta = \frac{R_i - R_s}{R_s} \times 1000$$

where

$R_i = O^{18}/O^{16}$ or C^{13}/C^{12} ratio of the sample,
$R_s = O^{18}/O^{16}$ or C^{13}/C^{12} ratio of the standard,
$\delta =$ the deviation in per mil (parts per thousand) of oxygen or carbon isotopic values of the sample from an arbitrary standard.

Oxygen and carbon isotopic values are reported relative to PDB Standard (Craig, 1957). The following equation (Clayton *et al.*, 1968, p. 426) defines the relation between the oxygen isotopic values given in one standard relative to the other

$$\delta_{(SMOW)} = 1.03037\,\delta_{(PDB)} + 30.37.$$

Corrections were made for valve mixing and δC^{13} according to Craig (1957). In calculating the oxygen isotopic data the fractionation factor of the phosphoric acid reaction with various carbonates (Sharma and Clayton, 1965) was taken into account. Fractionation factors of 1.01110 and 1.01025 were used for dolomite and calcite, respectively. All the values plotted in Figure 5 are calculated as dolomite, which means that samples with more than 50% calcite and the first of the three extractions made on a sample (if the gas can be regarded as CO_2 evolved only from the calcite without contam-

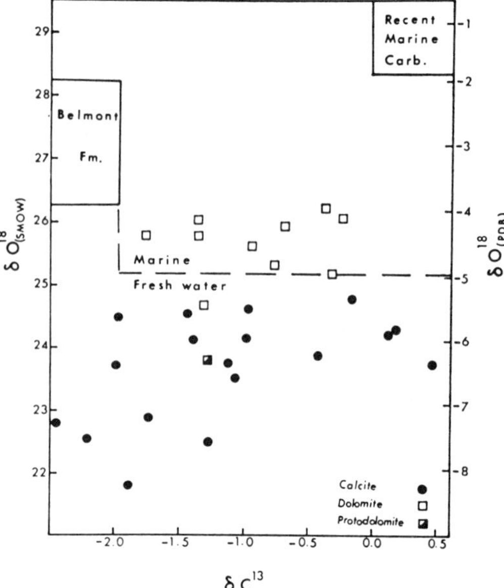

Fig. 5.—Oxygen isotope vs. carbon isotopic composition of the Mifflin Member. Dolomitic samples show higher oxygen isotopic values than calcitic samples. Compared to the Recent marine carbonates, Mifflin Samples have lower isotopic values indicating equilibration with meteoric water. The line dividing the Mifflin samples into marine and fresh water is based on data from Keith and Weber (1964). The box representing isotopic composition of the Recent marine carbonates is from Gross (1964).

ination with CO_2 from dolomite) should be 0.85‰ more positive.

The Mifflin dolomite has an average oxygen isotopic value of –4.47‰ with standard deviation of ± 0.5‰, whereas the Mifflin limestone has an average oxygen isotopic value of –6.43‰ with standard deviation of ± 0.8‰.

Interpretation of Results.—The use of oxygen isotope ratios for recognition of the origin of dolomite, like any other criterion used for solving this problem, has reached a controversial state that cannot be resolved definitely until the processes responsible for deposition of dolomite are fully understood. Since the work of Epstein *et al.* (1963), many authors have attempted to find a satisfactory answer to the fractionation problem; however, none is available so far.

Accepting 5–7‰ fractionation between coexisting calcite and dolomite, the results of the isotopic analysis of the Mifflin Member can be interpreted as follows. If δC^{13} of these samples is plotted against δO^{18}, as illustrated in Figure 5, all the dolomites except one sample have reasonably similar oxygen isotopic compositions whereas the limestones have lighter and more variable oxygen isotopic values. Moreover, for the same samples the carbon isotopic values are distributed within a narrower range. No distinction between the two facies can be made on the basis of the carbon isotopes. Isotopic composition of Recent Marine carbonates and Belmont Formation (Bermuda) are plotted in Figure 5 for comparison. The isotopic value of the Belmont will be used later to estimate the maximum time required for dolomitization of the Mifflin Member whereas the isotopic value of the Recent carbonates is shown for comparison with that of the Mifflin to illustrate the lower isotopic composition of the latter owing to diagenesis.

Keith and Weber (1964), using analyses of 500 samples of marine and freshwater carbonates and fossils from different geologic ages, proposed the line of $\delta C^{13} = -2\%_0$, as a line separating 80% of the marine deposited carbonates from those of freshwater origin. From their data a similar line can be drawn at $\delta O^{18} = -5\%_0$, separating more than 98% of the marine carbonates from those of freshwater origin, i.e., samples with δC^{13} and δO^{18} values heavier than $-2\%_0$ and $-5\%_0$ respectively are of marine origin.

All the dolomites and most of the calcite samples have δC^{13} values above $-2\%_0$, suggesting deposition in a marine environment. However, based solely on δO^{18} values, only dolomite samples fall in this category. It appears that dolomites have exchanged less isotopically with respect to calcites. Their oxygen isotopic values still fall within the region suggested for ancient marine carbonates in contrast with the limestones which have exchanged more and have isotopic values similar to those of fresh water carbonates (Fig. 6). The high degree of scatter in the data is probably due to permeability differences in the rocks, resulting in different degrees of exchange.

The dolomite collected along a joint in a quarry at Sun Prairie has a much lower oxygen isotopic value than other dolomites, and as will be shown later is a protodolomite with diffuse ordering peaks. Its lower oxygen isotopic value, diffuse ordering peak, and occurrence along joints are indicative of a late stage of dolomitization postdating the jointing.

Lack of 5–7‰ oxygen isotopic fractionation between coexisting calcite and dolomite in the Mifflin Samples demonstrates that these minerals were not coprecipitated and that the dolostones are not of primary origin. The presence

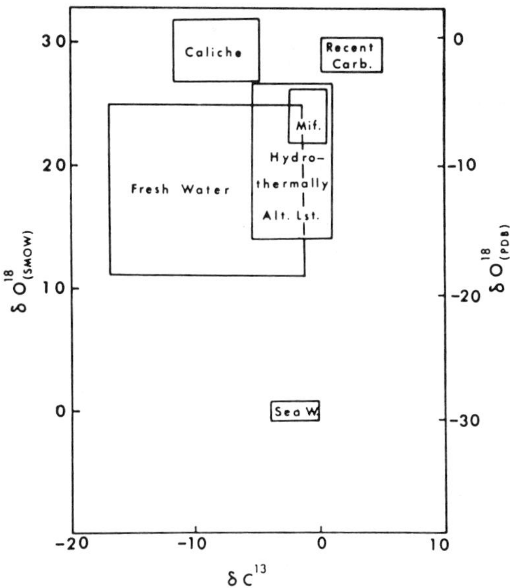

Fig. 6.—Isotopic distribution of carbonates precipitated in different environments. Although the Mifflin isotopic values are overlapped by hydrothermal and fresh water fields, several lines of evidence (see text) do not support an hypothesis of hydrothermal dolomitization. (Constructed on the basis of data collected by Degens, 1967).

of ore deposits in the study area and overlapping of the isotopic composition of the Mifflin samples by that of hydrothermally altered limestones (Fig. 6) suggests the possibility of hydrothermal dolomitization. If the present isotopic composition of the Mifflin samples was obtained during equilibration of the rocks with hydrothermal solutions, then the Mifflin limestones, which have lighter oxygen isotopic values than dolostones, represent equilibration with solutions of higher temperature. This conclusion contradicts the result of experimental dolomitization, which suggests higher efficiency of dolomitization at higher temperatures. Several other lines of evidence are also available to support the unlikelihood of hydrothermal dolomitization. Hall and Friedman (1969) studied isotopic variation within dolostones and limestones of the Champlainian Series in the Mississippi Valley zinc-lead district and concluded that such values in both facies show systematic changes immediately adjacent to ore bodies. Re-examination of their data, however, shows that only limestones display systematic changes, whereas dolomites fail to show such a trend. Their data suggest that the dolomitization occurred prior to mineralization and that the already formed dolomite was more resistant to alteration and was not markedly affected by the later hydrothermal solutions.

The outline map of the zinc-lead district in the upper Mississippi Valley (Heyl *et al.*, 1959, p. 7) shows that mineralization covers an area with the city of Platteville in the center. If dolomitization were due to hydrothermal solutions, the outlines of dolomitization and mineralization should coincide. However, the facies distribution of the Mifflin Member shows that the dolomite facies occurs only to the east of Platteville; a distribution inconsistent with the hypothesis of hydrothermal dolomitization. A possibility exists that such a distribution can result from permeability differences between rock occurring in the east and that in the west, but the presence of ore deposits in the calcitic facies precludes the possibility of significant differences in permeability between the two rock types. Also, because major shifts in dolostone-limestone boundary are associated with unconformity surfaces and marked differences in lithology, it is unlikely that the distribution of the facies boundary can be explained by hydrothermal dolomitization. The shift in the dolostone-limestone boundary and its juxtaposition with marked lithological changes is regarded as an exceptionally important feature bearing upon the mechanism of dolomitization and is discussed in detail later.

In summary, isotopic data from the Mifflin Member indicate isotopic exchange with meteoric water prior to and after dolomitization. Moreover, dolomitization appears to have occurred relatively early to maintain the isotopic values of the dolostones in the region suggested for ancient marine carbonates (Keith and Weber, 1964). Exchange reactions with meteoric water continued after dolomitization, and the different exchange rates of calcite and dolomite in equilibrating with interstitial water (Epstein *et al.*, 1963) resulted in the fractionation between the calcite and dolomite facies of the Mifflin Member.

X-Ray Diffraction

X-ray diffraction techniques were used for determination of the percentage of calcite and dolomite in the Mifflin samples and detection of mineralogical modifications introduced by diagenetic processes. Samples were crushed and ground for a few seconds by an electric grinder with a porcelain disk and the less than 44 micron fractions were used for X-ray analysis. A small amount of each sample was mixed with fluorite of the same grain size as an internal standard and was mounted on a glass slide with diluted Duco-cement. The analyses were made on a Norelco X-ray unit using nickel-filtered, CuK_α radiation.

The X-ray diffractograms all show calcite and/or dolomite; some samples contain quartz. The absence of other diffraction peaks is due in part to the lack of accessory minerals in quantities sufficient for detection at the sensitivity level employed. A calcite peak of 3.03 Å corresponding to 29.40° 2θ, and a dolomite peak of 2.88 Å corresponding to 30.96° 2θ, were used for determination of calcite and dolomite, respectively. Because the calcite and dolomite peaks are sharp and no other minerals were present in substantial amount, the peak height ratio of dolomite to calcite (2.88 Å dolomite peak-height/3.03 Å calcite peak-height) was used in conjunction with Tennant and Berger's (1957) graph to determine percentage of calcite and dolomite in the sample.

Substitution of Mg for Ca in calcite lattice results in a decreased calcite lattice constant owing to the smaller ionic radius of Mg. This reduction can be determined by means of X-ray diffraction patterns. Chave (1952) and Goldsmith *et al.* (1958a, 1961) have shown the effects of $CaCO_3$–$MgCO_3$ solid solutions on lattice constants of biogenic and synthetic calcites, respectively. Goldsmith and Graf (1958a, p. 680) assumed a linear relationship between the lattice constant (d_{211}) and substitution of $MgCO_3$ in calcite. Milliman *et al.* (1971) have shown that coralline algae contain considerably more magnesium than indicated by the peak shift in the X-ray pattern; local concentration of Mg in the algae is thought to be the cause of the discrepancy. They relate the asymmetry of the X-ray pattern and its tailing toward higher 2θ values to these local concentrations of Mg in the calcite.

High Mg-calcites are not present and sharp and symmetrical X-ray patterns indicate that carbonate minerals in the samples are of single phase. Determination of mole percent $MgCO_3$ in calcite and dolomite samples of the Mifflin Member using Goldsmith *et al.*'s (1958a) curve, reveals that all the calcites have less than 2 mole percent $MgCO_3$ in solid solution, whereas dolomite samples can be grouped into two distinct categories. One group has close to stoichiometric ratios of 49–50 mole % $MgCO_3$ and the other group has values ranging between 46 to 49%. The degree of ordering (Goldsmith and Graf, 1958b) for stoichiometric dolomite is high as compared with Ward's Natural Science Establishment Standard dolomite, whereas

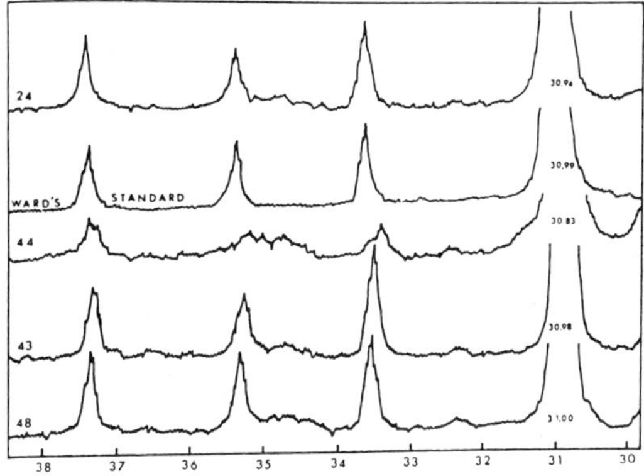

FIG. 7.—X-ray diffraction pattern of dolomitic facies of the Mifflin Member. Sample 44 has diffuse ordering peaks and is a protodolomite, whereas the others are well ordered dolomites. Degree of ordering is indicated by the relative height of the {01·5} peak (35.25° 2θ) to that of the {11·0} peak (37.25° 2θ). Corresponding rhombohedral indices are {221} and {101} respectively. The 2θ value of the main dolomitic peak {104} is given for each sample.

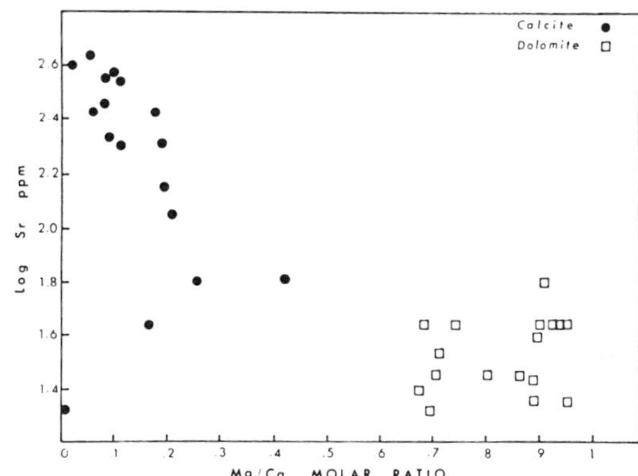

FIG. 8.—Strontium vs. Mg/Ca molar ratio of the Mifflin samples. Note increasing Sr concentration in the samples as Mg content decreases. Mean values of Sr are 229 ppm for limestone and 37 ppm for dolostone. These values are appreciably lower than Sr concentrations reported for Recent carbonates, suggesting diagenetic alteration of the Mifflin carbonates by meteoric water.

the protodolomites (Graf and Goldsmith, 1956) show a very low degree of ordering and are completely diffuse in some cases (Fig. 7).

Protodolomites with isotopic values similar to the limestone are present only outside the main dolomitized area and do not have a systematic stratigraphic relation to the Mifflin Member. These protodolomites occur as mottled dolomite along fractures. X-ray data, isotopic values, and geographic and stratigraphic settings of the protodolomites suggest a much later stage and different process of dolomitization than that deduced for the main dolomitic facies. The conditions of dolomitization of the protodolomites were not investigated.

Chemical Analysis

Examination of the composition of carbonate sediments through geologic time reveals that Recent carbonates are dominated by aragonite and Mg-calcite, whereas older rocks with few exceptions contain only calcite and/or dolomite. This variation is the result of dolomitization and other diagenetic processes resulting in the inversion of aragonite and Mg-calcite to calcite. These processes are enhanced in the presence of water (Fyfe and Bischoff, 1965). During inversion of aragonite and Mg-calcite to calcite, equilibration occurs between the new phase and interstitial solution resulting in modification of trace element distribution and isotopic composition of the sediments. These

modifications provide valuable information regarding diagenetic history of the sediments. Oxygen and carbon isotopic composition of the Mifflin samples was discussed in the previous section; the effect of diagenetic modifications on trace element distributions are discussed here.

If the ratio of Sr/Ca in a sample is measured and the temperature at which the sediment was deposited is known or can be estimated, the Sr/Ca ratio of the solution at the time of formation of the carbonate can be calculated and information regarding the environment of deposition can be obtained. Chemical analyses of Ca, Mg, Sr, and Na were obtained using a Jarrell Ash atomic absorption model 82-528.

Strontium.—Strontium values show an inverse relationship with percentage of dolomite in the sample, such that the Sr can be used as a discriminating factor (Fig. 8). Mean values of Sr are 228 ppm for limestone and 37 ppm for dolostone.

In Recent calcitic sediments Sr values have been reported to be about 1,000–1,200 ppm, but these values drop to about 400–700 ppm for ancient limestones (Kinsman, 1969). Kinsman and Holland (1969) show that concentration of Sr in the solid phase is primarily controlled by the ratio of Sr to Ca in the liquid phase and the temperature at which the sediment was deposited. Therefore, the Sr/Ca ratio in the solid can be determined by the distribution coefficient given as a function of temperature and Sr/Ca

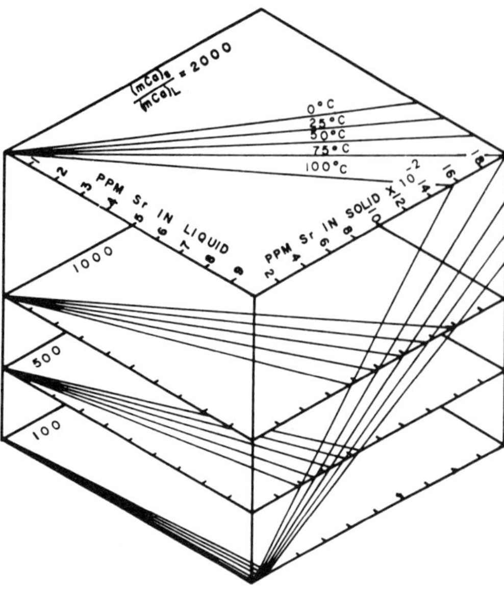

Fig. 9.—Diagrammatic representation of distribution of Sr in the solid as a function of $(mCa)_S/(mCa)_L$, Sr content of the liquid, and temperature. The diagram shows that as the ratio of $(mCa)_S/(mCa)_L$ decreases, i.e., the solution becomes more saturated with respect to calcite, the Sr content of the calcite precipitated from the solution decreases.

ratio of the liquid phase (for detailed discussion of trace element distribution coefficients, the reader is referred to the work of McIntire, 1963). Trace element distribution coefficient in any substance is given by

$$K_T = (M_i/C_j)_S / (M_i/C_j)_L \qquad (9)$$

where K_T = distribution coefficient at any specified temperature,
M_i = trace element concentration,
C_j = carrier element concentration,
S = solid,
L = liquid.

Rearranging the equation (9), trace element distribution in the solid is

$$(M_i)_S = \frac{(C_j)_S}{(C_j)_L} \times K_T (M_i)_L. \qquad (10)$$

Maintaining the ratio of the carrier elements in the solid and liquid constant, equation (10) can be rewritten as

$$(M_i)_S = \overset{*}{K_T}(M_i)_L, \qquad (11)$$

where $\quad \overset{*}{K_T} = \frac{(C_j)_S}{(C_j)_L} \times K_T.$

Variation of Sr content of calcite as a function of temperature, ratio of the Ca in the solid to Ca in the liquid and Sr content of the liquid is shown in Figure 9. It can be seen from Figure 9 that the Sr content in calcite is not only controlled by the Sr concentration of the liquid phase, but is dependent as well on the ratio of the Ca concentration in the solid to that in the liquid, e.g., a calcite formed from a solution with 8 ppm Sr and $(mCa)_S/(mCa)_L$ equal to 1,000 will contain 1,120 ppm Sr, whereas with the same Sr content but a $(mCa)_S/(mCa)_L$ of 100 the calcite will contain only 112 ppm Sr. Furthermore, it can be shown that Sr concentration in the calcite is inversely correlated with the degree of supersaturation of the liquid with respect to this mineral. The strontium value of pure calcite precipitated from sea water containing 8 ppm Sr at 25° C is equal to 1,120 ppm; a much higher concentration than observed values for the Mifflin calcites. Such a low Sr concentration can be attained either by precipitating calcite from highly supersaturated solutions (Fig. 9) $(mCa)_S/(mCa)_L \leq 100)$ or from solutions with much lower Sr content than that of sea water. The presence of principally subtidal marine fossils and a lack of evaporite minerals show that the Mifflin carbonates formed under normal marine conditions. In fact, based on mineralogical composition of their Recent analogs, many of the Mifflin skeletal materials were originally composed of aragonite containing up to 10,000 ppm Sr. Such a loss of Sr is attributed to diagenesis and neomorphism of the original sediment in the presence of meteoric water containing much lower Sr content than sea water.

The above conclusion cannot be made for the Mifflin dolomitic facies without knowledge of the distribution coefficient of Sr in dolomite. Unsuccessful attempts at precipitating dolomite at low pressure and temperature in the laboratory make it impossible to obtain a proper value for the Sr distribution coefficient for dolomite. However, an indirect approach was used by Behrens and Land (1972) to estimate the value of the distribution coefficient of dolomite. They proposed an approximate distribution coefficient for dolomite equal to one half of that of calcite, which implies that calcite precipitating from a solution will have twice as much Sr as co-precipitating dolomite, at equilibrium conditions.

The Mifflin limestone has Sr values six times greater than the Mifflin dolostone. This difference is regarded as resulting from non-equilibrium conditions prevailing at the time of development of limestone and dolostone and supports the conclusion reached by field observation and isotopic data that these minerals were not co-precipitated. The low Sr concentration of the Mifflin dolomites (average 37 ppm) suggests that fresh water diagenesis of these carbonates in part occurred prior to dolomitization. Analyses of Wisconsin ground water show very high concentrations of Sr (Skougstad and Horr, 1963) which can result in the formation of calcite extremely high in Sr (7,247 ppm). Calculations demonstrate that the presence of only two percent of calcite precipitated from such a ground water is sufficient to raise the total Sr content of the Mifflin limestone from 80 ppm Sr to the observed values. Consequently, much higher Sr content of the limestone than dolostone probably can be interpreted to be due to addition of secondary calcite (Badiozamani, 1972).

Sodium.—In Recent biogenic carbonates Na concentrations have been reported to be around 5,000–8,000 ppm (Harris and Pilkey, 1966) whereas the mean value of Na concentrations for the Mifflin limestone is 153 ppm and for the dolomitic facies of this member 190 ppm. The Na values of the Mifflin carbonates, therefore, confirm the equilibration of these carbonates with meteoric water and the loss of Na through diagenetic processes. Because the Na concentration of the dolomitic and calcitic facies of the Mifflin Member are almost identical (Fig. 10) and the Na partition coefficient for calcite and dolomite is not known, further interpretation based on the concentration of this element is not warranted.

Dolomitization of the Mifflin Member

The isotopic analyses indicate that exchange reactions occurred between Mifflin carbonates and meteoric water, the amount of exchange being less for the dolomitic facies than the calcitic one. If the isotopic value of the solution in contact with the rock or the rate of exchange were known, the time interval necessary for a known amount of exchange could be calculated. In the case of the Mifflin Member neither of these factors is known, but an indirect method can be used to estimate a rate of isotopic exchange. In Bermuda the average oxygen isotopic values for the Belmont Formation is –2.59‰ relative to PDB (Fig. 5). This is about 2‰ less than the average oxygen isotopic values of Recent marine carbonate. The age of the Belmont as given by Vacher (1971) is more than 300 × 10³ years. Assuming 300 × 10³ to

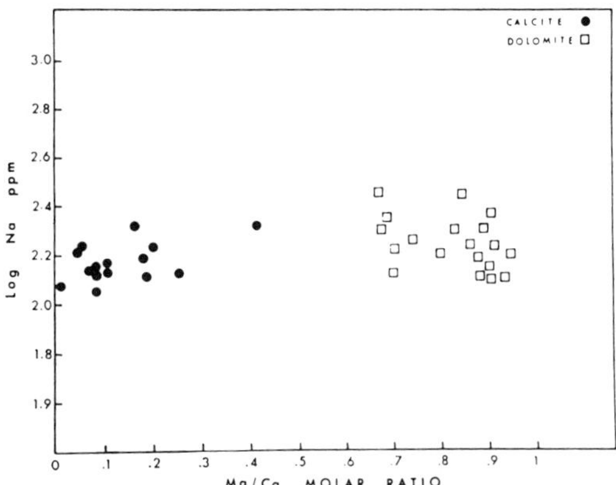

Fig. 10.—Sodium content vs. Mg/Ca molar ratio of the Mifflin Member. The Na concentration does not show any correlation with Mg/Ca ratio in the sample. The Na concentration of the Mifflin samples are much lower than reported values for Recent biogenic carbonates, suggesting diagenetic alteration of these carbonates by meteoric water.

be the minimum age of the Belmont, the minimum rate of exchange would be about 6.7‰ per 10⁶ years. Assuming equal porosity and permeability and equal ground water residence time for both facies, dolomitization must have been relatively early to maintain the oxygen isotopic values observed for the dolomitic facies. Using 6.7‰ per 10⁶ years rate of exchange of the Bermudian rocks, the maximum time of exchange for the Mifflin dolostone would be about 600 × 10³ years.

As previously discussed the low Sr and Na content of the Mifflin Member suggests neomorphism of the Mifflin carbonates and loss of Sr and Na owing to exchange with meteoric water.

Field observations indicate that the Mifflin Member was deposited in a broad, shallow-marine environment and the Mifflin sediment was exposed periodically to subaerial conditions. Paleontological investigation indicates a deepening of the environment of deposition basinward from the Wisconsin Arch. There are no pronounced lithological changes throughout the area of occurrence of the Mifflin carbonates and marine fossils in both facies are characteristic of a normal marine condition without evidence of a physically restricted or lagoonal type environment. Moreover, the presence of an open-marine fauna and a lack of mudcracks, algal mats, and evaporites and/or solution breccias, preclude supratidal deposition and dolomitization whereas there is a noteworthy ab-

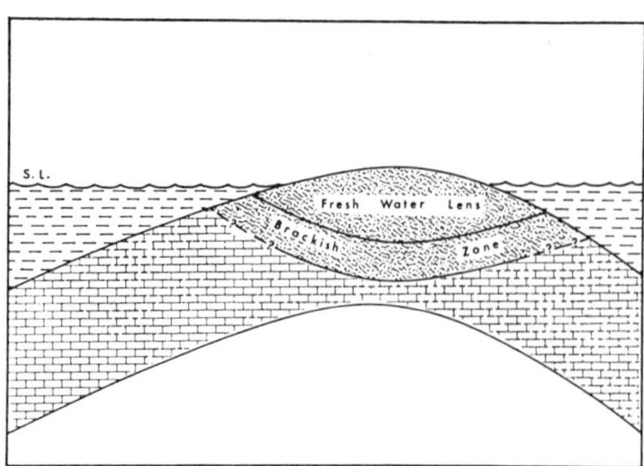

Fig. 11.—Diagrammatic representation of the dolomitization process of the Mifflin Member. This schematic representation shows the position of fresh water lens and underlying brackish zone after subaerial exposure of the Wisconsin Arch (see text).

sence of petrographic changes at the transition zone between the dolomitic and calcitic facies. These observations eliminate the possibility of the existence of a restricted lagoon (cf. Asquith, 1967) once believed responsible for Middle Ordovician dolomitization in Wisconsin.

Goodell and Garman (1969) proposed the cannibalization of the pre-existing strata as a mechanism for dolomitization, but this model is incapable of producing extensive masses of dolomite. In their calculation for the study area in the Bahamas, they require continuous deposition of carbonate sediments since early Cretaceous time at a maximum rate to provide 36,000 m of sediment necessary for production of Mg for dolomitization of the existing strata. They also had to assume that all the Mg produced by cannibalization was retained and used during the dolomitization process. The assumptions of continuous deposition and almost 100 percent consumption of Mg are not realistic and cannot be achieved. To cannibalize the pre-existing strata, there should be periods of non-deposition. Prevalence of non-depositional periods would result in reduction of the total thickness of the carbonates deposited in the Bahamas since early Cretaceous. Also, 100% dolomitization has not been observed even under extreme conditions where the solution is many times supersaturated with respect to dolomite. Hence, cannibalization cannot be used as a model to account for dolomitization of a large mass of dolostone. Hydrothermal dolomitization as was discussed in a previous section could not account for the Mifflin dolomitization either.

The available evidences strongly favor Dorag dolomitization as the principal mechanism of dolomitization of the Mifflin Member. The process would start after each regression and subaerial exposure of relatively high areas on the Wisconsin Arch. Dolomitization would proceed after the establishment of a lens of non-saline ground water and an underlying brackish zone (Fig. 11). If dolomitization were 100% complete, assuming a ground water flow rate of seven meters per year (Hanshaw *et al.*, 1965), the maximum rate of dolomitization in a rock with 10% porosity would be 0.85 m/1,000 yrs, if sea water and ground water with 1,300 and 30 ppm Mg, respectively, were mixed (for calculation see Hsu, 1966, p. 136). Using this rate and the 600×10^3 years (calculated from oxygen isotope data) as the maximum time required to dolomitize the Mifflin Member, the dolomitization front would move approximately 500 m in this period.

One hundred percent effective dolomitization and 30% sea water are the ideal optimum conditions for ground water dolomitization. Under normal conditions probably 10% effective dolomitization would be more realistic and the dolomite front would advance only 50 m, a distance sufficient to dolomitize the total thickness of the Mifflin carbonate. The Dorag dolomitization model is, therefore, capable of producing massive dolostone in shelf sediments at geologically reasonable rates. Moreover, the model can account for the close association of widespread dolomitic facies and shallow epicontinental shelves which have been reported by many authors (Prouty, 1948; Cloud and Barnes, 1957).

PALEO-ENVIRONMENTAL EXTRACTIONS

In the ideal model of Dorag dolomitization the dolostone-limestone boundary follows the lower margin of the ground water lens which produces an inclined boundary between the two rock facies. If detailed information about the boundary between the two facies is available, the position of the paleo-ground water lens may be ascertained. Any shift in this boundary will be a response to fluctuation of sea level; moreover, it may be possible to identify transgression and regression by translocation of the boundary between the two facies.

In the Middle Ordovician strata of southern Wisconsin, three major shifts in the boundary between dolostone and limestone have been reported (Agnew *et al.*, 1956). The three shifts are between the Pecatonica and the Mifflin, the Magnolia and the Quimbys Mill members of the Platteville Formation, and the Quimbys

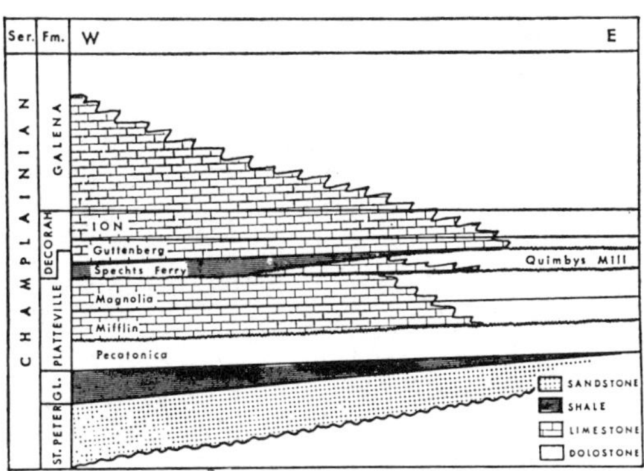

FIG. 12.—Diagrammatic cross section of the Champlainian Series in southwestern Wisconsin, showing three major shifts in the dolostone-limestone boundary. These shifts are between the Pecatonica-Mifflin, Magnolia-Quimbys Mill, and Quimbys Mill-Guttenberg members (modified after Agnew, 1956, and Ostrom, 1969).

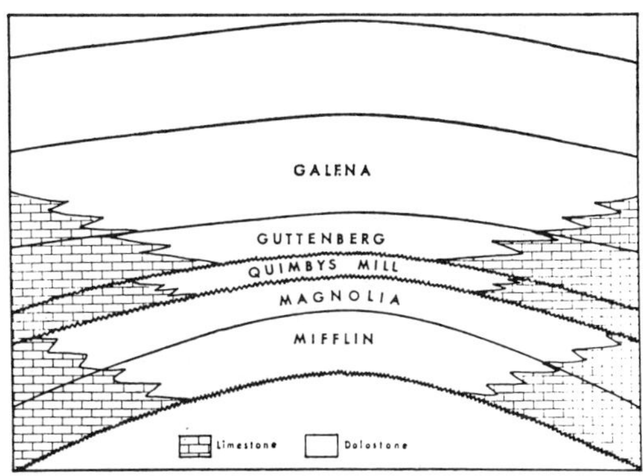

FIG. 13.—Postulated model for achieving three major shifts in the dolostone-limestone boundary in the Champlainian Series of Wisconsin. Each major shift represents a regression-transgression sequence. This interpretation is based on the fact that in the Dorag dolomitization model, the dolostone-limestone boundary follows the lower margin of the ground water lens which produces an inclined boundary between the two rock facies. The diagram, therefore, portrays the position of three successive paleoground water lenses at the time of subaerial exposure of these units.

Mill and the Guttenberg Member of the Decorah Formation (Fig. 12). Translocation of the facies boundary is accomplished also by major changes in lithology of the two adjacent members, e.g., the wavy and thin-bedded strata of the Mifflin and Magnolia give way to medium and thick-bedded strata of the Quimbys Mill. The latter are also very fine-grained, relatively unfossiliferous, and fracture conchoidally. Another translocation of the dolostone-limestone boundary occurs where the Quimbys Mill strata are replaced by thick to massive-bedded, fossiliferous and vuggy sediments of the Decorah and Galena formations.

If Dorag dolomitization has been operative, changes in the position of the facies boundary imply three major transgressions and regressions of the sea. These are represented in Figure 13 which diagrammatically illustrates the present distribution of the Middle Ordovician facies in southwestern Wisconsin, and shows the positions of successive ground water lenses during regressions and dolomitization.

Interpretations regarding the bathymetry of the environment of deposition can probably be made with reference to the dolostone-limestone boundary. Facies distribution of the Mifflin Member in southwestern Wisconsin (Fig. 14) shows an extension of a limestone tongue eastward from the general trend. This extension is interpreted in light of the proposed dolomitization model to be a relict of an embayment present at the time of deposition. After emergence of the Wisconsin Arch following Mifflin deposition this embayment remained connected

to the open sea. As a result, dolomitization was not operative in the embayment and the facies boundary followed its configuration (Badiozamani, 1972).

The details of the boundaries are shown based on available data and, therefore, are subject to change with additional information. A detailed study of the dolomite-calcite boundary as a function of time could result in recognition of higher-frequency sea level fluctuations of smaller magnitude during the Middle Ordovician.

In summary, the Dorag dolomitization model provides the most satisfactory explanation for the origin of the Mifflin and other Middle Ordovician dolostones in Wisconsin and adjacent states. Also, this model presents an alternative solution for the origin of dolostones along positive elements and associated with unconformities, where evidence for supratidal or Sabkha type dolomitization is lacking. The Dorag model has the advantage of being independent of extensive evaporation and high Mg/Ca ratios in solution, therefore is not limited geographically to low latitudes. These conditions and the fact that the mixing of fresh and sea water is readily attainable under many circumstances suggest that this type of dolomitization should be widespread and substantial volumes of ancient dolostones should be attributed to this category.

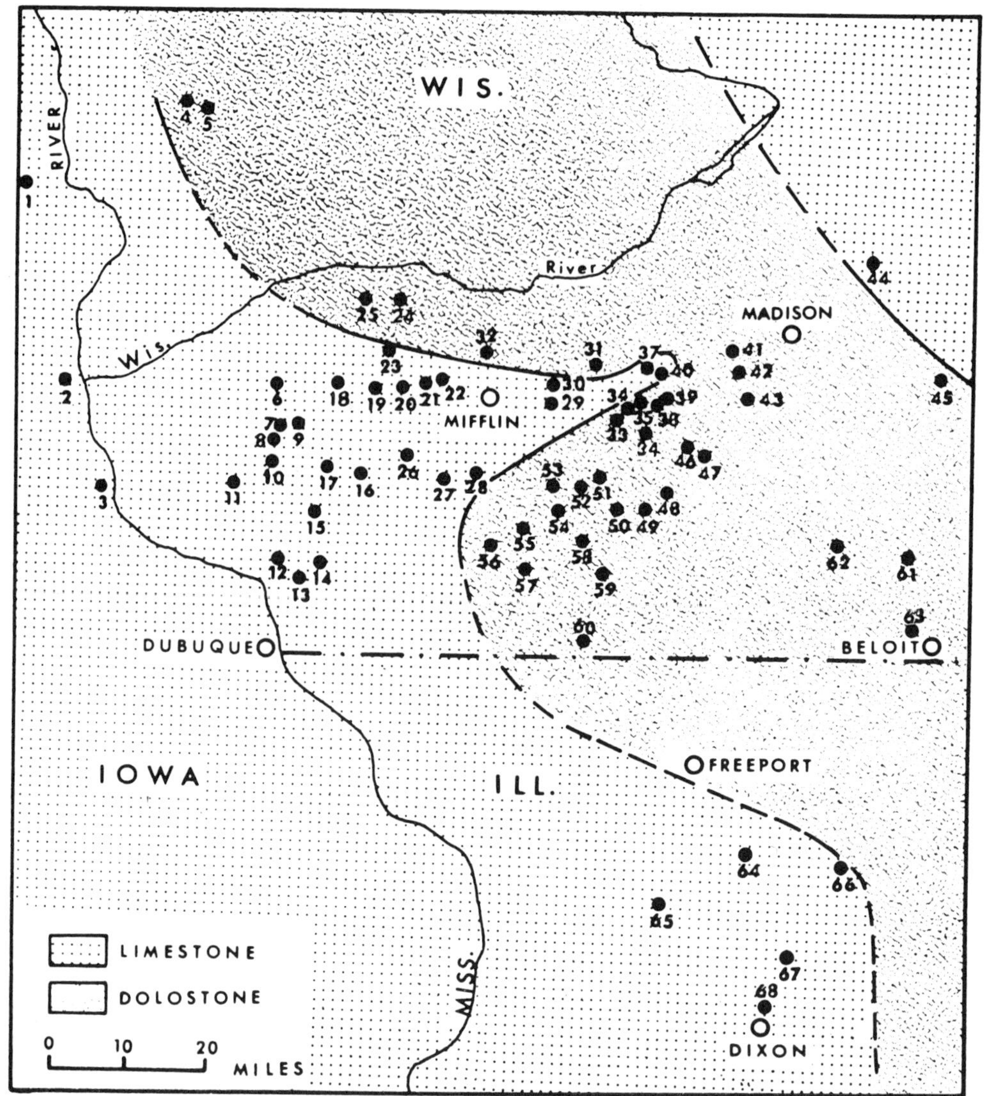

Fig. 14.—Facies distribution map of the Mifflin Member in southwestern Wisconsin. Black circles are collecting sites.

STATE OF IOWA

1. Quarry, Church, Allamakee County, SE¼ 29, 5N-1E.
2. Road cut, McGregor, Clayton County, SW¼ 35, 95N-3W.
3. Road cut, Guttenberg, Clayton County, SW¼ 5, 92N-2W.

STATE OF WISCONSIN

Vernon County

4. Quarry, Liberty Pole, SW¼ 21, 12N-4W.
5. Quarry, Liberty Pole, NW¼ 30, 12N-4W.

Grant County

6. Quarry, Fennimore, SW¼ 21, 6N-3W.
7. Quarry, Stitzer, SW¼ 8, 5N-2W.
8. Quarry, Stitzer, NW¼ 16, 5N-2W.
9. Quarry, Stitzer, SE¼ 19, 5N-2W.
10. Quarry, Lancaster, NE¼ 1, 4N-3W.
11. Road cut, Lancaster, SE¼ 1, NE¼ 19, 4N-3W.
12. Road cut, Potosi, NW¼ 7, 2N-2W.
13. Road cut, Dickeyville, SE¼ 15, 2N-2W.
14. Road cut, Platteville, NE¼ 12, 2N-2W.
15. Road cut, Ellenboro, NW¼ 34, 4N-2W.
16. Quarry, Arthur, NW¼ 12, 4N-1W.
17. Quarry, Arthur, NW¼ 7, 4N-1W.
18. Quarry, Montfort, E16, 6N-1W.

ACKNOWLEDGMENTS

The writer wishes to acknowledge, with appreciation, the assistance and stimulating discussion of L. L. Sloss, F. T. Mackenzie, and E. C. Dapples of Northwestern University. Both presentation and content were improved extensively by their critical reading of the manuscript. L. L. Sloss and F. T. Mackenzie supervised this work and gave valuable advice and suggestions. Thanks also go to P. W. Bretsky for his stimulating discussions in the early stages of the study and also for providing paleontological information. I am indebted to R. N. Clayton and Mrs. T. K. Mayeda from the University of Chicago who kindly made laboratory facilities available to the writer and guided isotopic analysis.

I wish to express my thanks to my wife, Manijeh, who encouraged and assisted me and shared the overburden of the work throughout the study. Thanks are also due to L. N. Plummer and M. D. Matthews for valuable comments and discussions.

I also want to thank R. M. Garrels of Northwestern University, L. S. Land of University of Texas, and R. F. Schmalz of the Pennsylvania State University for their cordial comments and reviewing the manuscript.

Financial assistance as well as computer time was made available by Northwestern University and additional aid was supplied by the National Iranian Oil Company for which I am indeed grateful. In its later stages, this work was supported by NSF Grant GA32120.

REFERENCES

AGNEW, A. F., AND HEYL, A. V., JR., 1946, Quimbys Mill, new member of Platteville Formation, Upper Mississippi Valley: Am. Assoc. Petroleum Geologists Bull., v. 30, p. 1585–1587.

———, ———, BEHRE, C. H., JR., AND LYONS, E. J., 1956, Stratigraphy of Middle Ordovician rocks in the zinc-lead district of Wisconsin, Illinois, and Iowa: U. S. Geol. Survey Prof. Paper 274-K, p. 251–312.

ALDERMAN, A. R., AND SKINNER, H. C. W., 1957, Dolomite sedimentation in the southeast of South Australia: Am. Jour. Sci., v. 255, p. 561–567.

ASQUITH, G. B., 1967, The marine dolomitization of the Mifflin Member Platteville limestone in southwest Wisconsin: Jour. Sed. Petrology, v. 37, p. 311–326.

BACK, W., AND HANSHAW, B. B., 1970, Comparison of chemical hydrogeology of the carbonate peninsulas of Florida and Yucatan: Jour. Hydrology, v. 10, p. 330–368.

BADIOZAMANI, K., 1972, The Dorag dolomitization model—application to the Middle Ordovician of Wisconsin: unpubl. Ph.D. dissertation, Northwestern University, Evanston, Illinois, 94 p.

←

Iowa County

19. Quarry, Montfort, SE¼ 19, 6N-1E.
20. Natural exposure, Cobb, NW¼ 22, 6N-1E.
21. Quarry, Cobb, NW¼ 19, 6N-2E.
22. Quarry, Edmund, SE¼ 15, 6N-2E.
23. Quarry, Highland, SE¼ 4, 6N-1E.
24. Quarry, Highland, SE¼ 34, 7N-1E.
25. Quarry, Highland, NE¼ 32, 7N 1E.
26. Road cut, Mifflin, NW¼ 35, 5N-1E.
27. Road cut, Mineral Point, NW¼ 15, 4N-8E.
28. Quarries, Mineral Point, NW¼ 5, 4N-3E.
29. Quarry, Ridgeway, SE¼ 26, 6N-4E.
30. Quarry, Ridgeway, NW¼ 14, 6N-4E.
31. Quarries, Barneveld, Center 10, 6N-5E.
32. Quarry, Dodgeville, W28, 7N-3E.

Dane County

33. Quarry, Daleyville, NW¼ 7, 5N-6E.
34. Quarry, Mt. Horeb, SE¼ 21, 6N-6E.
35. Quarry, Mt. Horeb, E21, 6N-6E.
36. Quarry, Mt. Horeb, W23, 5N-6E.
37. Quarry, Mt. Horeb, SW ¼ 35, 7N-6E.
38. Road cut, Mt. Horeb, NW¼ 25, 6N-6E.
39. Quarries, Mt. Horeb, NE¼ 30, 7N-6E.
40. Quarry, Mt. Horeb, SW¼ 18, 6N-7E.
41. Quarry, Middleton, SE¼ 27, 7N-8E.
42. Quarries, Madison, NW¼ 7, 6N-9E.
43. Quarries, Verona, Sec. 30, 6N-9E.
44. Quarries, Sun Prairie, NE¼ 34, 9N-11E.
45. Quarry, Cambridge, NE¼ 13, 6N-1E.
46. Quarry, New Glarus, NW¼ 34, 5N-7E.
47. Road cut, Bellville, NE¼ 36, 5N-7E.

Green County

48. Road cut, New Glarus, NE¼ 26, 4N-6E.
49. Quarry, Postville, NW¼ 2, 3N-6E.
50. Quarry, Postville, Center 5, 3N-6E.

Lafayette County

51. Quarry, Blanchardville, center 23, 4N-5E.
52. Quarry, Blanchardville, NW¼ 20, 4N-5E.
53. Quarry, Blanchardville, N24, 4N-4E.
54. Road cut, Yellowstone Lake, center 1, 3N-4E.
55. Quarry, Darlington, center N19, 3N-4E.
56. Quarry, Darlington, SE¼ 27, 3N-3E.
57. Quarry, Darlington, NE¼ 8, 2N-4E.
58. Quarries, Argyle, SW¼ 27, 3N-5E.
59. Quarry, Woodford, NW¼ 14, 2N-5E.
60. Quarries, South Wayne, NE¼ 15, 1N-4E.

Rock County

61. Quarries, Janesville, SE¼ 34, 3N-12E.
62. Quarries, Magnolia, NW¼ & SW¼ 15, 3N-10E.
63. Quarry, Beloit, E21, 1N-12E.

STATE OF ILLINOIS

64. Railroad cut, Leaf Rivery, Ogle County, NE¼ 33, 25N-10E.
65. Quarry, Carrol County, Brookville, NW¼ 21, 24N-7E.
66. Railroad cut, Oregon, Ogle County, SW¼ 7, 23N-10E.
67. Quarry, Dixon, Lee County, NW¼ 22, 22N-9E.
68. Quarry, Dixon, Lee County, SW¼ 33, 22N-9E.
69. Quarry, Troy Grove, Lasalle County, NE¼ 35, 35N-1E.

BAYS, C. A., 1937, The stratigraphy of the Platteville Formation: unpubl. Ph.D. dissertation, University of Wisconsin, Madison, Wisconsin, 217 p.

———, AND RAASCH, G. O., 1935, Mohawkian relations in Wisconsin: *in* Kansas Geol. Soc. 9th Ann. Field Conf. Guidebook, p. 296–301.

BEHRENS, E. W., AND LAND, L. S., 1972, Subtidal Holocene dolomite, Baffin Bay, Texas: Jour. Sed. Petrology, v. 42, p. 155–161.

BEVAN, A. C., 1926, The Glenwood beds as a horizon marker at the base of the Platteville Formation: Illinois Geol. Survey Rept. Inv. 9.

CALVIN, S., 1906, Geology of Winneshiek County: Iowa Geol. Survey, v. 16, p. 37–146.

CHAVE, K. E., 1952, A solid solution between calcite and dolomite: Jour. Geology, v. 60, p. 190–192.

CLAYTON, R. N., AND EPSTEIN, S., 1958, The relationship between O^{18}/O^{16} ratios in coexisting quartz, carbonate, and iron oxides from various geological deposits: Jour. Geology, v. 66, p. 352–371.

———, JONES, B. F., AND BERNER, R. A., 1968, Isotopic studies of dolomite formation under sedimentary conditions: Geochim. et Cosmochim. Acta, v. 32, p. 415–432.

CLOUD, P. E., AND BARNES, V. E., 1957, Early Ordovician sea in central Texas: Geol. Soc. America Mem. 67, p. 163–214.

COHEE, G. V., 1948, Cambrian and Ordovician rocks in Michigan Basin and adjoining areas: Am. Assoc. Petroleum Geologists Bull., v. 32, p. 1417–1448.

CRAIG, H., 1957, Isotopic standards for carbon and oxygen and correction factors for mass spectrometric analysis of carbon dioxide: Geochim. et Cosmochim. Acta, v. 12, p. 133–149.

DEFFEYES, K. S., LUCIA, F. J., AND WEYL, P. K., 1965, Dolomitization of Recent and Plio-Pleistocene sediments by marine evaporite waters on Bonaire Netherlands Antilles: *in* L. C. Pray and R. C. Murray (*eds.*), Dolomitization and limestone diagenesis, Soc. Econ. Paleontologists and Mineralogists, Spec. Pub. No. 13, p. 71–88.

DEGENS, E. T., 1967, Stable isotope distribution in carbonates, *in* G. V. Chilingar, H. J. Bissell, and R. W. Fairbridge (*eds.*), Carbonate rocks, 9B, Elsevier, Amsterdam, p. 193.

———, 1964, Oxygen and carbon isotope ratios in coexisting calcites and dolomites from Recent and ancient sediments: Geochim. et Cosmochim. Acta, v. 28, p. 23–44.

DEININGER, R. W., 1964, Limestone-dolomite transition in the Ordovician Platteville Formation in Wisconsin: Jour. Sed. Petrology, v. 34, p. 281–288.

ENGEL, A. E. J., CLAYTON, R. N., AND EPSTEIN, S., 1958, Variations in isotopic composition of oxygen and carbon in Leadville limestone (Mississippi, Colorado) and in its hydrothermal and metamorphic phases: Jour. Geology, v. 66, p. 374–393.

EPSTEIN, S., BUCHBAUM, R., LOWENSTAM, H., AND UREY, H. C., 1951, Carbonate-water isotopic temperature scale: Geol. Soc. America Bull., v. 62, p. 417–426.

———, GRAF, D. L., AND DEGENS, E. T., 1963, Oxygen isotope studies on the origin of dolomites: *in* Harmon Craig, S. L. Miller, and G. J. Wasserburg (*eds.*), Isotopic and cosmic chemistry, Amsterdam, North-Holland, p. 169–180.

FRIEDMAN, G. M., AND SANDERS, J. E., 1967, Origin and occurrence of dolostones: *in* G. V. Chilingar, H. J. Bissell, and R. W. Fairbridge (*eds.*), Carbonate rocks, 9A, Elsevier, Amsterdam, p. 267.

FRITZ, P., AND FONTES, J. C., 1966, Fractionment Isotopique pendant l'attaque acid des carbonates naturals. Role de la granulometrie: Comptes Rendus l'Acad. des Sci., Paris, v. 263, p. 1345–1348.

FYFE, W. S., AND BISCHOFF, J. L., 1965, The calcite-aragonite problem, *in* L. C. Pray and R. C. Murray (*eds.*), Dolomitization and limestone diagenesis, Soc. Econ. Paleontologists and Mineralogists Spec. Pub., No. 13, p. 3–13.

GARRELS, R. M., THOMPSON, M. E., AND SIEVER, R., 1961, Control of carbonate solubility by carbonate complexes: Am. Jour. Sci., v. 259, p. 24–45.

———, AND THOMPSON, M. E., 1962, A Chemical model for sea water at 25° C and one atmosphere total pressure: Am. Jour. Sci., v. 260, p. 57–66.

———, AND MACKENZIE, F. T., 1971, Evolution of Sedimentary rocks, Norton and Co., New York, 397 p.

GOLDBERG, E. D., 1957, Biogeochemistry of trace metals: *in* J. W. Hedgepeth (*ed.*), Treatise on marine ecology and paleoecology, v. 1, Geol. Soc. America Mem. 67, p. 345–357.

GOLDSMITH, J. R., AND GRAF, D. L., 1958a, Relation between lattice constants and composition of the Ca-Mg carbonates: Am. Mineralogist, v. 43, p. 84–101.

———, AND ———, 1958b, Structural and compositional variations in some natural dolomites: Jour. Geology, v. 66, p. 678–693.

———, ———, AND HEARD, H. C., 1961, Lattice constants of the Ca-Mg carbonates: Am. Mineralogist, v. 46, p. 453–457.

GOODELL, H. G., AND GARMAN, R. K., 1969, Carbonate geochemistry of Superior deep test well, Andros Island, Bahamas: Am. Assoc. Petroleum Geologists Bull., v. 53, p. 513–536.

GRAF, D. L., AND GOLDSMITH, J. R., 1956, Some hydrothermal syntheses of dolomite and protodolomite: Jour. Geology, v. 64, p. 173–186.

GROSS, H. G., 1964, Variation in the O^{18}/O^{16} and C^{13}/C^{12} ratios of diagenetically altered limestones in the Bermuda Island: Jour. Geology, v. 72, p. 170–193.

HALL, W. E., AND FRIEDMAN, I., 1969, Oxygen and carbon isotopic composition of ore and host rock of selected Mississippi Valley deposits: U. S. Geol. Survey Prof. Paper 650-C, p. C140–C148.

HANSHAW, B. B., BACK, W., AND RUBIN, M., 1965, Radiocarbon determinations for estimating ground water flow velocities in central Florida: Science, v. 148, p. 494–495.

———, AND ———, 1969, A geochemical hypothesis for dolomitization by ground water (abs.): Econ. Geology, v. 64, p. 349.

———, ———, AND DEIKE, R. G., 1971, A geochemical hypothesis for dolomitization by ground water: Econ. Geology, v. 66, p. 710–724.

HARRISS, R. C., AND PILKEY, O. H., 1966, Temperature and salinity control of the concentration of skeletal Na, Mn, and Fe, in Dendraster excentricus: Pacific Sci., v. 20, p. 235–238.

HELGESON, H. C., 1969, Thermodynamics of hydrothermal systems at elevated temperatures and pressures: Am. Jour. Sci., v. 267, p. 729–804.

HERSHEY, O. H., 1894, The Elk Horn Creek area of

St. Peter Sandstone in northwestern Illinois: Am. Geol., v. 14, p. 169–179.

HEYL, A. V., JR., AGNEW, A. F., LYONS, E. J., AND BEHRE, C. H., JR., 1959, The geology of the Upper Mississippi Valley zinc-lead district: U. S. Geol. Survey Prof. Paper 309, 310 p.

HOLLAND, H. D., 1964, Solubility of calcite in NaCl solutions between 50° and 200° C (*abs.*): Geol. Soc. America Spec. Paper 82, p. 94–95.

HSU, K. J., 1966, Origin of dolomite in sedimentary sequences: A critical analysis: Mineralium Deposita, v. 2, p. 133–138.

ILLING, L. V., WELLS, A. J., AND TAYLOR, J. C. M., 1965, Penecontemporary dolomite in the Persian Gulf: *in* L. C. Pray and R. C. Murray (*eds.*), Dolomitization and limestone diagenesis, Soc. Econ. Paleontologists and Mineralogists Spec. Pub., No. 13, p. 89–111.

JAMIESON, J. C., AND GOLDSMITH, J. R., 1960, Some reactions produced in carbonates by grinding: Am. Mineralogist, v. 45, p. 818–827.

KAY, M., 1928, Divisions of Decorah Formation: Science, New Ser., v. 67, p. 16.

———, 1929, Stratigraphy of the Decorah Formation: Jour. Geology, v. 37, p. 639–671.

———, 1935, Ordovician System in the Upper Mississippi Valley: *in* Kansas Geol. Soc. 9th Ann. Field Conf. Guidebook, p. 281–295.

KEITH, M. L., AND WEBER, J. N., 1964, Carbon and oxygen isotopic composition of selected limestones and fossils: Geochim. et Cosmochim. Acta, v. 28, p. 1787–1816.

KINSMAN, D. J. J., 1969, Interpretation of Sr^{++} concentrations in carbonate minerals and rocks: Jour. Sed. Petrology, v. 39, p. 486–508.

———, AND HOLLAND, H. D., 1969, the co-precipitation of cations with $CaCO_3$-IV. The co-precipitation of Sr^{++} with aragonite between 16° and 96° C: Geochim. et Cosmochim. Acta, v. 33, p. 1–17.

LAND, L. S., 1972, Contemporaneous dolomitization of Middle Pleistocene reefs by meteoric water, north Jamaica (*abs.*): Am. Assoc. Petroleum Geologists Bull., v. 56, p. 635.

———, AND EPSTEIN, S., 1970, Late Pleistocene diagenesis and dolomitization, north Jamaica: Sedimentology, v. 14, p. 187–200.

LANDES, K. K., 1946, Porosity through dolomitization: Am. Assoc. Petroleum Geologists Bull., v. 30, p. 305–317.

MATTHEWS, R. K., 1971, Diagenetic environments of possible importance to the explanation of cementation fabric in subaerially exposed carbonate sediments: *in* O. P. Bricker (*ed.*), Carbonate cements, The Johns Hopkins University Studies in Geology, No. 19, Johns Hopkins Press, p. 127–132.

MCCREA, J. M., 1950, On the isotopic chemistry of carbonates and a paleotemperature scale: Jour. Chem. Physics, v. 18, p. 849–857.

MCINTIRE, W. L., 1963, Trace element partition coefficient: A review of theory and application to geology: Geochim. et Cosmochim. Acta, v. 27, p. 1209–1264.

MCKINNEY, C. R., MCCREA, J. M., EPSTEIN, S., ALLEN, H. A., AND UREY, H. C., 1950, Improvements in mass spectrometers for the measurement of small differences in isotope abundance ratio: Rev. Sci. Instruments, v. 21, p. 724–730.

MILLIMAN, J. D., GASTNER, M., AND MULLER, J., 1971, Utilization of magnesium in Coralline Algae: Geol. Soc. America Bull., v. 82, p. 573–580.

MURRAY, R. C., 1960, Origin of porosity in carbonate rocks: Jour. Sed. Petrology, v. 30, p. 59–84.

NIER, A. O., 1947, A mass spectrometer for isotope and gas analysis: Rev. Sci. Instruments, v. 18, p. 398–411.

OSTROM, M. R., 1969, Champlainian Series (Middle Ordovician) in Wisconsin: Am. Assoc. Petroleum Geologists Bull., v. 53, p. 672–693.

OWEN, D. D., 1847, Preliminary report of the Geological Survey of Wisconsin and Iowa: U. S. General Land Office Rept., p. 160–173.

PROUTY, C. E., 1948, Trenton and Sub-Trenton stratigraphy of northwest belts of Virginia and Tennessee: Am. Assoc. Petroleum Geologists Bull., v. 32, p. 1596–1626.

RUNNELLS, D. D., 1969, Diagenesis, chemical sediments, and the mixing of natural waters: Jour. Sed. Petrology, v. 39, p. 1188–1201.

SCHUCHERT, C., AND BARRELL, J., 1914, A revised geologic time-table for North America: Am. Jour. Sci., 4th ser., v. 38, p. 1–27.

SHARMA, T., AND CLAYTON, R. N., 1965, Measurement of O^{18}/O^{16} ratios of total oxygen of carbonates: Geochim. et Cosmochim. Acta, v. 29, p. 1347–1353.

SHINN, E. A., GINSBURG, R. N., AND LLOYD, R. M., 1965, Recent supratidal dolomite from Andros Island, Bahamas: *in* L. C. Pray, and R. C. Murray (*eds.*), Dolomitization and limestone diagenesis, Soc. Econ. Paleontologists and Mineralogists Spec. Pub., No. 13, p. 112–123.

SKOUGSTAD, M. W., AND HORR, C. A., 1963, Occurrence and distribution of strontium in natural water: U. S. Geol. Survey Water Supply Paper. 1497-D, 97 p.

SLOSS, L. L., 1963, Sequences in the cratonic interior of North America: Geol. Soc. America Bull., v. 74, p. 93–114.

SONNENFELD, P., 1964, Dolomite and dolomitization: A review: Canadian Petroleum Geol. Bull., v. 12, p. 101–132.

STEIDTMANN, E., 1911, The evolution of limestone and dolomite, I: Jour. Geology, v. 19, p. 323–345.

TEMPLETON, J. S., AND WILLMANN, H. B., 1963, Champlainian Series (Middle Ordovician) in Illinois: Illinois State Geol. Survey Bull. 89, 260 p.

TENNANT, C. B., AND BERGER, R. W., 1957, X-ray determination of dolomite-calcite ratio of a carbonate rock: Am. Mineralogist, v. 42, p. 23–29.

UREY, H. C., LOWENSTAM, H. A., EPSTEIN, S., AND MCKINNEY, C. R., 1951, Measurement of paleotemperatures and temperatures of the Upper Cretaceous of England, Denmark and the southeastern United States: Geol. Soc. America Bull., v. 62, p. 399–416.

VACHER, H. L., 1971, Late Pleistocene sea-level history: Bermuda evidence: unpub. Ph.D. dissertation, Northwestern University, Evanston, Illinois, 153 p.

VAN TUYL, F. M., 1916, The origin of dolomite: Iowa Geol. Survey Ann. Rept. (for 1914), v. 25, p. 251–422.

VERNON, R. O., 1961, The geology and hydrology associated with a zone of high permeability

(Boulder zone) in Florida: Soc. Mining Engineers, Preprint 69-AG-12, 24 p.

WALTERS, L. J., JR., CLAYPOOL, G. E., AND CHOQUETTE, P. W., 1972, Reaction rates and δO^{18} variation for the carbonates-phosphoric acid preparation method: Geochim. et Cosmochim. Acta, v. 36, p. 129–140.

WEYL, P. K., 1960, Porosity through dolomitization: Conservation-of-mass requirements: Jour. Sed. Petrology, v. 30, p. 85–90.

19

Mg/Ca Ratio and Salinity: Two Controls over Crystallization of Dolomite[1]

ROBERT L. FOLK[2] and LYNTON S. LAND[2]

Abstract Carbonate precipitation is governed by both thermodynamic and kinetic considerations. The concentration and ionic composition of natural solutions influence not only the mineralogy of the precipitate, but also the crystal size and habit of both metastable and stable phases.

Under marine conditions sparry calcite does not precipitate because of poisoning of sideward crystal growth by Mg, which forces Mg-calcite and aragonite to assume a micritic, steep-rhombic, or fibrous habit. Blocky calcite can form only in low Mg/Ca waters, either from primary meteoric solutions, or where Mg has been removed from marine water by dolomite or clays.

In hypersaline environments precipitation is commonly rapid and, together with the high concentrations of foreign ions, it is difficult for dolomite to form because of the precise Ca-Mg ordering required. Instead, aragonite or Mg-calcite crystallizes. Dolomite can form only if the Mg/Ca ratio exceeds 5-10:1, and even then it is aphanitic and poorly ordered. At progressively reduced salinities dolomite is able to nucleate at still lower Mg/Ca ratios, approaching 1:1 in meteoric waters. The absence of foreign ions and slow crystallization, commonly under phreatic conditions, cause both dolomite and calcite to form exquisitely limpid, euhedral rhombs.

Thus, dolomite probably forms most readily by a *reduction* in salinity, particularly in a schizohaline environment (alternating between hypersaline and near-fresh conditions, as in a floodable sabkha or a phreatic mixing zone). Flushing marine saline waters with fresh water lowers salinity but maintains a high Mg/Ca ratio; crystallization is slower and the interfering effect of foreign ions is reduced.

INTRODUCTION

Many factors control the precipitation of the common sedimentary carbonate minerals. Among those known to be important are temperature, solution composition and concentration, rate of crystallization, and the presence and concentration of certain organic compounds (Sorby, 1879; Fyfe and Bischoff, 1965; Kitano and Hood, 1965; Pytkowicz, 1965; Berner, 1966; Lovering, 1969; and many others). But apparently the most important relations can be explained relatively simply by a diagram in which salinity is plotted against Mg/Ca ratio of the depositing solution (Figs. 1, 2; Folk and Land, 1972). Such a diagram can be used as a fundamental framework from which the other variables can be evaluated.

The relations among the various phases of Ca-CO$_3$ (calcite, magnesium-calcite, and aragonite) have been discussed elsewhere (Folk, 1973, 1974) and will only be summarized briefly here. Our emphasis in this paper is on dolomite.

The key idea is that dolomite is a very difficult mineral to form because of the precise Ca-Mg ordering required (Goldsmith, 1953). Ideal dolomite consists of carbonate sheets interlayered with alternating sheets consisting entirely of Ca atoms or entirely of Mg atoms. These ions have such similar properties that it is difficult to segregate them. We believe that segregation of ions to produce dolomite is accomplished more readily if the crystallization rate is very slow, or if solutions are dilute, with few interfering ions. Crystallization of dolomite is much more difficult if crystallization is rapid or if there is a high concentration of competing ions.

With a high concentration of Ca, Mg, and CO$_3$ ions, such as exists under evaporitic conditions, a highly oversaturated solution can lose energy by forming the improbable and difficult dolomite structure, or it can lose a smaller, but still substantial amount of energy by crystallizing as the simpler calcite or aragonite structures. In most cases the easier CaCO$_3$ mineral will form, and the Mg/Ca ratio in the remaining solution will rise to extreme values, as is observed in evaporitic environments (Kinsman, 1965). Only when extreme concentrations are reached is Mg finally forced to precipitate, as a poorly ordered, Ca-rich "protodolomite."

If the concentration of ions is much lower, such as in most meteoric waters, there is not nearly so much interference in crystal growth caused by lattice impurities. At slow rates of crystallization minerals can form that are more nearly in theoretical equilibrium with the surrounding solution; and in the case of dolomite the careful ordering of Ca and Mg which is required can be accomplished from solutions having Mg/Ca ratios near the theoretical value of 1:1.

We can draw a "bussing" analogy for the carbonates. First we have a new, comfortable, first class bus, called the "Dolomite Bus." But it has a single narrow door that is difficult to open. Behind it is a dirty second-class bus, the "Calcite Bus." It has several wide, readily opened doors.

[1]Manuscript received, November 11, 1973; accepted, April 27, 1974.

[2]The University of Texas at Austin.
We acknowledge the Geology Foundation, University of Texas, for helping with publication expenses.

247

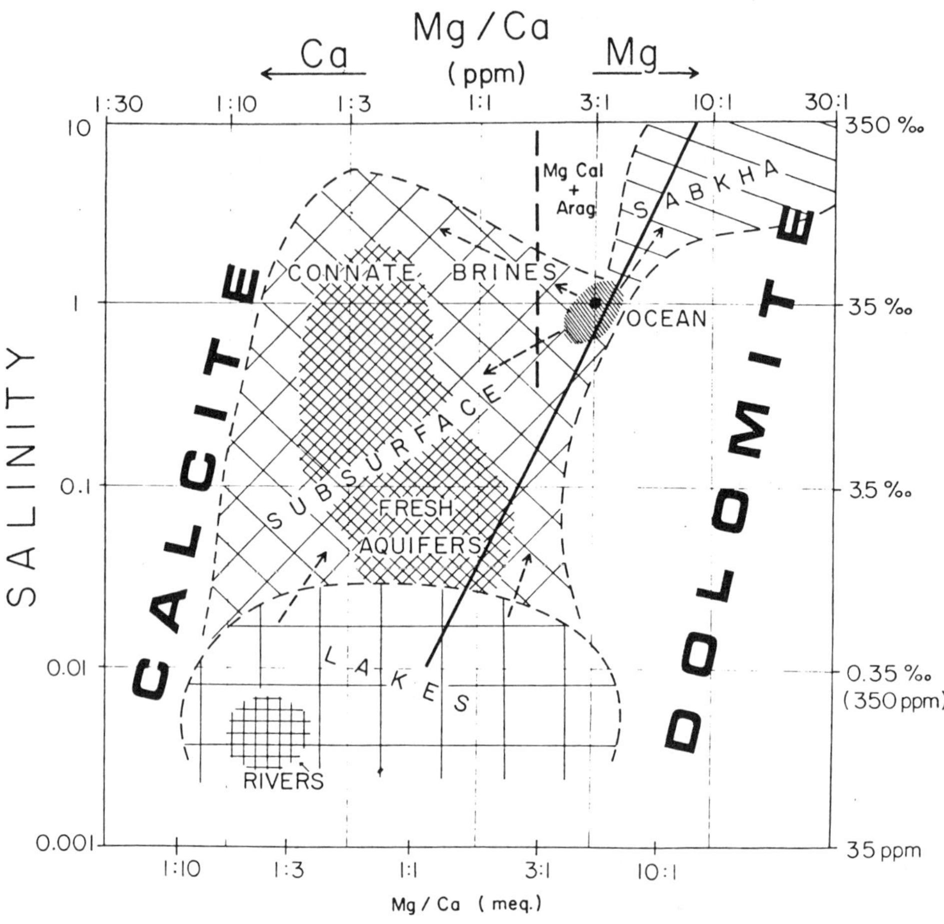

FIG. 1—Fields of occurrence of common natural waters plotted on graph of salinity versus Mg/Ca ratio. Sabkha waters are highly saline and Mg-rich. Subsurface waters vary from hypersaline to nearly fresh, but commonly show large excess of Ca over Mg (heavier shading shows most common values). Terrestrial fresh waters also are commonly Ca-rich. Some subsurface and lake waters, however, are anomalously high in Mg. Fields of preferred occurrence of dolomite, aragonite plus magnesian calcite, and calcite also are shown. At low salinities with few competing ions and slow crystallization rates dolomite can form at Mg/Ca ratios near 1:1. As salinity rises, it becomes more and more difficult for ordered dolomite structure to form, thus requiring progressively higher Mg/Ca ratios; in sabkha, Mg/Ca ratio of 5 or 10:1 is necessary before dolomite can crystallize owing to abundance of competing ions and rapid crystallization.

If there is a great mob waiting to get on the bus in a hurry and it starts to rain, they will all go on the uncomfortable "Calcite Bus" (high rate of precipitation or supersaturation means that calcite or aragonite will form because of its relative ease of crystallization). Almost nobody gets on the "Dolomite Bus," because it is too hard to get into in a hurry (high rate of carbonate precipitation means that the ordered dolomite structure is too difficult to develop). However, if only one person a minute comes by to get on a bus, he will get on the more comfortable "Dolomite Bus" (slow crys-

tallization, as in fresh water, means the difficult, ordered dolomite structure can form).

In the diagram here presented (Figs. 1, 2) the dolomite field is divided from the calcite field by a diagonal line indicating that, at low salinities (and low crystallization rates), dolomite can form readily at Mg/Ca ratios as low as 1:1; but as salinity and crystallization rate increase, the Mg/Ca ratio at which dolomite is first able to form rises until it must surpass 5:1 or 10:1 in hypersaline sabkhas. The line shows a generalized tendency and is a kinetic, not a thermodynamic, boundary;

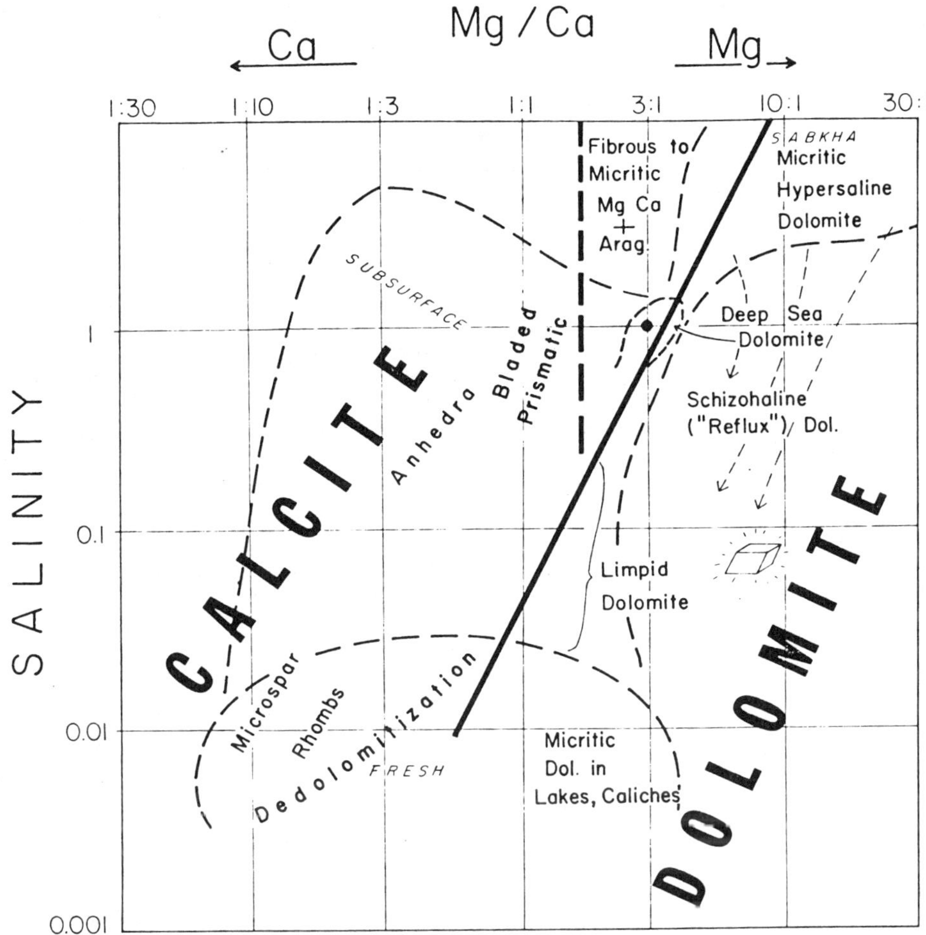

FIG. 2—Same diagram as Figure 1 emphasizing minerals and their morphologies. Hypersaline dolomite is micritic and not well ordered because of rapid crystallization and competition with other ions. Much dolomite is formed by dilution of sea or sabkha waters by fresh water (subsurface-mixing and schizohaline environments) causing tremendous drop in salinity but only slight change in high Mg/Ca ratio; water path drops almost vertically into dolomite field. Much of this type of dolomite is limpid—perfectly clear, gem-like, perfect rhombs. In "fresh" lake or cave waters dolomite can form at Mg/Ca ratios as low as 1:1, but if rate of precipitation is very rapid, *e.g.*, aided by organisms, then magnesian calcite or aragonite can form well into dolomite field. Rapid crystallization always forces dolomite/calcite boundary to right because of difficulty of ordering dolomite crystal.

In field of $CaCO_3$, at Mg/Ca ratios over 2:1, aragonite or magnesian calcite crystallizes in either micritic or fibrous habits because of side-poisoning effect of Mg ions. Left of that line calcite can form, but probably a strip-like field of crudely fibrous or bladed, prismatic to scalene calcite is present near 2:1 Mg/Ca boundary because side-poisoning effect of Mg here is executed only feebly. Bulk of calcite field is occupied by equant, mosaic polyhedral, or perceptibly anhedral calcite. In fresh waters of very low Mg/Ca ratio, calcite can form very uniformly sized 2-10-micron euhedra, often rhombs, on stream surfaces or in caliches; or calcite can form loafish rhombs of very uniform 5-10-micron size (microspar) by freshwater recrystallization of micrite; or calcite can replace dolomite (dedolomitization). None of the morphologic boundaries is exact.

it is not meant to be a precise separator, because temperature, presence of various foreign ions, *etc.*, can shift the precise position of this boundary.

DOCUMENTATION OF DIAGRAM

Information is accumulating rapidly on the chemistry of natural waters. Without making a comprehensive review of such information, we present here a summation of the data, beginning with the classic (but overplayed) hypersaline sabkha, progressing to deep-sea and normal-marine dolomite, then to dolomite in nearly fresh surface waters, and finally to dolomite precipitated in the subsurface.

Dolomite in Shallow, Hypersaline Waters

From the sheer mass of papers, hypersaline dolomite would win easily any popularity contest by an overwhelming margin. Discovery of dolomite forming as crusts in the supratidal zone (Shinn *et al.,* 1965), or in hypersaline tidal flats, sabkhas (Kinsman, 1965), and lagoons (Deffeyes *et al.,* 1964; Behrens and Land, 1972) touched off a breakneck stampede to a new dogma. The hypersaline point of view was stated strongly by Friedman and Sanders (1967); a more reasonable tack has been taken in the excellent review by Zenger (1972).

In such hypersaline environments, it is reasoned, precipitation of gypsum and consequent loss of Ca raises the Mg/Ca ratio of the brine to values of 5:1 or more (by weight), sometimes as high as 100:1. Total salinity can rise from about five times that of normal sea water (at which point gypsum precipitates) to values as high as 10:1 or more. Kinsman (1965) has shown that hypersaline dolomite begins to crystallize only when the Mg/ Ca ratio exceeds 5:1 to 10:1. Any carbonate crystallized below such high Mg/Ca ratios comes out as the easier-to-form aragonite or magnesian-calcite.

Dolomite in Normally Saline Bays and Deep Sea

Normal sea water has a Mg/Ca weight ratio of close to 3:1. Considerable controversy exists over whether dolomite can form in marine water of normal salinity, *e.g.*, in nonrestricted bays or in deep-sea sediments. We believe that if dolomite can form in both hypersaline environments and in fresh water, then it certainly must be able to form in sea water as well, although simple immersion for millions of years is probably insufficient to cause the reaction (Emery *et al.,* 1954). The controversy on the presumed existence of deep-sea dolomite has been reviewed recently by Zenger (1972). Blatt *et al.* (1972, p. 482-485) asserted that

theoretical calculations show that normal sea water should indeed be capable of producing dolomite, but that dolomite does not normally form because of the ordering difficulty. In ancient rocks, both Scholle (1971) and Mossler (1971) have proposed dolomite formation in marine waters of normal salinity. Modern examples of normal marine dolomite from the deep sea (Thompson *et al.,* 1968) are fraught with problems, commonly the presence of hydrothermal solutions.

Dolomite in Dilute Meteoric Waters

Most freshwater lakes and streams average between 50-500 ppm total dissolved salts, whereas most large rivers contain 50-150 ppm. Most of these surface waters have Mg/Ca ratios between about 1:10 and 1:3; thus, calcite should be the expected phase formed. However, some exceptional lakes have an excess of Mg over Ca. Müller (1970) and Müller *et al.* (1972) reported magnesian calcite, aragonite, and protodolomite forming in sediments of Lake Balaton, Hungary, with a total salinity under 500 ppm. But the Mg/Ca ratio in the lake varies from 1:1 to 3:1 and becomes even higher in the interstitial waters of the muds where protodolomite is present. From a survey of other lakes they found dolomite present where the Mg/ Ca ratio was highest (over 7:1). We propose that the rapid crystallization rate in lake-bottom sediments, possibly aided by organic withdrawal of CO_2, is responsible for dolomite being able to form only at such high Mg/Ca ratios. Dolomite also formed in the silts around glacial Lake Agassiz, probably through evaporation (Sherman and Thiel, 1939; Sherman *et al.,* 1962).

In caves, dolomite and other assorted magnesian carbonates can form if the Mg/Ca ratio exceeds 1:1 (Halliday, 1961; Moore, 1961; Holland *et al.,* 1964; Fischbeck and Müller, 1971; Thrailkill, 1971). Dolomite can form in other freshwater deposits such as spring tufas (Klähn, 1928).

In caliches calcite is again the normal product, as the Mg/Ca ratio of most soil waters is low. However, dolomitic caliche can form by attack of rain water upon Mg-rich source rocks (Sherman *et al.,* 1947).

We conclude that dolomite can form in surface fresh waters with Mg/Ca ratios as low as 1:1 providing crystallization rate is slow enough for ordering to take place. A more rapid crystallization rate requires a greater excess of Mg/Ca. A high total salinity is not in the least required; indeed, it tends to hamper crystallization of dolomite and favor calcite or aragonite instead.

Dolomite in Subsurface Waters

Subsurface waters show an extremely wide range in composition, from drinkable waters of as low as 100 ppm total salinity to very concentrated brines as high as six times seawater salinity (White, 1965). Because of a great loss of Mg relative to Ca in most subsurface brines, the Mg/Ca ratio tends to be low, between 1:2 and 1:4 (de Sitter, 1947; Dickey, 1966, 1969).

Subsurface waters of mainly meteoric derivation, though dilute, have Mg/Ca ppm ratios that approach a limit very close to 1:2 (molar ratios of 0.8 Mg to 1.0 Ca). This situation has been reported by Langmuir (1971) in Pennsylvania well waters drawn from Paleozoic carbonate rocks, by Holland *et al.* (1964) for cave waters in Paleozoic carbonates, by Hsü (1963) in Florida ground waters drawn from Tertiary carbonates, and by Hanshaw *et al.* (1971) in aquifers of Florida, Yucatan, and elsewhere in the world. Because these diverse waters all reach a limiting Mg/Ca value of 1:2 (in ppm) this is interpreted to be the point at which under subsurface conditions calcite and dolomite are both stable, consequently precipitation of either phase can occur with slight shifts in composition. Of course, for large amounts of dolomite to be formed, enough time, adequate total supply of Mg, and actively moving waters must be available (Hanshaw *et al.*, 1971).

In summary, these data on natural waters show that, in a hypersaline environment with high ion concentration and rapid crystallization, the Mg/Ca ratio must exceed 5:1 or 10:1 for dolomite to form. In normal marine waters (such as deep-sea sediments), dolomite probably forms at Mg/Ca values over 3:1. In some fresh water and low-salinity subsurface waters, dolomite can form at Mg/Ca ratios as low as 1:1 because of a lack of competing ions and a generally slower rate of crystallization. The lower the salinity, the easier it is for dolomite to order.

On the diagram (Fig. 1), this leads to a diagonal kinetic boundary line for the dolomite field. The fact that this boundary is diagonal is critically important. It means that the easiest way to produce dolomite is by *lowering* the salinity, reducing the concentration of competing ions, and in most cases slowing the rate of crystallization. Dolomite also can form at any given salinity by raising the Mg/Ca ratio, but that is not so common a process.

METHODS FOR PRODUCING DOLOMITE

The two most commonly accepted ways of forming dolomite are (1) with the hypersaline sabkha model, saline waters supplied either by supratidal flooding or by capillary upsucking; or (2) by

a supposed reflux of heavy brines through porous sediments beneath the saline basins, a theory now in some doubt (Hsü and Siegenthaler, 1969). Both these models rely upon using a brine that is always hypersaline. Hypersaline dolomite certainly exists, but we propose that it is much easier and of much greater quantitative importance to produce dolomite by dilution with fresh water. This can happen in several ways: (1) the schizohaline environment, wherein hypersaline brines are periodically mixed with fresh waters in a near-surface environment; (2) mixture in the shallow subsurface of evaporitic, highly saline brines with fresh water; (3) mixture of normal sea water or its connate equivalent with meteoric water, as in a salt-water/ freshwater-lens contact zone; and (4) phreatic meteoric water collecting Mg purged from magnesian-calcite during diagenesis.

Schizohaline[3] environment (Folk and Land, 1972; Siedlecka, 1972)—Many shallow hypersaline environments are subjected to episodic flushing by fresh water. A sabkha can be flooded with monsoon rains; or a shallow hypersaline bay may be flushed by storms or hurricanes. In either case the salinity drops suddenly and drastically to nearly fresh conditions, only to be built slowly back to hypersaline conditions as the normal evaporitic regime regains mastery. Such environments, characterized by wild swings between hypersaline and nearly fresh conditions, with essentially no times of normal marine salinity, are designated as "schizohaline" environments (Folk and Siedlecka, in press).

Why is a schizohaline environment such an ideal place to form dolomite? Let us assume a typical sabkha with perhaps five times normal salinity and a Mg/Ca ratio of double normal, say 7:1. Adding fresh water drops the salinity drastically, but the Mg/Ca ratio remains almost as high as it was initially because of the low concentration of total salts, including Ca and Mg, carried by the diluting water. On the diagram (Fig. 1) this is illustrated by a line dropping almost vertically and the composition of the water plunges deeply into the dolomite field.

For example, diluting one part normal sea water with 9 parts of average river water with about 100 ppm total dissolved solids (typically including 20 ppm Ca and 7 ppm Mg) decreases the salinity of the mixture to about one-tenth-normal sea water, but changes the Mg/Ca ratio only from 3:1 to 2.2:1. Diluting a typical hypersaline sabkha brine with 9 parts of river water would drop the Mg/Ca ratio only from 7:1 to 6:1. Water of such composi-

---

[3]"Schizo-" is Greek for "split"; name coined by Harry Mueller, graduate student, University of Texas at Austin.

tion in the subsurface then should continue to precipitate dolomite until the Mg/Ca ratio approaches 1:2.

Repetition of this process every several decades might allow significant quantities of dolomite to form. In such a dynamic environment of repeated flushing, effective transport of Mg ions through the system is insured. The large freshwater head developed during flooding may mix with and transport Mg-rich, diluted saline waters into the deep subsurface as a thick lens, and provide the "pump" necessary for massive diagenesis.

Subsidence or uplift—Even without the effect of flooding, gradual subsidence of hypersaline sabkha or lagoon sediments as deposition continues may allow penetration of meteoric waters from a landward freshwater table into the hypersaline-saturated sediments. A similar chemical revolution ensues. A fall in sea level will have the same effect. Maiklem (1971) showed, for example, that evaporative drawdown within a large basin may allow extensive flushing by fresh waters. Or uplift and emergence of sabkha sediments will raise the hypersaline system permanently into the vadose freshwater zone.

Mixing of fresh water with normal sea water—Many examples are known where a freshwater lens of huge size overlies sea water, with a zone of mixing in between characterized by dynamic water movement. Examples have been given by Cotecchia (1955) for Puglia, Italy; Kohout (1967) for Florida; Back and Hanshaw (1970) and Hanshaw *et al.* (1971) for Florida and Yucatan; Land (1973a, b) for the Pleistocene of Jamaica; and Badiozamani (1973) for the Ordovician of Wisconsin. In each case, mixing drastically lowers the salinity, but the Mg/Ca ratio remains high because the amounts of Ca and Mg added by fresh water are very small compared to the large amounts of these cations present in sea water. Mg/Ca ratios only can be lowered to typical subsurface values of 1:2 or 1:4 by crystallizing a large amount of dolomite or Mg-clays.

The failure to find gypsum in the supratidal flats of the Bahamas suggests that some of the dolomite crusts may be formed by wet-season rain-water dilution of sea water that is either of normal salinity or that has not yet evaporated down to the gypsum stage (see Bourrouilh, 1972, for another example); thus, the cause of the dolomite is, again, freshwater dilution.

Mg-purging—Incongruent dissolution of Mg-calcites can cause dolomitization at nearly constant salinity by increasing the Mg/Ca ratio of the pore waters (Land and Epstein, 1970; Folk and Siedlecka, in press). Only minor amounts of dolomite, restricted to Mg-calcite allochems or their internal pores, can be produced by this method, however.

LIMPID DOLOMITE: PRODUCT OF DILUTE WATERS

Freshwater dolomite is characteristically limpid (Folk and Land, 1972; Siedlecka, 1972; Land, 1973a). Crystals are readily identifiable with a low-power binocular microscope (or even hand lens or naked eye) because of their mirror-smooth faces, reflecting sunlight like tiny faceted gems. Crystals are usually perfectly brilliant, transparent, and water-clear, lacking any inclusions; they look like minute diamonds. Under the petrographic microscope they appear as perfect euhedra with geometrically exact faces, and again are extremely clear though some are zoned. At higher magnifications obtained by electron microscope they appear as near-perfect crystals with absolutely plane crystal faces, lacking any imperfections, inclusions, growth steps, or any other sort of blemish.

The significance of limpid dolomite was first realized in rocks of Bear Island (Siedlecka, 1972, 1973; Folk and Siedlecka, in press). Bear Island lies between Spitsbergen and Norway, and consists mainly of Upper Carboniferous to Permian carbonate rocks. The Upper Carboniferous *Fusulina* limestone formation consists mainly of very finely crystalline subhedral dolomite with ghosts of pellets and laminae. It probably represents a hypersaline, sabkha-type depositional environment. However, fractures and bird's-eye vugs in this fine dolomite are filled with coarse, sparry calcite and 0.05-0.10-mm crystals of limpid dolomite. The calcite is younger and is a product of phreatic freshwater diagenesis, on the basis of its coarse crystal size and the fact that large crystals of calcite extend right up to the pore walls without any wall-lining crust of bladed to fibrous calcite. The Permian *Spirifer* limestone formation is largely biosparite, but limpid dolomite crystals 0.05-0.10 mm in diameter are present locally together with sparry calcite cement.

Limpid dolomite also has been observed in Hell and other localities on Grand Cayman Island, west of Jamaica (Folk *et al.*, 1973). There the dolomite is in the Tertiary Bluff limestone.[*] Etching reveals limpid dolomite lining cavities that later were filled with poikilotopic sparry calcite. This limestone has been affected greatly by freshwater flushing; much of it is overlain by freshwater swamps, and surface outcrops are characterized by grotesque phytokarst. Large crystals of limpid dolomite up to 5 mm across fill cavities and fractures in the rocks of the Dolomite Alps, Italy.

The obvious physical perfection is not imaginary, for limpid dolomite dissolves much more slowly in acid; thus, limpid dolomite is apparently also more noble chemically than ordinary dolomite. A piece of medium-to-coarse crystalline dolomite rock left in a beaker of weak HCl for several hours shows extensive corrosion of normal dolomite with formation of skeletal relics. In sharp contrast, limpid dolomite in the same beaker is very little affected, retaining its sharp and planar faces, only a slight clouding and dulling of luster being observed. Apparently limpid dolomite dissolves at rates of $1/10$-$1/100$ that of ordinary dolomite. Modern "protodolomites" dissolve at even faster rates.

The low rate of chemical attack probably results from a more perfect stoichiometry and ionic ordering of limpid dolomite, the result of slow, careful precipitation from dilute solution. Slow crystallization leads to unusually perfect crystals which consequently dissolve slowly, and in the case of dolomite this probably explains the difficulty workers have experienced and the differing results they have obtained in trying to determine the K_{sp} of the mineral from dissolution experiments and in the interpretation of isotopic fractionation factors.

Morphology of Calcite, Magnesian-Calcite, and Aragonite

Since the work of Leitmeier (1910, 1915) it has been recognized that the Mg^{++} ion favors the precipitation and stabilization of aragonite over calcite. In both natural waters and in experimental work (*e.g.*, Usdowski, 1963) a Mg/Ca ratio of about 2:1 appears to be a commonly observed limit for calcite precipitation. Because of its small size the presence of the Mg ion prevents growth of calcite by blocking sideward growth of the calcite crystal lattice, but c-axis growth proceeds at more rapid rates (Folk, 1973). Thus aragonite is the easiest phase to form; any calcite that does manage to crystallize is loaded with Mg-ion impurities (average about 10-18 percent $MgCO_3$) and assumes a crystal habit that is either anhedral micrite, micrite of steep rhombic habit, or fibrous. Accumulated sidewise misfit caused by Mg usually prevents sideward growth of magnesian calcite crystals beyond a few microns, though these can coalesce into large polycrystals.

Thus, fibrous to micritic aragonite and magnesian calcite are the phases formed from marine water with its high-Mg content; they can crystallize either as beachrock cement, as shallow submarine cement of hardgrounds, in reefs, or as deep-sea cement.

Once the Mg/Ca ratio drops below about 2:1, ordinary low-magnesian calcite can form as equant, sparry crystals. Consequently sparry calcite is the typical cement formed by meteoric waters or dilute subsurface waters where the Mg/Ca ratio ranges generally from 1:1 to 1:10. In very dilute solutions without competing cations such as Na^+ or K^+, calcite in the near-surface zone can form uniformly sized unit rhombohedra in the 2-10-micron range. Such rhombs are found in caliches, in lake sediments, and as precipitates on stream surfaces (Folk, 1971, 1973), and form as a common cement in vadose carbonate sands (such as in dune rocks). Experiments by Usdowski (1963) also show that calcite rhombs form in low-salinity waters with Mg/Ca 1:4 or less.

In the phreatic zone (Land, 1970; Thorstenson *et al.*, 1972) sparry calcite forms large, commonly poikilotopic crystals that extend to the allochem boundaries without any intervening crust or fringe of finer fibrous or bladed calcite. Presumably this is the result of slow precipitation from cavities that are completely water-filled.

In recent years it has become dogmatic that sparry calcite is the result of freshwater diagenesis. Indeed, much of it does form that way, but we propose that a substantial quantity of sparry calcite may form from buried marine fluids. As sea water is buried with sediments and mixed with meteoric water, it also undergoes extensive diagenesis. Particularly important is the great loss of Mg relative to Ca; the Mg/Ca ratio is completely reversed from 3:1 in sea water to values typically between 1:2 and 1:4 in brines, even highly saline ones. Magnesium is apparently lost through reaction to form dolomite and is snatched out by clays to form chlorite or montmorillonite (Folk, 1973).

The great importance of this process is in the removal of the poisoning effect of the Mg^{++} ion (and perhaps SO_4^- as well), so that coarse sparry calcite can crystallize at depth.

Thus blocky, sparry calcite can form either by (1) surficial meteoric waters that are initially poor in Mg or (2) subsurface solutions, either connate or mixed with meteoric water, that have lost much Mg relative to Ca by reaction to form dolomite or clays.

Conclusions

Dolomite, because of the difficulty of ordering required for crystallization, can form most easily by slow crystallization. At more rapid crystallization rates, aragonite and magnesian calcite crystallize because they are simpler structures. Dolomite is favored similarly when solutions are dilute because few impurities disrupt the precise ordering of the lattice.

Dolomite can form from any natural solution providing the Mg/Ca ratio is over approximately 1:1, even in lakes or subsurface waters of very low total salinity. An important way to precipitate dolomite is to dilute sea water or sabkha-evaporitic water with fresh water. Dilution allows the Mg/Ca ratio to remain very high, but slows the crystallization rate and reduces the concentration of competing ions. Two ideal sites where such a mixing mechanism can take place (schizohaline environments) are floodable sabkhas or inundatable shallow lagoons where salinity undergoes rapid fluctuation between hypersaline and nearly fresh conditions; another site is the subsurface zone where sea water or evaporitic waters come into contact with a wedge or lens of meteoric water and salinity reduction occurs. In both cases Mg is *supplied* by saline waters, but precipitation is *permitted* only by dilution with fresh waters.

Dolomite formed from dilute solutions is characteristically limpid (euhedral, perfectly clear with plane mirrorlike faces) and is much more resistant to solution than ordinary dolomite.

The Mg/Ca ratio also controls the morphology of calcite. Abundant Mg favors formation of fibrous crystals (aragonite or Mg-calcite), whereas a lack of Mg allows formation of blocky, sparry calcite. Calcite rhombs form in very dilute, near-surface waters or Mg- and SO_4-free subsurface brines.

REFERENCES CITED

Back, W., and B. B. Hanshaw, 1970, Comparison of chemical hydrogeology of the carbonate peninsulas of Florida and Yucatan: Jour. Hydrology, v. 10, p. 330-368.

Badiozamani, K., 1973, The Dorag dolomitization model—application to the Middle Ordovician of Wisconsin: Jour. Sed. Petrology, v. 43, p. 965-984.

Behrens, E. W., and L. S. Land, 1972, Subtidal Holocene dolomite, Baffin Bay, Texas: Jour. Sed. Petrology, v. 42, p. 155-161.

Berner, R. A., 1966, Chemical diagenesis of some modern carbonate sediment: Am. Jour. Sci., v. 264, p. 1-36.

Blatt, H., G. Middleton, and R. Murray, 1972, Origin of sedimentary rocks: Englewood Cliffs, New Jersey, Prentice-Hall, 634 p.

Bourrouilh, F., 1972, Diagenese recifale: Calcitisation et dolomitisation leur repartition horizontale dans un atoll souleve ile Lifou, Territoire de Nouvelle Caledonie: Paris, Office Recherche Sci. et Tech. Outre-Mer, Cah., ser. geol., v. 4, no. 2, p. 121-148.

Cotecchia, V., 1955, Influenza dell'acqua marine sulle falde acquifere in zone costiere, con particulare riferimento alle ricerche d'acqua sotteranea in Puglia: Geotecnica, v. 2, no. 3, p. 105-128.

Deffeyes, K. S., F. J. Lucia, and P. K. Weyl, 1964, Dolomitization; observations on the Island of Bonaire, Netherlands Antilles: Science, v. 143, p. 678-679.

De Sitter, L. U., 1947, Diagenesis of oil-field brines: AAPG Bull., v. 31, p. 2030-2040.

Dickey, P. A., 1966, Patterns of chemical composition in deep subsurface brines: AAPG Bull., v. 50, p. 2472-2478.

———— 1969, Increasing concentration of subsurface brines with depth: Chem. Geology, v. 4, p. 361-370.

Emery, K. O., J. I. Tracey, Jr., and H. S. Ladd, 1954, Geology of Bikini and nearby atolls: U.S. Geol. Survey Prof. Paper 260-A, 265 p.

Fischbeck, R., and G. Müller, 1971, Monohydrocalcite, hydromagnesite, nesquehonite, dolomite, aragonite, and calcite in speleothems of the Fraenkische Schweiz, Western Germany:

Contr. Mineralogy und Petrology, v. 33, p. 87-92.

Folk, R. L., 1971, Caliche nodule composed of calcite rhombs, *in* O. P. Bricker, ed., Carbonate cements: Johns Hopkins Univ. Studies Geology, no. 19, p. 167-168, 376 p.

———— 1973, Carbonate petrography in the post-Sorbian age, *in* R. N. Ginsburg, ed., Evolving concepts in sedimentology: Johns Hopkins Univ. Studies Geology, no. 21, p. 118-158.

———— 1974, The natural history of crystalline calcium carbonate; effect of magnesian content and salinity: Jour. Sed. Petrology, v. 44, p. 40-53.

———— and L. S. Land, 1972, Mg/Ca vs salinity; a frame of reference for crystallization of calcite, aragonite, and dolomite (abs.): Geol. Soc. America Abs. with Programs, v. 4. no. 7, p. 508.

———— and A. Siedlecka, in press, Sedimentary and diagenetic fabrics of the schizohaline environment, exemplified by late Paleozoic rocks of Bear Island, Svalbard: Sed. Geology.

———— H. H. Roberts, and C. H. Moore, 1973, Black phytokarst from Hell, Cayman Islands, British West Indies: Geol. Soc. America Bull., v. 84, p. 2351-2360.

Friedman, G. M., and J. E. Sanders, 1967, Origin and occurrence of dolostones, *in* G. V. Chilinger, H. J. Bissell, and R. W. Fairbridge, eds., Carbonate rocks: Amsterdam, Elsevier, Developments in Sedimentology 9A, p. 267-348.

Fyfe, W. S., and J. L. Bischoff, 1965, The calcite-aragonite problem, *in* L. C. Pray and R. C. Murray, eds., Dolomitization and limestone diagenesis: Soc. Econ. Paleontologists and Mineralogists Spec. Pub. 13, p. 3-13.

Goldsmith, J. R., 1953, A simplexity principle and its relation to ease of crystallization: Jour. Geology, v. 61, p. 439-451.

Halliday, W. R., 1961, More dolomite speleothems: Natl. Speleolog. Soc. News, v. 19, no. 11, p. 143.

Hanshaw, B. B., W. Back, and R. G. Deike, 1971, A geochemical hypothesis for dolomitization by ground water: Econ. Geology, v. 66, p. 710-724.

Holland, H. D., T. V. Kirsipu, J. S. Heubner, and U. M. Oxburgh, 1964, On some aspects of the chemical evolution of cave waters: Jour. Geology, v. 72, p. 36-67.

Hsü, K. J., 1963, Solubility of dolomite and composition of Florida ground waters: Jour. Hydrology, v. 1, p. 288-310.

———— and C. Siegenthaler, 1969, Preliminary experiments on hydrodynamic movement induced by evaporation and their bearing on the dolomite problem: Sedimentology, v. 12, p. 11-26.

Kinsman, D. J. J., 1965, Gypsum and anhydrite of recent age, Trucial Coast, Persian Gulf, *in* J. L. Rau, ed., Second symposium on salt, v. 1: Cleveland, Ohio, Northern Ohio Geol. Soc., p. 302-326.

Kitano, Y., and D. W. Hood, 1965, The influence of organic material on the polymorphic crystallization of calcium carbonate: Geochim. et Cosmochim. Acta., v. 29, p. 29-41.

Klähn, H., 1928, Susswasserkalkmagnesiagesteine and Kalkmagnesiasusswasser: Chemie der Erde, v. 3, p. 453-587.

Kohout, F. A., 1967, Ground-water flow and the geothermal regime of the Floridian Plateau: Gulf Coast Assoc. Geol. Socs. Trans., v. 17, p. 339-354.

Land, L. S., 1970, Phreatic versus vadose meteoric diagenesis of limestones: evidence from a fossil water table: Sedimentology, v. 14, p. 175-185.

———— 1973a, Contemporaneous dolomitization of middle Pleistocene reefs by meteoric water, North Jamaica: Bull. Marine Sci., v. 23, p. 64-92.

———— 1973b, Holocene meteoric dolomitization of Pleistocene limestones, North Jamaica: Sedimentology, v. 20, p. 411-424.

———— and S. Epstein, 1970, Late Pleistocene diagenesis and dolomitization, North Jamaica: Sedimentology, v. 14, p. 187-200.

Langmuir, D., 1971, The geochemistry of some carbonate ground waters in central Pennsylvania: Geochim. et Cosmochim. Acta, v. 35, p. 1023-1045.

Leitmeier, H., 1910, Zur Kenntnis der Carbonate, pt. I, Die Di-

morphie des kohlensauren Kalkes: Neues Jahrb. Mineralogie, v. 1, p. 49-74.
———— 1915, Zur Kenntnis der Carbonate, pt. 2: Neues Jahrb. Mineralogie, Beilageband, v. 40, p. 655-700.
Lovering, T. S., 1969, The origin of hydrothermal and low temperature dolomite: Econ. Geology, v. 64, p. 743-754.
Maiklem, W. R., 1971, Evaporative drawdown—a mechanism for water level lowering and diagenesis in the Elk Point basin: Bull. Canadian Petroleum Geology, v. 19, p. 487-503.
Moore, G. W., 1961, Dolomite speleothems: Natl. Speleolog. Soc. News, v. 19, no. 7, p. 82.
Mossler, J. H., 1971, Diagenesis and dolomitization of Swope Formation (Upper Pennsylvanian), southeast Kansas: Jour. Sed. Petrology, v. 41, p. 962-970.
Müller, G., 1970, High-magnesian calcite and proto-dolomite in Lake Balaton (Hungary) sediments: Nature, v. 226, p. 749-750.
———— G. Irion, and U. Forstner, 1972, Formation and diagenesis of inorganic Ca-Mg carbonates in the lacustrine environment: Naturwissenschaften, v. 59, p. 158-164.
Pytkowicz, R. M., 1965, Rates of inorganic calcium carbonate nucleation: Jour. Geology, v. 73, p. 196-199.
Scholle, P. A., 1971, Diagenesis of deep-water carbonate turbidites, Upper Cretaceous Monte Antola flysch, northern Apennines, Italy: Jour. Sed. Petrology, v. 41, p. 233-250.
Sherman, G. D., Y. Kanehiro, and C. K. Fujimoto, 1947, Dolomitization in certain calcareous soils: Hawaii Agric. Experiment Station Rept., 1944-46, 52 p.
———— F. Schultz, and F. J. Alway, 1962, Dolomitization in soils of the Red River Valley, Minnesota: Soil Sci., v. 94, p. 304-315.
———— and G. A. Thiel, 1939, Dolomitization in glaciolacustrine silts of Lake Agassiz: Geol. Soc. America Bull., v. 50, p. 1535-1552.
Shinn, E. A., R. N. Ginsburg, and R. M. Lloyd, 1965, Recent supratidal dolomite from Andros Island, Bahamas, *in* L. C. Pray and R. C. Murray, eds., Dolomitization and limestone diagenesis: Soc. Econ. Paleontologists and Mineralogists Spec. Pub. 13, p. 112-123.
Siedlecka, A., 1972, Sedimentary and diagenetic fabrics of some late Paleozoic rocks of Bear Island, Svalbard (abs.): Geol. Soc. America Abs. with Programs, p. 662.
———— 1973, Length-slow chalcedony and relicts of sulphate—evidence of evaporite environments in the Upper Carboniferous and Permian beds of Bear Island, Svalbard: Jour. Sed. Petrology, v. 42, p. 812-816.
Sorby, H. C., 1879, On the cause of the production of different secondary forms of crystals: Min. Mag., v. 3, p. 111-113.
Thompson, G., V. T. Bowen, W. G. Melson, and R. Cifelli, 1968, Lithified carbonates from the deep-sea of the equatorial Atlantic: Jour. Sed. Petrology, v. 38, p. 1305-1312.
Thorstenson, R. C., F. T. Mackenzie, and B. L. Ristvet, 1972, Experimental vadose and phreatic cementation of skeletal carbonate sand: Jour. Sed. Petrology, v. 42, p. 162-167.
Thrailkill, J., 1971, Carbonate deposition in Carlsbad Caverns: Jour. Geology, v. 79, p. 683-695.
Usdowski, H. E., 1963, Der Rogenstein des norddeutschen unteren Buntsandsteins, ein Kalkoolith des marinen Faziesbereichs: Fortschr. Geologie Rheinland u. Westfalen, v. 10, p. 337-342.
White, D. E., 1965, Saline waters of sedimentary rocks, *in* A. Young and J. E. Galley, eds., Fluids in subsurface environments: AAPG Mem. 4, p. 342-366.
Zenger, D. H., 1972, Dolomitization and uniformitarianism: Jour. Geol. Education, v. 20, p. 107-124.

20

Reprinted from *Geology* **6**:556–559 (1978), courtesy of the Geological Society of America

Diagenetic dolomite formation related to Paleozoic paleogeography of the Cordilleran miogeocline in Nevada

J. B. Dunham*
E. R. Olson
Department of Earth Sciences
University of California
Riverside, California 92521

ABSTRACT

Many dolomite units in the early to middle Paleozoic Cordilleran miogeocline of Nevada formed through diagenetic replacement of subtidally deposited $CaCO_3$ by $CaMg(CO_3)_2$. Across the miogeocline, the consistent temporal and spatial distribution of limestone on the west and dolomite on the east indicates that there was a paleogeographic control on diagenetic dolomite formation. Dolomitization models involving hypersaline brines are inadequate to account for the origin of regionally extensive replacement-dolomite formations in Nevada that lack associated evaporite-mineral suites.

Several workers have recently proposed that the mixing of marine pore water with meteoric-derived ground water would lead to dolomitization of $CaCO_3$ in the subsurface, without precipitation of evaporites. We propose that subtidally deposited $CaCO_3$ in east-central Nevada became dolomitized in the subsurface as a result of dilution of marine pore water by fresh ground water derived from subaerially exposed tracts in the eastern part of the miogeocline. $CaCO_3$ sedimentary deposits in west-central Nevada were never intruded by fresh water because they were not proximal to areas of freshwater recharge; consequently, these sediments escaped dolomitization. The freshwater-seawater–mixing model is compatible with the regional extent of replacement-dolomite formations in Nevada and resolves the relation between dolomitization and paleogeography.

INTRODUCTION

Lower to middle Paleozoic carbonate rocks of the Cordilleran miogeocline have been interpreted as continental shelf deposits that represent shallow-water depositional environments to the east and deeper-water environments on the outer shelf to the west (Stewart and Poole, 1974; Matti and McKee, 1977). Within these shelf carbonates there is observed a consistent pattern in the temporal and spatial distribution of limestone and dolomite. The lower to middle Paleozoic carbonate rocks east of central Nevada are predominantly dolomite, whereas those to the west are predominantly limestone (Fig. 1). The consistent temporal and spatial pattern of deeper-water environments to the west, coupled with the consistency of the limestone-dolomite distribution, strongly suggests a paleogeographic control on limestone-dolomite distribution.

It can be shown, however, that many regionally extensive dolomite units in Nevada formed through the diagenetic replacement of subtidally deposited $CaCO_3$ by $CaMg(CO_3)_2$. Subtidal deposition is demonstrated by a variety of primary depositional fabrics and by the distribution of benthic faunas that include corals, stromatoporoids, brachiopods, and pelmatozoans. In these rocks it is evident that the original sediment was composed of $CaCO_3$; dolomite formation was a postdepositional diagenetic process rather than a penecontemporaneous or primary process.

The purpose of this paper is to discuss the areal distribution of Nevada diagenetic dolomites in light of current ideas on the mechanism of replacement-dolomite formation. Observations in Nevada demand that any model explaining the limestone-dolomite distribution must account for postdepositional replacement of large volumes of $CaCO_3$ sediment deposited under normal to nearly normal marine conditions (Berry and Boucot, 1970, p. 89). The model must also account for dolomitization without accompanying precipitation of evaporite minerals and for the apparent relation between limestone-dolomite distribution and paleogeography.

GREAT BASIN DOLOMITE

Many regionally extensive dolomite units in Nevada formed by diagenetic replacement of subtidally deposited $CaCO_3$ by $CaMg(CO_3)_2$. Ross (1977, p. 31) noted that dolomite replacement of corals and brachiopods in dolomite units of Cincinnatian age in Nevada indicates that some of the dolomite must have formed by replacement of $CaCO_3$. Miller

---

*Present address: Union Oil Company Research Center, P.O. Box 76, Brea, California 92621.

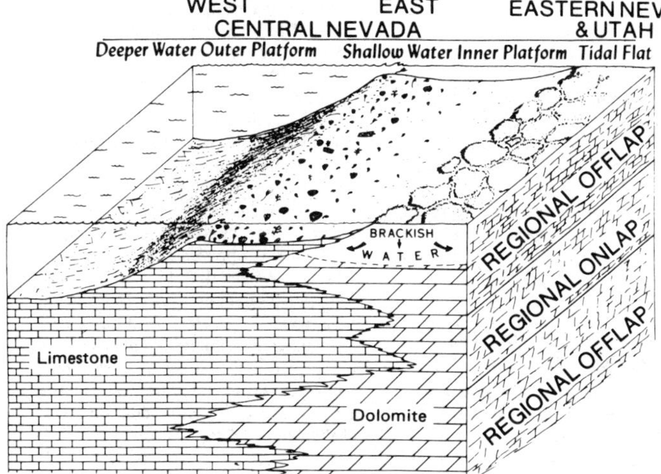

Figure 1. Block diagram illustrating generalized paleogeography of Paleozoic Cordilleran miogeocline. Exposed tracts of eastern pàrt of miogeocline are areas of freshwater recharge. Lateral migration of recharge areas with regional onlap and offlap resulted in migration of limestone-dolomite boundary with time. Limestones in west-central Nevada escaped dolomitization because of distance from areas of recharge. Horizontal dimension represents tens to hundreds of kilometres; vertical dimension represents tens to hundreds of metres.

and Walch (1977) discussed Upper Ordovician through Lower Devonian carbonate rocks of the southern Great Basin and noted that the distribution of fossils suggests that most or all of the dolomite formed by postdepositional replacement of $CaCO_3$. Dunham (1977) described dolomite replacement of corals and stromatoporoids, as well as pellets, oolites, and oncolites, within the Ordovician-Silurian Hanson Creek Formation of central Nevada. Berry and Boucot (1970) described subtidal benthic faunas from the Lower Silurian through lower-Middle Silurian Laketown Dolomite of eastern Nevada and western Utah.

In the southern Great Basin, Zenger (1972) noted that $CaMg(CO_3)_2$ has replaced $CaCO_3$ in the Middle to Upper Devonian Lost Burro Formation, although evidence for supratidal deposition is lacking. In central Nevada, Potter (1975, p. 73) described a Middle Devonian dolomite unit occupying the stratigraphic posi-

tion of the middle Denay Limestone in the Hot Creek Range. The presence of brachiopods, corals, and bryozoans in the unit indicates a subtidal environment of deposition and a replacement origin for the dolomite. Kendall (1975) described dolomite replacement of pelmatozoan columnals and other fossil debris in the Lower to Middle Devonian "Sadler Ranch Formation" in the Sulphur Spring Range of central Nevada.

This summary of representative rock units points out that many regionally extensive dolomite units in the miogeocline of Nevada formed by diagenetic replacement of subtidally deposited $CaCO_3$.

DOLOMITIZATION MODELS

Folk and Land (1975, p. 63) noted that many models for dolomitization involve hypersaline brines. Precipitation of gypsum removes Ca^{++} from the brine; this raises the Mg/Ca ratio of the liquid to extreme levels and results in the replacement of $CaCO_3$ by $CaMg(CO_3)_2$. Land and others (1975) also noted, however, that such models require precipitation of large quantities of gypsum in order to produce significant quantities of dolomite. Within Paleozoic rocks of the Great Basin, evaporite minerals, their pseudomorphs,

or large formations of collapse breccias that would indicate the former presence of evaporites are extremely rare. The lack of vast quantities of evaporite within the eastern part of the Cordilleran miogeocline precludes dolomitization of subsurface sediment by reaction with hypersaline brine.

A model for subsurface dolomitization that does not require evaporite precipitation is diagenesis by intersititial sea water diluted by meteoric ground water (Hanshaw and others, 1971; Badiozamani, 1973; Folk and Land, 1975; Land and others, 1975). The proponents of this model suggested that precise Ca-Mg cation-ordering required for stoichiometric dolomite-crystal growth would be favored in the presence of dilute solutions, where crystallization rates would be slow and where there would be a low concentration of "foreign ions" competing for lattice sites (Folk and Land, 1975).

The chemical and physical properties of dolomite crystals formed from dilute solutions are distinct from those of crystals formed from brines. Slow growth from dilute solutions produces relatively large, well-ordered dolomite crystals, containing low concentrations of foreign ions. By contrast, rapid crystallization from brine produces very small (less than 5-μm diameter), poorly ordered Ca-rich dolo-

257

mite crystals, containing high concentrations of foreign ions (Land and others, 1975, p. 1623). The marked physical and chemical differences between recent hypersaline dolomites and most ancient dolomite rocks led Land and others (1975) to conclude that many, if not most, ancient dolomites can be explained by the freshwater-seawater–mixing model.

Evidence for subsurface diagenetic replacement of subtidally deposited $CaCO_3$ by dolomite crystals coarser than 5 μm, coupled with lack of evaporites, strongly suggests that many Nevada dolomites formed from dilute solutions, rather than from reaction with hypersaline brines.

RELATION BETWEEN DOLOMITE AND PALEOGEOGRAPHY

Many recent subaerially exposed carbonate terranes in humid climates are underlain by large lenses of fresh water. Thus a mechanism for introduction of fresh water into subsurface sediment exists in certain modern settings. Back and Hanshaw (1970) reported that within the sediments underlying both the Florida and Yucatan peninsulas, a lens of fresh water and a thicker zone of brackish water overlie an extensive body of salt water. Kohout (1967) described percolation of fresh water, derived from the karst region of central Florida, through the offshore sediments of the eastern continental shelf. The meteoric water mixes with sea water and is finally discharged by upward leakage through confining beds or through the issuance of submarine springs on the continental shelf and slope. Manheim (1967, p. 89) noted that JOIDES drilling on the continental shelf and slope off Florida revealed fresh and brackish waters in sedimentary strata as far as 120 km seaward of the Florida coast.

It is thus an observation that freshwater lenses not only underlie subaerially exposed carbonate terranes, but also are capable of extending laterally for some distance into adjacent submarine sediment of the continental shelf. Direct subaerial exposure of a body of carbonate sediment is not required for intrusion of meteoric water to take place. Rather, it is the proximity of a body of sediment to areas of freshwater recharge that is required.

Regarding the Paleozoic Cordilleran miogeocline of Nevada, we suggest that much of the replacement dolomite formed in response to dilution of marine pore water by fresh water in the subsurface diagenetic environment. Intrusion of fresh water into submarine-deposited sediment took place as a result of lateral extension of freshwater lenses that had developed beneath subaerially exposed tracts of the eastern part of the miogeocline. Large tracts of the shallow-water inner platform would have become exposed to rainfall during periods of regional offlap (Fig. 1). As such, these tracts would have acted as regions of freshwater recharge. Paucity of evaporites in the many Paleozoic tidal-flat sequences of the eastern part of the miogeocline indicates that the prevailing paleoclimate was humid—a Bahama type of climate as opposed to a Persian Gulf type. We believe that this paleoclimate was humid enough to provide adequate fresh water to charge a shelf aquifer of regional extent. Subtidally deposited sediment, proximal to areas of freshwater recharge, was intruded by fresh water in the subsurface; pervasive dolomitization of the original $CaCO_3$ sediments was the result. By this model, the position of dolomite-to-limestone transitions from east to west in Nevada mark the maximum westward lateral extent of freshwater lenses in the subsurface. Because freshwater intrusion took place in the subsurface, after deposition of the original sediment, there is no reason to expect that the limestone-dolomite boundary would correspond with a boundary between differing depositional environments of the original sediment.

The limestone-dolomite boundary represents a boundary between sediments in which freshwater-seawater mixing occurred (now dolomite rocks) and sediments in which freshwater-seawater mixing did not occur (now limestone rocks). Thus, it is the position of the freshwater lens at a given period of time that determines the position of the mineralogic transition in the subsurface. The position of the freshwater lens is determined by the position of areas of freshwater recharge. Therefore, although the dolomite is diagenetic and the mineralogic transition seldom corresponds with changes in depositional environment, the position of the mineralogic transition is controlled by the position of exposed tracts of the carbonate platform. The position of exposed tracts and thus of freshwater lenses is paleogeographically controlled.

By this model, changes in regional paleogeography should produce changes in the position of the limestone-dolomite boundary. The model predicts that regional regressions would result in a westward shift of the limestone-dolomite boundary because marine offlap would produce a westward shift in position of subaerially exposed carbonate terranes (Fig. 1). Conversely, regional transgressions would flood formerly exposed tracts; this would push areas of recharge, and thus the limestone-dolomite boundary, eastward. Such shifting of the mineralogic transition has been documented by Johnson (1971, 1974) for Devonian onlap-offlap cycles of western North America. For example, Johnson (1971, Fig. 4) showed the limestone-dolomite boundary lying in east-central Nevada during the Middle Devonian. Johnson (1971, Fig. 6) showed that the Taghanic onlap, spanning the Middle-Upper Devonian boundary, shifted the limestone-dolomite boundary several hundred kilometres eastward by Late Devonian time. It is this close correlation between changing mineralogic distribution and changing paleogeography that argues for dolomite formation in the shallow subsurface soon after burial—rather than dolomite formation after deep burial of the carbonate sediment, which would nullify any influence of paleogeography.

Johnson (1974, p. 466) noted that an even greater onlap in mid-Mississippian time covered much of the transcontinental arch and left only small emergent areas. This maximum onlap resulted in deposition of a very broad sheet of $CaCO_3$ sediments, extending well into the continental interior. Johnson attributed the limited dolomitization of this sheet in the continental interior to an absence of contemporaneous hypersalinity. The paucity of mid-Mississippian dolomite may equally reflect the reduced catchment area for freshwater recharge during maximum onlap.

DISCUSSION

We believe that much of the replacement dolomite making up a large volume of rocks in the eastern part of the Cordilleran miogeocline formed as a result of mixing of fresh and marine waters in the subsurface. This model obviates any requirement for precipitation of vast quantities of evaporites and is entirely appropriate for dolomitization of rocks containing fossils that indicate deposition in normal sea water. Dolomitization by freshwater-seawater mixing is a diagenetic process and is not directly related to the depositional environment of the original sediment.

Figure 1 depicts operation of the mixing

model. Sediment containing corals, brachiopods, and stromatoporoids is deposited in a normal marine environment. Pore waters just below the sediment-water interface, and for some metres below, would be of near normal marine character. At a depth of a few tens of metres, however, the pore water would become brackish, and dolomitization would proceed through slow solution of calcite and precipitation of dolomite. The large tongues of dolomite reflect lateral migration of the freshwater lens with time. The lens, and thus the limestone-dolomite boundary, migrates eastward when regional onlap floods recharge areas.

Figure 1 also illustrates the relation between dolomite and paleogeography. Shallow-water depositional environments to the east either became subaerially exposed during regional offlap or were in proximity to areas of freshwater recharge. Deeper-water areas of the platform to the west never became subaerially exposed during the offlaps that affected the shallow shelf. Thus, no areas of freshwater recharge existed in west-central Nevada. The sediments of the outer shelf were located beyond the greatest lateral extent of the eastwardly derived freshwater lenses throughout early to middle Paleozoic time; consequently, these sediments never became dolomitized.

We have tested this model on a local scale in Nevada (Dunham and Olson, 1977). Limestone and dolomite of the Ordovician-Silurian Hanson Creek Formation were examined. The isotopic and trace-element compositions of the dolomite were consistent with values proposed for dolomite formed from dilute solutions (Land and others, 1975, p. 1623). This model could be tested on a much larger scale by application of petrographic and geochemical methods described by Badiozamani (1973) and by Land and others (1975).

Supratidal dolomite certainly exists in the eastern part of the miogeocline. Operation of the freshwater-seawater-mixing model requires an exposed catchment area for freshwater recharge. Freshwater-seawater mixing explains dolomitization of the many areally extensive formations in Nevada that were deposited in a subtidal environment.

REFERENCES CITED

Back, W., and Hanshaw, B. B., 1970, Comparison of chemical hydrology of the carbonate peninsulas of Florida and Yucatan: Journal of Hydrology, v. 10, p. 330-368.

Badiozamani, K., 1973, The Dorag dolomitization, application to the Middle Ordovician of Wisconsin: Journal of Sedimentary Petrology, v. 43, p. 965-984.

Berry, W.B.N., and Boucot, A. J., 1970, Correlation of the North American Silurian rocks: Geological Society of America Special Paper 102, 289 p.

Dunham, J. B., 1977, Depositional environments and paleogeography of the Upper Ordovician-Lower Silurian carbonate platform of central Nevada, *in* Stewart, J. H., and others, eds., Paleozoic paleogeography of the Western United States: Society of Economic Paleontologists and Mineralogists, Pacific Section, Pacific Coast Paleogeography Symposium 1, p. 157-164.

Dunham, J. B., and Olson, E. R., 1977, Petrologic and geochemical data from the Upper Ordovician-Lower Silurian Hanson Creek Formation of central Nevada: A reflection on the dolomitization process: Geological Society of America Abstracts with Programs, v. 9, p. 959.

Folk, R. L., and Land, L. S., 1975, Mg/Ca ratio and salinity: Two controls over crystallization of dolomite: American Association of Petroleum Geologists Bulletin, v. 59, p. 60-68.

Hanshaw, B. B., Back, W., and Deike, R. G., 1971, A geochemical hypothesis for dolomitization by groundwater: Economic Geology, v. 66, p. 710-724.

Johnson, J. G., 1971, Timing and coordination of orogenic, epeirogenic, and eustatic events: Geological Society of America Bulletin, v. 82, p. 3263-3298.
———— 1974, Shorelines of epeiric seas: American Journal of Science, v. 274, p. 465-470.

Kendall, G. W., 1975, Some aspects of Lower and Middle Devonian stratigraphy in Eureka County, Nevada [M.S. thesis]: Corvallis, Oregon State University, 199 p.

Kohout, F. A., 1967, Groundwater flow and the geothermal regime of the Floridian plateau: Gulf Coast Association of Geological Societies Transactions, v. 17, p. 339-354.

Land, L. S., Salem, M. R., and Morrow, D. W., 1975, Paleohydrology of ancient dolomites: Geochemical evidence: American Association of Petroleum Geologists Bulletin, v. 59, p. 1602-1625.

Manheim, F. T., 1967, Evidence for submarine discharge of water on the Atlantic continental slope of the southern United States, and suggestions for further search: New York Academy of Science Transactions, ser. 2, v. 29, p. 839-853.

Matti, J. C., and McKee, E. H., 1977, Silurian and Lower Devonian paleogeography of the outer continental shelf of the Cordilleran miogeocline, central Nevada, *in* Stewart, J. H., and others, eds., Paleozoic paleogeography of the Western United States: Society of Economic Paleontologists and Mineralogists, Pacific Section, Pacific Coast Paleogeography Symposium 1, p. 181-215.

Miller, R. H., and Walch, C. A. 1977, Depositional environments of Upper Ordovician through Lower Devonian rocks in the southern Great Basin, *in* Stewart, J. H., and others, eds., Paleozoic paleogeography of the Western United States: Society of Economic Paleontologists and Mineralogists, Pacific Section, Pacific Coast Paleogeography Symposium 1, p. 19-38.

Potter, E. C., 1975, Paleozoic stratigraphy of the northern Hot Creek Range, Nye County, Nevada [M.S. thesis]: Corvallis, Oregon State University, 129 p.

Ross, R. J., 1977, Ordovician paleogeography of the Western United States, *in* Stewart, J. H., and others, eds., Paleozoic paleogeography of the Western United States: Society of Economic Paleontologists and Mineralogists, Pacific Section, Pacific Coast Paleogeography Symposium 1, p. 19-38.

Stewart, J. H., and Poole, F. G., 1974, Lower Paleozoic and uppermost Precambrian Cordilleran miogeocline, Great Basin, Western United States, *in* Dickinson, W. R., ed., Tectonics and sedimentation: Society of Economic Paleontologists and Mineralogists Special Publication 22, p. 28-57.

Zenger, D. H., 1972, Significance of supratidal dolomitization in the geologic record: Geological Society of America Bulletin, v. 83, p. 1-12.

ACKNOWLEDGMENTS

Reviewed by J. G. Johnson, and J. C. Matti. Research supported by the donors of the Petroleum Research Fund, administered by the American Chemical Society.

MANUSCRIPT RECEIVED MAR. 13, 1978
MANUSCRIPT ACCEPTED JUNE 23, 1978

Part VI

LATE DIAGENETIC
DOLOMITIZATION

Editors' Comments
on Papers 21 and 22

21 FREEMAN
Sedimentology and Dolomitization of Muschelkalk Carbonates (Triassic), Iberian Range, Spain

22 LOVERING
The Origin of Hydrothermal and Low Temperature Dolomite

Under the preferred term *late diagenetic dolomite,* we have chosen to follow generally the definition of the second edition of the *Glossary of Geology* (AGI, 1980): "Deep seated diagenesis, occurring a long time after deposition, when the sediment is more or less compacted into a rock . . . it represents a transition from diagenesis to metamorphism." Thus, our usage corresponds to the "epigenetic" or "catagenic" dolomite of Chilingar et al. (1979, p. 426). Late diagenetic dolomites commonly are considered to be deep-burial (mesogenetic) in the sense of Choquette and Pray (1970) and generally are to be contrasted, in a spatial sense, with the shallow-burial environment (eogenetic) of early diagenetic dolomites. However, inasmuch as lithification may occur relatively early in the history of carbonate sediments (for example, marine and/or meteoric cementation) late diagenetic dolomitization in the temporal sense may similarly occur during the eogenetic stage! Also, epigenetic dolomites, included here as a type of late diagenetic dolomite—equivalent to the epigenetic or structurally controlled dolomite of Friedman and Sanders (1967) and to the "T-dolostone" of Dunbar and Rodgers (1957)—could well be telogenetic (Choquette and Pray, 1970). Narkiewicz (1979) provides a recent discussion of mesogenetic and telogenetic dolomites in Poland. Hydrothermal dolomites must also be considered; many epigenetic dolomites are, in fact, hydrothermal. We appreciate the difficulty in distinguishing hydrothermal and metamorphic dolomites.

A pertinent question concerns the relative significance of early versus late diagenetic dolomitization in the geologic record. The carbonate literature is replete with references to early diagenetic (eogenetic) dolomitization, and this concept is reflected in the majority of models presented thus far in this book. It seems that the majority of researchers favor early, shallow-burial dolomitization. Part of this belief stems from the deduction that because seawater is the most readily available source of the Mg^{2+} ions required for dolomitization, it would be most accessible during early diagenesis or in the eogenetic stage, and the discovery of modern, predominantly marginal marine dolomites. By invoking uniformitarianism, it has been natural that the more observable and more easily conceived shallow-burial processes would be emphasized. However, it appears very likely that the role of late diagenetic dolomitization has been underestimated (Mattes and Mountjoy, 1980; Zenger and Dunham, 1980).

Various field and petrographic criteria have been used to distinguish early and late diagenetic dolomites in the geologic record (for example, Papers 2 and 3). Hewett's (1928, 1931) classic studies on dolomite and associated ore deposits, particularly in the Great Basin of the United States, demonstrated the association of dolomites and fracture and fault systems, which indicates a postorogenic timing of dolomitization. The association of base-metal ores and late diagenetic (epigenetic) dolomites is well known (see Beales and Hardy, 1980). In concluding that many dolomites of the Rocky Mountain region are of late diagenetic (postorogenic) origin, Miller (1966) showed an impressive relation between dolomite occurrences and various structural features such as the belt of overthrusting, anticlines, domes, faults, and the Mississippian-Triassic unconformity. Other workers also have recognized dolomitization that appears to have been genetically related to their position beneath unconformities for example, beneath the pre-Triassic unconformity in Leicestershire and Derbyshire, England (Chilingar et al., 1979, pp. 493-497; Parsons , Paper 3)—although this type may account for relatively little ancient dolomitization. Zenger (1976) ascribed the origin of dolomite "dikes" in the uppermost Precambrian in the northern Inyo Mountains of California to epigenetic dolomitization. Based on a relationship between fractures and dolomites in the Swiss Alps, Spörli (1968) concluded that dolomitization was effected by the expulsion of formation waters during postdepositional deformation of limestone strata.

Such examples of postorogenic, epigenetic dolomitization are much less equivocal than late diagenetic, stratiform dolomites, which can be very difficult to distinguish from early diagenetic types. In fact, some dolomites are probably a bit of both insofar as their diagenetic histories have been long and complex. Studies involving oxygen isotopes (for example, Choquette, 1971; Engel et al., 1958;

Fritz, 1971) have been applied to the problem of timing, depleted values of ^{18}O suggesting the involvement of warm to hot, relatively deep subsurface waters. Fluid inclusion analyses (for example, Roedder, 1968, 1971) commonly demonstrate that dolomites associated with ore deposits formed at higher temperatures, and often through reaction with more saline waters, than did "coexisting" calcite. Zenger (1981) used the collective evidence of coarse crystallinity, pervasive and nearly obliterative dolomitization, very negative values of $\delta^{18}O$, and fluid inclusion data to conclude that at least the last diagenetic event in the dolomitization of the stratiform Little Falls Dolomite (Upper Cambrian, New York State) was related to hot, subsurface waters. Chilingar et al. (1979, pp. 493-511) evaluated field, microscopic, stable isotope, and fluid inclusion criteria used to distinguish early and late diagenetic dolomites.

We are pleased to include here Freeman's outstanding paper (Paper 21) on late diagenetic (postorogenic) dolomitization of Triassic Muschelkalk carbonates in Spain. Aware of the peritidal characteristics of the Muschelkalk, its high proportion of dolomite, and the association with the superjacent evaporitic Keuper beds, Freeman expected to find evidence of penecontemporaneous, tidal-flat dolomitization. Surprisingly, he found that the dolomite bore no relationship to sedimentary facies but was, instead, joint controlled. Freeman suggested that meteoric water descending through the overlying Keuper evaporites unexpectedly attained a high Mg^{2+}/Ca^{2+} ratio because of the introduction of Mg^{2+} ions provided by diagenesis of Keuper chlorites. Such fluids, coupled with higher temperatures provided by the geothermal gradient prior to unloading, were believed to be responsible for the dolomitization of the Muschelkalk dolomites. Earlier, Freeman (1966) described petrographic criteria that indicated a late diagenetic origin for dolomite in the Plattin and Joachim (Ordovician) carbonate units in Arkansas and has felt that late diagenetic dolomitization generally is much more important than we would believe (written communication, 1970). This view is supported by Miller (1966), Mattes and Mountjoy (1980), and Wanless (1979).

Certainly there appear to be fewer geochemical restraints on late diagenetic, deep burial dolomitization than on early diagenetic dolomitization. For example, the concentration of nonhydrated or only partially hydrated Mg^{2+} ions increases in solutions at higher temperatures, consequently decreasing this inhibiting effect on crystallization (Murata, Friedman, and Craemer, 1972, p. 4). Since, in the subsurface, conditions required for equilibrium are stable over longer periods of time than on the earth's surface, time and temperature will permit dolomitization of carbonates provided that the reacting solutions (groundwater) have a higher magnesium content than solutions that are at equilibrium with calcite and dolomite. Usdowski (1968, pp. 28-31) has pointed out that most pore waters contain more Mg^{2+} than corresponds to this equilibrium. The excellent study by Lovering (Paper 22) deals with this subject and represents independent work that supported Usdowski's findings. Emphasizing hydrothermal dolomitization, Lovering concentrated on its occurrence in mining districts. He pointed out that the Mg^{2+}/Ca^{2+} ratios in solutions associated with ore deposits range from high values, representing fluids capable of dolomitizing precursor carbonates at lower temperature, to low values representing those that could not dolomitize at temperatures within the range of formation of associated mesothermal ore deposits. He presented evidence that limestones would be converted to dolomite where they are in contact with circulating groundwaters saturated with respect to calcium, provided they have a Mg^{2+}/Ca^{2+} ratio appropriate to the temperature: 6.7 at 20°C to 0.03 at 420°C. The general conclusions of Lovering and of Usdowski (1968) are also supported by the findings of Murata, Friedman, and Craemer (1972), who contended that at temperatures of 60°C, solutions with Mg^{2+}/Ca^{2+} ratios ranging from 0.33 to 2.00 are at equilibrium with dolomite. Lovering concluded that the locally widespread hydrothermal dolomite in the vicinity of mineralized districts was caused by groundwater heated to the appropriate temperature relatively early in the igneous cycle to which the ore deposits were genetically related.

Several models have been proposed for late diagenetic dolomitization involving primarily the lateral and/or upward movement of subsurface waters, with the necessary magnesium possibly derived from the hot fluids expressed by compaction from shales or carbonates. For example, Brown (1959) concluded that connate waters derived from the compaction of argillaceous sediments had Mg^{2+}/Ca^{2+} values that were sufficiently high to dolomitize porous, oolitic carbonates along permeability trends in Upper Cambrian strata of Montana and Wyoming. Illing's (1959) model for late diagenetic dolomitization of permeable Upper Devonian Leduc reefs in the Alberta Basin involved the movement of Mg-enriched compaction waters expressed primarily from shales. Mattes and Mountjoy (1980) accepted the Illing model as a feasible one to account for late diagenetic dolomitization of the Miette buildup (Devonian) in the Alberta Rockies. Earlier, Jodry (1969), arguing against seepage refluxion, also applied this model to dolomitization of Silurian reefs in Michigan, although the compacting sediments were believed to be carbonate muds. We have much more to learn about late diagenetic dolomitization, specifically with regard to the processes involved, as well as its distribution in time and space.

For example, our impression is that there is relatively little of such dolomitization in the Tertiary. M. Esteban (oral communication, 1980), however, has pointed out the possibility of deep-burial, pervasive dolomitization of foram-algal reefs of Teritary age in southern Brazil and southwestern Asia.

REFERENCES

Bates, R. L., and J. A. Jackson, ed. 1980, *Glossary of Geology,* 2d ed., American Geological Institute, Falls Church, Va., 749p.

Beales, F. W., and J. L. Hardy, 1980, Criteria for the Recognition of Diverse Dolomite Types with an Emphasis on Studies on Host Rocks for Mississippi Valley-Type Ore Deposits, in *Concepts and Models of Dolomitization,* ed. D. H. Zenger, J. B. Dunham, and R. L. Ethington, Soc. Econ. Paleontologists and Mineralogists Spec. Pub. 28, Tulsa, Ok., pp. 197-213.

Brown, C. W., 1959, Diagenesis of Late Cambrian Oölitic Limestone, Maurice Formation, Montana and Wyoming, *Jour. Sed. Petrology* **29**:260-266.

Chilingar, G. V., D. H. Zenger, H. J. Bissell, and K. H. Wolf, 1979, Dolomites and Dolomitization, in *Diagenesis in Sediments and Sedimentary Rocks,* ed. G. Larsen and G. V. Chilingar, Elsevier, Amsterdam, pp. 423-536.

Choquette, P. W., 1971, Late Ferroan Dolomite Cement, Mississippian Carbonates, Illinois Basin, U.S.A., in *Carbonate Cements,* ed. O. P. Bricker, The Johns Hopkins University Press, Baltimore, Md., pp. 339-346.

Choquette, P. W., and L. C. Pray, 1970, Nomenclature and Classification of Porosity in Sedimentary Carbonates, *Am. Assoc. Petroleum Geologists Bull.* **54**:207-250.

Dunbar, C. O., and J. Rodgers, 1957, *Principles of Stratigraphy,* John Wiley, New York, 356p.

Engel, A. E. J., R. N. Clayton, and S. Epstein, 1968, Variations in Isotopic Composition of Oxygen and Carbon in Leadville Limestone (Mississippian, Colorado) and in Its Hydrothermal and Metamorphic Phases, *Jour. Geology* **66**:374-393.

Freeman, T., 1966, Post-lithification Dolomite in the Joachim and Plattin Formations (Ordovician), Northern Arkansas (Abstract), *Geol. Soc. America Spec. Paper 87,* p. 59.

Friedman, G. M., and J. E. Sanders, 1967, Origin and Occurrence of Dolostones, in *Carbonate Rocks: Origin, Occurrence, and Classification,* ed. G. V. Chilingar, H. J. Bissell, and R. W. Fairbridge, Elsevier, Amsterdam, pp. 267-348.

Fritz, P., 1971, Geochemical Characteristics of Dolomites and the ^{18}O Content of Middle Devonian Oceans, *Earth and Planetary Sci. Letters* **11**:277-282.

Hewett, D. F., 1928, Dolomitization and Ore Deposition, *Econ. Geology* **23**:821-863.

Hewett, D. F., 1931, Geology and Ore Deposits of the Goodsprings Quadrangle, Nevada, *U.S. Geol. Survey Prof. Paper 162,* Washington, D.C., 172p.

Illing, L. V., 1959, Deposition of Some Upper Paleozoic Carbonate Sediments in Western Canada, *Fifth World Petroleum Cong. Proc.* (sec. 1), pp. 23-52.

Jodry, R. L., 1969, Growth and Dolomitization of Silurian Reefs, St. Clair County, Michigan, *Am. Assoc. Petroleum Geologists Bull.* **53**:957-981.

Mattes, B. W., and E. W. Mountjoy, 1980, Burial Dolomitization of the Upper Devonian Miette Buildup, Jasper National Park, Alberta, in *Concepts and Models of Dolomitization,* ed. D. H. Zenger, J. B. Dunham, and R. L. Ethington, Soc. Econ. Paleontologists and Mineralogists Spec. Pub. 28, pp. 259-297.

Miller, D. N., Jr., 1966, Diagenesis of Sedimentary Rocks, in *Symposium on Recently Developed Geologic Principles and Sedimentation of the Permo-Pennsylvanian of the Rocky Mountains,* Wyoming Geol. Assoc. 12th Ann. Conf., Casper, Wy., pp. 17-21K.

Murata, K. J., I. Friedman, and M. Craemer, 1972, Geochemistry of Diagenetic Dolomites in the Miocene Marine Formations of California and Oregon, *U.S. Geol. Survey Prof. Paper 724-C,* Washington, D.C., 12p.

Narkiewicz, M., 1979, Telo- and Mesogenetic Dolomites in Subsurface Upper Devonian to Lower Carboniferous Sequences of Southern Poland, *Neues Jahrb. Geologie u. Paläontologie Abh.* **158**:180-208.

Roedder, E., 1968, Temperature, Salinity, and Origin of the Ore-Forming Fluids at Pine Point, Northwest Territories, Canada, from Fluid Inclusion Studies, *Econ. Geology* **63**:439-450.

Roedder, E., 1971, Fluid-Inclusion Evidence on the Environment of Formation of Mineral Deposits of the Southern Appalachian Valley, *Econ. Geology* **66**:777-791.

Spörli, B., 1968, Syntectonic Dolomitization in the Helvetic Nappes of Central Switzerland, *Geol. Soc. America Bull.* **79**:1839-1846.

Usdowski, H.-E., 1968, The Formation of Dolomite in Sediments, in *Recent Developments in Carbonate Sedimentology in Central Europe,* ed. G. Müller and G. M. Friedman, Springer-Verlag, Berlin, pp. 21-32.

Wanless, H. R., 1979, Limestone Response to Stress: Pressure Solution and Dolomitization, *Jour. Sed. Petrology* **49**:437-462.

Zenger, D. H., 1976, Dolomitization and Dolomite "Dikes" in the Wyman Formation (Precambrian), Northeastern Inyo Mountains, California, *Jour. Sed. Petrology* **46**:457-462.

Zenger, D. H., 1981, Stratigraphy and Petrology of the Little Falls Dolostone (Upper Cambrian), East-Central New York, *New York State Museum, Map and Chart Ser. 34,* Albany, N.Y., 138p.

Zenger, D. H., and J. B. Dunham, 1980, Concepts and Models of Dolomitization—An Introduction, in *Concepts and Models of Dolomitization,* ed. D. H. Zenger, J. B. Dunham, and R. L. Ethington, Soc. Econ. Paleontologists and Mineralogists Spec. Pub. 28, Tulsa, Ok., pp. 1-9.

21

Reprinted from *Am. Assoc. Petroleum Geologists Bull.* **56**:434–453 (1972)

Sedimentology and Dolomitization of Muschelkalk Carbonates (Triassic), Iberian Range, Spain[1]

TOM FREEMAN[2]

Abstract Triassic rocks of the Iberian Range consist of a succession of continental sandstones (Buntsandstein), peritidal carbonates and shales (Muschelkalk), and continental claystones and evaporites (Keuper). This "Germanic" platform facies closely resembles equivalent rocks in the Balearic Islands; consequently, any hinge line separating platform from geosynclinal facies must lie an undetermined distance southeast of the Balearics.

Dolostone, which comprises 75 percent of Muschelkalk carbonates, bears no relation to sedimentary facies; to the contrary, it is demonstrably joint controlled. This fact, together with the near absence of "dedolomite," indicates a late and sustained magnesium/calcium ratio in excess of that expected from an evaporite-bearing assemblage.

Water samples from Keuper "salinas" (salt-producing evaporating pans) show that meteoric water descending through the Keuper evolves into a brine with Mg:Ca ratios locally 4:1, in spite of the abundant gypsum and anhydrite. X-ray analysis suggests chlorite as a source of the magnesium, the solubility of which probably reflects diagenetic fixing in the Keuper evaporite basin. The composition of Keuper water, aided prior to unloading by geothermal gradient, is thought to be responsible for Muschelkalk dolomitization.

Introduction

The Triassic of the Iberian Peninsula is areally subdivisible into three facies (Fig. 1), the broadest of which consists of Buntsandstein sandstones, Muschelkalk carbonates, and Keuper shales and evaporites.[3] Local additions to this "Germanic" section are (1) an interval of evaporite-bearing shale, the "Röt," that occurs between the Buntsandstein sandstones and Muschelkalk carbonates in the eastern part of the study area, and (2) an interval of evaporite-bearing shale, the "middle Muschelkalk" of Virgili (1958), that is present within Muschelkalk carbonates of the Catalanide Range. No evidence of an unconformity was discovered within the Buntsandstein-Muschelkalk-Keuper sequence. A single isolated corrosion surface (hardground), reflecting lithification and solution during Muschelkalk deposition, is discussed under the section on dolomitization.

The Spanish succession, Buntsandstein-Muschelkalk-Keuper, broadly resembles the threefold subdivision of the Germanic Triassic, and therefore has been similarly assigned to the "platform facies" of interregional maps. However, more specific statements concerning the depositional environment of the Spanish Triassic have not been made, owing to a lack of petrologic studies in the peninsula. Before Perconig (1968) cataloged the microfacies of the Triassic and Jurassic sediments of Spain, there did not exist a single published photomicrograph of the Muschelkalk of the Iberian Peninsula. Earlier studies, for example those of Riba (1959) and Virgili (1958), although excellent in quality, were largely classical in approach, treating the areal geology and stratigraphy and, to a lesser degree, the paleontology. The illustrating by Perconig (1968) of a few scattered samples of the Spanish Muschelkalk only served to invite the "much needed paleogeographic analyses" to which Rutten referred (1969, p. 408).

The diagenesis of the Muschelkalk invited study because of its proximity to overlying Keuper evaporites. This succession of carbonates and evaporites seemed a likely place to discover and study both the products of early reflux dolomitization and the products of later calcitization. Such proved *not* to be the case, however; dolomitization occurred *after* lithification and hard-rock fracturing (at least some dolomite precipitation probably was post-mountain building), and only a trace of calcitized dolomite was discovered. The late dolomite, together with the near

[1] Manuscript received, April 28, 1971; accepted, June 25, 1971.

[2] Department of Geology, University of Missouri-Columbia.

I am pleased to acknowledge the hospitality of Carmina Virgili, Catedrático del Departamento de Estratigrafia, Universidad de Madrid, who generously supplied space, technical assistance, and encouragement. Particularly helpful were discussions with Louis Sánchez de la Torre, of the Universidad de Oviedo. Field expenses were provided by Chevron Overseas Petroleum, Inc., and travel expenses in conjunction with sabbatical leave were provided by the University of Missouri-Columbia. Finally, I thank R. G. C. Bathurst, of the University of Liverpool, who provided facilities during the final phase of this work, and to Ian West, of that university, who gave valuable counsel concerning the petrography of evaporites.

[3] The names Buntsandstein-Muschelkalk-Keuper, as applied here, are lithic extrapolations of the Germanic section (*i.e.,* sandstone-carbonate-shale) and might or might not be temporal equivalents of their nomenclatural counterparts elsewhere.

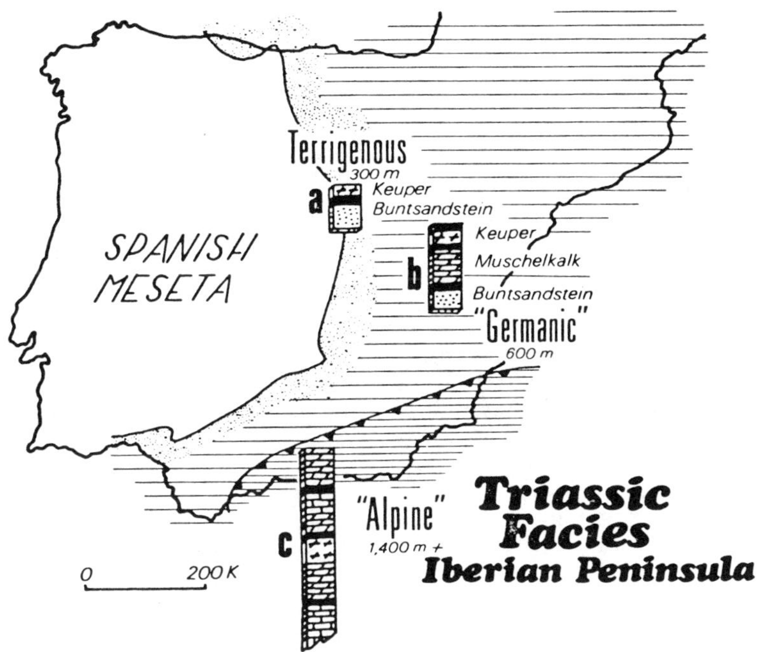

Fig. 1—General limits of Triassic facies, Iberian Peninsula. Map after Querol (1969). Sections generalized (*a*) from Simancas (1968); (*b*) from Riba (1959); and (*c*) from Perconig (1962).

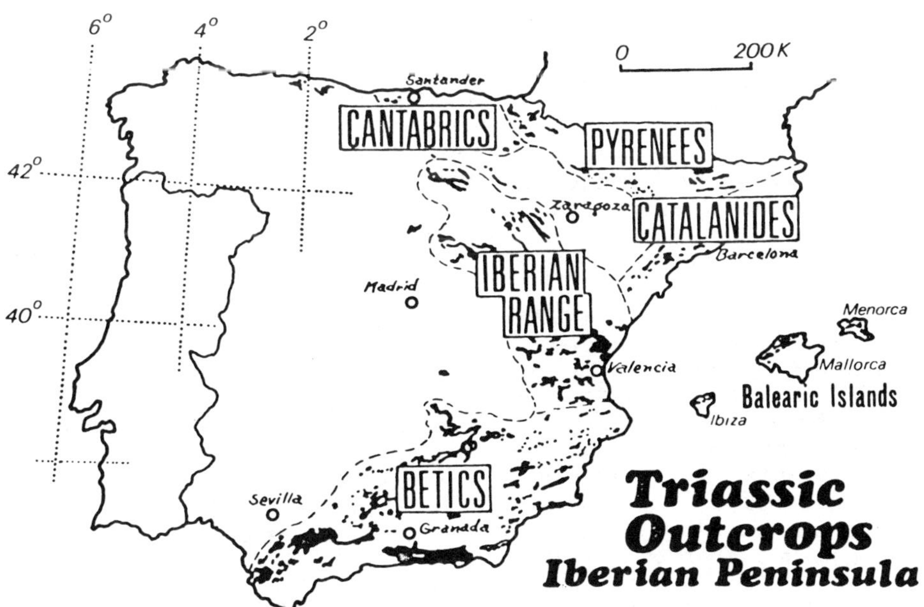

Fig. 2—Geography of Triassic outcrops, Iberian Peninsula, reflecting Tertiary mountain building and subsequent sedimentation.

absence of "dedolomite," reflects a late and sustained magnesium/calcium ratio in excess of that expected from groundwater reaction with evaporites. An important aim of this study is to reconcile this apparent anomaly.

Sedimentology

General Statement

Superimposed on the distribution of the Spanish Triassic is the fabric of early Tertiary ("Alpine") mountain building and subsequent continental sedimentation. The result is a group of mountain ranges to which exposures of Triassic and younger Mesozoic rocks are confined (Fig. 2), flanked by intermontane basins filled largely with Cenozoic deposits. Inasmuch as the Iberian Range trends down the Triassic depositional dip, this seemed a likely place to study the Muschelkalk in an area of probable maximum sedimentologic variation. However, the petrology of the Muschelkalk along the length of the Iberian Range proved to be surprisingly uniform. Figure 3 illustrates the few changes that do occur southeastward; namely, (1) an increase in thickness, (2) an increase in the amount of limestone at the expense of dolostone, and (3) a decrease in the amount of chert.

In the Iberian Range the Muschelkalk does not lend itself to vertical subdivision as it does in the Catalanides (Virgili, 1958), nor do the facies show a neat lateral compartmentalization. Certain samples can be assigned with assurance to various depositional environments, but a precise reconstruction of their sedimentologic distribution, with a chronology of bathymetric and hydrographic changes, would require considerably more time than that allotted to this investigation. The discussion that follows is presented only in an effort to document the general sedimentologic setting of the Muschelkalk in the Iberian Range, including a brief analysis of its Balearic Islands equivalent.

Facies Analysis

Dolomitization has obscured the original texture of Muschelkalk carbonates to such a degree that original grain sizes of carbonate detritus are difficult to determine. For this reason, the upper size-limit of "lime mud" suggested by Dunham (1962) proved to be more applicable than limits proposed in other classifications. Dunham's classification (1962) is adapted to the carbonates of this report (see "depositional facies" of Figure 3), even though there is an overprint of dolomitization on 75-80 percent of the carbonates. In Figure 3 those parts of the rock column in which limestones persist are indicated by line patterns at the left of the sections. The terms "laminated" and "nodular" (Fig. 3) refer to lime mudstones in which the fabric is thought to be a more diagnostic label than the texture. (A nomenclature combining fabric and texture was deliberately avoided for clarity of illustration).

Lime grainstones.—Within the Muschelkalk of the Iberian Range lime grainstones are exceedingly scarce. Even so, they are significant in that they attest to at least local shallow-water deposition. The oolite especially reflects a within-wave-base environment of generation. Oolite occurs at 7 and 25 ft above the base of section 10 (Fig. 3), referred to as 10-7 and 10-25 respectively[4] (Fig. 4 is an example), and lime grainstones consisting of fossil fragments are at 3-23, 3-64, 6-64, and 6-170. These six rather thin intervals comprise the only apparent Muschelkalk carbonates that reflect effective winnowing by waves or currents. The absence of lime grainstones at the base of Muschelkalk sections suggests that either (1) high-energy beach conditions were confined to the largely subaerial Buntsandstein (assuming partial contemporaneity), or (2) shoaling conditions were well east of the area of study and the Muschelkalk of the Iberian Range was deposited across a broad lagoon within which bathymetry dictated relatively quiet conditions. From a consideration of both lithology and paleontology, I am inclined to believe the latter, although the stratigraphy of the Buntsandstein-Muschelkalk does suggest partial contemporaneity (*i.e.,* conformability within the Buntsandstein-Muschelkalk-Keuper interval, with the Muschelkalk pinching out westward; Fig. 1).

Lime wackestones—Because of dolomitization, it is difficult to distinguish between original lime wackestones and lime mudstones of the Muschelkalk. The most useful clue appears to be the general heterogeneity of the former, best described as oatmeal texture. On close examination, flecks of varicolored material prove to be ghosts of probable skeletal debris. These lime wackestones, which comprise 14 percent of the total carbonate measured, host recognizable burrows, pellets, intraclasts, and sparse macrofossils (small snails and clams suggestive of restricted conditions). Desiccation cracks and laminations are lacking. The lone example of a pelagic fauna (Fig. 5) was found within a bed of dolomitized lime wackestone.

Evidently, Muschelkalk lime wackestones accumulated in a subtidal environment, frequented

---

[4] This scheme of sample designation is followed throughout this paper: the number of the measured section from which the sample was collected is followed by a number corresponding to the sample's position expressed in feet above the base of that section.

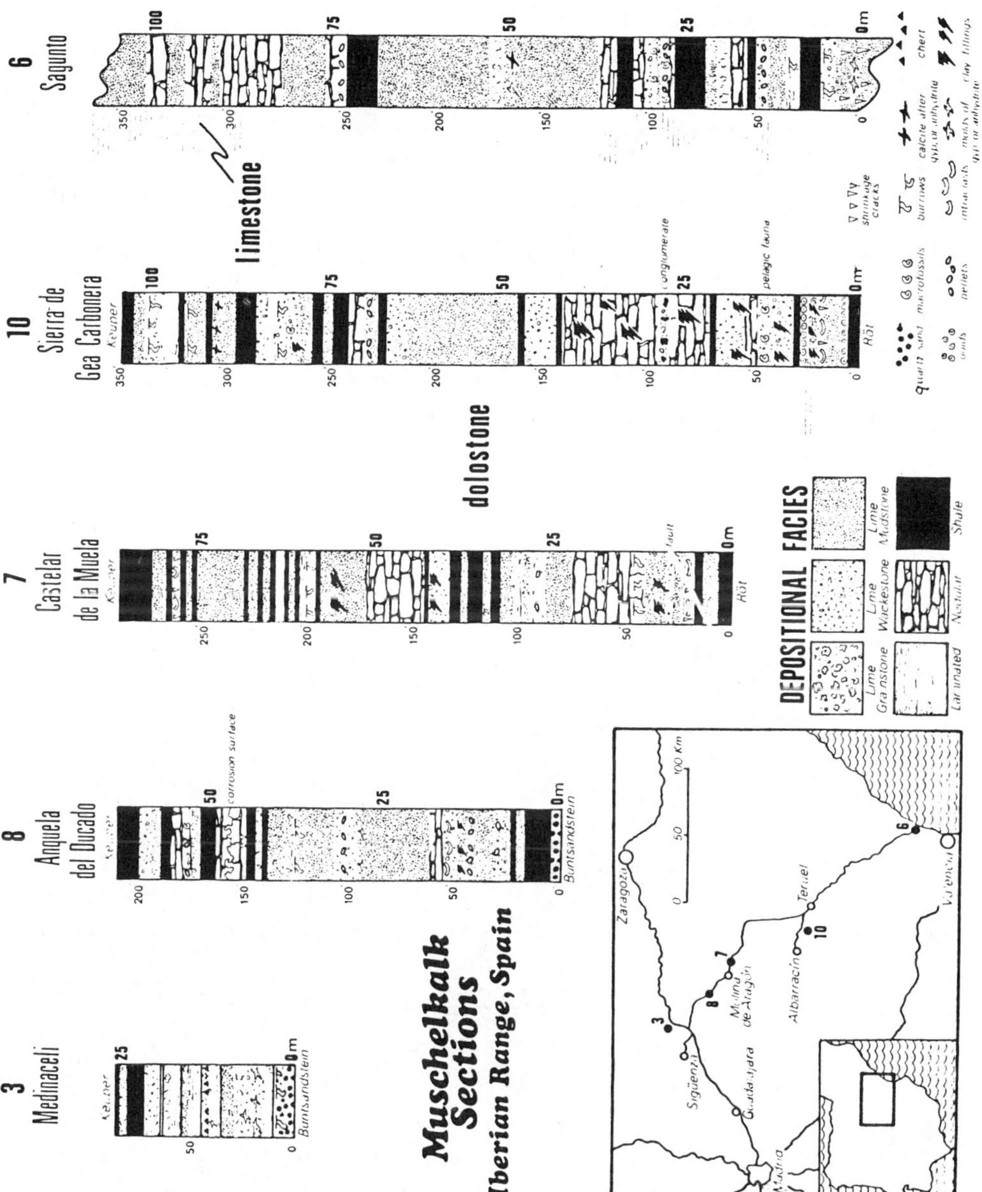

Fig. 3—Measured and described sections of Muschelkalk, Iberian Range, Spain. Intervals of dominant limestone (with shale) are marked by horizontal ruling at left of sections; remainder is dominantly dolostone (with shale). Carbonates, although largely dolomitized, are represented as *depositional facies* with rock nomenclature of Dunham (1962).

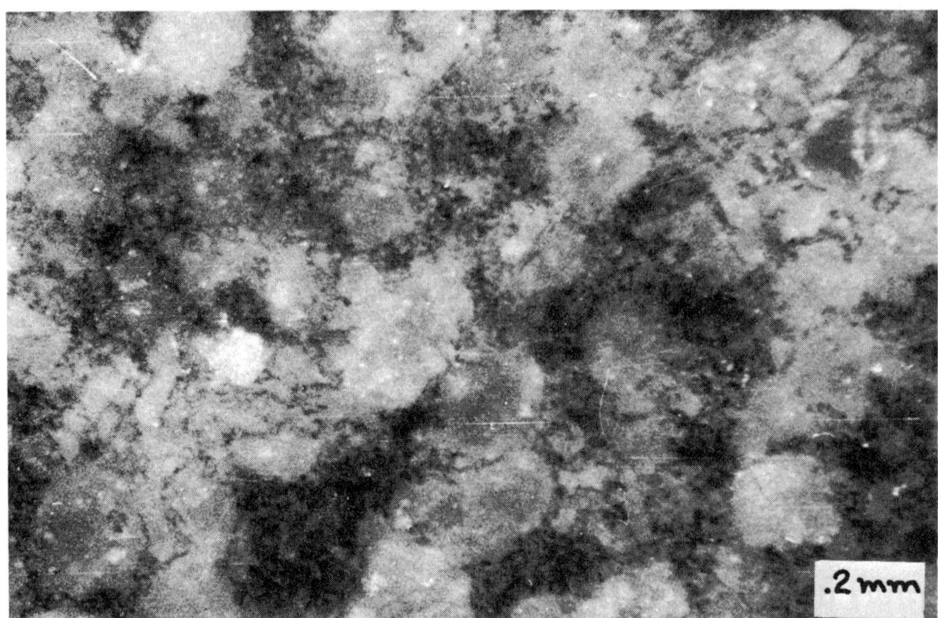

FIG. 4—Photomicrograph of dolomitized oolite grainstone from 7 ft above base of section 10 (*i.e.*, sample 10-7).

FIG. 5—Photomicrograph of dolomitized lime wackestone (10-52) with lone example of pelagic fauna found in this study.

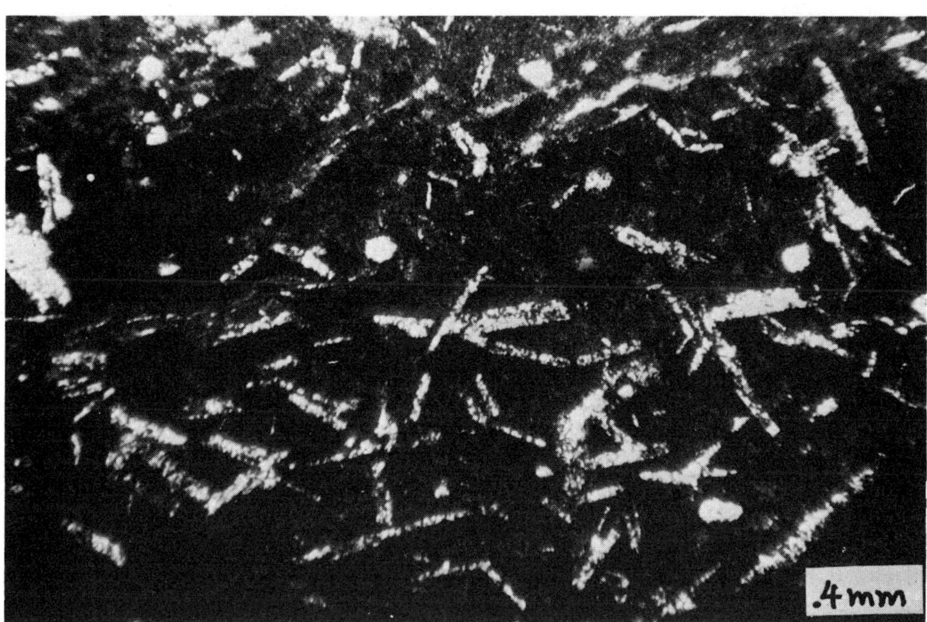

Fig. 6—Photomicrograph of pseudomorphs of ferroan calcite-after-gypsum in dolostone. Sample 10-307. Probable intertidal deposit.

by burrowing and masticating organisms, which was at times subject to disrupting waves and currents.

Lime mudstones—Unlike the "oatmeal" texture of the lime wackestones, the texture of Muschelkalk dolomitized lime mudstones is one of uniformity in color and crystallinity. Generally, the crystal size of the dolomite of the mudstones is smaller than that of the dolomitized grainstone and wackestone, but important exceptions preclude this characteristic from being a criterion for classification.

The lime mudstones are generally lighter in color than the lime wackestones, and this, coupled with the near absence of recognizable fossils, hints at a greater degree of subaerial exposure for the former. Comparison with modern examples from the Bahamas (Shinn *et al.*, 1969) suggests a tidal-flat environment, probably the intertidal zone. Consistent with this interpretation is the local presence of pseudomorphs and molds after gypsum and anhydrite (Fig. 6), which, with one exception, are not present in the lime wackestones. In recent deposits of the Trucial Coast (Persian Gulf), gypsum crystals of similar size and orientation are characteristic of the intertidal zone (Shearman, 1966, p. B210).

Solution of scattered evaporite crystals has resulted in the development of many of the tiny vugs in some lime mudstones (Fig. 7). However,

the vugs generally are irregular in shape, so the full importance of this process is difficult to assess. In chert nodules from the same bed as that of sample 3-38a, lutecite was produced by the replacement of anhydrite by SiO_2 (Fig. 8); therefore, some of the vugs in sample 3-38a can reasonably be assumed to be products of general anhydrite solution following local silicification. However, the density of vugs in sample 3-38a is greater than that of lutecite crystals in sample 3-38b; therefore, it appears that there is still another mechanism of vug formation in addition to that of evaporite solution. Whether the development of this vuggy (moldic?) porosity is restricted to surface exposures, or whether it persists in the subsurface, is unknown. That evaporites did influence the precipitation of SiO_2 in these rocks is suggested by the length-slow chalcedony that occurs in association with the microcrystalline chert. Pittman and Folk (1970) claimed that this variety of chalcedony reflects the former presence of evaporites.

Laminated lime mudstones—The "laminated" facies (Fig. 3) is a special group of lime mudstones which, in addition to being characteristically laminated, are lighter in color than the nonlaminated lime mudstones. The laminations (Fig. 9), especially the crinkled ones (Fig. 10), are suggestive of algal-mat activity, and the lighter color is perhaps the result of prolonged subaerial oxi-

Fɪɢ. 7—Hand specimen of vuggy dolostone; sample 3-38a. Undetermined number of vugs were produced
by solution of evaporites (Fig. 8).

Fɪɢ. 8—Photomicrograph of lutecite-after-anhydrite (arrow) in chert. Sample (3-38b) was collected lateral
to sample 3-38a of Figure 7. Former vug is lined with length-slow chalcedony.

FIG. 9—Hand specimen of laminated "birds-eye" dolomitized lime mudstone, sample 7-106, suggestive of supratidal zone.

dation. According to Shinn *et al.* (1969), both algal laminae and a high degree of oxidation are indicative of the supratidal zone. Desiccation of these Muschelkalk laminites, another product of subaerial exposure, produced mudcracks in surprisingly few samples, however. Figure 11 is a convincing, although admittedly atypical, example of mudcracks in the laminated facies. That the clasts are in fact mudcrack polygons, rather than post-depositional pull-aparts, can be shown by details of internal sedimentation (Fig. 12).

Unlike the lime mudstones of the intertidal zone, these supratidal laminites lack evidence of evaporite mineralization.

Nodular lime mudstones—A third variant in the realm of lime mudstones is the "nodular" facies. Figure 3 shows that 17 percent of the total rock measured, including a few feet of Röt and Keuper in certain sections, is shale. In addition to the clay comprising these shale intervals, there is abundant clay in the matrix of the "nodular" facies (Fig. 13). Flow structures within the matrix, together with streamlining of the lime mud pseudo-clasts, reflect a history of alternate deposition of clay-rich and clay-poor lime muds, followed by compaction and/or sliding that caused the less competent clay-rich zones to flow, pulling apart and shaping the more competent clay-poor lime mud laminae. This interpretation agrees with that suggested by Wilson (1969), but

similar nodular carbonates have been explained by concretionary segregation (Hallam, 1967) and by subaqueous shrinkage cracking followed by differential consolidation (Garrison and Fischer, 1969). In support of my interpretation I should mention that in the Muschelkalk "nodular" facies are abundant fucoids that I interpret as the record of burrowing organisms which inhabited lime mud sediments consisting of alternate clay-poor and clay-rich laminae. A back-filling of clay-poor lime mud through an adjacent clay-rich lamina understandably would weather out in the bold relief that characterizes these fucoids.

For want of a modern analogue, the distribution of clay and clay shale in Muschelkalk carbonates is difficult to interpret. I doubt that they reflect climatic or eustatic changes, because none seems to be regionally persistent (Fig. 3). They might, instead, record deposition in bathymetric lows, reflecting either flocculation in higher salinities (layered solutions) or water quieter than that in which the subtidal lime wackestones accumulated.

Synthesis

From the preceding discussion of sedimentology, a reasonable synthesis is presented in Figure 14. The general sedimentologic setting of the Muschelkalk of the Iberian Range is one of a lagoon-tidal flat complex that underwent suffi-

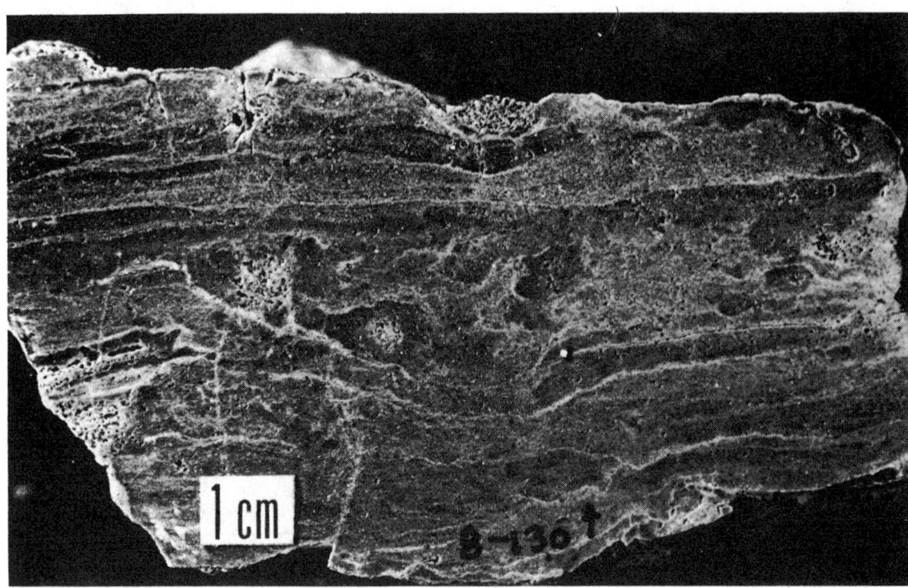

Fig. 10—Hand specimen of laminated dolomitized lime mudstone, sample 8-130, believed to reflect crinkling of algal mats in supratidal zone.

cient evaporation (at least in the intertidal zone) to permit diagenesis of gypsum and anhydrite. This evaporite setting and, perhaps more important, the near absence of pelagic faunal elements suggest that a possible bank or reef complex, with its attendant petroleum-bearing potential, might separate the Iberian Muschelkalk from basin deposits on the east. For this reason, I examined the Muschelkalk of the Balearic Islands.

Balearic Island Triassic—The Balearic Island Triassic generally is believed by geologists to be "alpine" (geosynclinal) or at least "transitional," in nature (Hollister, 1942). If the Balearic Island Muschelkalk were in fact geosynclinal, then a possible barrier would lie northwest of the islands' original position. (The amount of northwestward thrusting, by way of a commonly postulated extension of the Betic zone, would have to be assessed before the position of such a barrier could be reasoned). If, on the other hand, the Balearic Island section proved to be like that of the Iberian Range, any possible barrier would have to be well southeast of the islands.

The geologic maps of Hermite (1879) were used as guides to Triassic sections on the Balearic Islands of Mallorca and Menorca.

Fig. 11—Hand specimen of mudcracked limestone, 7-283, reflecting desiccation in supratidal zone.

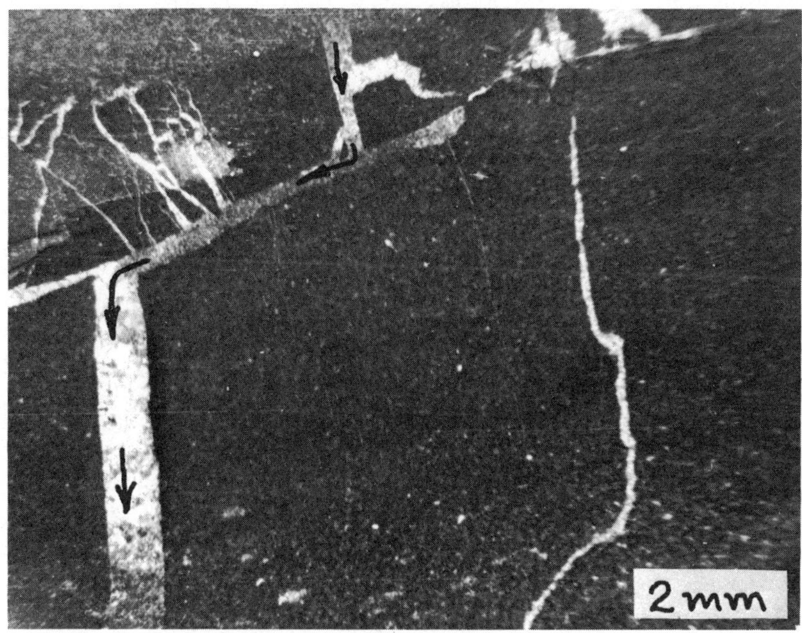

FIG. 12—Photomicrograph of sample 7-283 (Fig. 11) showing details of post-desiccation internal sedimentation (arrows).

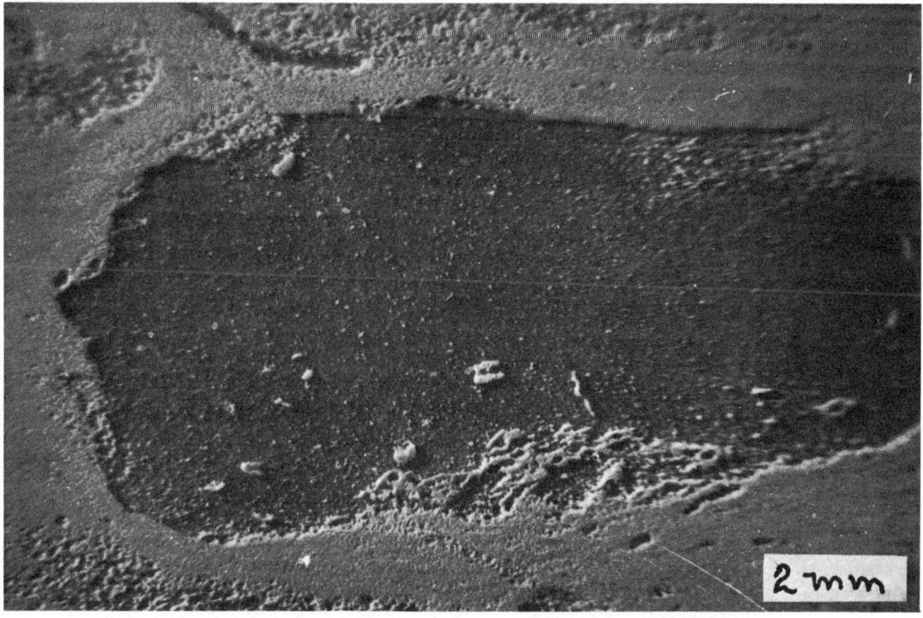

FIG. 13—Photomicrograph of etched surface of limestone 6-57, nodular lime mudstone, showing clayey matrix.
One calcareous nodule is etched into negative relief.

Along the northwest coast of Mallorca there is a mountain range with associated thick exposures of Triassic and Jurassic rocks. Along that part of the coast nearest the town of Estallenchs is a complete section of southeastward-dipping Muschelkalk with recognizable Röt and Buntsandstein below and a covered dip-slope reentrant of postulated Keuper above (the Keuper appears to be thin here owing to probable northwestward thrusting). The Muschelkalk of Mallorca can be described as follows:

1. The thickness, an estimated 85 m, is no greater than that measured in eastern sections of the Iberian Range (*i.e.*, there is *no suggestion of hinge-line thickening* between the island and the mainland).

2. Surprisingly, the progressive southeastward increase in limestone at the expense of dolostone (Fig. 3) is interrupted on Mallorca by an *entirely dolomitized section.*

3. The rock types (*i.e.*, the facies of Figure 3) of the Iberian Range Muschelkalk can be matched in essentially all aspects with their Mallorcan counterparts. There is *no sedimentologic change* between Mallorca and the mainland. The exposures of Muschelkalk on Menorca, most of which are thin because of the low relief of the island, serve to corroborate the observations made on Mallorca. The only difference between the Muschelkalk sections of the two islands is an appreciable quantity (40 percent?) of limestone on Menorca.

The Balearic Island Muschelkalk is so similar to that of the Iberian Range that one wonders why the former has been considered "Alpine" in origin. Evidently, this claim is based on a few alpine-type ammonites collected by Schwartz in the 19th century (referred to by Hollister, 1942, p. 81) and their sedimentologic implications. However, judging from the lithic similarities between the Balearics and the peninsula, the boundary between platform and geosynclinal facies should be portrayed an undetermined distance southeast of the islands (Fig. 15, arrows).

Therefore, the evidence dictates that any barrier reflected in the lithology and paleontology of the Iberian Range Muschelkalk must lie well southeast of the Balearic Islands and, consequently, beyond the reach of present economic exploitation.

DOLOMITIZATION

Terminology

Although the nomenclatural treatment of dolomite by Sander (1936) was largely petrographic (with an awkward grouping under "primary" of both detrital and pore-fill dolomites), his terms "primary" and "secondary" have since acquired temporal significance (*i.e., early* and *late*). Consequently, an ambiguity has arisen from this mixture of petrography and genesis.

Friedman and Sanders (1967), who most recently tried to put the terminology of dolostones in order, proposed (p. 267) that "most dolostone deposits in the geologic record owe their origin to hypersaline brines," which react with the deposit either synsedimentologically ("capillary concentration" or "refluxion"; p. 338) or as "remnants of sea water trapped with the sediments at the time of their deposition" (p. 335). Concerning the latter, they suggested that hypersalinity can result from "subsurface processes not altogether understood, by which waters are concentrated by diffusion, membrane filtering, or other processes" (p. 334), but there was no mention of the possibility of dolomitization by groundwater to which there has been important addition of remobilized ions. With only surface brines and trapped remnants thereof as possible dolomitizing fluids, there remained only *time* as a basis for epigenesis, and they proposed that structural control be used to assess this. Their classification follows:

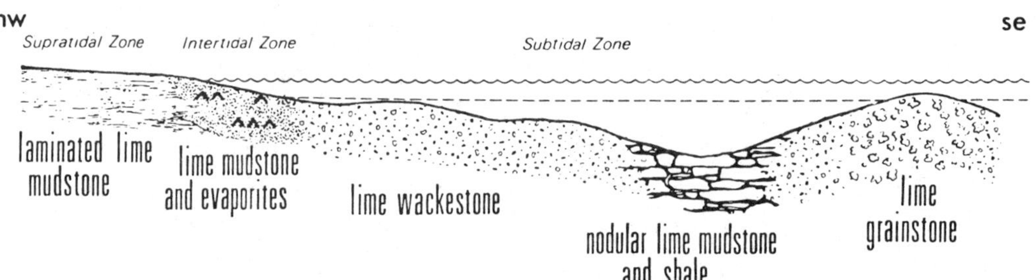

FIG. 14—Diagrammatic interpretation of Muschelkalk sedimentation, Iberian Range. Extent of any one specific facies is not intended to express its relative abundance within Muschelkalk, *e.g.*, there is no quantitatively important accumulation of lime grainstone.

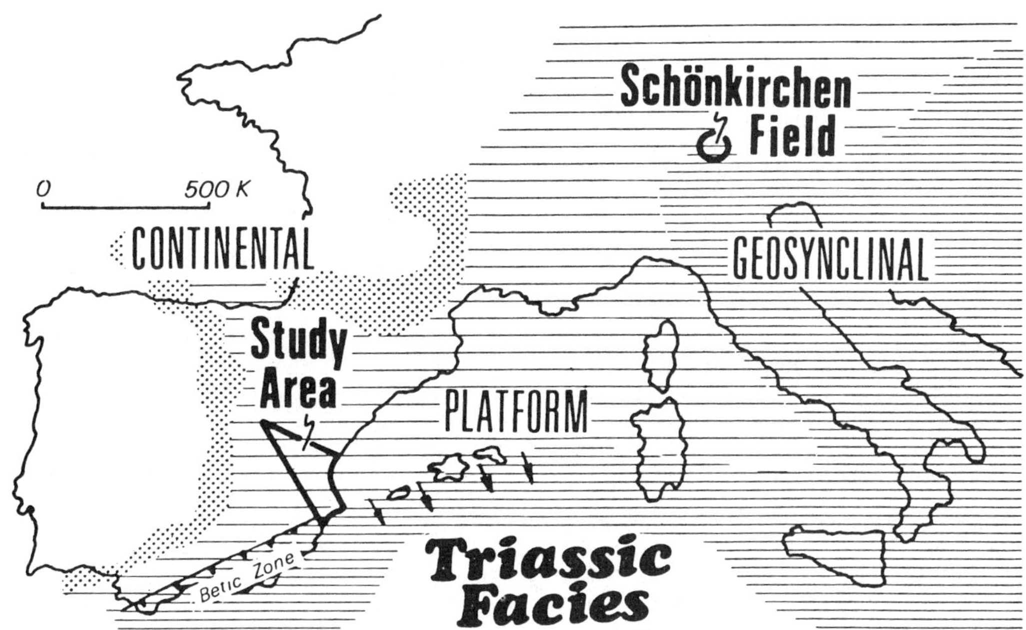

FIG. 15—Triassic facies, southwestern Europe (after Kent, 1969). Observations made in Balearic Islands dictate moving boundary between platform and geosynclinal facies undetermined distance southeast of islands (arrows). Notice that Schönkirchen field (Austria) occupies position relevant to this hinge line.

1. Syngenetic, ". . . formed penecontemporaneously in its environment of deposition . . ." (*i.e.*, precipitated?);

2. Detrital;

3. Diagenetic, ". . . formed by replacement of calcium carbonate during or after consolidation of the sediment";

4. Epigenetic, ": . . formed by replacement of limestone with the dolomite being localized by post-depositional structural elements" (*i.e.*, post-lithification?).

Perhaps a more rational basis for a *genetic* classification is the geochemical setting which the particular dolomite reflects. To be sure, the setting is difficult to determine, but it is an end to which any study of dolomite should strive. During the course of study a more descriptive nomenclature, with various levels of refinement, should be made available to the petrographer, and it is hoped that the classification presented in Figure 16 will serve both needs.

The degrees of decision under "detrital dolomite" (Fig. 16) are difficult to attain even when working with Holocene sediments, not to mention their lithified counterparts. Even so, various studies suggest that these subcategories are real, and my classification, theoretical as it might seem, anticipates more sophisticated petrographic techniques. Differences between "syngenetic" and "epigenetic" will become clear in the history of epigenesis that follows.

Genesis

Consideration of syngenesis—The most convincing arguments for synsedimentary dolomitization of the rock record have been based on sedimentary associations (*c.g.*, Laporte, 1964; West *et al.*, 1968). Superficially, one might be guided by similar diagnostic sedimentary features to interpret Muschelkalk dolostones as synsedimentary in origin. However, on close inspection, the distribution of these sedimentary features proves not to coincide with the distribution of dolostone. For example, lime grainstones are thoroughly dolomitized (Fig. 4), whereas my candidate for a supratidal deposit (Fig. 11), curiously enough, is *limestone;* and distribution of Muschelkalk dolostone is not stratiform but cross-cutting (Fig. 17). Moreover, the precipitation of $CaSO_4$, considered by some to be essential in the supratidal model of dolomitization, occurred demonstrably earlier than at least some dolomite precipitation (Fig. 18). In short, there is nothing about their sedimentology, geometry, and petrography that points to a synsedimentary ("supratidal") origin for Muschelkalk dolostones. On the contrary, the distribution of these dolostones relative to hard-rock fractures (Fig. 19) is a convincing argument for late dolomitization. Is this late dolomite syngenetic (Fig. 16) possibly the product of refluxing from the Keuper sedimentary regime, or is it epigenetic, probably later than Keuper sedimentation and reflective of possible subsurface evolution of dolomitizing waters? The time of lithification (the maximum possible age of hard-rock fractures) becomes important in the discussion of whether the hard-rock fractures could have existed at the time of Keuper sedimentation and have influenced the course of refluxing waters.

As mentioned, the entire Buntsandstein-Keuper interval appears to be conformable, and at only one locality (at 160 ft in section 8, Fig. 3) did I discover a corrosion surface reflecting lithification and probable subaerial exposure of the Muschelkalk contemporaneous with its accumulation. In the absence of intra-Triassic unconformities, it is difficult to envision Muschelkalk carbonates sufficiently lithified to support joints at the time of Keuper sedimentation. Although a growing number of examples attests to submarine cementation of lime sands (Fischer and Garrison, 1967; Taft *et al.*, 1968; Land and Goreau, 1970; Shinn, 1969, 1970), evidence to date suggests that the lithification of lime mud requires subaerial exposure. Inasmuch as most of the Muschelkalk is lime mud, it is unlikely that hard-rock fractures in the Muschelkalk existed at the time of any possible Keuper refluxing.

A Keuper refluxing model for Muschelkalk dolomitization is further complicated by the relative impermeability of the Keuper. A hydrostatic head is sufficient for moving water through the Keuper (meteoric water is flowing in, through, and out of the Keuper today), but it is doubtful that simply the higher density of a surface brine at sea level would effect downward migration through this claystone. Unfortunately, not enough is known about the hydrology of the reflux model in various media.

Indications are that Muschelkalk dolostones are neither supratidal nor reflux in origin, and therefore are not related to marine brines. What then is the age of the dolomitization, and what was the origin of the dolomitizing water?

Interpretation of epigenesis—Dolomite both exhibits a distribution relation to joints (Fig. 19) and occurs as druse within joints and vugs. Still, the age of these fractures is problematic; specifically, it is not known whether they are older or younger than Tertiary mountain building. The presence of dolomite druse in certain vugs suggests at least some post-mountain dolomite precipitation. Figure 20 is a diagram of a sample collected from 70 ft above the base of section 10, where the beds are vertical. The orientation of the sample represented in Figure 20 is the same as it was in the field; the plane of the illustration is the plane of vertical stratification. Some vugs are filled with a combination of red detrital clay and calcite cement, and from geopetal structures it can be seen that sedimentation of the clay occurred after the beds acquired their present 90° dip (*i.e.*, post-mountain building). Between the vug wall and its clay-calcite contents is a druse of dolomite cement. An argument for post-structure dolomite would be more convincing were the dolomite druse between the clay and the calcite cement, but such is not the case. However, the vug "plumbing system" is not likely to have remained devoid of sediment and cement during the entire period of mountain building. For this reason I suspect that both the formation of the vugs and their subsequent filling by dolomite-clay-calcite were post-mountain building events.

Calcitization—Because of the abundant gypsum and anhydrite in the overlying Keuper, one might expect a considerable amount of "dedolomite" in the Muschelkalk. Surprisingly, dedolomite was discovered at only a single locality (Fig. 21). The late dolomite described earlier, together with the near absence of dedolomite, indicates a late and sustained magnesium/calcium ratio in excess of that expected from water in association with calcium sulfate-bearing rocks. For this reason water samples were collected and analyzed in an effort to measure the amount of magnesium that might be available to the Muschelkalk today.

For centuries the Spaniards have utilized the evaporite content of the Keuper in the manufacture of salt. Groundwater issuing from the Keuper drains into salinas, where evaporation promotes the precipitation of NaCl. Water was collected from six salinas, each of which is within a Keuper valley with an undetermined thickness of Keuper below. Figure 22 is a diagram illustrating the geologic-physiographic setting of all six water samples. Calcium and magnesium contents, which were determined by the atomic absorption method, are shown as Ca/Mg ratios in Figure 23. The high magnesium content, locally four times that of calcium, seems anomalous when the abundant calcium available to these waters from Keuper gypsum and anhydrite is considered. However, after a study of the petrography and distribution of Muschelkalk dolomites and dedolomites, the ionic composition of these water samples comes as no surprise; if it is assumed that these compositions are reflected in Muschelkalk diagenesis, they might explain both the late dolomitization and the absence of dedolomite in all but a single locality. Incidentally, that single locality where dedolomite was observed (Fig. 21)

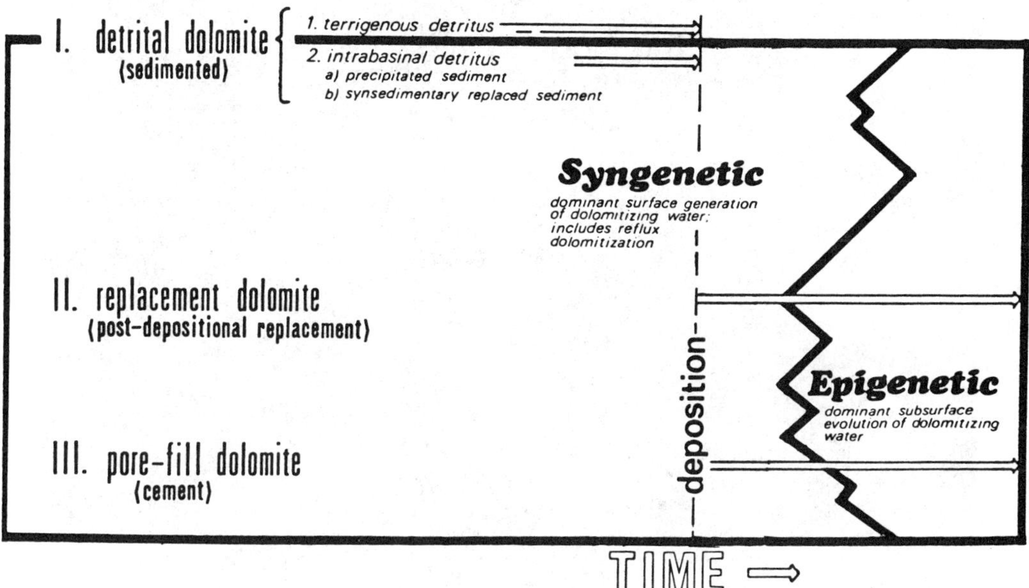

Fig. 16—Classification of products and geneses of dolomitization.

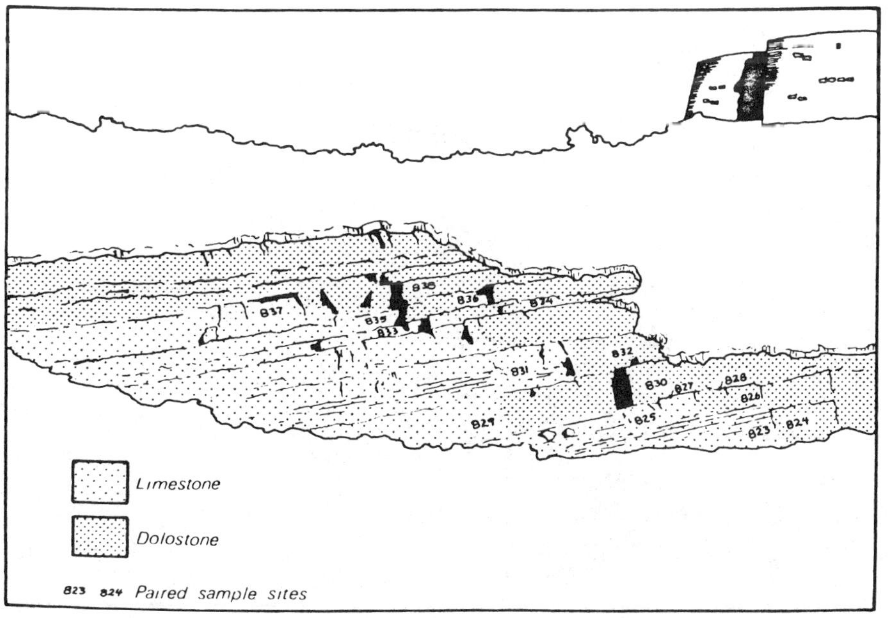

Fig. 17—Diagram made from field photograph of Muschelkalk in quarry near Sagunto showing discordant relation
of limestone and dolostone.

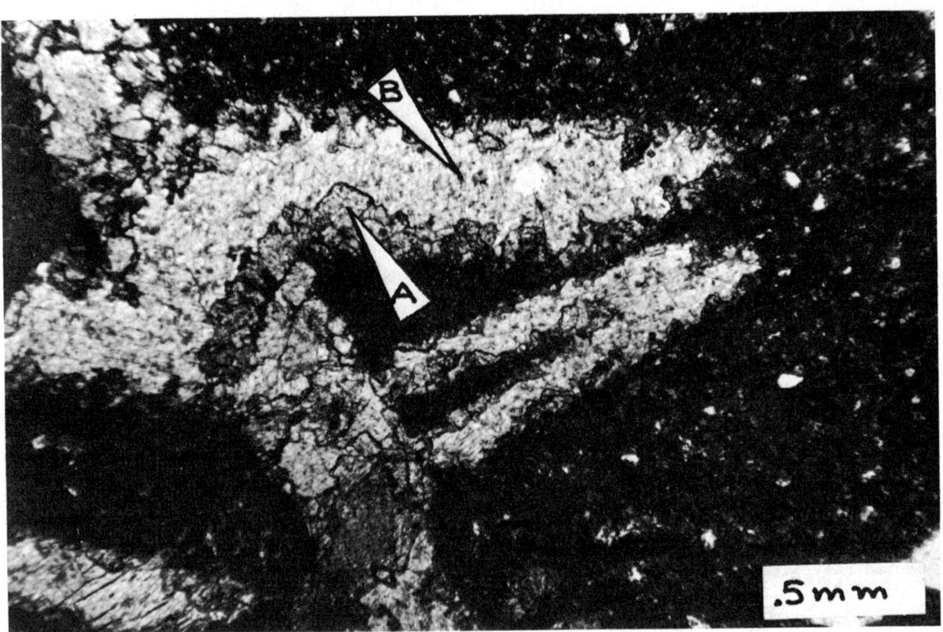

Fig. 18—Photomicrograph of gypsum mold filled with first-generation dolomite cement *(A)* and second-generation calcite cement *(B)*. Dolomite precipitation followed *solution*, not formation, of evaporite.

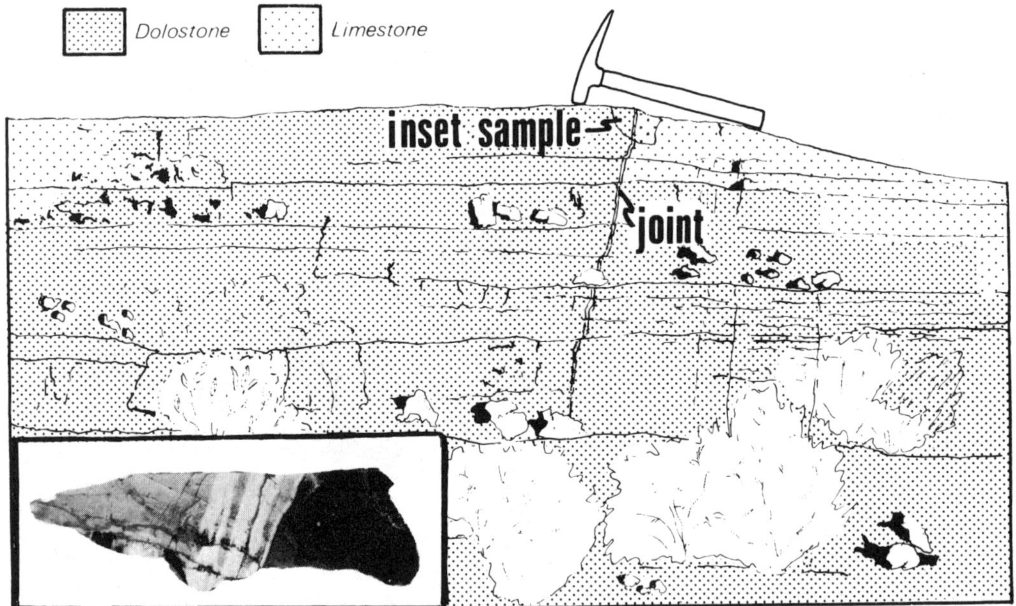

Fig. 19—Diagram from field photograph showing distribution of limestone-dolostone relative to joint. Near Sagunto. Sample taken from across "welded" joint was stained to corroborate field observation (insert).

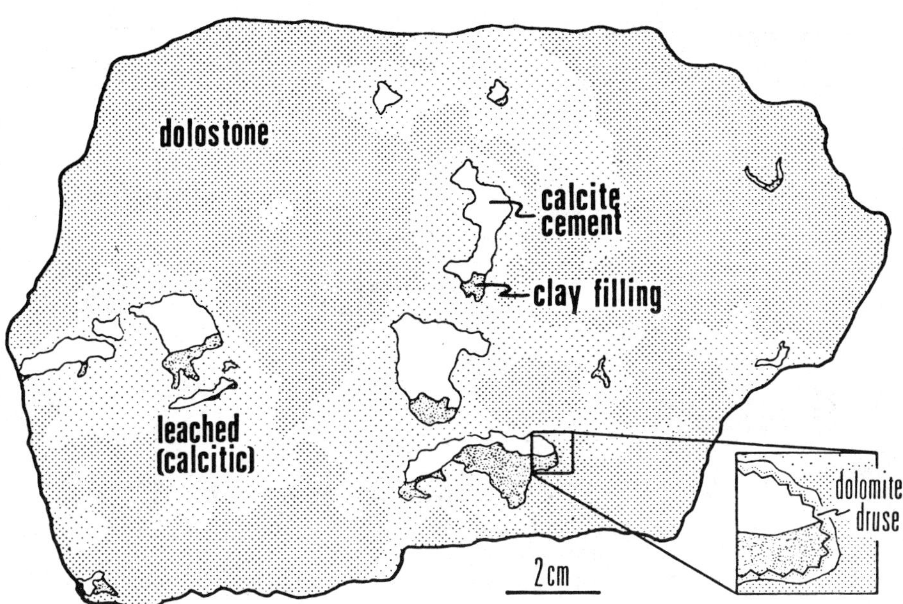

Fɪɢ. 20—Drawing of hand specimen (sample 10-70) showing geopetal structures of detrital clay and carbonate cements. Inset shows dolomite druse that is post-vug but pre-clay in origin. Plane of drawing is plane of stratification, which, in field, is vertical.

is very near Domeño, where the highest Ca/Mg ratio was found!

Surely the high magnesium values, relative to those of calcium, explain the occurrence of aragonite, which is so common within the Keuper[5] (Fig. 24). Murray (1954), in a notable effort to explain the occurrence of aragonite versus calcite in cave deposits, concluded that $CaCO_3$ precipitated from waters with Mg/Ca ratios in excess of 1.0 tends to be aragonite, whereas $CaCO_3$ precipitated from waters with lesser Mg/Ca values is typically calcite (Table 1).

Source of magnesium—Marfil Pérez's (1970) study of the Keuper shows that the only possible important source of magnesium in the Keuper is chlorite, which, according to her analyses, is the most abundant clay mineral in the formation. Figure 25 shows the occurrence of chlorite in a single sample of Keuper clay, and Figure 26 illustrates a second analysis of that same sample after treatment with ethylene glycol. The shifting of the 2ϕ spacing is an expression of the displacement of water in the chlorite by the ethylene glycol. The water in turn reflects a deficient interlayer Mg content, which is probably the result of earlier leaching. However, just how much magnesium has been leached, and thereby made available for Muschelkalk dolomitization, would be difficult to determine.

Conclusions

Highly soluble magnesium, presumably fixed diagenetically in the Keuper and held loosely in the chlorite, probably is responsible for (1) the anomalously low Ca/Mg ratios from this evaporite-rich formation, (2) the precipitation within the Keuper of aragonite rather than calcite, (3) the late dolomitization of Muschelkalk limestones, and (4) the near absence of dedolomite in the Muschelkalk.

The ionic composition of Keuper water (Table 2) is suggestive of seawater, but it is more logically meteoric in origin. For this water to be entombed seawater, the rate of its bleeding away from the Keuper would have to be slower than the rate of physiographic dissection (Fig. 22), which is not tenable, and there is no apparent hydrologic head that could force connate water up from deeper reservoirs. Instead, "dehydrated seawater," stored in the mineral phases of the Keuper, has been remobilized and made effective in Muschelkalk diagenesis. Granted, the concentrations presented in Table 2 are produced partly

[5] The name "aragonite" is taken from the region of Aragon, which includes a large part of the area of study. See Marfil Pérez (1970) for data on the distribution of aragonite in the Keuper. See Sánchez de la Torre and Agueda Villar (1970) for a lithofacies map of the Keuper in the western part of the present study area.

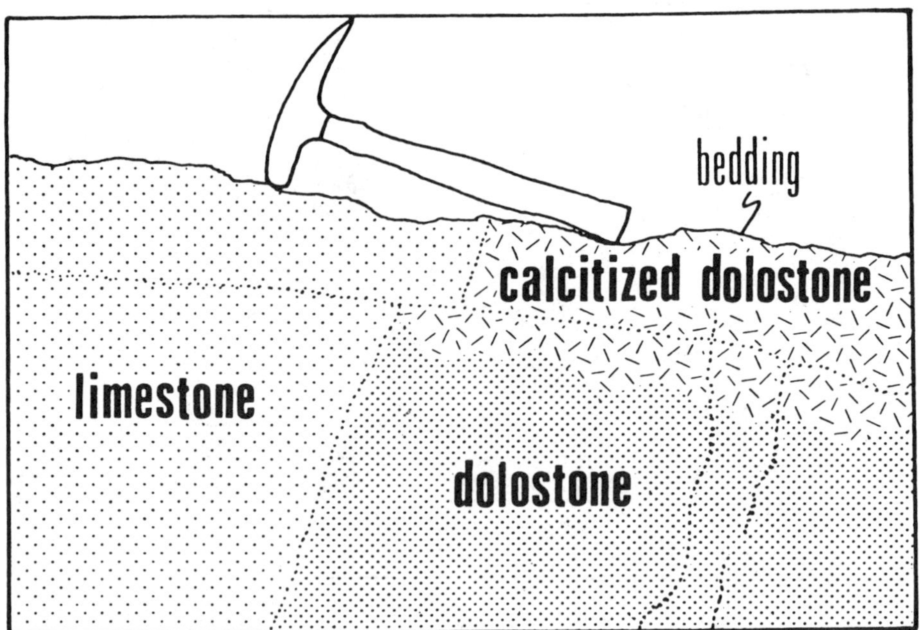

Fig. 21—Diagram showing joint-controlled dolostone (right) that is partly calcitized (upper-right). Sample from near Domeño.

by surface evaporation, but the salt-manufacturing industry is effective only because the waters are brines in the subsurface. Nowhere in the areas of salinas is the groundwater potable.

Inasmuch as Muschelkalk dolomitization is related to the subsurface evolution of a high Mg/Ca brine, the term "epigenesis" applies (Fig. 16). However, the latest event that can be detected petrographically is the precipitation of *calcite*, not dolomite (*e.g.*, Figs. 18, 20). Therefore, I suggest that the dolomitizing potential of Keuper water was aided prior to unloading by geothermal gradient, and that subsequent lowering of the temperature, through erosion, shifted the geochemical environment from the stability field of dolomite to that of calcite. An example of this mechanism, documented by fluid inclusion analyses, can be found in examining Tables 5 and 8 of Schmidt (1962), where the temperature ranges of dolomite and calcite in the Miami-Picher district are 76°-100°C and 52°-79°C respectively. The temporal relation, dolomite-calcite, is suffi-

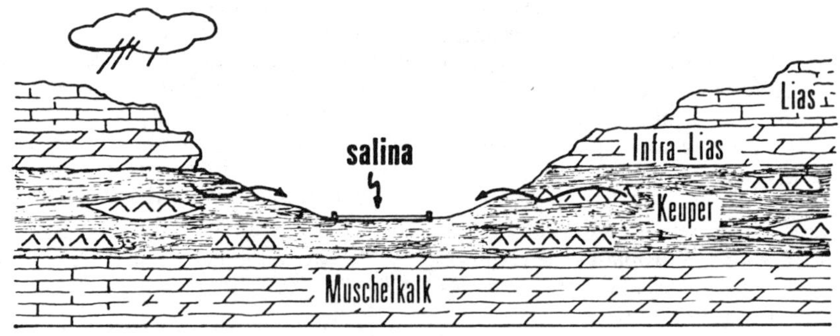

Fig. 22—Diagram showing geologic-physiographic setting of Keuper salinas. Most water supplied to salinas, rather than surface runoff, oozes from Keuper (arrows).

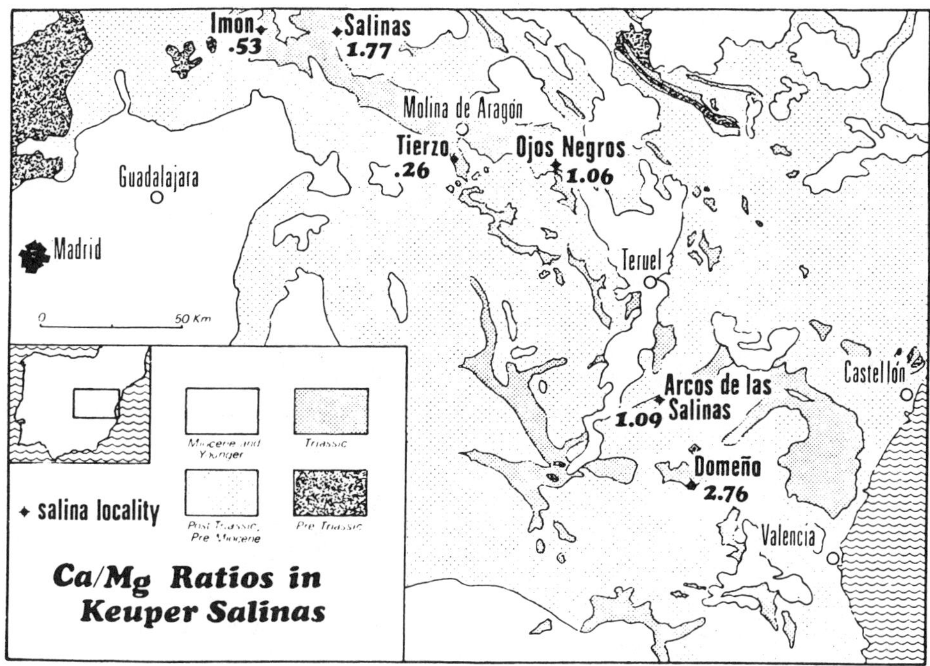

FIG. 23—Ca/Mg ratios in water samples from Keuper salinas. Base map from España Instituto Geológico y Minero (1958).

FIG. 24—Aragonite crystals from Aragon region, within area of report.

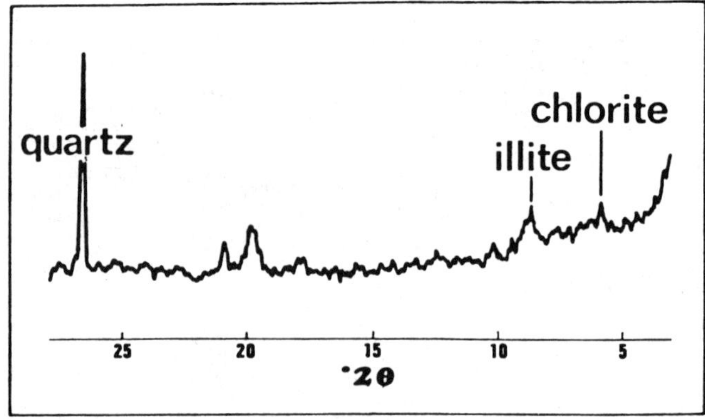

FIG. 25—X-ray diffractogram showing presence of chlorite in sample of Keuper clay.

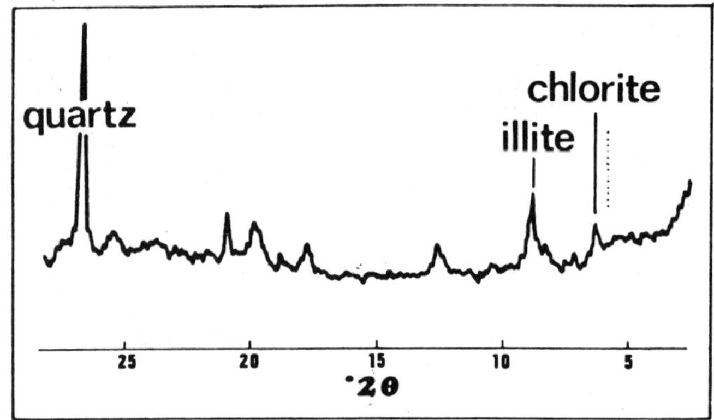

FIG. 26—Same sample as that represented in Figure 25 after treatment with ethylene glycol.

Table 1. Precipitation of Aragonite and Calcite from Cave Waters*

Analyses of Cave Waters (ppm)			Associated Mineral Precipitate
Sulfate	Calcium	Magnesium	
38	27	27	aragonite
	26	24	aragonite
	22	27	aragonite
27	31	30	aragonite
8	46	35	calcite
27	28	15	calcite
15	64	33	calcite
15	31	28	calcite
35	27.5	22	calcite
	23.6	13.4	calcite

* From Murray (1954).

Table 2. Ionic Composition of Waters from Keuper

Sample Locality	Concentration (ppm)			
	Chloride	Magnesium	Calcium	Sodium
Imón	108,500	1,025	905	65,500
Salinas	61,500	725	2,120	37,500
Tierzo	88,000	1,025	440	49,000
Ojos Negros	67,500	1,075	1,880	41,500
Arcos de las Salinas	155,300	1,250	2,250	82,500
Domeño	350	101	460	245

ciently fundamental to suggest that an important part of vug and fracture cementation is, in general, an unloading process.

REFERENCES CITED

Dunham, R. J., 1962, Classification of carbonate rocks according to depositional texture, in Classification of carbonate rocks: Am. Assoc. Petroleum Geologists Mem. 1, p. 108-121.

España Instituto Geológico y Minero, 1958, Edición del mapa geológico de España y Portugal peninsulares, Baleares y Canarias.

Fischer, A. G., and R. E. Garrison, 1967, Carbonate lithification on the sea floor: Jour. Geology, v. 75, p. 488-496.

Friedman, G. M., and J. E. Sanders, 1967, Origin and occurrence of dolostones, chap. 6 in Developments in sedimentology—9A, Carbonate rocks—origin, occurrence and classification: Amsterdam, Elsevier Pub. Co., p. 267-348.

Garrison, R. E., and A. G. Fischer, 1969, Deep-water limestones and radiolarites of the Alpine Jurassic, in Depositional environments in carbonate rocks: Soc. Econ. Paleontologists and Mineralogists Spec. Pub. 14, p. 20-56.

Hallam, A., 1967, Sedimentology and palaeogeographic significance of certain red limestones and associated beds in the Lias of the Alpine region: Scottish Jour. Geology, v. 3, pt. 2, p. 195-219.

Hermite, H., 1879, Etudes géologiques sur les îles Baléares. Première partie Majorque et Minorque: Paris, Impre. Becquet, 362 p.

Hollister, J. S., 1942, La posicion de las Baleares en las orogenias varisca y alpina: p. 71-102 in Publicaciones alemanas sobre geologia de España, t. 1: Consejo Superior Investigaciones Científicas, 168 p.

Kent, P. E., 1969, The geological framework of petroleum exploration in Europe and North Africa and the implications of continental drift hypotheses, in The exploration for petroleum in Europe and north Africa: London, Inst. Petroleum, p. 3-17.

Land, L. S., and T. F. Goreau, 1970, Submarine lithification of Jamaican reefs: Jour. Sed. Petrology, v. 40, p. 457-462.

Laporte, L., 1964, Supratidal dolomitic horizons within the Manlius Formation (Devonian) of New York: Geol. Soc. America Guidebook no. 2, Miami Mtg., p. 59-66.

Marfil Pérez, R., 1970, Estudio petrogenético del Keuper en el sector meridional de la Cordillera Ibérica: Estudios Geol., v. 26, p. 113-161.

Murray, J. W., 1954, The deposition of calcite and aragonite in caves: Jour. Geology, v. 62, p. 481-492.

Perconig, E., 1962, Sur la constitution géologique de l'Andalousie occidentale, en particulie du bassin du Guadalquivir, in Livre a la mémoire du Professeur Paul Fallot: Soc. Géol. France, t. 1, p. 229-256.

—— 1968, Microfacies of the Triassic and Jurassic sediments of Spain: International Sedimentary Petrographic Series, v. 10, Leiden, E. J. Brill, 63 p.

Pittman, J. S., and R. L. Folk, 1970, Length-slow chalcedony: a new testament for vanished evaporite minerals (abs.): Geol. Soc. America Abs. with Programs, v. 2, p. 654-655.

Querol, R., 1969, Petroleum exploration in Spain, in The exploration for petroleum in Europe and North Africa: London, Inst. Petroleum, p. 49-71.

Riba, O. A., 1959, Estudio geológico de la Sierra de Albarracin: Sec. Geomorfología de Barcelona, Inst. Estudios Turolenses, Monografías 16, 283 p.

Rutten, M. G., 1969, The geology of western Europe: Amsterdam, Elsevier Pub. Co., 520 p.

Sánchez de la Torre, L., and J. A. Agueda Villar, 1970, Paleogeografia del triásico en el sector occidental de la Cordillera Ibérica: Estudios Geol., v. 26, p. 423-430.

Sander, B. K., 1936, Contributions to the study of depositional fabrics; rhythmically deposited Triassic limestones and dolomites; translated by E. B. Knopf: Am. Assoc. Petroleum Geologists (pub. 1951), 207 p.

Schmidt, R. A., 1962, Temperatures of mineral formation in the Miami-Picher district as indicated by liquid inclusions: Econ. Geology, v. 57, p. 1-20.

Shearman, D. J., 1966, Origin of marine evaporites by diagenesis: Inst. Mining and Metallurgy Trans., v. 75, sec. B, Bull. no. 717, p. 208-215.

Shinn, E. A., 1969, Submarine lithification of Holocene, carbonate sediments in the Persian Gulf, in Lithification of carbonate sediments, 1: Sedimentology, v. 12, p. 109-144.

—— 1970, Submarine formation of bored surfaces (hardgrounds) and possible misinterpretation in stratigraphic applications (abs.): Am. Assoc. Petroleum Geologists Bull., v. 54, p. 870.

—— R. M. Lloyd, and R. N. Ginsburg, 1969, Anatomy of a modern carbonate tidal-flat, Andros Island, Bahamas: Jour. Sed. Petrology, v. 39, p. 1202-1228.

Simancas, R., 1968, Geologia de los alrededores de Grado del Pico, Segovia: Thesis, Madrid Univ., 125 p.

Taft, W. H., F. Arrington, A. Haimovitz, C. MacDonald, and C. Woolheater, 1968, Lithification of modern marine carbonate sediments at Yellow Bank, Bahamas: Bull. Marine Sci., v. 18, p. 762-828.

Virgili, C., 1958, El triásico de los Catalanides: España Bol., Inst. Geol. y Minero, t. 69, 856 p.

West, I. M., A. Brandon, and M. Smith, 1968, A tidal flat evaporite facies in the Visean of Ireland: Jour. Sed. Petrology, v. 38, p. 1079-1093.

Wilson, J. L., 1969, Microfacies and sedimentary structures in "deeper water" lime mudstones, in Depositional environments in carbonate rocks: Soc. Econ. Paleontologists and Mineralogists Spec. Pub. 14, p. 4-19.

22

Reprinted from *Econ. Geology* **64**:743–754 (1969)

The Origin of Hydrothermal and Low Temperature Dolomite[1]

T. S. LOVERING

Abstract

Laboratory data and field evidence provide compositions of water that are in equilibrium with dolomite and calcite from temperatures of about 20° C to 400° C. From these data it is possible to construct a chart showing the stability fields of calcite and dolomite in dilute solutions as a function of temperature and Ca/Mg molar ratios; this chart indicates that most non-saline groundwaters have insufficient magnesium in them to allow the formation of dolomite unless they are abnormally hot, or have passed through a magnesium silicate environment. Nevertheless, in some areas present groundwaters contain substantial amounts of magnesium and the Ca/Mg molar ratios are such as to indicate that limestone would be converted into dolomite at higher temperatures. If such areas were above or near intrusive centers or shallow hearths where magma laid quiescent for some time, the rocks and their contained waters should have heated up as the temperature gradient between surface and hot intrusion became stabilized, and a hyperthermal area such as we find in many parts of the world today would result. A convective circulation would start in permeable rocks and this would provide large volumes of magnesia for conversion of limestone to dolomite, where the ratio of calcium to magnesia was appropriate. Age relations of hydrothermal dolomite in various Cordilleran districts show that the dolomitic alteration in nearly all is quite early, and this would suggest that it formed in response to the initial rise of magma in the district well before ores as such were deposited at the horizon of the dolomite. Most of the puzzling features of the hydrothermal dolomite associated with ore deposits are readily explained if the source of the magnesia is primarily groundwater, the composition of which varies from district to district. The difficulty of providing an adequate source of magnesia from magmatic differentiation in rocks that contained very little magnesia is avoided. It also explains the occurrence of hydrothermal dolomite in areas remote from ore deposits but in which there are deep reaching fractures or other permeable channels that could conduct hot groundwater from depth.

Introduction

THE origin of dolomite has been debated for well over a century, and there is still no widespread agreement as to the chemistry involved in its formation. New evidence from the laboratory and the field suggests that we may be approaching a solution of the problem. Evidence will be presented to indicate that limestones are converted to dolomite wherever they are in contact with circulating groundwater saturated with respect to calcite and having a molar calcium/magnesium ratio appropriate to the temperature; this molar ratio ranges from about 0.15 at 20° C to 35.0 at 420° C. The writer concludes that the variable but locally widespread hydrothermal dolomite in areas near ore deposits or mineralized districts was caused by groundwater heated to the appropriate temperature early in the igneous cycle to which the ore deposits were genetically related.

A summary of the development of ideas prior to 1914 concerning the origin of different types of dolomite is presented in an excellent review article by Van Tuyl (1914). Van Tuyl's paper contains more than 250 bibliographic references and is a scholarly review of the evidence for and against the various theories of origin that had been proposed up to the date of its publication. He was, however, concerned chiefly with the mode of origin of extensive beds of dolomite rather than with smaller areas of dolomite formed by hydrothermal solutions. These smaller areas were the subject of an excellent paper by Hewett in 1928, which will be considered later. Van Tuyl classifies the theories of the origin of dolomite in a systematic way as follows:

1. Primary deposition theories:
   a. The chemical theory
   b. The organic theory
   c. The elastic theory

<hr>

[1] Low temperature is a term used to designate those underground solutions that are relatively cool as compared with hydrothermal solutions. The term "relatively cool" as employed by the writer means temperatures below those that will allow organic life to exist, or below the boiling point of water at geologically low pressures, approximating a maximum temperature of 100° C.

2. Alteration theories:
   a. The marine alteration theory
   b. The ground water alteration theory
   c. The pneumatolytic alteration theory
3. Leaching theories:
   a. The marine leaching theory
   b. The surface leaching theory

Van Tuyl concludes that most stratified dolomites result from the alteration of limestone or limy sediments (p. 397). He notes the possibility of direct chemical precipitation of dolomite from the sea on a small scale and the possibility that impure dolomitic limestones of minor importance may represent clastic deposits. He finds little or no evidence that organisms give rise to other than very slightly magnesian limestones, and he discounts the probability that marine and surface leaching of such limestones produce dolomite, but this latter conclusion has been questioned by several sedimentologists since. Goodell and Garman, 1969, who report results of drilling a deep test well in the Bahamas, found that magnesian calcites occur at shallow depths; they conclude that the dolomites which dominate below 400 feet all formed by cannibalization of more unstable carbonate phases during uplifts when they were exposed to subaerial conditions. However, as discussed further under "Experimental work," it would be magnesium that was leached during such subaerial stages to enrich the groundwaters in Mg^{2+} while the calcite was removed mechanically.

The evidence available suggested to Van Tuyl that the alteration of limestone to dolomite has largely occurred beneath the sea and that dolomitization by ground water is very local and very imperfect. In discussing the time and place of dolomitization, Van Tuyl notes that in the imperfectly altered limestone, dolomite follows original lines of weakness rather than joints or fractures, the presence of perfect rhombs of dolomite in partly altered limestone suggests the latter had not yet solidified when dolomitization took place; the widespread extent and uniform composition of many dolomites show they were formed by an agent capable of operating uniformly over wide areas such as the ocean; that an adequate source of magnesium for transforming extensive limestone formations into dolomite is only to be found in the sea, which contains many times as much of this constituent as ordinary ground water (413 ppm Ca, 1294 ppm Mg according to Culkin *in* Riley and Skirrow, 1965, p. 122); that many dolomites are directly and regularly overlain by pure limestone or by thick shale beds, which Van Tuyl interpreted to mean that they were formed before the overlying beds were deposited. He then notes that "dolomites of minor importance such as those associated with ore deposits" and others clearly related to fractures,

must have been formed through the agency of unusual ground water. Because present day ground water and river water are almost uniformly low in magnesium and contain much more Ca than Mg, he concludes that such waters could not dolomitize limestone. In the case of mineralizing springs and mineralizing solutions related to ore deposition, it was conceivable to Van Tuyl that magnesium would be present in sufficient proportions to accomplish local dolomitization and doubtless most vein dolomites were so formed.

Returning again to the eight theories of deposition, Van Tuyl's work suggests that chemical precipitation may be a minor source, that clastic dolomite is present in few localities, that there is no evidence for the organic theory, that the marine leaching theory, the surface leaching theory, and the pneumatolytic alteration theory are highly improbable and lack evidence to support them. This leaves only the two alteration theories as worthy of consideration for extensive masses of bedded dolomite. Van Tuyl, in common with many other early investigators, tried in vain to make dolomite in the laboratory and his work casts doubt on the reported synthesis of dolomite in the laboratory by earlier investigators. In fact, we are still waiting for evidence of dolomite precipitated in the laboratory at room temperature. Van Tuyl notes, however, several instances where the evidence indicates that dolomitization has taken place recently in closed lagoons, where the sea water is concentrated, and also in the outer part of coral reefs facing the open sea. He was puzzled, however, over the chemistry of dolomitization by solutions so low in magnesium as those of sea water and the inability of all experimenters to precipitate dolomite in similar or much more concentrated solutions in the laboratory.

The idea that dolomite formed by alteration of calcite sediments or limestone through the agency of sea water was first suggested by Dana in 1843. Dana, however, ascribed the dolomitization to recurrent heating of the sea water by igneous agencies and this, in a sense, anticipated the theory of hydrothermal origin of some dolomites. The possibility of dolomitization of sediments in modern times has received attention recently and the evidence for the process is persuasive. It is interesting to note that the chemical evidence recently acquired suggests that the waters may become highly super-saturated with magnesium before dolomitization starts, but once it begins, the ratio of Ca to Mg rises continually until the Ca to Mg value is between 0.3 and 0.2, equivalent to molar ratios of 0.18 and 0.12, see Kinsman (1966). Deffeyes, Lucia and Weyl (1965, p. 86) suggest that the Ca/Mf ratio approximates 1/30 before reaction takes place and that as calcium carbonate is dolomitized, the ratio rises steadily to approximately 1 to 5.

Chemical analysis of the waters in lagoons showed that the dolomite was precipitated from sea water concentrated to the point where calcium-carbonate and gypsum had been precipitated, thus lowering the Ca/Mg ratio. Somewhat similar results were obtained by Kinsman (1965, 1966), who studied dolomitization and the development of evaporites in the lagoonal sediments in the southern part of the Persian Gulf, where he found temperatures of about 34 degrees at depths of 1 to 4 feet. Evaporation losses from the upper sediment surface are balanced by inward movement of lagoonal waters and as the brine moves inward, gypsum is first precipitated, thus lowering the ratio of Ca to Mg to values near 0.09 or less. Dolomitization of the dominantly organic sediments then begins and produces early diagenetic prelithification dolomite and very locally magnesite also.

Illing, Wells and Taylor (1965), who also studied dolomitization of recent sediments in the Persian Gulf, report that oceanic pore water in carbonate sand at the edge of a low swamp beach (a sebkha) on the Persian Gulf has a Ca/Mg ratio of about 0.33; but this ratio decreases across the lagoonal flats in the calcite sediments to nearly 0.08, and then rises in a short space as one proceeds farther inland and dolomite appears in the sediments. The salinity and the ratio of Ca to Mg in the pore solution gradually increases landwards, and for some distance dolomitization of $CaCO_3$ takes place where the Ca/Mg weight percent ratio is about 0.16. The ratio of Ca/Mg rises to 0.33 (a molar ratio of 0.19) in the much more concentrated pore solution on the landward side of the lagoonal flats. In the area where Ca/Mg = 0.33 the pore water is concentrated to about 15% chloride.

These three papers agree that the brines become more concentrated than sea water before dolomitization begins and that the critical ratio of Ca to Mg (weight percent) at a temperature of about 35° C is apparently no more than 0.33 and no less than 0.16, and it would seem that a ratio of about 0.33 weight percent or 0.20 molar percent was the approximate ratio for dolomitization to end in solutions of moderately concentrated sodium-chloride brine, at temperatures near 35° C.

Formation of dolomite in South Australia at still lower temperatures is recorded by Skinner (1963). According to Mrs. Skinner (p. 461) calcian dolomite (or protodolomite") precipitates together with magnesian calcite at a temperature of about 18° C. Where only calcite precipitates, the Ca/Mg ratio is more than 0.27 weight percent (0.16 molar percent);

where calcium-carbonate and dolomite co-precipitate, the ratio of Ca/Mg is 0.27 or less.

The classic paper on hydrothermal dolomitization is undoubtedly that of Hewett's (1928). He notes that dolomite as a characteristic alteration of limestone adjacent to ore deposits has been known for many years and was probably first reported by Schmidt in 1875. Since then it has been observed in many places both in ore deposits of the Mississippi Valley type and associated with undoubtedly hydrothermal ore deposits in the Rocky Mountains and in Europe. Many of these occurrences are described by Hewett and need not be discussed in detail here. In most of the districts where the limestones have been altered to dolomite near ore, the chief metals mined are zinc and lead but in a few of them, copper is important also. The limestone walls are altered to nearly pure dolomite in many districts but not in all. The zone of transition between limestone and dolomite is commonly narrow and sharply defined. Traces of small fossils and even of larger fossils are commonly destroyed. In these districts the dolomite commonly contains more iron than the original limestone, and it is also noteworthy that the porosity of the dolomite is little different from that of the limestone unless subsequently leached, although a mole per mole replacement of magnesium for calcium should result in an increase in porosity of nearly 13 percent. Sulfides commonly replace dolomitized limestone but are also deposited in open spaces within breccias made up of dolomitized limestone. It is clear that at such places dolomitization distinctly precedes deposition of sulfide even though some dolomite crystallizes in open spaces after deposition of the sulfides.

The fact that some ore bodies have an envelope of dolomitized limestone and others do not suggests that the process of ore deposition and dolomitization do not coincide in space and time, and raises the question, therefore, whether the source of magnesium is the same as the source of the metals. The metals are usually ascribed to igneous emanations in the typical hydrothermal veins of the Rocky Mountain province but as Hewett notes, the relations to these ore deposits clearly indicate pre-ore hydrothermal solutions were the agent of dolomitization. No igneous rocks are known in the Mississippi Valley lead-zinc areas which could supply such hydrothermal solutions, and this is also true of the Upper Silesian lead-zinc deposits of Poland. Furthermore, in the hydrothermally dolomitized limestones of many of the Rocky Mountain deposits, the dolomitization is widespread not only near the intrusions but as much as 60 miles away (Banks, 1967). As many of the intrusions in mineralized regions are highly sericitized, Hewett suggests that the magnesium removed during such alteration might be the source of magnesium for dolomitized limestones, but at the same time, he points out that about ten volumes of igneous rock would have to be sericitized to supply one volume of dolomite. He then gives instances where many cubic kilometers of hydrothermal dolomite have been formed and recognized the difficulty of finding an adequate hydrothermally altered source rock for a replacement of this magnitude. Hewett notes that where magnesite appears to replace limestone near intrusives, there is commonly a zone of dolomite between the limestone and the magnesite which lies next to the intrusive rock. Hewett reflects the general agreement of earlier students of hydrothermal dolomite that magnesia must be derived from deep sources closely related to the intrusives which, however, show a wide range in composition; intermediate types such as monzonite are the most common. He considers altogether five sources of magnesium. 1. The waters of the sea. 2. The shell of sedimentary rocks. 3. The underlying crystalline rocks such as gneisses. 4. The shallow bodies of intrusive rocks. 5. The deeper magma sources.

1. Hewett states (and it is now generally agreed) that sea water produces dolomite either by precipitation or by replacement which accounts for the so-called syngenetic dolomites that show 11 to 13% porosity.

2. Hewett accepts Siebenthal's (1916) explanation of the origin of the dolomitic replacement of limestone associated with the zinc and lead deposits of the Mississippi Valley type ore deposits. Siebenthal pointed out that the deep well waters of the region derived from Cambrian and Ordovician dolomite contain much more magnesium in proportion to calcium than do waters from the Mississippian limestone. The molar ratio of Ca²⁺ to Mg²⁺ is approximately 1.0 (requiring solutions at a temperature of about 100° C to dolomitize a limestone; see fig. 1, T.S.L.) and the total of Ca + Mg ranges up to several hundred parts per million. Siebenthal ascribed the magnesium needed to dolomitize the limestone to solution of bedded dolomites at depth and transportation by deeply migrating ground water. Hewett accepted this explanation for the Mississippi Valley type of ore deposits but rejected it for Goodsprings, Nevada, and Iglesias, Sardinia, because he felt the structural situation was unfavorable to artesian circulation and to a long distance lateral migration of magnesia in ground water. Hewett also noted that elimination of magnesia from limestone or dolomite occurs wherever it is replaced by iron, silica, or other substances and that the resulting magnesia entering either hydrothermal or normal ground water circulations would then be available to dolomitize rocks at higher levels.

3. The possibility that a "basic front" emanating from the metamorphism of schists and gneisses with the elimination of pore water was regarded by Hewett as most unlikely and rejected altogether in favor of other sources of magnesium.

4. Hewett states that the relationship of altered and unaltered shallow bodies of intrusive rocks to dolomite shows that pre-dolomitization intrusions are "either sparse or absent"; in my own experience the term "sparse" could be omitted. Widespread hydrothermal dolomite associated with mining districts seems to be everywhere earlier than associated igneous rocks and therefore, they could not have been the source of the dolomitizing fluids.

5. Hewett appeals to deeper magma sources as the most probable parents of dolomitizing solutions and strongly advocates the hydrothermal alteration of deep igneous rocks as the most likely source of magnesium for hydrothermal dolomite. He thus concludes that the dolomites associated with ore deposits have two types of genesis—(1) the magnesian ground waters that he and Siebenthal assume account for dolomitization of the Mississippi Valley type of deposit and, (2) the hydrothermal solutions that have altered igneous rocks in mining districts and leached Mg from them at depth and were thus able to carry magnesium to higher levels.

Experimental work on the solution and synthesis of dolomite.—Since the time of Van Tuyl's and Hewett's papers, dolomite has been successfully synthesized at high temperatures and some of the chemistry of its formation has been clarified. Work done by Holland and his associates (1966 and 1967) shows that the fields of calcite, dolomite, and magnesite are well defined at high temperatures and that the molar ratio of Ca²⁺ to Mg²⁺ in the solutions, i.e., their activities, is of paramount importance in determining the boundaries of these fields at different temperatures. The work of Rosenburg, Burt and Holland (1967) shows that the molarity of the solution is also of increasing importance at lower temperatures in the range they studied (420° to 290° C) but that the difference between 0.16 molar and 1 molar solutions is less than the difference between 1 molar and 2 molar solutions. Indeed, the sparse experimental data shown for the 0.16 and 0.5 molar solutions are compatable with identical Ca²⁺/Mg²⁺ ratios for calcite-dolomite stability in the 1.0 molar solutions.

Rosenburg et al. (1967, p. 395) derived equations for the calcite-dolomite boundary (1.1) and the dolomite-magnesite boundary (1.2) for 1 molar Ca/Mg solutions as shown below:

where $T = 295°\text{–}420°$ C:

$$\log (M \text{ Ca}^{2+}/M \text{ Mg}^{2+}) = 3.15 - 1.14 \times 10^3/°K \ldots \ldots 1.1;$$

$$\log (M \text{ Ca}^{2+}/M \text{ Mg}^{2+}) = 3.21 - 1.66 \times 10^3/°K \ldots \ldots 1.2.$$

These equations plot as straight lines on semilog paper, see Figure 1; if the T-C relations at lower temperatures as described for South Australia and the Persian Gulf are plotted, it is plain that for dilute chloride solutions having concentrations of

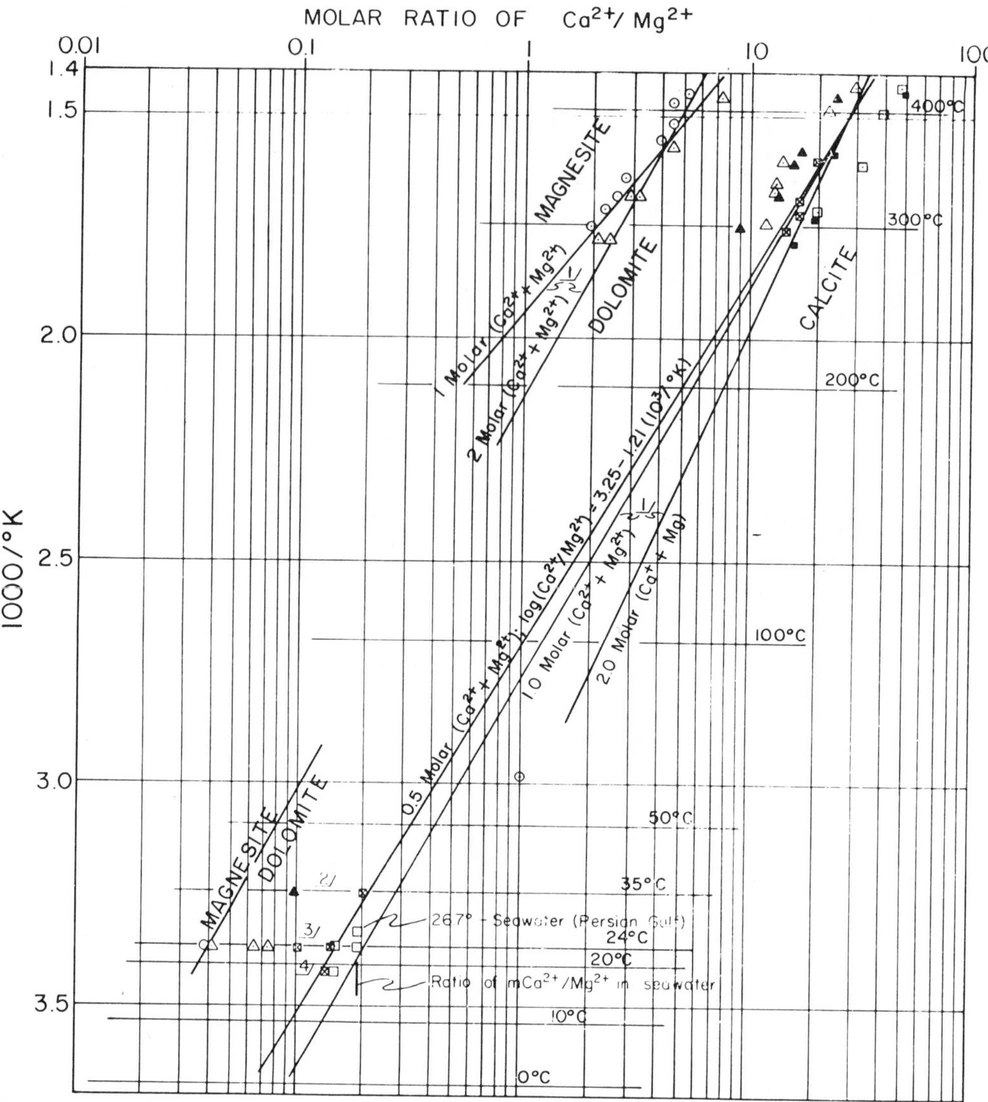

Figure 1. Molar ratios of Ca²⁺/Mg²⁺ chloride solutions precipitating, replacing, or in equilibrium with calcite, dolomite and magnesite, 10°C to 450°C.

<u>EXPLANATION</u>

▫ Calcite only
■ Dolomite replaced by calcite (0.5 Molar)
⊠ Calcite and dolomite or protodolomite
△ Dolomite only
▲ Calcite replaced by dolomite (0.5 Molar)
○ Magnesite

1/ Data at 290°C – 420°C from Rosenberg and Holland (1964 & 1967)
2/ Data at 35°C from Illing, Welk and Taylor (1965)
3/ Data at 24°C from von der Borch (1965)
4/ Data at 18°C from Skinner (1963)

Open symbols above 290° represent 1.0 *M* (Ca+Mg) solutions; open symbols below 35° represent natural brines less than 0.2 *M* (Ca+Mg).

Table 1 (of table 5, and Figure 2 von der Borch, 1965)

	pH range Apr.-Oct.	Weight percent Mg/Ca	*ionic ratio Ca^{2+}/Mg^{2+}
Aragonite+magnesian calcite	8.2-8.5	2.5-4.0	0.24-0.5
Magnesian calcite	7.8-8.4	2.5-5.0	0.24-0.12
Protodolomite+magnesian calcite	7.6-8.9	4 -6	0.15-0.10
Ordered dolomite	8.4-9.8	8 -9	0.075-0.067
Dolomite+magnesite	7.9-10.4	(16)	0.038
Aragonite+hydromagnesite	8.4-8.9	(20)	0.030

*Calculated by T. S. L.

$Ca + Mg$ less than 1 M, the experimental data at high temperature can be used to generate a line separating the calcite and dolomite fields at low temperatures also if we assume than M Ca/M Mg $\simeq M$ Ca^{2+}/M Mg^{2+} as is assumed for the experimental points.

The data on the composition of the solutions and the solid phases in the Coorong area of South Australia is excellent (Skinner, 1963; von der Borch, 1965), but the temperature data are meager. Von der Borch (1965, p. 795) shows the approximate fields of stability of the diagenetic and precipitating minerals as a function of the Mg/Ca ratio (in weight percent of the total salinity) and states that the average summer temperature is about 75° F (24° C). These data are shown in Table 1 and Figure 1 in the equivalent M Ca^{2+}/M Mg^{2+} ratios.

Aragonite and hydromagnesite formed in a saline lake where the weight percent of Mg/Ca = 20 ($Ca^{2+}/Mg^{2+} = 0.03$), but in another place brine in contact with a dolomite-magnesite pair had a molar ratio of $Ca^{2+}/Mg^{2+} = 0.04$; the aragonite-hydromagnesite pair probably represents a metastable phase. Kinsman (1966) found the same pair precipitated early in the sequence of evaporite minerals formed at the shore edge of the Trucial Coast of the Persian Gulf. Aragonite and gypsum formed as the ratio of Ca^{2+}/Mg^{2+} changed from about 0.25 to 0.09, when magnesite appeared. Inland from this point the Ca/Mg ratio increased steadily to about 0.33, and dolomite was the only magnesian mineral precipitated. The air temperatures there average 28° C (ranging from 5°–47° C), and the ground water at 2 to 4 feet depth in the sebkha averages 34° C (annual range 26°–39° C) about 5° or 6° warmer than the sea. It would seem that the isolated aragonite-hydromagnesite pair of von der Borch also represents a transient metastable recent precipitate from magnesian brines having little calcium.

Studies of Persian Gulf sebkha sediments on the west coast of the Qatar Peninsula (Illing, Wells, and Taylor, 1965) about 20 miles southeast of Bahrein Island, and about 200 miles northwest of the Trucial Coast area of Kinsman show clear evidence of dolomitization of calcareous sediment which precipitated in a lagoonal evaporite environment. The mean annual temperature at Bahrein is (H.O. Pub 62, 1960) 79° ($\approx 26°$ C) and as in so many other areas of the world, this temperature corresponds to the average surface temperature in the latter part of October or the first of November, see Lovering and Goode, 1960, p. 21 and 40.

The mean annual temperature of the coastal waters of the Persian Gulf on the west coast of the Qatar Peninsula as interpreted from the charts for February (66° F), May (81° F), August (93° F), November (79° F) of U.S.N. Hydrographic Office Pub 62 (1960) would closely approximate 80° F, or 26.7° C. This is the temperature assumed in Figure 1 for the precipitation of calcite at or below sea level, where the Ca^{2+}/Mg^{2+} molar ratio is that of sea water or 0.19. Illing, Wells, and Taylor (1965) record only a few isolated temperatures in their studies but show the Ca/Mg ratios (weight percent) in the ground water of the sebkha and the composition of the sediments that the water saturates as determined in a series of pits dug across the flats from the shore to the interior edge of the sebkha. The writers were obviously impressed with the high temperatures measured in early November (33° to 34° C) in the moist sediment at the surface and 36° C about 8 inches below the surface in one pit; they speculate that the temperatures were much hotter on the inner landward portions and suggest a steady summer temperature of about 40° C. It is quite possible, however, that here as on the Trucial Coast the temperature averages only 5 or 6° C above the adjoining sea water giving an average temperature of about 32° C. The C^{14} age of two samples (7 to 20 cm and 23 to 32 cm) indicate that in them the dolomite at a depth of about 12 cm is approximately 2,500 years old and at a depth of 28 cm is about 3,400 years old. Comparison with the dating of the Coorong dolomites as given by von der Borch, Rubin, and Skinner (1964) shows the time span of sediment formation is comparable.

Coorong:	Age (yrs B.P.)	Depth cm.
	300 ± 250	0.0
	700 ± 250	7.6
	770 ± 250	15.2
	1,780 ± 250	30.5
	2,030 ± 250	33.0

The ease with which aragonite and hydromagnesite precipitate as contrasted with dolomite suggest that mean annual temperatures are more meaningful for reactions converting calcite to dolomite than any single brine temperature, but as will be shown below, the pH of the water may be more important than a temperature change of several degrees.

The difficulty of synthesizing dolomite at low temperatures (25°–100° C) has led to a variety of experimental and theoretical attacks on the problems of dolomite-calcite equilibria and solubility, most of which result in precise but inconsistent figures. The thermodynamic calculations all must assume that the ionic strength of the solution *in equilibrium* with the dolomite can be measured or is known. Solubility product calculations for saline solutions mean little except as modified by correction factors for free ions, and activity coefficients.

Holland (1967) has given an excellent and detailed summary of experimental work on the solubility of calcite and dolomite. The solubility of calcite, which has been well investigated, increases notably with increasing pressure of CO_2 but diminishes rapidly with increasing temperature. Its solubility also increases with the content of NaCl present in the solution until high NaCl concentrations are reached, and up to 300° C at least the solubility of $CaCO_3$ in NaCl solutions is also retrograde with increasing temperature.

Solubility data on dolomite are conflicting. The inability of experimenters to precipitate dolomite at room temperature makes all such solubility determinations suspect. Garrels et al. (1960) found a large variation in solubility that seemed clearly a function of sample preparation (grinding time, fineness) and derived equilibrium constants (K_{dol}) ranging from 3.6×10^{-17} to 4.6×10^{-20} with the latter figure favored. Other investigators (see, Barnes and Back, 1969) have calculated value of K_{dol} ranging from 1×10^{-17} to 6.6×10^{-19}. The generally accepted value of K_{dol}; is about 2.5×10^{-17} which approximates the square of $K_{cal} = 5.1 \times 10^{-9}$, the value calculated by Garrels and Dreyer (1953); Krauskopf (1967 p. 79) uses a value of 4.5×10^{-9}.

If the equilibrium reaction for dolomite is a simple ionic substitution according to the equation:

$$2\ CaCO_3 + Mg^{2+} \rightleftharpoons CaMg(CO_3)_2 + Ca^{2+},\ \text{and}$$

$$a_{Ca^{2+}}/a_{Mg^{2+}} = 1,\ \text{then}\ a_{Ca^{2+}} = a_{Mg^{2+}};$$

$$\text{and}\ K_{cal-dol} = (a_{Ca^{2+}})^2.$$

The fact that $K_{cal-dol} \simeq [K_{cal}]^2$ does not, however, prove that the value of a_{Ca}/a_{mg} is actually unity! Hsu (1963) assumed that Florida ground water was in equilibrium with dolomite and limestone, and calculated the value of $K_{cal-dol}$ from the Ca and Mg content of the water. The values ranged from 1.9×10^{-15} to 5.7×10^{-16} with a mean value of 1.3×10^{-15} for 8 determinations. As this value seemed too low, Hsu calculated $K_{cal-dol}$ on the assumption that it was the square of the value of the K_{cal}. The mean of the 8 values for this calculation was 1.9×10^{-17} which Hsu regards as acceptable. Similar discrepant results are reported by Holland et al. (1964) in their study of cave waters in limestone and dolomite terranes. Ie seems probable that equilibrium with dolomite was not attained in either case. Both observation of nature and laboratory work show that calcite is much more readily dissolved and precipitated than dolomite. Furthermore, evidence is accumulating that suggests the solution of magnesian calcite and of dolomite in nature is incongruent, and takes place according to the equation:

$$(Ca \cdot Mg)\ CO_3 \rightarrow CaCO_3 + Mg^{2+}$$

$$Ca\ Mg\ (CO_3)_2 \rightarrow CaCO_3 + Mg^{2+} + CO_3^{=}.$$

Land (1967) in some long duration experiments found that magnesian calcites dissolved incongruently at 25° C, the Mg going into solution preferentially. This agrees with the findings of Goodell and Garman (1969). The work of Yanat'eva (1945) showed that dolomite also dissolves incongruently.

The activity coefficients of Ca, Mg, and CO_3 in sea water have been calculated by Garrels and Thompson (1962) as follows: $^aCa^{2+} = 0.28$, $^aMg^{2+} = 0.36$, $^aCO_3^{=} = 0.20$; but only 9 percent of the $CO_3^{=}$, 87 percent of the Mg^{2+}, and 91 percent of the Ca^{2+} are present as free ions. The molality of Ca and of Mg in ocean water at 20° C and a salinity of 3.5 percent (chlorinity 19.3%) according to Riley and Skirrow (1965, p. 122) are 0.0103 M Ca and 0.0540 M Mg, but the molality of $CO_3^{=}$ is a function of temperature, pH, and concentration of CO_2.

The inorganic carbon content of the oceans changes appreciably below the thermocline which is 75 to 100 meters below the surface. Above the thermocline the average inorganic carbon content is $0.20 g/cm^2$, and below itis given as 7.25 g/cm^2 (Riley and Skirrow, 1965, p. 229).

Above the thermocline the CO_2 content is equivalent to $0.73\ g/10^5$ cc. This content would yield a solution having a molarity of $1.67 \times 10^{-3}\ M\ CO_2$, but the $CO_3^{=}$ ion makes up only a small part of the dissociation products. The actual percentage depends chiefly on temperature and pH. At a temperature of 24° C, a pH of 7.5 and a chlorinity of 21‰, 3.4 percent of the CO_2 is present as $CO_3^{=}$, at a pH of 8.0 some 10 percent is $CO_3^{=}$, and at pH 8.4 nearly 22 percent of the CO_2 is represented by $CO_3^{=}$, (Riley and Skirrow, 1965, p. 668).

In the Persian Gulf and South Australia where calcite and dolomite are now forming the pH ranges from 7.5 to more than 10. The pore water in the calcite-aragonite lagoonal sediments of the Persian Gulf has a pH of 7.5 (Illing et al. 1965). Using the molality values given for ocean water at a pH of 7.5 and a temperature of 26° C the solubility product for calcite can be calculated, using the values given by Garrels and Thompson for free ions and activities:

$$K_{cal} = [Ca^{2+}] [CO_3^=] = CaCO^\circ_3$$

$$K_{cal} = [0.28 \times 0.87 \times 0.01] [0.20 \times 0.09$$
$$\times (1.67 \times 10^{-3} \times 0.033)] = 2.5 \times 10^{-9}.$$

Obviously this figure is close to the accepted values of 4.5 to 5.1×10^{-9}; if the pH were assumed to be 7.8, the value would be $K_{cal} = 5.1 \ 10^{-9}$. Although the pH was measured in the lagoonal calcite-aragonite sediments, the CO_2 content of the pore water was not given by Illing, but is commonly appreciably more than in the overlying lagoonal waters. It is very probable that it was as much as twice the figure used in the calculation, and if so the calculated value of K_{cal} would be 5×10^{-9}.

For salt water in equilibrium with both dolomite and calcite the molar ratio of $Ca^{2+}/Mg^{2+} \simeq 0.15$ (see Table 1 and Fig. 1); using the approximate figure of 0.054 for the molarity of Mg^{2+} in salt water in equilibrium with both calcite and dolomite, the molarity of Ca would have decreased from 0.01 M (by precipitation of calcite and gypsum) to 0.0081 M. Calculating the value of $K_{cal-dol}$ on the data from the Coorong lagoon (von der Borch, 1965) we assume that the value of $K_{cal} = 5 \times 10^{-9}$ and that the amount of $CO_3^=$ present when dolomite plus calcite are in equilibrium with the solution is determined by $K_{cal} = 5 \times 10^{-9}$. At the dolomite-calcite equilibrium point we assume that $M \ Ca/M \ Mg = 0.15$ (Table 1 and Figure 1) and therefore $M \ Ca^= = 0.15 \times 0.054 = 0.0081 \ M$ Ca. Using this figure we can calculate the value of $M \ CO_3^=$ present:

$$[Ca^{2+}] [CO_3^=] = 5 \times 10^{-9},$$
$$[CO_3^=] = 5 \times 10^{-9}/8.1 \times 10^{-3} = 6.2 \times 10^{-5}$$

To calculate the value of $K_{cal-dol}$ from the data on the Coorong lagoonal lakes we assume that at equilibrium $Ca^{2+} = 0.0081 \ M$, $Mg^{2+} = 0.054 \ M$, $CO_3^= = 6.2 \times 10^{-5}$ at a temperature of 24° C:

$$K_{cal-dol} = [K_{cal}] [Mg^{2+}] (CO_3^=)$$

$$K_{cal-dol} = [5 \times 10^{-9}] [0.87 \times 0.36 \times 0.054]$$
$$[0.09 \times 0.20 \times 6.2 \times 10^{-5}]$$

$$K_{cal-dol} = 1.17 \times 10^{-17} \text{ at } 24° C.$$

The values given above for K_{cal} and $K_{cal-dol}$ calculated from data on sea water—or sea water modified by precipitation—presumably in equilibrium with calcite-aragonite or calcite-dolomite agree very well with published data. There seems to be no persuasive reason to assume that the ratio Ca^{2+}/Mg^{2+} is unity when dolomite plus calcite plus sea water are in equilibrium.

In the Coorong area both carbonates precipitate from the lagoonal waters actively in spring and summer but not during the winter; here the average summer temperature of 75° F (24° C) as given by von der Borch (1965) seems a reasonable one to use for plotting the reaction series given by Alderman and von der Borch (1963). This has been done in Figure 1. The dolomite-protodolomite and calcite points show a most remarkable coincidence with the projected fields of dolomite and calcite as determined at high temperature, see Figure 1, but the magnesite-dolomite point does not correspond to a similar projection of the boundary of the high temperature dolomite and magnesite fields. Undoubtedly the appearance of other phases such as hydromagnesite, and huntite makes this point on the diagram unusable for defining stability fields at the temperature of 24° C.

After critically studying all the data, I conclude that the Ca/Mg molar ratio in dilute solutions where (Ca + Mg) is 0.5 molar or less and is in equilibrium with calcite and dolomite is closely approximated between 10° C and 420° C by the line shown in Figure 1 passing through the dolomite-calcite point of the von der Borch and Alderman reaction series at 24° C for 0.5 molar (Ca + Mg). Its equation is:

$$\log: (M \ Ca^{2+}/M \ Mg^{2+})$$
$$= 3.25 - 1.21 \ (10^3/°K) \ldots\ldots\ldots 1.3.$$

Hewett's conclusion that hydrothermal dolomite in Cordilleran ore deposits is related to intrusive rocks in many areas and represents magnesium derived by the hydrothermal alteration of igneous rocks or dolomite at depth (Hewett 1928) has some—but not enough—field evidence to support it. In the Goldsprings district, where Hewett (1931) studied the hydrothermal dolomitization problems most thoroughly, he concluded that the Big Devil Rhyolite and the related glassy rhyolite northwest of the Sultan mine were responsible for the alteration of the Monte Cristo Limestone to dolomite. In both places the intrusive was separated from the normal limestone by an area of strongly fractured and bleached dolomite adjacent to the intrusion. Hewett concludes (p. 47) "It is clear that the effect of the intrusion has been to convert the limestone to dolomite and to add a little silica." Although the close association of the dolomite and the intrusion suggests a genetic relationship, it should be emphasized that the dolomite had been broken before or by the intrusion of the latite and furthermore, that neither intrusion shows the sericitic alteration to which Hewet ascribed the source of the magnesium. The Big Devil Rhyolite is fine-grained orthoclase, quartz, and biotite, with some sodic plagioclase, whereas the intrusive rhyolite near the Sultan mine is mostly glassy but also contains orthoclase and biotite. Biotite, however, (which was unaltered) is one of the earliest minerals to be sericitized wherever sericitic alteration is present. It would seem that the field relations are better explained by ground water dolomitization and subsequent bleaching related to the removal of carbonaceous material in the dolomite during the intrusion of the rhyolite.

Experimental work, natural waters, and chemical relations: In considering the origin of hydrothermal dolomite, the relation of the boundary between the dolomite and calcite fields to temperature is instructive (see Fig. 1). Wherever the molar ratio of ionic calcium to ionic magnesium is 1 or less, any temperature above 100° C is sufficient to make dolomite (or magnesite) the stable phase in equilibrium with the solution. Similarly, even with a calcium to magnesium molar ratio of 2 to 1, dolomite would be stable above 137° C, and at as much as 5 to 1, dolomite would be the stable phase above a temperature of 200° C.

Hot water in geothermal areas not associated with ore deposits, such as the Niland brines near the Salton Sea and waters of the geothermal areas in New Zealand, readily dissolve constituents of the rocks through which they pass. The ratios of calcium to magnesium vary greatly, depending on the environment of the thermal waters. The brine of a well in the CO_2 enriched part of the Salton Sea geothermal system of Imperial Valley, California, contains 2.38% solids and is listed by White, Hem, and Waring (1963, p. 42) as containing 580 ppm of Mg and having a weight ratio of Ca/Mg = 0.67, or M Ca/M Mg = 0.40.

Although most of the thermal waters listed by White, Hem and Waring (1963) have calcium to magnesium ratios greater than 10, some of the analyses of saline waters similar to oil field brines from both cold and warm springs show weight percent ratios of Ca/Mg = 1.1 to 6 (p. 36, 38–39). These springs include waters from crystalline rocks, limestone, dolomite and basalts but the lowest ratio is that of a spring in a serpentine area where the weight ratio of Ca to Mg is 0.025. But in it, the magnesium is only 58 ppm and the Na content is more than 9,000 ppm. Sulfate-bicarbonate waters associated with oil fields show Ca/Mg weight percent ratios ranging from 0.8 to 5 (table 14, p. 34) whereas the oil field waters and other deep well brines high in sodium- and calcium-chloride show Ca/Mg ratios ranging from about 2.5 to about 10; several of these brines are extremely high in magnesium, running nearly 2,000 ppm in the Michigan Basin, weight percent ratio Ca/Mg of 8. Although the analyses of ground waters from dolomite show Ca/Mg weight

ratios ranging from 0.4 to 2.5 the total solids in the waters reported are relatively small (229 to 1,370 ppm), and Mg ranges from 14 to 86 ppm (from 0.4 to 3.6 millimoles Mg). Nearly all of the chemical analyses of ground waters from granite, rhyolite, and similar rock types are quite dilute and the magnesium content commonly ranges from less than 1 ppm to about 10 ppm. In such dilute solutions the ratios of Ca to Mg lose significance but range from more than 10 to as little as 0.55.

The solubility of dolomite is less than that of calcite at all temperatures where it has been measured (Holland, 1967) and the solubility of calcite decreases with increasing temperature both in sodium-chloride solutions and in bicarbonate solutions. It would seem that dolomite would not furnish much magnesium to hydrothermal solutions at temperatures above 150° C, whereas basic silicate rocks such as serpentines or basalt could readily yield a substantial amount to hot solutions. Such solutions or connate brines associated with oil fields could carry large amounts of magnesium.

Rosenburg, Burt, and Holland (1967) succeeded in establishing calcite-dolomite equilbrium in solutions above 290° with concentrations as low as 0.16 molar and Ca/Mg = 15; (0.15 M Ca + 0.01 M Mg) equivalent to a concentration of 243 ppm Mg, but they report no syntheses at lower temperatures, presumably because the rate of reaction was very slow. It seems probable that failure to synthesize dolomite at room temperatures in the laboratory is primarily caused by this reluctance to crystallize at low temperatures even in highly supersaturated solutions. The approximation that reaction rates double with every increase of 10° C also implies that the reaction rate decreases with lowering temperatures similarly. Furthermore, since the crystallization requires the collision of two ions in comparatively low energy states in a solution before nucleation begins, it is obvious that the cooler the solution and the more dilute it is, the less probable is collision and nucleation. Once nucleation has started, however, collision of low energy ions with the growing crystal phase is much more probable and supersaturation should be reduced in a much shorter time than is required in the initial nucleation. To a large extent geologic time offsets dilution and lowering temperatures; but it is apparent that as the degree of supersaturation approaches zero, the time required for the initial nucleation of a crystal in a solution containing none must approach infinity.

As the solubility of calcite in hot water falls off rapidly with increasing temperature, presumably—but not certainly——that of dolomite does also except in a very concentrated brine, whether the water is saline or saturated with CO_2. In a 0.2 molar NaCl solution it amounts to about 350 ppm $CaCO_3$ at 100° C, 220 ppm at 150° C, and 100 ppm at 200° C; in water charged with carbon dioxide at a pressure of one atmosphere the solubility of calcite is decreased; 220 ppm at 100° C, 90 ppm at 150° C, and 40 ppm at 220° C, but increases materially in the presence of both CO_2 and NaCl (Holland, 1967). In solutions saturated with respect to dolomite and calcite it is very probable that the solubility of dolomite is appreciably less than that of calcite above 100° C. If this is true, a 0.2 molar NaCl solution at 100° C would have no more than 350 ppm of Ca. At this temperature according to Figure 1, dolomite should form in solutions if the molar ratio of Ca/Mg is unity or less. Allowing for a certain degree of supersaturation, it would seem that the minimum concentration of magnesium and calcium at which dolomite and calcite would precipitate at 100° C might approximate Ca/Mg = 400 ppm/243 ppm (a molar ratio of 1.0). Unless the solutions were appreciably more concentrated than this or moving through the area for a long time, it seems unlikely that much dolomite would form before the solutions were depleted; and if the solution were cooling as it moved it would very shortly be undersaturated and have a composition wholly in the calcite field.

Although most non-saline ground waters have insufficient magnesium in them to allow the formation of hydrothermal dolomite unless they are abnormally hot, or had passed through a magnesian-silicate environment and so became enriched in Mg, ground water containing substantial amounts of magnesium are present in many areas especially where connate water or oil field brines are present. If such areas were above or near intrusive centers or shallow hearths where magma lay quiescent for some time, the rocks and their contained water would inevitably heat up as the temperature gradient between surface and hot intrusion became stabilized. As shown by Jaeger (1957) this would occur in a relatively short time geologically. Such conditions probably occur in many places in the world today, viz. in Yellowstone Park, in the Larderello geothermal area of Italy, at The Geysers in California, and in the New Zealand geothermal area.

Many mining camps show evidence of intermittent intrusion of magma to different depths below the surface; some of the intrusions were accompanied by hydrothermal alteration that clearly preceded other intrusions which in turn, still were earlier than ore deposits associated with the camp. Perhaps the East Tintic district and the Main Tintic district in Utah give us some of the best examples of this type of relationship. In the hyperthermal[3] area caused by the early intrusives that preceded the deposition of ore here by a substantial amount, ground water cells heated by the conduction of heat into the rocks would form convection cells above or in the neighborhood of an intrusion as has been suggested by the work of Jaeger (1967), Lovering (1965), and White (1968). Such a convection cell would cause hydrothermal dolomite to replace limestone if the ratio of magnesium to calcium was appropriate and there was a sufficient concentration of magnesium. The presence of saline waters in the East Tintic district, which have a calcium to magnesium molar ratio of 2.4 to 1 and have temperatures above 65° C at the present time (Lovering and Morris, 1965), suggest that dolomitic alteration by such geothermal cells is a distinct possibility at temperatures of 150° C. Similarly some boreholes in the Salton Sea area, another hyperthermal area which is well known, where the temperature of the highly saline water reaches 350° C at depth, contain water having substantial amounts of magnesium in a ratio to calcium that would cause dolomitic alteration if limestone were present. In the shallow Niland well in the CO_2 enriched part of this field, the water contains 320 ppm Ca and 580 ppm Mg (a Ca/Mg ratio of 0.33) a molar concentration that would seem to be quite adequate to cause dolomitic alteration if circulated through limestone at quite moderate temperatures (White, Waring and Hem, 1963).

The age relations of hydrothermal dolomite to ore deposits in various Cordilleran mining camps show that nearly all of the hydrothermal dolomite is quite early but that locally some dolomite crystallized from solutions at various times and after ore deposition.

The composition of the fluid inclusions as reported by Roedder (1967) shows that the fluid inclusions associated with the Mississippi Valley type ore deposits and the hydrothermal ore deposits of Cordilleran lead-zinc veins varies greatly in both the ratio of calcium to magnesium and in their concentrations. At Rico, Colorado, according to Roedder (1967, p. 548–549), fluid inclusions occur in sphalerite that formed at temperatures of about 275° C; they contain about 2.5% solids and have a molar ratio of Ca/Mg of 1.5/1; fluid inclusions from some Asian fluorites show Ca/Mg weight percent of 1/1 (molar 0.6) the vapor bubbles in ZnS and CaF_2 from the Mississippi Valley districts homogenize at 100° to 150°, and their molar Ca/Mg ratios range from about 3 to 4.2

[3] "Hyperthermal" is a term used for areas where the geothermal gradient and heat flow are much above the average; in general this refers to a temperature gradient of more than 150° C/km and a heat flow of more than 8 microcalories per second per square centimeter, but the exact limits are not yet defined.

and the inclusion liquids contain 1.5 to 2% dissolved solids. These ratios contrast with the fluid inclusions in gold-quartz veins which have molar ratios of Ca/Mg of 18/1. Clearly the ratios of calcium to magnesium in solutions associated with ore deposits range in composition from those that could be expected to dolomitize limestone at low temperatures to others that could not dolomitize limestone at temperatures in the range of mesothermal ore deposits.

The puzzling features of hydrothermal dolomite associated with ore deposits seem readily explained if the source is primarily ground water, the composition of which varies greatly from district to district; it is difficult to believe that a process of magmatic differentiation which resulted in hydrothermal fluids that precipitated ores of striking similarity in different camps, would be so different in composition that in places they would cause a tremendous amount of hydrothermal dolomite to form and in other places virtually none.

Where the ore depositing epoch is completely separated from the formation of the hydrothermal dolomite, the presence of the dolomite in regions remote from ore as well as around hydrothermal conduits later followed by the ore depositing solutions is a most reasonable relationship, if related to ground water. Whether the ore depositing fluid was magmatic or not, the formation of hydrothermal dolomite at an early stage by magnesian ground water would be expected where ground water became heated and started to circulate through a limy environment. If the ground water had a highly saline composition and also contained appreciable amounts of the ore metals as in the Salton Sea area, hydrothermal dolomite could form as soon as the appropriate ratios existed at the temperature of the solution in contact with limestone, whereas ore deposition would have to wait the introduction of sulfur in a reduced form. Both processes could conceivably be simultaneous although the field occurrences thus far studied indicate that wherever extensive dolomitization occurred, it preceded the ore stage. Nevertheless, some minor dolomitization or the precipitation of dolomite and related carbonates does occur immediately before, during, and after sulfide deposition in some localities.

The variation of O^{18}/O^{16} isotope ratios in hydrothermal dolomite has been noted by many workers—notably Engel, Clayton, and Epstein (1958) and Lovering, McCarthy, and Friedman (1963)—who have generally ascribed the variation in these ratios to the differences in temperature of the magma-derived dolomitizing solutions; it is also possible, however, that they represent variations in the composition and temperature of heated ground water.

The isotopic composition of such ground water would be a function of its temperature and isotopic exchange with minerals traversed enroute to the site of dolomitizing reactions. The shorter the time that such solutions have been in contact with silicates and carbonate minerals at high temperatures, the more nearly the oxygen isotope ratios would reflect the equilibrium between heated local meteoric water and any hydrothermal dolomite formed by it. Changes in the isotopic composition of ground water genetically related to dolomite could thus be engendered by heat, magmatic emenations or time of contact with oxygen-bearing minerals at depth. It is quite clear from the work of Engel et al. (1958) and others that several generations of dolomite exist in the vicinity of the ore deposits of Gilman, Colorado, and that the oxygen isotopes vary systematically in the later generations of dolomite as hydrothermal conduits are approached. On the other hand, the widespread hydrothermal dolomite in the areas adjacent to these ore deposits show a relatively uniform isotopic composition which could well reflect the equilibrium between Laramide ground water and the limestones at an early stage of convective circulation.

The theory outlined above to explain the formation of hydrothermal dolomite seems to harmonize with the field and theoretical studies; it would explain the fact that hydrothermal dolomite makes up huge volumes of rocks in some districts but is completely lacking near other replacement ore bodies that show strong affiliation with those associated with hydrothermal dolomite; it emphasizes a fact that is gradually beginning to emerge, the formation of most ore deposits and the alteration of their host rocks are not simultaneous events representing a simple short one-stage process but reflect rather a relatively complex sequence of disparate chemical events taking place over an appreciable bit of geological time measurable in tens of thousands or hundreds of thousands of years.

REFERENCES

Alderman, A. R., and von der Borch, C. C., 1963, A dolomite reaction series: Nature, v. 198, p. 465–466.

Banks, N. G., 1967, Geology and geochemistry of the Leadville limestone and its supergene, hydrothermal and metamorphic derivatives: Unpub. Ph.D. thesis, Univ. of Calif. at San Diego.

Barnes, Ivan, and Back, William, 1964, Dolomite solubility in ground water: Article 160 *in* U. S. Geol. Survey Prof. Paper 475-D, p. D179–D188.

Culkin, Frederick, 1965, Properties of sea water, *in* Chemical Oceanography; Riley, J. P., and Skirrow, Geoffrey, ed., N. Y. Acad Press, p. 122.

Dana, J. D., 1843, On the analogies between the modern igneous rocks and the so-called primary formations, and the metamorphic changes produced by heat in the associated sedimentary deposits: (Art. XIV), Am. Jour. Sci., v. 45, no. 1, p. 104–129.

Deffeyes, K. S., Lucia, F. J., and Weyl, P. K., 1965, Dolomitization of recent and Plio-Pleistocene sediments by marine evaporite waters on Bonaire, Netherland Antilles, *in* Soc. Econ. Paleontologists and Mineralogists Spec. Pub., 13, p. 71–88.

Engel, A. E. J., Clayton, R. N., and Epstein, S., 1958, Variations in isotopic composition of oxygen and carbon in Leadville limestone (Mississippian, Colorado) and in its hydrothermal and metamorphic phases: Jour. Geology, v. 66, p. 374–393.

Garrels, R. M., Thompson, M. E., and Sevier, Raymond, 1960, Stability of some carbonates at 25° C and one atmosphere pressure: Am. Jour. Sci., v. 258, p. 402–418.

——, and Dreyer, R. M., 1952, Mechanism of limestone replacement at low temperatures and pressures: Geol. Soc. America Bull., v. 63, p. 325–380.

——, and Thompson, M. E., 1962, A chemical model for sea water at 25° C and one atmosphere total pressure: Am. Jour. Sci., v. 260, p. 57–66.

Goodell, H. G., and Garman, R. K., 1969, Carbonate geochemistry of Superior deep test well, Andros Island, Bahamas: Am. Assoc. Petroleum Geologists Bull., v. 53/3, p. 513–536.

Hewett, D. F., 1928, Dolomitization and ore deposition: Econ. Geol. v. 23, no. 8, p. 821–863.

——, 1931, Geology and ore deposits of the Goodsprings quadrangle, Nevada: U. S. Geol. Survey Prof. Paper 162, 172 p.

Holland, H. D., 1967, Gangue minerals in hydrothermal deposits, *in* Geochemistry of Hydrothermal Mineral Deposits: Barnes, H. L., ed., New York, Holt, Rinehart, and Winston, Inc., p. 382–436.

——, Kirsipu, T. V., Heubner, J. S., and Oxburgh, U. M., 1964, On some aspects of the chemical evolution of cave waters: Jour. Geology, v. 72, p. 36–67.

Hsu, K. H., 1963, Solubility of dolomite and composition of Florida ground water: Jour. Hydrology (Amsterdam), v. 1, p. 288–310.

H. O. Pub. 62, 1960, USN Sailing directions for the Persian Gulf, 5d., U. S. Govt. Printing Office, 347 p.

Illing, L. V., Wells, A. W. J., and Taylor, J. C. M., 1965, Penecontemporaneous dolomite in the Persian Gulf, *in* Pray, L. C., and Murray, R. C., eds, Dolomitization and limestone diagenesis—a symposium: Soc. Econ. Paleontologists and Mineralogists Spec. Pub. 13, p. 89–111.

Jaeger, J. C., 1957, The temperature in the neighborhood of a cooling intrusive sheet: Am. Jour. Sci., v. 255, no. 4, p. 306–318.

Kinsman, D. J. J., 1965, Dolomitization and evaporite development including anhydrite, in lagoonal sediments, Persian Gulf (abs): Geol. Soc. of America Spec. Paper 82.

——, 1966, Gypsum and anhydrite of Recent age, Trucial Coast Persian Gulf, *in* Second Symposium on Salt, v. 1, Northern Ohio Geol. Soc., Cleveland, Ohio, p. 302–326.

Krauskopf, Konrad, 1967, Introduction to Geochemistry. New York, McGraw-Hill, 721 p.

Land, L. S., 1967, Diagenesis of some skeletal carbonates: Jour. Sed. Petrology, v. 37, p. 914–930.

Lovering, T. S., 1965, Some problems of geothermal exploration: Mining Eng., v. 17, no. 9, p. 95–99; also in Mining Eng. Trans., v. 232, no. 3, p. 274–281.

——, McCarthy, J. H., Friedman, Irving, 1963, Significance of O^{18}/O^{16} and C^{13}/C^{12} ratios in hydrothermally dolomitized limestones and manganese carbonate replacement ores of the Drum Mountains, Juab County, Utah, *in* Geol. Survey Research, 1963, U. S. Geol. Survey Prof. Paper 475-B, p. B1–B9.

——, and Morris, H. T., 1965, Underground temperatures and heat flow in the East Tintic district, Utah: U. S. Geol. Survey Prof. Paper 504-F, 28 p.

——, and Goode, H. D., 1963, Measuring geothermal gradients in drill holes less than 60 feet deep, East Tintic district, Utah: U. S. Geol. Survey Bull, 1172.

Riley, J. P., and Skirrow, Geoffrey, ed., 1965, Chemical Oceanography, London and New York, Academic Press, v. 1, 712 p.

Roedder, Edwin, 1967, Fluid inclusions as samples of ore fluids, *in* Geochemistry of Hydrothermal Mineral Deposits: Barnes, H. L., ed., New York, Holt, Rinehart, and Winston, Inc., p. 515–574.

Rosenburg, P. E., Burt, D. M., and Holland, H. D., 1967, Calcite-dolomite-magnesite stability relations in solutions—the effect of ionic strength: Geochim. et Cosmochim. Acta, v. 31, no. 3, p. 391–396.

Schmidt, Adolf, 1875, On the forms and origin of the lead and zinc deposits of southwest Missouri: St. Louis Acad. Sci. Trans., v. 3, p. 246–252.

Siebenthal, C. E., 1916, Origin of the zinc and lead deposits of the Joplin Region, Missouri, Kanasas and Oklahoma: U. S. Geol. Survey Bull. 606, 283 p.

Skinner, H. C. W., 1963, Precipitation of calcian dolomites and magnesian calcites in the southeast of South Australia: Am. Jour. Sci., v. 261, no. 5, p. 449–472.

Van Tuyl, F. M., 1914, The origin of dolomite: Iowa Geol. Survey Ann. Rept., v. 25, p. 253–421.

von der Borch, Christopher, 1965, Distribution and primary geochemistry of modern carbonate sediments in the Coorong area South Australia, Geochim. et Cosmochim. Acta, v. 29, p. 781–799.

von der Borch, Christopher, Rubin, Meyer, and Skinner, B. J., 1964, Modern dolomite from South Australia: Am. Jour. Sci., v. 262, p. 1116–1118.

White, D. W., 1968, Environments of generation of some base metal ore deposits: Econ. Geol., v. 63, no. 4, p. 301–335.

White, D. W., Hem, J. D., and Waring, G. A., 1963, Chemical composition of subsurface waters: *in* Data of Geochemistry: Fleisher, Michael, ed., 6 ed., U. S. Geol. Survey Prof. Paper 440-F, p. F1–F67.

Yanat'eva, O. K., 1954, Solubility of dolomite in water in the presence of carbon dioxide: Izvest. Akad. Nauk, Otdel. Khim. Nauk, no. 6, p. 1119–1120.

Part VII

OTHER MODELS AND CONCEPTS OF DOLOMITIZATION

Editors' Comments
on Papers 23 Through 30

23 HARRIS
Dolomitization Model for Upper Cambrian and Lower Ordovician Carbonate Rocks in the Eastern United States

24 VON DER BORCH, LOCK, and SCHWEBEL
Ground-Water Formation of Dolomite in the Coorong Region of South Australia

25 WOLFBAUER and SURDAM
Origin of Nonmarine Dolomite in Eocene Lake Gosiute, Green River Basin, Wyoming

26 SPOTTS and SILVERMAN
Organic Dolomite from Point Fermin, California

27 KAHLE
Possible Roles of Clay Minerals in the Formation of Dolomite

28 GEBELEIN and HOFFMAN
Algal Origin of Dolomite Laminations in Stromatolitic Limestone

29 LINDHOLM
Detrital Dolomite in Onondaga Limestone (Middle Devonian) of New York: Its Implications to the "Dolomite Question"

30 SHINN
Selective Dolomitization of Recent Sedimentary Structures

Many of the better known and more general concepts of dolomitization have been discussed previously in this book. However, in an attempt to convey some appreciation for the range of ideas considered through the years, we present this Part VII. In many cases, the concepts presented here overlap to some extent with others in this part and also with some described in previous papers. For example, the lacustrine model presented by Wolfbauer and Surdam (Paper 25) involves an evaporative concentration mechanism similar to that proposed for dolomitization in marine tidal flats; the possible contribution of Mg^{2+} from algal stromatolitic laminae, as discussed by Gebelein and Hoffman (Paper 28), could also serve as a late diagenetic model and is relevant to a consideration of selective dolomitization (see Shinn, Paper 30).

Evaporitic dolomitization of tidal-flat sediments was the theme of Part IV. There Zenger (Paper 16) suggested that many ancient dolomites represent sudtidal accumulations (see also Asquith, 1967; Choquette and Steinen, 1980; Mossler, 1971; Schmidt, 1965; Cullis, Paper 1; Parsons, Paper 3). Recognizing that environments for intertidal and supratidal dolomitization are relatively restricted in Holocene carbonates, Harris (Paper 23) questioned the then (and possibly still) popular notion that the peritidal zone is the exclusive site for regional dolomitization by evaporitic waters. In a study of the impressively extensive and thick, shallow-water Upper Cambrian and Lower Ordovician strata of the eastern United States (90 percent of the carbonates being dolomite) as his case, he presented convincing arguments for penecontemporaneous subtidal dolomitization wherein a shoreward-increasing salinity gradient in an epeiric setting (Irwin, 1965; Shaw, 1964) could provide a mechanism responsible for widespread subtidal dolomitization. He pointed out that significant salinity gradients can occur without physical barriers even in the much narrower epicontinental seas of today.

Dolomite, often actually protodolomite (and usually not abundant), has been reported from many Holocene lacustrine environments such as the Great Salt Lake (Eardley, 1938), the Dead Sea (Ingerson, Paper 8), Lake Balkash, the Soviet Union (Graf, 1960, p. 23), and Deep Spring Lake, California (Jones, 1961; Peterson, Bien, and Berner, 1963). Müller, Irion, and Förstner (1972) studied inorganic Ca-Mg carbonates in twenty-five lakes of differing hydrochemistry in both humid and arid zones of central and eastern Europe, Turkey, Afghanistan, and East Africa. "Poorly ordered, Ca-rich proto-dolomite" is a secondary mineral only in the lake sediments with high-Mg calcite in association with lake or pore waters having Mg/Ca ratios greater than 7. A startling piece of news was the report by Alderman and Skinner (Paper 9) of dolomite ("plastic dolomitic sediments") precipitating in a shallow lagoon, known as the Coorong, and nearby permanent and intermittent alkaline lakes in South Australia. In the ensuing years much has been written about this region of relatively widespread Holocene dolomitization, including the question of the ordering of the dolomite and its possible primary origin. The significance of groundwater in the formation of this dolomite was established by von der Borch, Lock, and Schwebel (Paper 24). The Holocene stratigraphy of the coastal plain inland from the Coorong defines a regressive cycle of sedimentation for

303

the area wherein restricted marine and lagoonal carbonates pass upward into ephemeral lake deposits that contain protodolomite as well as ordered dolomite. Today the coastal zone is the principal area in which seaward-flowing groundwaters discharged from unconfined aquifers beneath the coastal plain are evaporated; this situation explains why the bulk of the dolomite now forming is near the coast. However, groundwater discharge zones are also present, resulting in ephemeral lakes at some inland localities, in interdune corridors remote from the present lagoon, and landward of the seawater-groundwater mixing zone. These workers concluded that dolomite ideally forms from ions supplied by fresh groundwaters, whereas aragonite, Mg calcite, and protodolomite assemblages form from marine or mixed water. Although the basic interpretive picture has not changed appreciably since 1975, for later and more detailed aspects the reader is referred to the papers of von der Borch and Lock (1979) and Muir, Lock, and von der Borch (1980).

The Eocene Green River Formation of the central Rocky Mountains has long been studied and recognized as representing, for the most part, lacustrine facies (for example, Bradley, 1929); an alkaline playa lake model was proposed by Eugster and Surdam (1973). The formation of dolomite in the mud flats surrounding this ancient lake was described by Wolfbauer and Surdam (Paper 25). Considering various possible origins for the dolomite in the Laney Shale Member, they concluded that protodolomite formed by penecontemporaneous replacement of $CaCo_3$ precursors in the mud flats adjacent to the lake. The dolomitizing fluids were derived from evaporative concentration of groundwater into Mg-rich brines following regressive phases in an arid climate.

Is it possible that there has been a significant control on dolomitization, either directly or indirectly, by organic matter? Much remains to be learned in this regard. In his 1964 review, Sonnenfeld concluded that both primary and secondary dolomites are phytogenic, or plant derived. He postulated that primary dolomites appear to be dependent on the metabolism of bacteria or of higher plants. Magnesium for secondary dolomitization could be derived from weathering products of ultrabasic rocks, and Sonnenfeld believed that organic compounds derived from plant decomposition could aid in the process. However, one must wonder about the extensive stratiform dolomites of replacement origin that developed in the Upper Precambrian and early Paleozoic prior to the advent of land plants. In describing heavy- and light-carbon dolomites and their relationship to organic matter, Murata, Friedman, and Craemer (1972) pointed out that certain organic acids promote the entry of Mg^{2+} into the calcite lattice and thus could conceivably play a role in dolomitization. Davies and Ferguson (1975) concluded that protodolomite formation associated with organic matter in the Great Barrier Reef of Australia may result from the high alkalinity provided by photosynthesis or decaying organic material. Also subscribing to alkalinity as the major control in dolomitization, Lippmann (1973, pp. 182-187) considered as organic those dolomites incorporating their CO_3^{2-} from bacterial reduction of sulfate and the accompanying oxidation of organic matter. Most recently, Mansfield (1980) described the organic production of crystallographically ordered dolomite found in uroliths in the bladder of a Dalmatian dog; his geochemical considerations seemed to implicate the causal effect of anaerobic bacteria. Whether his contention that the Dalmatian bladder is a microcosm of widespread carbonate environments is, at least to us, open to question. As a representative of concepts considering dolomitization as related to organic matter or organic activity we present the well-known paper by Spotts and Silverman (Paper 26). They described dolomite rhombs closely associated with tar as cement in sandstones in the Miocene Monterey Formation exposed in sea cliffs near Los Angeles, California. Carbon isotope analysis of those crystals indicate very depleted $\delta^{13}C$ values and further suggest, at least in part, a derivation of the carbon from oxidized organic matter. Although the dolomite is not abundant there, the possible broader influence of organic matter on dolomitization must be a consideration; we are, to a large extent, unknowledgeable in this area.

Important in any consideration of dolomitization is the source of Mg^{2+} ions. The most obvious contributor is seawater with its characteristic Mg^{2+}/Ca^{2+} molar ratio of about 5.2, although scant to no dolomitization has been recognized in association with exposure to normal seawater. No doubt this potentially large reservoir of Mg^{2+} ions has been one of the reasons for the popularity of shallow-burial (eogenetic) models of dolomitization. Other sources of Mg^{2+} ions have been referred to in our earlier comments in this book—for example ultrabasic rocks (Sonnenfeld, 1964) and basinal shales (Illing, 1959; Mattes and Mountjoy, 1980). Lovering (Paper 22) pointed out that groundwaters, especially where connate water or oil-field brines are involved, contain sufficient magnesium for dolomitization, given the requisite high temperature. High-magnesium calcite, commonly in the form of skeletal material such as echinoderms or coralline algae, has been suggested as both a source of Mg^{2+} and as a nucleation site for dolomitization (for example, Fairbridge, Paper 7; Peterson, 1962, pp. 26-27; Schlanger, 1957). In connection with a unique model of "solution-cannibalization" for dolomitization of subsurface carbonates beneath the Bahama platform, Goodell and Garman (1969) concluded that incongruent dissolution of unstable high-magnesium calcite during uplift would produce dense magnesium-rich brines that would percolate downward—a type of seepage refluxion mechanism. They believed that dolomite growth began on abrasion particles of high-magnesium calcite and on high-magnesium crystallites in skeletal fragments. Although this model may be workable, Land (1973, p. 86) contended that it fails to provide sufficient magnesium for appreciable dolomitization.

In the context of this problem of source of magnesium, we present two papers that describe different concepts of dolomitization. Kahle (Paper 27), recognizing that a review of the geologic literature seemed to suggest a positive correlation between the percentages of dolomite and insoluble residue, proposed that clay minerals could play a significant role in the formation of some dolomites. In addition to several magnesium-bearing varieties such as montmorillinite, chlorite and vermiculite, which may serve as a direct source of Mg^{2+} ions, clays can also be considered membranes that could act as barriers to ionic migration (salt filtration). In addition, clays can provide sites of nucleation for dolomitization and also could enter into chemical reactions involving calcite, which, in turn, would produce dolomite as one of the products.

In Paper 28, Gebelein and Hoffman were impressed by certain thick Proterozoic and Lower Paleozoic carbonate sequences in the Canadian Northwest Territories and central Appalachians, respectively, in which interlaminated limestone and dolomite were ubiquitous. They sought in vain to find an exact modern mineralogical analog to these mm-rhythmites; the closest counterpart appeared to be alternate algal- and sediment-rich laminae in peritidal algal stromatolites, which, unfortunately, contained little or no dolomite. Basing their arguments on comparative field observations in both modern sediments and ancient rocks and on laboratory experimentation, they demonstrated that the dolomite in ancient laminites is secondary and represents the algal-rich member of the modern couplets that are Mg^{2+} rich. They proposed that the magnesium ions are complexed organically in the algal mat during deposition, but possibly long after lithification, decomposition of the organic matter releases this excess magnesium for dolomitization in the microenvironment of the algal laminae. Although their experimentation using modern algal material produced no dolomite (but rather 17 to 20 mole percent Mg calcite), they suggested that dolomitization must be a longer-term process that results from replacement of carbonate material in contact with the relict algal mat laminae and/or selective replacement of the high-

Mg calcite within these algal layers. Interestingly, their model involved aspects of other concepts, such as late diagenetic dolomitization, organic influence on dolomitization, and selective dolomitization. As these authors admitted, such a model would be relevant only for certain ancient interlaminated stromatolitic limestones and dolomites; many dolomite successions have no such stromatolitic laminations. One point worth mentioning is that their model could provide an explanation for late diagenetic dolomite that conforms to primary sedimentary bedding, a characteristic that we believe has been all too rapidly invoked as evidence for penecontemporaneous to early diagenetic dolomitization.

Three papers appearing in 1962 collectively emphasized the possible significance of detrital (terrigenous) dolomite in the geologic record. Deffeyes and Martin (1962) radiocarbon-dated dolomite in Florida Bay sediments, earlier described as organically related, authigenic dolomite by Taft (1961), as being older than 35,000 years. Because modern sedimentation in Florida Bay began less than 4000 years ago, these authors proposed that Taft's crystals might, in fact, represent detrital material derived from older carbonate rocks. The situation is, however, not clear inasmuch as many of these dolomite crystals are euhedral and there are delicate clusters of rhombs that suggested to Taft that transportation was not involved in their accumulation. Is there a possibility of a terrigenous origin of grains followed by some degree of overgrowth development? This disagreement brings out the important general point that without recognition of detrital nuclei, it may be difficult to determine a terrigenous origin of dolomite (for example, Kopp and Sledz, 1980). Amsbury (1962) described detrital dolomite in Lower Cretaceous fluvial deposits in Texas. This dolomite occurred as fragments of microcrystalline rock or as rounded-to-angular central cores of dolomite crystals with epitaxial overgrowths. The distribution of well-rounded detrital dolomite grains, ranging in size from granules to medium-grained sand, in various Cretaceous sandstones of the Uinta Basin were described by Sabins (1962) as being controlled by the availability of dolomite detritus rather than by depositional environment. Sabins presented a comparison among his detrital, primary, and secondary dolomite "grains." To represent the "detrital dolomite" concept or model, we include the oft-referred-to contribution by Lindholm (Paper 29). Insofar as the cloudy interior of dolomite grains in the Devonian Onondaga Limestone of New York displayed no crystal form, Lindholm suspected that the grains may represent overgrowths on a rounded-to-subrounded detrital core. The positive correlation between quartz and the abundance and size of associated dolomite grains makes his argument for a detrital origin all the more convincing. Lindholm envisioned as a source of the detrital dolomite either ancient dolomite (making the dolomite grains in the Onondaga "terrigenous") or reworked penecontemporaneous dolomite from adjacent and coeval marginal marine environments, possibly involving transportation by wind. The fine-grained detritus could possibly account for some "primary" dolomite grains reported in the literature. If the detrital grains served as nuclei around which diagenetic overgrowths resulted by total replacement of the surrounding calcareous sediment, there could result the more typical coarsely crystalline texture of diagenetic dolomites.

One of the main contributors to our knowledge of Holocene dolomite is E. A. Shinn, who coauthored a paper on supratidal dolomite (Paper 11) in the Bahamian tidal flats. In this part we include as an example of many other cases of preferred dolomitization his 1968 paper concerning selective dolomitization of modern structures in tidal-flat carbonates in the lower Florida Keys. Shinn referred to the one-to-four-inch thick, pelleted, dolomitic crust similar to that in the Bahamas described in Paper 11. The crust contains nondolomitic limestone nodules (taken to represent the polygonal-shaped mud between desiccation cracks) embedded in a lithified dolomitic sediment. As mud chips are eroded today they are incorporated into more porous and permeable sediment, which is, presumably, more susceptible to the effect of transmitted Mg^{2+}-rich brines. Shinn pointed out that ancient analogs are common in the geologic record. Certainly rock, fabric, or structural selectivity has played a part in many reported cases of dolomitization. Murray (1969) and Murray and Lucia (1967) also noted a permeability control on dolomitization at Bonaire and in the Mississippian Turner Valley Formation of Alberta, respectively. It has become fairly well accepted that finer matrices, such as aragonite or calcite mud, are more susceptible to dolomitization than are allochems or spar (for example, see Pray and Choquette, 1966; Füchtbauer, Paper 33) although Mattes and Mountjoy (1980, p. 274) reported that within fenestral mudstones associated with the Miette buildup (Devonian, Alberta), calcite spar in voids, fractures, bioclasts, and laminae may be replaced by dolomite in contrast to adjacent unaltered micrite. Selective dolomitization of Mg-rich calcite has been mentioned by many workers, including Inden and Koehn (1979) and Schlanger (1957); however, Friedman and Sanders (1967, p. 296) reported examples of high-magnesium calcite that resisted dolomitization.

The papers in this part represent but some of the less-noted concepts of dolomitization. As an example of an intriguing recent proposal, Sibley (1980) has suggested a possible climatic control on dolomitization: in humid climatic regimes, which should have interstitial waters with relatively high P_{CO_2}, early solution or conversion of metastable carbonates to low-magnesium calcite could inhibit dolomitization. It is, perhaps, fitting to close with the concluding statement of Zenger and Dunham (1980, p. 7) in their introduction to the recently published Special Paper 28 of the Society of Economic Paleontologists and Mineralogists:

> Yet, we have hardly mentioned all the types of dolomite such as inland lacustrine, deep water, spelean, terrigenous, metamorphic, etc. Dolomite has been found as a fluvial precipitate(?) along northern California Coast Range streams draining serpentinite terrain (Barnes and O'Neil, 1971), in the teeth of living echinoids (Schroeder and others, 1969), in volcaniclastic sequences, in part as a replacement product of montmorillonite (R. C. Surdam, 1978, oral communication), and even forming in the bladder of a Dalmatian (Mansfield, 1979). Certainly some of these dolomite occurrences are minor, perhaps even freakish. There is still value in attempting to determine which of the more plausible mechanisms account for the greater portion of the dolomitization we observe, but clearly there is no one model and no one unique solution to the complexity of this enigmatic mineral.

REFERENCES

Amsbury, D. L., 1962, Detrital Dolomite in Central Texas, *Jour. Sed. Petrology* **32:**5-14.

Asquith, G. B., 1967, The Marine Dolomitization of the Mifflin Member Platteville Limestone in Southwest Wisconsin, *Jour. Sed. Petrology* **37:**311-326.

Barnes, I., and J. R. O'Neil, 1971, Calcium-Magnesium Carbonate Solid Solutions from Holocene Conglomerate Cements and Travertines in the Coast Range of California, *Geochim. et Cosmochim. Acta* **35:**699-718.

Bradley, W. H., 1929, Algae Reefs and Oolites of the Green River Formation, *U.S. Geol. Survey Prof. Paper 154-G*, Washington, D.C., pp. 203-223.

Choquette, P. W., and R. P. Steinen, 1980, Mississippian Non-Supratidal Dolomite, Ste. Genevieve Limestone, Illinois Basin: Evidence for Mixed-Water Dolomitization, in *Concepts and Models of Dolomitization*, ed. D. H. Zenger, J. B. Dunham, and R. L. Ethington, Soc. Econ. Paleontologists and Mineralogists Spec. Pub. 28, Tulsa, Ok., pp. 163-196.

Davies, P. J., and J. Ferguson, 1975, Dolomite and Organic Material, *Nature* **255:**472-473.

Deffeyes, K. S., and E. L. Martin, 1962, Absence of Carbon-14 Activity in Dolomite from Florida Bay, *Science* **136:**782.

Eardley, A. J., 1938, Sediments of Great Salt Lake, Utah, *Am. Assoc. Petroleum Geologists Bull.* **22:**1305-1411.

Eugster, H. P., and R. C. Surdam, 1973, Depositional Environment of the Green River Formation of Wyoming: A Preliminary Report, *Geol. Soc. America Bull.* **84:**1115-1120.

Friedman, G. M., and J. E. Sanders, 1967, Origin and Occurrence of Dolostones, in *Carbonate Rocks: Origin, Occurrence, and Classification,* ed. G. V. Chilingar, H. J. Bissell, and R. W. Fairbridge, Elsevier, Amsterdam, pp. 267-348.

Goodell, H. G., and R. K. Garman, 1969, Carbonate Geochemistry of Superior Deep Test Well, Andros Island, Bahamas, *Am. Assoc. Petroleum Geologists Bull.* **53:**513-536.

Graf, D. L., 1960, Geochemistry of Carbonate Sediments and Sedimentary Carbonate Rocks, Part I, Carbonate Mineralogy, Carbonate Sediments, *Illinois Geol. Survey Circ. 297,* Urbana, Il., 39p.

Illing, L. V., 1959, Deposition and Diagenesis of Some Paleozoic Carbonate Sediments in Western Canada, *Fifth World Petroleum Cong. Proc.* (sec. 1), pp. 23-52.

Inden, R. F., and H. Koehn, 1979, Dolomitization of Offshore Carbonate Deposits in Hammett Shale, Lower Cretaceous, Texas (Abstract), *Am. Assoc. Petroleum Geologists Bull.* **63:**472.

Irwin, M. L., 1965, General Theory of Epeiric Water Sedimentation, *Am. Assoc. Petroleum Geologists Bull.* **49:**445-459.

Jones, B. F., 1961, Zoning of Saline Minerals at Deep Spring Lake, California, in *Geol. Survey Research, 1961: Short Papers in the Geologic and Hydrologic Sciences,* U.S. Geol. Survey Prof. Paper 424-B, Washington, D.C., pp. 199-202.

Kopp, O. C., and J. J. Sledz, 1980, Detrital (?) Dolomite in an Upper Cambrian Conglomerate: Its Recognition, Potential Sources, and Significance (Abstract), *Geol. Soc. America Abs. with Programs* **12:**464-465.

Land, L. S., 1973, Contemporaneous Dolomitization of Middle Pleistocene Reefs by Meteoric Water, North Jamaica, *Bull. Marine Sci.* **23:**64-92.

Lippmann, F., 1973, *Sedimentary Carbonate Minerals,* Springer-Verlag, Berlin, 228p.

Mansfield, C. F., 1979, Possible Biogenic Origin for Some Sedimentary Dolomite (Abstract), *Am. Assoc. Petroleum Geologists Bull.* **63:**490.

Mansfield, C. F., 1980, A Urolith of Biogenic Dolomite—Another Clue in the Dolomite Mystery, *Geochim. et Cosmochim. Acta* **44:**829-839.

Mattes, B. W., and E. W. Mountjoy, 1980, Burial Dolomitization of the Upper Devonian Miette Buildup, Jasper National Park, Alberta, in *Concepts and Models of Dolomitization,* ed. D. H. Zenger, J. B. Dunham, and R. L. Ethington, Soc. Econ. Paleontologists and Mineralogists Spec. Pub. 28, Tulsa, Ok., pp. 259-297.

Mossler, J. H., 1971, Diagenesis and Dolomitization of Swope Limestone (Upper Pennsylvanian) Southeast Kansas, *Jour. Sed. Petrology* **41:**962-970.

Muir, M., D. Lock, and C. von der Borch, 1980, The Coorong Model for Penecontemporaneous Dolomite Formation in the Middle Proterozoic McArthur Group, Northern Territory, Australia, in *Concepts and Models of Dolomitization,* ed. D. H. Zenger, J. B. Dunham, and R. L. Ethington, Soc. Econ. Paleontologists and Mineralogists Spec. Pub. 28, Tulsa, Ok., pp. 51-67.

Müller, G., G. Irion, and U. Förstner, 1972, Formation and Diagenesis of Inorganic Ca-Mg Carbonates in the Lacustrine Environment, *Naturwissenschaften* **59:**158-164.

Murata, K. J., I. Friedman, and M. Craemer, 1972, Geochemistry of Diagenetic Dolomites in Miocene Marine Formations of California and Oregon, *U.S. Geol. Survey Prof. Paper 724-C,* Washington, D.C., 12p.

Murray, R. C., 1969, Hydrology of South Bonaire, N.A.—A Rock Selective Dolomitization Model, *Jour. Sed. Petrology* **39:**1007-1013.

Murray, R. C., and F. J. Lucia, 1967, Cause and Control of Dolomite Distribution by Rock Selectivity, *Geol. Soc. America Bull.* **78:**21-36.

Peterson, M. N. A., 1962, The Mineralogy and Petrology of Upper Mississippian Carbonate Rocks of the Cumberland Plateau in Tennessee, *Jour. Geology* **70:**1-31.

Peterson, M. N. A., G. S. Bien, and R. A. Berner, 1963, Radiocarbon Studies of Recent Dolomite from Deep Spring Lake, California, *Jour. Geophys. Research* **68:**6493-6505.

Pray, L. C., and P. W. Choquette, 1966, Genesis of Carbonate Reservoir Facies (Abstract), *Am. Assoc. Petroleum Geologists Bull.* **50:**632.

Sabins, F. F., Jr., 1962, Grains of Detrital, Secondary, and Primary Dolomite from Cretaceous Strata of the Western Interior, *Geol. Soc. America Bull.* **73:**1183-1196.

Schlanger, S. O., 1957, Dolomite Growth in Coralline Algae, *Jour. Sed. Petrology* **27:**181-186.

Schmidt, V., 1965, Facies, Diagenesis, and Related Reservoir Properties in the Gigas Beds (Upper Jurassic), Northwestern Germany, in *Dolomitization and Limestone Diagenesis,* ed. L. C. Pray and R. C. Murray, Soc. Econ. Paleontologists and Mineralogists Spec. Pub. 13, Tulsa, Ok., pp. 124-168.

Schroeder, J. H., E. J. Dwornik, and J. J. Papike, 1969, Primary Protodolomite in Echinoid Skeletons, *Geol. Soc. America Bull.* **80:**1613-1616.

Shaw, A. B., 1964, *Time in Stratigraphy,* McGraw-Hill, New York, 365p.

Sibley, D. F., 1980, Climatic Control of Dolomitization, Seroe Domi Formation (Pliocene), Bonaire, N.A., in *Concepts and Models of Dolomitization,* ed. D. H. Zenger, J. B. Dunham, and R. L. Ethington, Soc. Econ. Paleontologists and Mineralogists Spec. Pub. 28, Tulsa, Ok., pp. 247-258.

Sonnenfeld, P., 1964, Dolomites and Dolomitization: A Review, *Bull. Canadian Petroleum Geology* **12:**101-132.

Taft, W. H., 1961, Authigenic Dolomite in Modern Carbonates Along the Southern Coast of Florida, *Science* **134:**561-562.

von der Borch, C. C., and D. Lock, 1979, Geological Significance of Coorong Dolomites, *Sedimentology* **26:**813-824.

Zenger, D. H., and J. B. Dunham, 1980, Concepts and Models of Dolomitization—An Introduction, in *Concepts and Models of Dolomitization,* ed. D. H. Zenger, J. B. Dunham, and R. L. Ethington, Soc. Econ. Paleontologists and Mineralogists Spec. Pub. 28, Tulsa, Ok., pp. 1-9.

23

Reprinted from *U.S. Geol. Survey Jour. Res.* **1**:63–78 (1973)

DOLOMITIZATION MODEL FOR UPPER CAMBRIAN AND LOWER ORDOVICIAN CARBONATE ROCKS IN THE EASTERN UNITED STATES

By LEONARD D. HARRIS, Knoxville, Tenn.

Abstract.—Existing models for dolomitization emphasize that penecontemporaneous dolomitization can occur in both subtidal and supratidal environments if the necessary chemical and physical factors favorable for the development of magnesium-rich hypersaline waters exist. Holocene shallow-water hypersaline environments that have the potential to produce dolomite without deposition of more soluble evaporite minerals are found in Shark Bay, Australia, and on the Great Bahama Bank. These hypersalinity systems are characterized by near-vertical isosalinity layers of increasing concentration landward from the open ocean and show little or no relationship to bottom topography. I suggest that a similar but larger scale epicontinental salinity system covering the Late Cambrian and Early Ordovician carbonate continental shelf produced a broad wedge of subtidally deposited dolomite in the eastern half of the United States from Mexico to Canada. The distribution of limestone and dolomite in this region is closely keyed to the salinity gradient and accounts for the natural progression from a normal marine limestone facies through a transition zone to a highly saline dolomite facies phase. Algal stromatolite mats and domes occupied low-energy niches in both the limestone and dolomite facies, whereas stratiform algal stromatolites were confined to the areas of moderate energy within the dolomite facies.

In the past, construction of environmental models that account for the large-scale deposition of dolomite in ancient widespread carbonate environments has defied resolution because the feasibility of the suggested processes could not be confirmed experimentally nor compared to natural modern sedimentary conditions. Recently some progress concerning the origin of dolomite has been gained through the discovery of dolomite in Holocene carbonate rocks and the subsequent study of its environmental parameters. The major contribution from these studies is the general acceptance of the idea that penecontemporaneous dolomitization is common where sediments in high intertidal or supratidal areas contain pore fluids consisting of concentrated magnesium-rich brines, as for example along wide salt flats, termed "sabkhas" in the Persian Gulf (Illing and others, 1965) and carbonate mudflats in the Bahamas (Shinn and others, 1965). The model of supratidal penecontemporaneous dolomitization has been widely used in interpreting ancient carbonate environments; however, it seems to be most applicable to relatively restricted geographic areas (commonly less than 10,000 sq mi) where thin carbonate units can be demonstrated to change from subtidal and intertidal limestone to supratidal dolomite. It is less applicable as a principal site of dolomitization in regionally extensive megacarbonate environments (approximately 1 million square miles) such as the Late Cambrian and Early Ordovician, where nearly 90 percent of the carbonate is dolomite. Some widespread uniform process, other than the supratidal model, must have been operating to have dolomitized thousands of square miles of carbonate as much as 3,000 feet thick. This report questions the currently popular usage of the intertidal and supratidal environments as exclusive sites for large-scale regional dolomitization, and offers an alternate hypothesis involving a magnesium-rich subtidal environment.

UPPER CAMBRIAN AND LOWER ORDOVICIAN CARBONATE ROCKS

A paleogeographic and lithofacies map of the Late Cambrian and Early Ordovician shows that during this time a vast continental shelf, about 2,000 miles long and from 100 to 1,000 miles wide, occupied much of the eastern half of the United States (fig. 1). Maximum development of the shelf and its attendant epicontinental sea occurred late in Cambrian time and persisted with apparent continuous sedimentation until late in Early Ordovician time. Facies of Late Cambrian and Early Ordovician age show an almost classic relationship to near-shore clastic sediments containing subordinate dolomite interfingering with offshore carbonate rocks. A multitude of stratigraphic names has been used to subdivide this ancient sequence of strata into stratigraphic units of differing ranks. Because the nomenclatural complexities are not critical to this study the reader is referred to the latest summary report (Holland, 1971). One major subdivision, the Knox Group, has been widely recognized in the southern and central Appalachians within the predominantly carbonate sequence, and I suggest that many of the general features of the Knox can be applied on a shelf-wide basis.

The Knox Group, as much as 3,000 feet thick, occupies most of the middle segment of the offshore carbonate sequence, crops out in Georgia, Alabama, eastern Tennessee, and southwestern Virginia, and lies in the subsurface of middle

.312

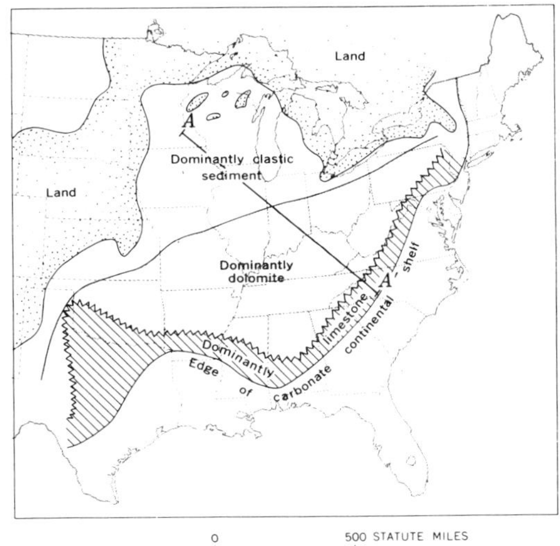

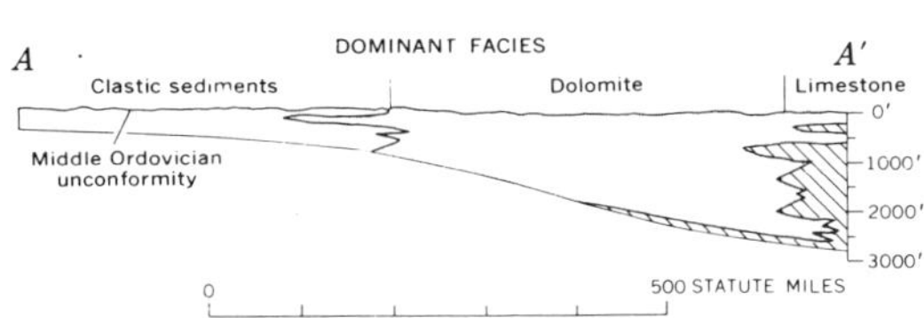

Figure 1.—Generalized paleogeographic map of Late Cambrian and Early Ordovician time, showing regional distribution of facies of the Knox Group and equivalent rocks on the ancient continental shelf. Based on Chenoweth (1968), Donaldson and Page (1964), Folk (1959), Lochman-Balk (1971), Perry (1964), Rodgers (1968), Swartz (1948), and Williams (1969).

and western Tennessee, Kentucky, West Virginia, Ohio, and parts of Indiana and Illinois. Throughout most of this area the Knox is dolomite, but in easternmost Tennessee and southwestern Virginia it is divided into a western dolomite facies and an eastern limestone facies. Recently Laporte (1971) summarized the view of sedimentologists concerned with the change of limestone to dolomite in Upper Cambrian and Lower Ordovician carbonate strata in the eastern United States. They suggest that the change is a response to the environment of deposition and indicate that the limestone has a subtidal origin, whereas the dolomite had an intertidal and supratidal origin. In their view each facies characteristically contains recognizable stratigraphic criteria imposed by environmental controls: subtidal limestone units, although exhibiting regional change and variation, tend to be uniform and may commonly show great lateral persistence. In contrast, variable conditions of sedimentation in the intertidal and supratidal dolomite regime produce abrupt and frequent

changes in lithologic units, so that correlation of dolomite beds even between adjacent outcrops is nearly impossible. These apparent clear-cut differences between limestone and dolomite facies of Cambrian and Ordovician age as summarized by Laporte have not been corroborated by the studies of economic geologists in eastern Tennessee, who concentrated on details of stratigraphy without preconceived ideas of environmental constraints.

More than 40 years ago, when geologists realized that commercial quantities of zinc were confined regionally to a 200-foot zone in the upper part of the dolomite phase of the Knox Group of eastern Tennessee, zinc mining companies initiated intensive foot-by-foot stratigraphic studies in search of key beds to aid in exploration and in mine development. Detailed studies of mine workings demonstrated that, rather than a series of laterally discontinuous lithologic units, almost any distinctive stratum, such as a 1-inch sandstone or shale bed, a chert nodule zone, minor bedding or color variation, and so on, could be correlated from mine to mine in the Mascot-Jefferson City zinc district. Continuing surface and subsurface studies of approximately the upper 900 feet of the Knox in Tennessee have resulted in the careful documentation of an average of one key bed per approximately 11 feet of vertical section, many of which have regional significance (Oder and Miller, 1945; Crawford, 1945; Oder and Ricketts, 1961; Harris, 1969; Stagg and Fischer, 1970). In the past 20 years thousands of core holes have penetrated parts of the predominantly dolomite facies of the Knox in search for economic zinc deposits throughout much of Tennessee and in parts of Kentucky, Alabama, and Virginia. In every area, where enough closely spaced drilling has been done to develop a stratigraphic sequence, laterally extensive key beds can be recognized and can be correlated for considerable distances. In central and northern Virginia, far removed from the Tennessee zinc district, Young and Wedow (1962) and Herbert and Young (1956) experienced no difficulty in recognizing and using thin key units in rocks of the Beekmantown Dolomite, the lateral equivalent of the upper part of the Knox. The widespread uniformity of multiple key units in the dolomite facies of the Knox Group and equivalent rocks, rather than being characteristic of discontinuous intertidal-supratidal sedimentation, is more characteristic of the uniform subtidal sedimentation of Laporte (1971). Consequently many of the criteria summarized by Laporte and used to distinguish between subtidal, intertidal, and supratidal environments need to be critically analyzed.

PRIMARY FEATURES AS ENVIRONMENTAL INDICATORS

Identification of ancient intertidal and supratidal sediments is based on the recognition of features similar to those found on modern carbonate tidal flats. Unfortunately, there is a tendency to ignore the possibility that many of the primary features found on Holocene tidal flats may not necessarily be limited to that environment. In fact, few primary features are infallible. Both shrinkage cracks and algal stromatolites are excellent examples of "infallible guides" used to identify positively the intertidal and supratidal environments. Shrinkage cracks are common features of modern intertidal and supratidal areas; however, Burst (1965) has experimentally confirmed that shrinkage cracks can form subaqueously, in response to an increase in salinity, in mud that contains a small quantity of swelling clay. In the Knox Group, shrinkage cracks are found in about equal abundance in both the limestone and dolomite facies. Detailed facies studies and insoluble residue contents suggest that cracked beds show a close affinity for niches within a subtidal environment where clay-mineral abundance, usually greater than 5 percent, is sufficient to react to major fluctuations in salinity. Major changes in salinity are undoubtedly common along the transition zone between masses of normal marine and penesaline water. Consequently, shrinkage cracks, although useful in many places as indicators of subaerial exposure, may just as commonly indicate fluctuations in salinity within subtidal environments.

Holocene algal stromatolites

Algal stromatolites are laminated structures formed by the trapping and binding of sediment by an algal film. They are organosedimentary features in which only the gross structure of the algal films is preserved; cellular and other detailed structure are not preserved. Logan, Rezak, and Ginsburg (1964) suggested that the form of a Holocene stromatolite is related to environmental conditions, and does not lend itself to the limited definition of form genera or form species. They proposed to abandon all generic and specific names, substituting a descriptive nomenclature based on the arrangement of certain geometric forms. Their classification contains two basic forms—the hemispheroid and the spheroid in three major arrangements: (1) laterally linked hemispheroids (LLH), (2) discrete, vertically stacked hemispheroids (SH), and (3) discrete spheroids (SS). Usefulness of algal stromatolites as environmental indicators was stressed by their suggesting that LLH and SH forms were restricted to intertidal areas and SS forms restricted to subtidal areas of shallow agitated water. Other studies of Holocene algal stromatolites (Ginsburg, 1960; Monty, 1965; Kendall and Skipwith, 1968; Gebelein, 1969; and G. R. Davis, 1970) have not fully confirmed either the form distribution or environmental parameters proposed by Logan, Rezak, and Ginsburg (1964), but generally all agree that there are three dominant geometric forms—mats, hemispheroids, and spheroids. Investigations of intertidal algal stromatolites by Kendall and Skipwith (1968) in the Persian Gulf and by G. R. Davis (1970) in Shark Bay, Western Australia, indicate that the dominant geometric growth form is a subparallel mat rather than a hemispheroid or a spheroid. The mat may be altered mechanically to many subforms depending upon its position within the intertidal and supratidal area. A summary of Holocene stromatolite reports suggests that the intertidal and supratidal areas can be subdivided into as many as five distinct zones (fig. 2).

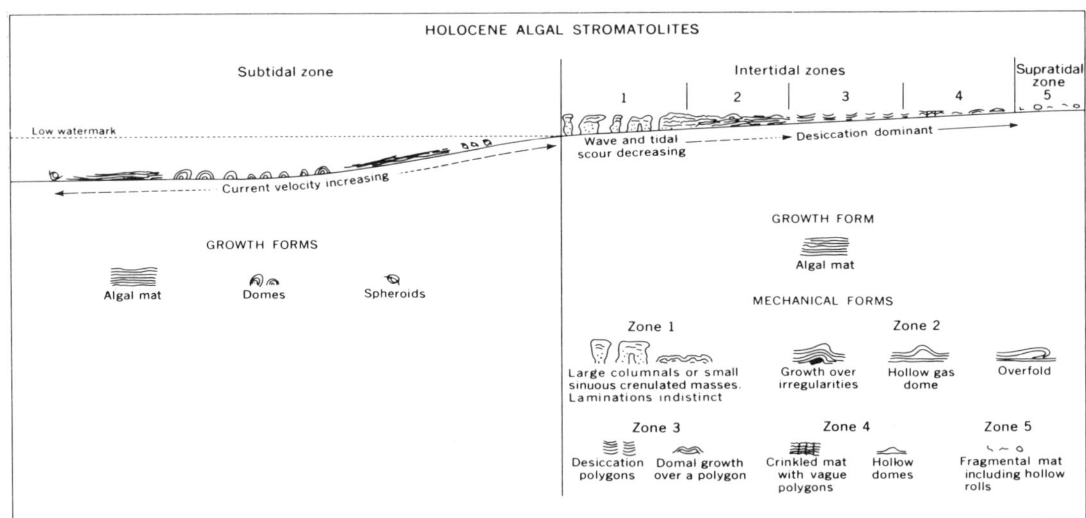

Figure 2.—Form distribution of Holocene algal stromatolites. Based on Black (1933); Ginsburg (1960); Logan, Rezak, and Ginsburg (1964); Kendall and Skipwith (1968); Gebelein (1969); and G. R. Davis (1970).

Zone 1.—A continuous algal mat may develop almost to the low watermark in a protected intertidal mudflat environment, but varieties of forms ranging from sinuous domes or ridges to columnar structures result where waves and tides have increased erosive force. Logan (1961) attributed the development of columnar forms to upward growth of algal stromatolites by progressive stacking of hemispheroidal laminae (SH-V). However, their development seems much more complex, and, as suggested by Logan (1961, figs. 2 and 3) and G. R. Davis (1970, p. 199), they appear to result from scour and beach-rock formation in conjunction with algal growth. Active algal mats are not mechanically resistant to moderate erosive forces (Gebelein, 1969, p. 56) and a slight increase in erosive energy scours the mat irregularly. Once the mat is breached, beach rock, which results from cementation caused by alternate wetting and drying in the intertidal zone, may begin to line the walls of channels. The protection on the walls afforded by the beach-rock formation causes erosion to proceed in a dominantly vertical direction, thereby inducing the development of characteristic SH-V columnar forms of Logan, Rezak and Ginsburg (1964). Along the low watermark, beach rock is sculptured by erosion into columnar structures that grade toward the high watermark into sinuous mounds or ridges, which may or may not be indurated, and finally into unbroken algal mat. Within the beach-rock environment the growth of algal mat is restricted to the top of structures where erosive forces are less. Continued growth of the mat adds to the development of sculptured forms, but its effect is minimal as suggested by the restriction of crudely defined laminations to the upper part of the structures.

Zone 2.—Smooth, nearly flat algal mats characteristically occupy diurnally flooded intertidal areas of low energy. Irregular domelike structures result from growth over irregularities, obstructions, or from erosion and slight desiccation. Superficially these domelike structures resemble the laterally linked hemispheroidal (LLH) structures of Logan, Rezak, and Ginsburg (1964), but they lack symmetry, and their mechanical origin is clearly discernible.

Zone 3.—Desiccation of the algal mat produces a typical polygonal pattern in which individual polygons are saucer shaped and exhibit a wide range in size depending upon the grain size of local sediment and the moisture distribution.

Zone 4.—The algal mat in this zone has a poorly defined polygonal pattern but exhibits abundant hollow domes and a crinkled surface produced by trapped gases. Locally this zone may change to a relatively flat mat similar to that of zone 2.

Zone 5.—Above the high watermark, in the supratidal zone, algal growth is inhibited, as shown by detrital fragments and spheroidally coiled pieces of algal mat.

Subtidal Holocene algal stromatolites have not received the same amount of study as the intertidal forms because they are more difficult to observe and classify. Although these forms were noted by Ginsburg (1960) and Monty (1965) in Florida and the Bahamas, a thorough study of their distribution and environmental parameters did not emerge until Gebelein

(1969) published a paper on subtidal algal stromatolites of Bermuda. Gebelein found that algal development is favored in the lee of islands and on wide, shallow shelf areas where reduced wave and current scour takes place. Form distribution is closely keyed to current velocity and sediment supply (fig. 2). Areas of moderate current velocity and ample sediment supply develop algal mats, similar to those described in zone 2 of the intertidal zone, in water as much as 35 feet deep. In areas of reduced velocity where sediment supply is small, algal mats differentiate into discrete small domes; where sediment supply is large, they differentiate into large domes. Both domes tend to elongate in the direction of current movement. Lamination within small domes is distinct, convex, regular, and symmetrical; in contrast, lamination within domes is convex, indistinct, and irregular in thickness.

To summarize, Logan, Rezak, and Ginsburg (1964) based their classification of algal stromatolites on the assumption that hemispheroids or spheroids were the basic growth forms of algal stromatolites. Other studies (Kendall and Skipwith, 1968; Gebelein, 1969; G. R. Davis, 1970) indicated that there are three growth forms—mats, hemispheroids, and spheroids. The basic structural element of each form is a simple lamina. Logan, Rezak, and Ginsburg (1964) showed that development of laminae and form differentiation are controlled by physical, chemical, and biological constraints of the environment. Thus, wide variation in type and intensity of environmental constraints between subtidal and intertidal settings results in the development of distinctive algal stromatolite forms keyed to the particular environment (fig. 2). Natural growth of hemispheroidal forms is favored in subtidal areas where sediment supply is ample and current velocity is low. Spheroidal forms are more likely to develop in subtidal areas where current velocity alternates between low and moderate. Algal mats grow in both subtidal and intertidal environments because the controlling physical constraints of moderate current velocity and sediment supply are similar. The distinguishing characteristic between subtidal and intertidal algal mats is the greater diversity of forms mechanically produced in the intertidal zone in conjunction with erosion, beach rock formation, gas generation, and desiccation. Because Logan, Rezak, and Ginsburg (1964) did not recognize the algal mat as a distinct form and did not distinguish between intertidally mechanically produced forms and subtidal growth forms, they unintentionally limited the usefulness of algal stromatolites as environmental indicators of ancient carbonate depositional environments.

Algal stromatolites of the Knox Group

Preservation of algal stromatolites in the Knox Group of Tennessee and southwestern Virginia ranges from vague outlines in dolomite to minute detailed structures in chert. Because of intense weathering, the residuum from the Knox contains vast quantities of chertified algal stromatolites. On the basis of detailed study of bedrock and residual products, ancient algal structures in the Knox are limited to algal mats, stratiform layers, and growth domes, with only a rare spheroidal structure (fig. 3). By analogy with Holocene algal stromatolite zonation (fig. 2) the absence of mechanically produced structures, such as polygonal forms and mechanical domes, suggests that intertidal and supratidal environments were absent or poorly developed during deposition of the Knox.

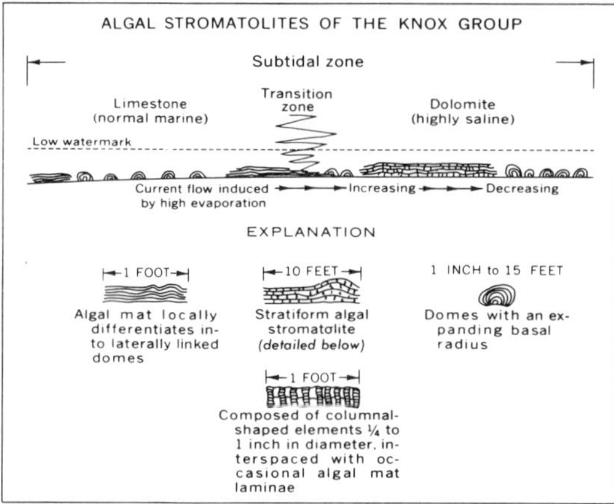

Figure 3.—Forms and distribution of Late Cambrian and Early Ordovician algal stromatolites in the Knox Group. Comparison of the small-scale algal mat structure with the large-scale stratiform algal structures suggests many similarities; the main difference is size.

Beds consisting of near-parallel paper-thin laminae, similar to those described (Black, 1933, p. 170–171) as the result of rhythmic alternation of Holocene carbonate sediments and dark algal films, are common in the Knox Group. Unfortunately, rhythmic alternation in organic content does not in itself identify laminated algal structures. Periodic changes in salinity, which are not unlikely in the highly saline environment suggested for the Knox, could cause mass mortality in phytoplankton (Emery and Stevenson, 1957, p. 687), which in turn would result in the deposition of a lamina high in organic matter, resembling a primary algal lamina. The lack of suitable criteria to differentiate between laminated algal sediments and organic lamination from other processes limits the number of such beds that can be assigned to an algal origin.

Paper-thin laminated beds in the Knox Group without exception have been found to have a clay content ranging from 10 to more than 25 percent, strongly suggesting that the lamination primarily resulted from rhythmic changes in clay content. That rhythmic changes in amounts of clay alternate with algal mat development is implied by the fact that locally some paper-thin laminated beds are found to differentiate into LLH structures.

The most abundant algal stromatolite in the Knox Group is in regionally persistent stratiform units in which the basic structural element is a crude column exhibiting paper-thin and vertically stacked hemispheroidal laminae. Thousands of the individual basic columnar elements, from ¼ inch to 1 inch in diameter, are in laterally continuous layers from 1 inch to 3 feet thick. Silicified specimens in which many primary structures are preserved in detail, suggest that the layering results from periodic interruption in upward growth of individual columnal elements. These interruptions are recorded as laminae which form natural partings that present-day weathering tends to accentuate. In all respects the growth layers imitate algal mat development—they form laterally continuous subparallel stratiform layers that locally differentiate into a series of hemispheroids several feet in height and more than 10 feet in diameter. The major difference between the laminae of an algal mat and the layers in the columnar structures is a matter of scale, plus the fact that each growth layer is composed of thousands of individual units (fig. 3).

Stratiform algal stromatolites in massive units from 3 to 140 feet thick are confined to the dolomite facies of the Knox Group of eastern Tennessee and southwestern Virginia. Detailed studies indicate that many of these thick stratiform zones persist laterally for more than 50 miles. Intercalated within these zones are thin beds of oolite, intraformational round-pebble conglomerate, and dololutite. Lateral facies changes from algal to nonalgal material involving individual beds within the zone can be transitional or abrupt. Where transitional, the stratiform algal beds change laterally to algal debris, then to oolite and dololutite, and finally to dololutite containing abundant growth domes; where abrupt, they change directly to dololutite with thin beds of oolite, intraformational round-pebble conglomerate, and growth domes. Distribution of insoluble residues across the facies boundaries shows a direct relationship between the amount of clay residue and rock type. Clay content of stratiform algal stromatolites, oolite, and some debris beds is less than 1 percent; in contrast, the clay content of all dololutite beds exceeds 5 percent. The low clay content of the algal beds, oolites, and algal debris suggests that these rocks were deposited in an environment where the energy level was sufficiently great to prevent the mechanical accumulation of clay, whereas dololutite with higher clay content accumulated in areas of lower energy.

The laterally continuous nature of stratiform algal stromatolite deposits suggests that uniformly widespread subtidal environmental conditions must be responsible for their development. Low clay content and presence of oolite and round-pebble intraformational conglomerate implies that stratiform algal deposits developed in areas of moderate current activity where scouring action of the current prevented growth between columnals. Periodic reduction in current velocities encouraged limited growth of laterally continuous algal laminae. Exceptional conditions exist in present-day hypersaline environments where water lost to high evaporation is continuously replaced by oceanic water (Emery and Stevenson, 1957, p. 686). The continuous inflow of marine water to replace water lost to evaporation in hypersaline environments may be the mechanism which accounts for the widespread development of laterally persistent stratiform algal zones and their restriction to the dolomite facies of the Knox Group.

Unlike the SH-C domes of Logan, Rezak, and Ginsburg (1964), which are defined as discrete structures in which succeeding hemispheroidal laminae develop without an increase in the basal radius, algal growth domes in the Knox Group develop with an increasing basal radius. Growth domes are distinct and nearly symmetrical structures markedly different from the randomly disposed and asymmetrical mechanical domes developed in Holocene intertidal settings. Normally, growth domes in the Knox are arranged in laterally continuous layers composed of hundreds or thousands of individual domes. The size of individual domes within a layer ranges from a few inches up to 15 feet in diameter. The initial spacing of individual domes within a layer is critical and determines the ultimate size and configuration of the mature structures. If growth is initiated at widely spaced intervals, individual elements develop without interference into large, wide-spaced domes; but if development starts at more closely spaced intervals, the growing domes soon impinge on one another and compound forms with complex growth patterns develop.

That both limestone and dolomite facies contain domes demonstrates that their growth was not dependent on variations in salinity. However, their growth does appear to be keyed to subtidal zones with low current velocities, where they are associated with low-energy lithologies such as calcilutite and dololutite containing more than 5 percent insoluble clay.

MODELS OF DOLOMITIZATION

Studies of Holocene massive carbonate deposits have emphasized that dolomite is formed as a penecontemporaneous replacement of calcium carbonate and that the process of dolomitization is operative in supratidal areas in which magnesium-rich hypersaline solutions are developed (Illing and others, 1965; Shinn and others, 1965). Similarly, studies of ancient evaporite deposits have emphasized that subtidal dolomite can be an early primary deposit (Irwin, 1965) or can form late in the evaporite cycle as a consequence of refluxion of magnesium-rich brines (Adams and Rhodes, 1960). The unifying theme emerging from these studies is that dolomitization can occur in both supratidal and subtidal environments, where high evaporation rate and low fresh-water dilution produce magnesium-rich brines. Salinity systems with the potential to develop and maintain the necessary chemical and physical factors favoring widespread subtidal dolomitization through a considerable time span have been described for the deepwater barred basin and the shallow epicontinental sea model (fig. 4).

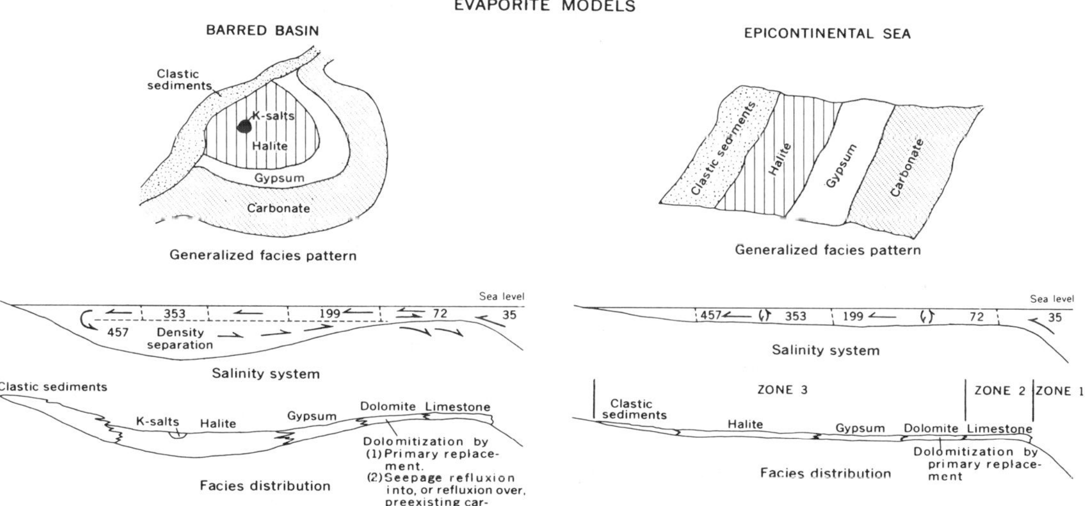

Figure 4.—The salinity systems, facies distributions, and generalized facies patterns of the barred-basin and epicontinental sea evaporite models. Based on data from King (1947), Scruton (1953), Adams and Rhodes (1960), Shaw (1964), Irwin (1965), and Hite (1968, 1970). The values given in the salinity systems are parts per thousand.

Barred basin

The barred-basin evaporite model (Ochsenius, 1877; King, 1947), which is operative in warm arid areas where evaporation exceeds dilution by fresh and oceanic water, is generally a relatively deep topographic depression separated from the open ocean by a shallower physical barrier. Commonly the barrier has been a shelf covering thousands of square miles (Hite, 1970, p. 50) that nearly or completely surrounds the basin. As sea water moves across the shallow shelf toward the basin, a strong horizontally increasing salinity gradient is produced in response to high evaporation rates (Lang, 1937). Continued evaporation establishes a circulatory system in the basin because of density stratification and density flow. Dense brine in the farthest reaches of the basin sinks and flows back beneath the less dense influxing current toward the bottom of the basin; some of this brine is refluxed back over the barrier to the sea (King, 1947, p. 474). The combination of chemical and physical factors controlling the circulation in a barred basin produces a distinctive and somewhat circular facies pattern in which chemical and biochemical carbonates occur on the fringing shelf and the more soluble salts occur near the center (fig. 4). A large part of the shelf carbonate is thought to have been dolomitized by seepage refluxion through permeable beds (Adams and Rhodes, 1960) or by refluxion of the brines over the carbonate shelf (Hite, 1970, p. 53).

Continental shelf

Although several shallow-water sedimentation models have been proposed to explain sedimentary patterns in a variety of carbonate and evaporite sequences, only relatively recently has it been recognized that certain elements were common to all shallow-water models, enabling Shaw (1964) and Irwin (1965) to describe a generalized model for epicontinental sedimentation. Shaw and Irwin emphasize that an epicontinental sea environment is characterized by the great width of the shelf (hundreds of miles) where bottom depositional slopes are on the order of less than 1 foot per mile, and where circulation, except near the edge, is restricted by the shallow depth of the sea. According to Shaw and Irwin, lateral variation in biochemical, chemical, and physical factors from shore to deeper water results in the development of three major zones of sedimentation. These sedimentary zones, in contrast to the nearly circular sedimentary pattern of a barred basin, tend to form a series of broad bands roughly parallel to the shore (fig. 4). Their three zones include:

1. An open ocean zone of low energy, hundreds of miles wide, where the depth of water restricts the hydraulic energy. Sediments are largely derived from zone 2 (see below) and consist of fine-grained organic debris. Bottom conditions tend to be reducing, thus preserving introduced organic matter and producing dark sediments.

2. A zone of high energy, tens of miles wide extending from the point where waves first strike the bottom to the limit of tidal action. Physical, chemical, and organic activity within this zone causes an orderly shoreward progression of facies ranging from biogenetic sediments to oolites and coated grains to pellets to mud.

3. A zone of low energy, free from the effects of oceanic waves and tides, extending from zone 2 to shore. In areas of high evaporation and low fresh-water runoff, salinities should increase progressively shoreward, initiating chemical precipitation keyed to that gradient. Precipitation begins with calcium carbonate and progresses consecutively through dolomite to gypsum, halite, and other more soluble salts.

Wind-generated currents within a shallow epicontinental sea tend to homogenize the water mass to a depth of at least 65 feet, thereby preventing the development of vertical density stratification. Instead, a horizontal gradient is established with increasing salinity and density toward the shore (fig. 4). The higher density water mass is prevented from free refluxion to the open ocean by the lack of sufficient hydrologic head to overcome the frictional resistance of the gently sloping sea bottom (Scruton, 1953). This frictional entrapment of highly saline water is in contrast to the hydrologic system of the deeper barred basin, where circulation is dependent upon a density differential between influxing surface currents and refluxing bottom currents. The existence at the present time of hydrologic condition similar to the epicontinental sea model but on a smaller scale seems well established (Emery and Stevenson, 1957), particularly by the long-term measurements (13 years) in Shark Bay, Western Australia (Logan and Cebulski, 1970), and the consistencies in data obtained through a 29-year time span for the Great Bahama Bank (Black, 1933; Smith, 1940; Newell and others, 1959; Cloud, 1962).

Shark Bay, Western Australia, with a semiarid subtropical climate is a shallow-water embayment open to the ocean, whereas the Great Bahama Bank with a humid subtropical climate is a shallow-water carbonate shelf surrounded by the open ocean. Despite these environmental differences, the hydrologic systems of both areas are similar in that high salinities are formed in response to high evaporation, small tidal influx, and low fresh-water dilution. Surface and bottom salinity in Shark Bay shows a gradient from normal marine salinity at the bay mouth to hypersalinity in the bay head (Logan and Cebulski, 1970, fig. 6). Cross sections drawn through the long axis of the bay show no horizontal density stratification, nor do isohalines show any relationship to bottom topography (fig. 5). Instead, the structure of the high-salinity water mass is nearly vertical, with increasing salinity and density toward the bay head. The vertical isosalinity zones, which are uniformly oxygenated from top to bottom, result from continuous homogenization by currents and wave action (Logan and Cebulski, 1970, p. 14). Surface

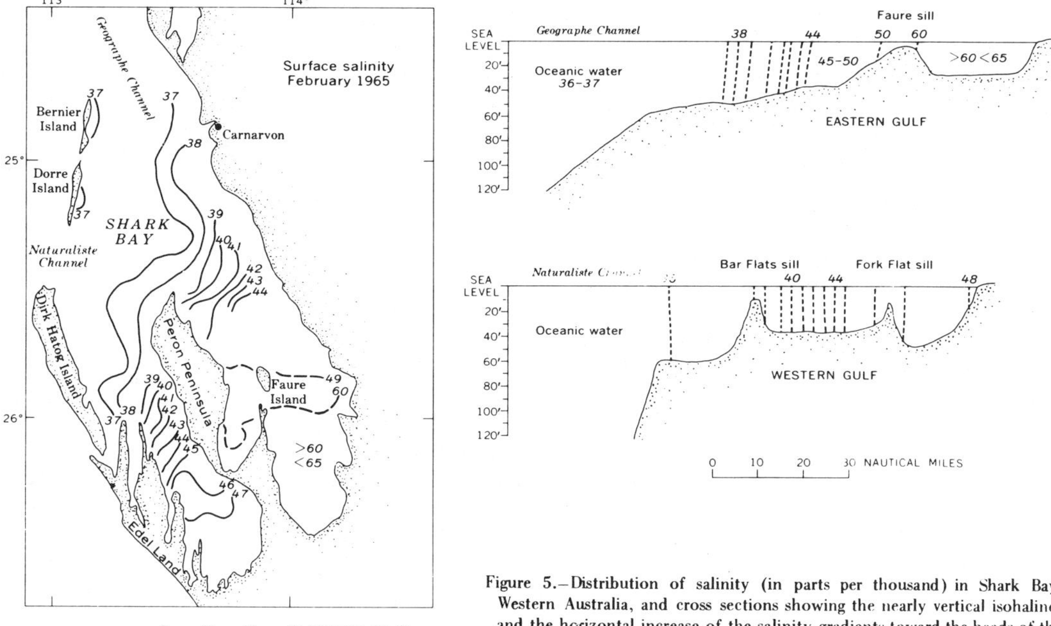

Figure 5.—Distribution of salinity (in parts per thousand) in Shark Bay, Western Australia, and cross sections showing the nearly vertical isohalines and the horizontal increase of the salinity gradients toward the heads of the two gulfs. From Logan and Cebulski (1970).

and bottom salinity gradients (Cloud, 1962, table 25) for part of the Great Bahama Bank show that the structure of the Bahama high-salinity mass is defined by steeply inclined isohalines, similar to the configuration at Shark Bay (fig. 6).

Circulation in Shark Bay and on the Great Bahama Bank is sluggish, having an estimated residual time for a particular water mass to complete a circuit from open ocean and back to be on the order of 5 months in Shark Bay (Logan and Cebulski, 1970, p. 21) and 3 months in the Bahamas (Smith, 1940, p. 156). An equilibrium condition between evaporation rate and oceanic exchange has been accomplished in Shark Bay by the constant slow movement of tidal currents bringing oceanic water in on east-facing shores and a reflux of higher salinity water out along west-facing shores (Logan and Cebulski, 1970, p. 21). Tidal currents play a minor role as a controlling factor in the circulation on the Great Bahama Bank. The direction of tidal flooding is such that its effect is largely canceled out; thus, residual current directions are determined by seasonal changes in prevailing winds (Smith, 1940, p. 15). During the summer months when prevailing winds are from the east, continuing evaporation results in the development of a high-salinity water mass with its apex off the northwest shore of Andros Island (fig. 6). Water with salinities greater than 39 parts per thousand is carried westward by persistent residual currents to cover much of the bank. Apparently, 40 to 60 inches of rain, mainly during the summer months, tends to prevent the salinity of the Bahama water mass from attaining concentrations greater than 45 parts per thousand. During the winter months the wind shifts to the northwest and oceanic water of lower salinity is carried shoreward to dilute the high-salinity mass. Thus, the salinity system developed on the bank is dynamic and tends to move and deteriorate seasonally.

DOLOMITIZATION

Studies of Holocene dolomitization, especially that of Illing, Wells, and Taylor (1965), emphasize that magnesium-ion concentrations are sufficiently great early in the evaporite cycle to initiate dolomitization before large concentrations of gypsum or anhydrite are deposited. The relationship is well illustrated by a plot of Illing, Wells, and Taylor's data from the Persian Gulf (fig. 7). Incipient dolomitization begins subtidally as suggested by the combination of minor dolomite and abundant magnesian calcite. Supratidally, where dolomitization is more complete, dolomite is abundant and magnesian calcite is sparse. The mole Mg-Ca ratio of pore water increases to about 19 at the seaward edge of the supratidal area, where gypsum first appears in the sediments; thereafter it decreases as magnesium is subtracted by the process of dolomitization. The relatively slight increase in chlorinity across the supratidal sabkha suggests that the system is near equilibrium and that

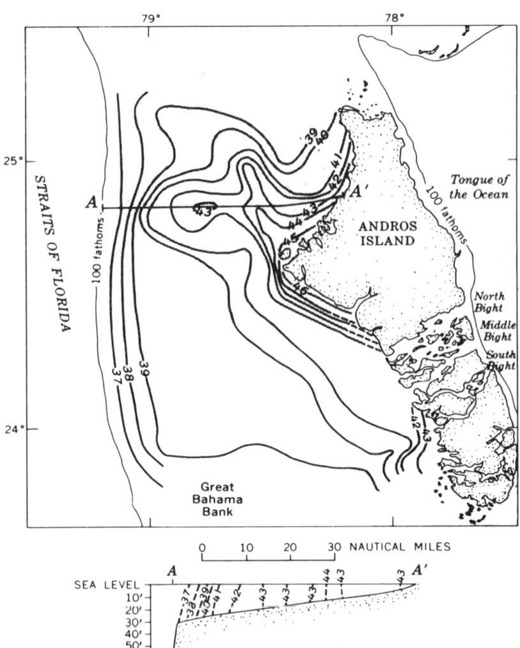

Figure 6.—Distribution of salinity (in parts per thousand) on the Great Bahama Bank, and cross section showing the nearly vertical isohalines and the horizontal increase of the salinity gradient shoreward. Shaded areas contain more than 3 percent magnesium carbonate. Data from Newell and Rigby (1957) and Cloud (1962).

dolomitization is proceeding rapidly, accompanied by deposition of small amounts of gypsum. Accordingly, dolomitization occurs early in the evaporite cycle during the transition from deposition of calcium carbonate to gypsum. Significantly, many ancient evaporite deposits have a marked deficiency of magnesium relative to its abundance in sea water (Stewart, 1963, p. 40), lending support to the concept that magnesium in ancient evaporite basins was extracted early in the process of dolomitization, thereby depleting residual brines so that only small quantities of magnesium salts were available for precipitation and refluxion. The natural progression of facies from limestone to dolomite to gypsum to halite to potassium salt is simply a reflection of the varying concentration of salts within a salinity system.

The process of magnesium concentration in sediments appears to be operative subtidally within a present-day shallow carbonate environment. This is suggested by the comparison of the distribution of salinity on the Great Bahama Bank (Cloud, 1962, table 25) to distribution of magnesium (Newell and Rigby, 1957, fig. 13). Areas of greatest magnesium in the bottom sediment appear to be closely allied to the configura-

tion of the Bahama salinity system; in contrast, sediments with low magnesium content are deposited where oceanic water with lower salinities circulates more freely, such as north of and southwest of Andros Island and in the area of the bights near the center of Andros (figs. 6, 8). Cloud (1962, p. 94) suggests that most high-magnesium calcite in the bottom sediment of the bank is probably related to the skeletal content of the sediment. However, the amount of magnesium in skeletal material is thought to be dependent upon temperatures and salinity of the environment as well as the genetic effect of the organism (Chave, 1954; Lowenstam, 1954; Chilingar, 1962). Magnesium in the Bahamas may well be all concentrated in skeletal material, but this simply reflects the high-salinity environment within which the material formed. Magnesium distribution superimposed on a facies map of the Great Bank (Imbrie and Purdy, 1962, fig. 10) demonstrates that no clear-cut relation exists between facies and magnesium content (fig. 8). The lack of control by facies lends credence to the inference that the distribution of magnesium in the bottom sediments is related to the configuration of the salinity system on the Great Bahama Bank.

The epicontinental-sea salinity system of Shaw (1964) and Irwin (1965) and its shoreward horizontally increasing salinity gradient has the potential to develop the necessary chemical and physical factors to promote widespread subtidal dolomitization. I believe that such a system is responsible for the regional development of subtidal dolomite during Late Cambrian and Early Ordovician time.

LATE CAMBRIAN AND EARLY ORDOVICIAN SEDIMENTATION MODEL

Massive evaporite deposits attributed to deepwater deposition have not been described in the Knox Group. Minor amounts of anhydrite in vugs, small veins, and thin laminae have been observed in Tennessee and also in Lower Ordovician carbonate rocks in Texas (Barnes, 1959, p. 57) and Illinois (Saxby and Lamar, 1957, p. 14). The limited amounts of anhydrite and the broad-band facies distributions (figs. 1 and 4) tend to rule out the barred-basin model as a means of explaining deposition of dolomite in the Knox. Instead, the epicontinental-sea salinity system of Shaw (1964) and Irwin (1965) and its shoreward increasing salinity gradient is postulated as a more likely environment for the deposition of this facies. Characteristically, the deposition of limestone and dolomite in the Knox was related to the epicontinental salinity gradient, so that a shoreward succession exists from normal marine limestone on the east through a transitional zone to a highly saline dolomite phase on the west (fig. 9). Subsurface information (summarized by Lochman-Balk, 1971) shows that the Knox gradually thins northwestward toward Illinois and is replaced by a dominantly clastic sequence (fig. 1). Primary characteristics of outcropping Upper Cambrian and Lower Ordovician rocks of Wisconsin (Ostrom, 1970; R. A. Davis, 1970) compared to the primary characteristics of equivalent

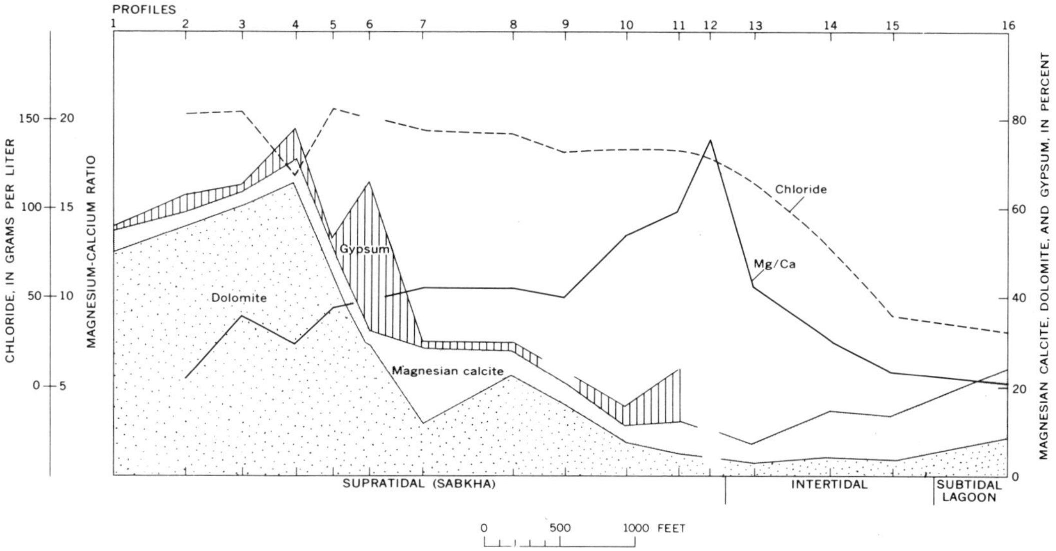

Figure 7.—Profiles to a depth of 58 cm across the Sabkha Faishakh on the Persian Gulf, showing changes in chlorinity, Mg-Ca ratio, and percentage of magnesian calcite, dolomite, and gypsum. The process of dolomitization appears to intensify progressively from incipient dolomitization subtidally to massive dolomitization in the high supratidal area. Adapted from Illing, Wells, and Taylor (1965).

age rocks of the Knox Group suggest that in near-shore areas the ancient salinity system did not continue to increase (fig. 10). Apparently fresh-water dilution from the bordering land lowered salinities to the point that a more diverse fauna developed, as well as contributed toward the preservation of minor amounts of limestone.

Reconstruction of the configuration of the original carbonate shelf of the Knox Group is hampered by numerous thrust faults of regional extent in eastern Tennessee (fig. 11). These faults with displacements on the order of miles obscure regional facies trends by burying large segments of the original depositional shelf and bringing into juxtaposition the facies which accumulated on separate parts of the shelf. This is especially true in regard to the limestone facies. Before faulting, the original width of the area of limestone deposition was on the order of 100 miles (fig. 1). Mass transport by faults has buried the easternmost edge of the original shelf where sediments reflecting high energy were deposited. Consequently, only that part of the limestone sequence deposited under a low-energy condition is preserved in eastern Tennessee. Despite the fact that parts of the record are obscure, generalized facies trends can be ascertained in a series of stratigraphic sections extending from the easternmost outcrop belt in eastern Tennessee to deep wells in eastern Kentucky (fig. 9). A summary of selected primary features of the limestone facies, the transition zone, and the dolomite facies clearly indicates that many of the same primary elements are

found throughout all facies of the Knox Group (fig. 12), emphasizing the regional impress of the overall subtidal environment. However, physical and chemical differences within the subtidal environment tended to restrict the abundance and even the presence of few elements to particular facies.

Where the limestone facies is present, it is characterized by dark-gray calcilutite containing thin zones of paper-thin laminated beds, angular intraclasts, oolite, arenite, and domal algal stromatolites. This combination of lithologies and algal domes suggests that the eastern limestone facies was deposited in water of normal salinity near wave base but not below the zone of photosynthesis. The energy level as suggested by the abundance of dark-gray calcilutite was normally low, although intraclasts and arenite beds record an occasional short-term increase in energy from strong storm currents. That a low-energy level was maintained throughout deposition of the limestone sequence is probably related to the fact that much of the ocean's energy from tides and wave action was largely dissipated before reaching that part of the facies now preserved in eastern Tennessee.

The transition zone between the dominantly limestone facies on the east and the dolomite facies on the west is confined to a relatively narrow geographic belt (fig. 9). Prouty (1947; 1948, p. 1618) thought that this abrupt facies change was related to a structural barrier, but my stratigraphic work shows no unusual thickening or thinning across the so-called

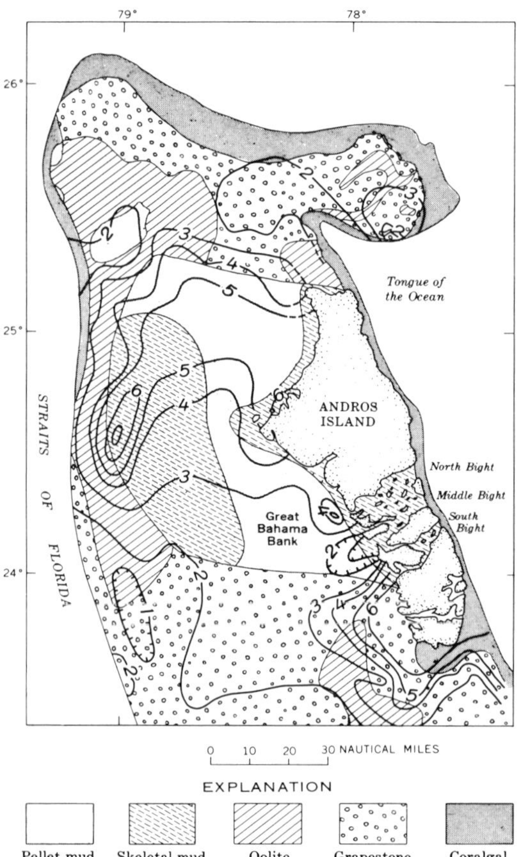

Figure 8.—Map of part of the Great Bahama Bank, showing the distribution of facies and magnesian carbonate content of sediments, in percent. From Newell and Rigby (1957) and Imbrie and Purdy (1962).

barrier. I believe that the transition zone, which has been partly telescoped by Appalachian thrusting, marks the location of a relatively constant geographic zone of change from normal marine water to penesaline water within an epicontinental sea. Surface and subsurface studies (Swartz, 1948; Donaldson and Page, 1964; Perry, 1964; Barnes, 1959; Chenoweth, 1968; Williams, 1969) suggest that this same transition zone within carbonate rocks equivalent to the Knox Group persists from New York to Texas (fig. 1), thus implying that the change from normal marine water to penesaline water existed completely around the edge of the ancient continental shelf.

The transition zone characteristically contains about equal proportions of dolomite and limestone, apparently reflecting the gradual lateral shift of the eastern part of the high-salinity water mass where magnesium concentration was large enough to promote dolomitization. The facies within the transition zone contains a higher proportion of oolite in relation to intraclasts and records the first stratiform algal stromatolites in dolomite. The transition facies apparently accumulated near the limit of oceanic wave and tidal effect, but under the influence of induced currents produced by water loss to evaporation.

The dolomite facies of the Knox Group contains a diversity of rock types ranging from dololutite to oolite, produced by widely different levels of depositional energy within a high-salinity environment. Forcefully emphasizing that although dolomitization may be controlled by the concentration of magnesium ions within an epicontinental sea, it does not follow that concentration of magnesium can only occur in low-energy environments. Despite the fact that a dolomite facies was deposited largely beyond the reach of oceanic waves and tidal currents, circulation within the high-salinity water mass was not static. Instead, high evaporation induced a constant shoreward current flow, which had a profound effect upon the kind and lateral distribution of facies. These induced currents were uniform and widespread, and caused the deposition of laterally persistent lithologic units as little as 1 inch thick. Where these currents were moderate, oolite and massive-bedded stratiform algal stromatolite as much as 140 feet thick accumulated; where current velocities were low, dololutite with domal algal stromatolites accumulated. Because of the similarity in energy levels, the distribution of dololutite, oolite, and stratiform algal stromatolite of the highly saline environment closely resembles the distribution pattern of like lithologic elements in the normal marine limestone phase of Shaw (1964) and Irwin (1965), where oolite and reefs accumulated in the higher energy zones at the shelf edge and calcilutite accumulated in areas of lower energy.

SUMMARY

Approximately 90 percent of the Upper Cambrian and Lower Ordovician carbonate strata is dolomite. This is in sharp contrast to Holocene carbonate deposition sites where most of the carbonate strata are limestone and only a small percentage of the total sediment is dolomite, mainly confined to the supratidal area. The epicontinental-sea sedimentation model and its shoreward increasing salinity gradient is postulated to be the likely mechanism responsible for widespread subtidal dolomitization during Late Cambrian and Early Ordovician time. Comparison of the epicontinental sedimentation model to present-day carbonate depositional sites suggests that Holocene sites represent only the normal marine limestone phase of the epicontinental model.

Present-day examples of the epicontinental salinity system, although measured only by tens of miles, illustrate that appreciable salinity increase can occur without physical

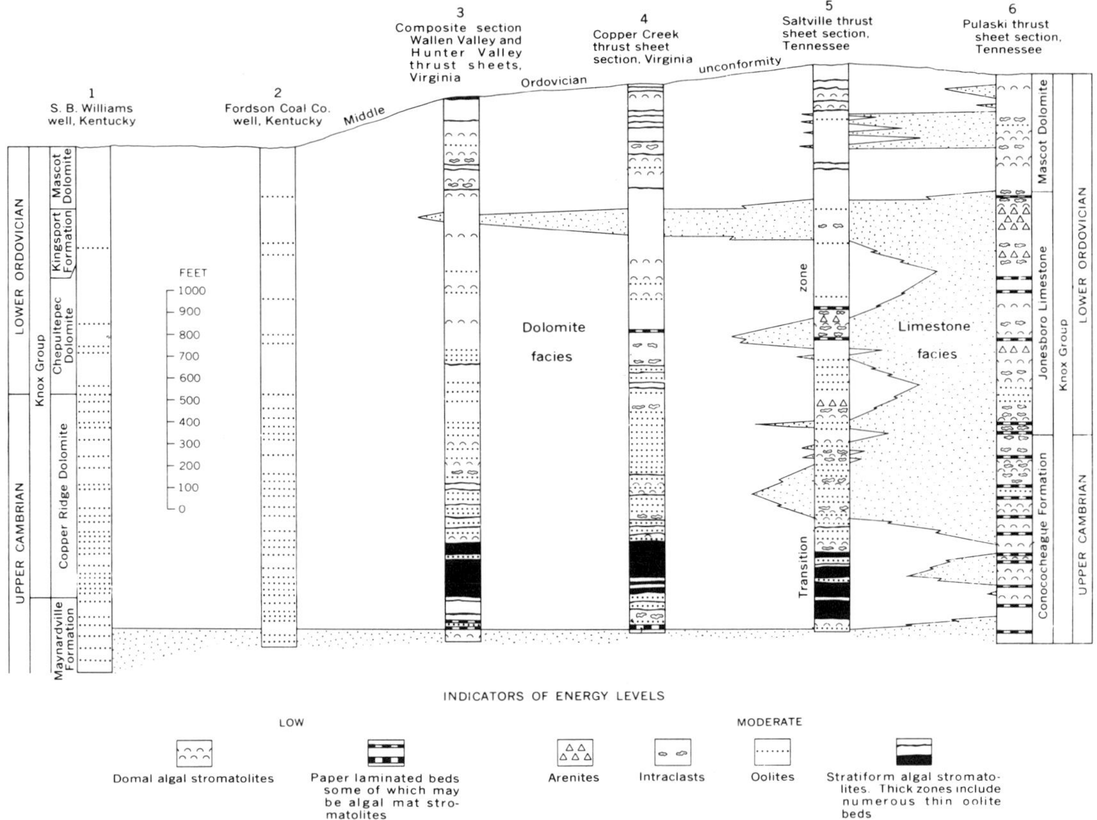

Figure 9.—Sections showing the distribution of energy-indicative beds in the Upper Cambrian and Lower Ordovician carbonate strata in wells in Kentucky and surface sections in Virginia and Tennessee. Algal stromatolites could not be identified in well cuttings from Kentucky. (See fig. 11 for line of section.)

barriers. Such a system developed on an ancient, shallow continental shelf, hundreds of miles wide, could easily concentrate sea water to the point of gypsum saturation. This would result in the development of subtidal dolomite on a regional scale. Increased fresh-water runoff from land or seasonal shifts in wind direction, as in the Bahamas, would contribute to a seasonal deterioration of the ancient salinity system. Resolution of calcium sulfate during the period of lower salinity would account for only trace amounts of gypsum now found in the rocks of Late Cambrian and Early Ordovician age.

Sedimentation within an epicontinental sea, as typified by the Knox Group of Late Cambrian and Early Ordovician age in the southern Appalachians, is closely related to a westward-increasing salinity gradient, so that a lateral transition from normal marine limestone facies to a transition zone to a dolomite facies exists. Domal algal stromatolites are abundant throughout both the limestone and dolomite facies of the Knox, indicating that variation in salinity has little effect on their growth, but that water depth during deposition of either facies was not below the zone of photosynthesis. Limestone deposition was restricted to a rather narrow band at the oceanic bordering edge of the epicontinental sea, where continual flushing by wave and tidal currents prevented extensive buildup in salinity. The transition zone developed near the limit of tidal exchange, thus it occupies a relatively fixed geographic position throughout deposition of the Knox. Although the dolomite facies was deposited in high-salinity water beyond the reach of oceanic and tidal currents, energy levels within that environment were not uniformly low. Instead, the diversity in rock type, from dololutite to oolite, suggests that energy levels ranged from low to moderate. The

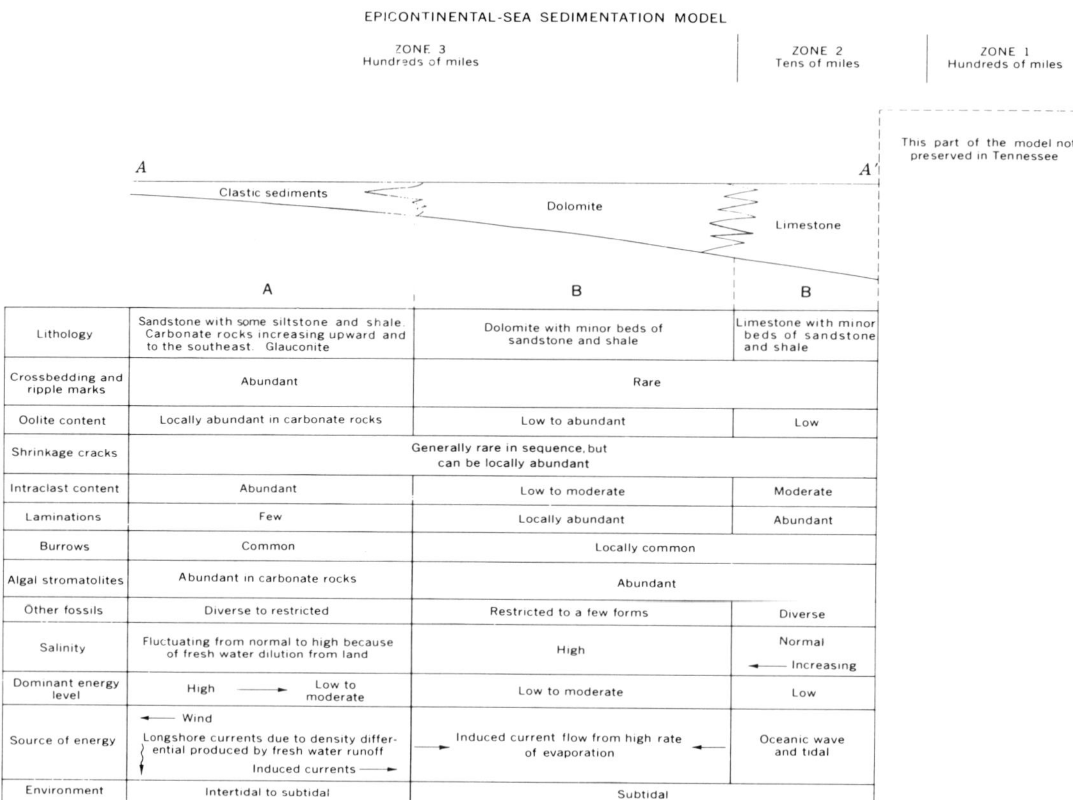

	A	B	B
Lithology	Sandstone with some siltstone and shale. Carbonate rocks increasing upward and to the southeast. Glauconite	Dolomite with minor beds of sandstone and shale	Limestone with minor beds of sandstone and shale
Crossbedding and ripple marks	Abundant	Rare	
Oolite content	Locally abundant in carbonate rocks	Low to abundant	Low
Shrinkage cracks	Generally rare in sequence, but can be locally abundant		
Intraclast content	Abundant	Low to moderate	Moderate
Laminations	Few	Locally abundant	Abundant
Burrows	Common	Locally common	
Algal stromatolites	Abundant in carbonate rocks	Abundant	
Other fossils	Diverse to restricted	Restricted to a few forms	Diverse
Salinity	Fluctuating from normal to high because of fresh water dilution from land	High	Normal ◄── Increasing
Dominant energy level	High ──► Low to moderate	Low to moderate	Low
Source of energy	◄── Wind Longshore currents due to density differential produced by fresh water runoff Induced currents ──►	Induced current flow from high rate of evaporation	Oceanic wave and tidal
Environment	Intertidal to subtidal	Subtidal	

Figure 10.—Summary of primary characteristics and interpretive environments of the rocks of the Knox Group of Tennessee (B) compared with similar parameters of equivalent age rocks in Wisconsin (A). Cross section is line $A-A'$ of figure 1. (Wisconsin data from Ostrom, 1970, and R. A. Davis, 1970).

distribution of the different rock types was related to circulatory patterns induced where a constant shoreward current flow developed from replacement of water by evaporation loss. Where velocities of the induced current were moderate, oolite and stratiform algal stromatolite development is favored; where velocities were low, dololutite with domal algal stromatolites accumulated.

REFERENCES CITED

Adams, J. E., and Rhodes, M. L., 1960, Dolomitization by seepage refluxion: Am. Assoc. Petroleum Geologists Bull., v. 44, no. 12, p. 1912–1920.

Barnes, V. E., 1959, Stratigraphy of the pre-Simpson Paleozoic subsurface rocks of Texas and southeast New Mexico, V. 1–2: Texas Univ. Pub. no. 5924, 836 p.

Black, Maurice, 1933, The algal sedimentation of Andros Island, Bahamas: Royal Soc. London Philos. Trans. ser. B, vol. 222, p. 165–192.

Burst, J. F., 1965, Subaqueously formed shrinkage cracks in clay: Jour. Sed. Petrology, v. 35, no. 2, p. 348–353.

Chave, K. E., 1954, Aspects of the biochemistry of magnesium—[Pt.] 1, Calcareous marine organisms: Jour. Geology v. 62, no. 3, p. 266–283.

Chenoweth, P. A., 1968, Early Paleozoic (Arbuckle) overlap, southern Mid-continent, United States: Am. Assoc. Petroleum Geologists Bull. v. 52, no. 9, p. 1670–1688.

Chilingar, G. V., 1962, Dependence on temperature of Ca/Mg ratio of skeletal structure of organisms and direct chemical precipitation out of sea water: Southern California Acad. Sci. Bull., v. 61, p. 45–60.

Cloud, P. E., Jr., 1962, Environment of calcium carbonate deposition west of Andros Island, Bahamas: U.S. Geol. Survey Prof. Paper 350, 138 p.

Crawford, Johnson, 1945, Structural and stratigraphic control of zinc deposits in East Tennessee: Econ. Geology, v. 40, p. 408–415.

Davis, G. R., 1970, Algal-laminated sediments, Gladstone embayment, Shark Bay, Western Australia, in Carbonate sedimentation and environments, Shark Bay, Western Australia: Am. Assoc. Petroleum Geologists Mem. 13, p. 169–205.

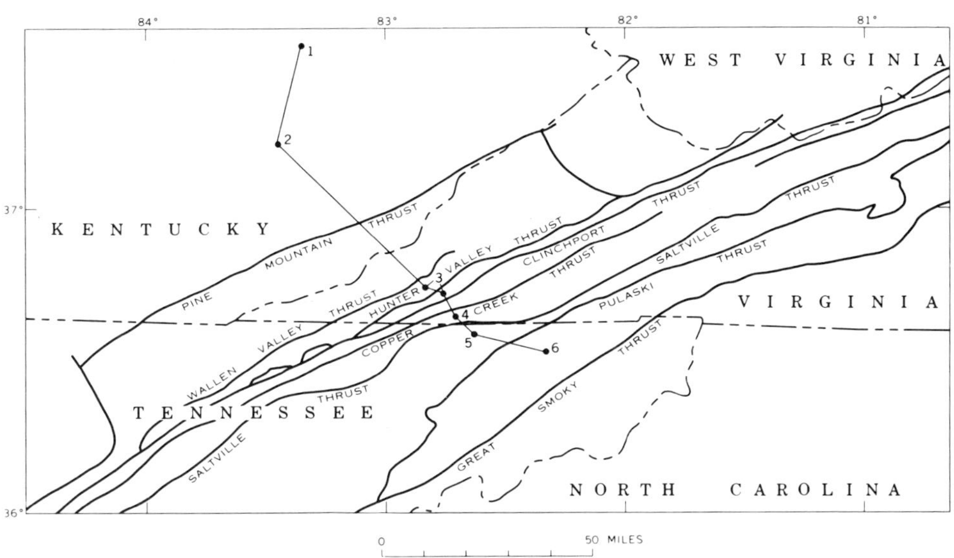

Figure 11.—Map showing the location of wells and measured sections in figure 9 and the trace of major thrust faults in the area.

Davis, R. A., Jr., 1970, Prairie Du Chien Group in the Upper Mississippi Valley, *in* Field trip guidebook for Cambrian-Ordovician geology of western Wisconsin—Geol. Soc. America, Ann. Mtg., Milwaukee, Wis., 1970: Wisconsin Geol. and Nat. History Survey Inf. Circ. 11, p. 35–44.

Donaldson, Alan, and Page, Ronald, 1964, Stratigraphic reference sections of Elbrook and Conococheague Formations (Middle and Upper Cambrian) and Beekmantown Group (Lower Ordovician) in Berkeley and Jefferson Counties, West Virginia: West Virginia Acad. Sci. Proc., v. 35, p. 149–171.

Emery, K. O., and Stevenson, R. E., 1957, Estuaries and lagoons—[Pt.] 1, Physical and chemical characteristics, *in* Chap. 23 *of* Hedgpeth, J. W., ed., Ecology: Geol. Soc. America Mem. 67, p. 673–693.

Folk, R. L., 1959, Thin-section examination of pre-Simpson Paleozoic rocks, *in* V. 1 *of* Barnes, V. E., Stratigraphy of the pre-Simpson Paleozoic subsurface rocks of Texas and southeast New Mexico: Texas Univ. Pub., no. 5924, p. 95–130.

Gebelein, C. D., 1969, Distribution, morphology, and accretion rate of Recent subtidal algal stromatolites, Bermuda: Jour. Sed. Petrology, v. 39, no. 1, p. 49–69.

Ginsburg, R. N., 1960, Ancient analogues of Recent stromatolites: Internat. Geol. Cong., 21st, Copenhagen, 1960, Rept., pt. 22, p. 26–35.

Harris, L. D., 1969, Kingsport Formation and Mascot Dolomite (Lower Ordovician) of East Tennessee, *in* Papers on the stratigraphy and mine geology of the Kingsport and Mascot Formations (Lower Ordovician) of East Tennessee: Tennessee Div. Geology Rept. Inv. 23, p. 1–39.

Herbert, Paul, Jr., and Young, R. S., 1956, Sulfide mineralization in the Shenandoah Valley of Virginia: Virginia Div. Geology Bull. 70, 58 p.

Hite, R. J., 1968, Distribution and geologic habitat of marine halite and associated potash deposits, *in* Seminar on sources of mineral raw materials for the fertilizer industry in Asia and the Far East: U.N. ECAFE Mineral Resources Devel. Ser. 32, p. 307–326.

——— 1970, Shelf carbonate sedimentation controlled by salinity in the Paradox basin, southeast Utah, *in* Rau, J. L., and Dellwig, L. F., eds., Third symposium on salt, V. 1: Cleveland, Ohio, Northern Ohio Geol. Soc., p. 48–66.

Holland, C. H., ed., 1971, Cambrian of the New World, V. 1: New York and London, Interscience Publishers, 456 p.

Illing, L. V., Wells, A. J., and Taylor, J. C. M., 1965, Penecontemporary dolomite in the Persian Gulf, *in* Dolomitization and limestone diagenesis—A symposium: Soc. Econ. Paleontologists and Mineralogists Spec. Pub. 13, p. 89–111.

Imbrie, John, and Purdy, E. C., 1962, Classification of modern Bahamian carbonate sediments, *in* Classification of carbonate rocks—A symposium: Am. Assoc. Petroleum Geologists Mem. 1, p. 253–272.

Irwin, M. L., 1965, General theory of epeiric clear water sedimentation: Am. Assoc. Petroleum Geologists Bull., v. 49, no. 4, p. 445–459.

Kendall, C. G., and Skipwith, P. A., 1968, Recent algal mats of a Persian Gulf lagoon: Jour. Sed. Petrology, v. 38, no. 4, p. 1040–1058.

King, R. H., 1947, Sedimentation in Permian Castile Sea: Am. Assoc. Petroleum Geologists Bull. v. 31, no. 3, p. 470–477.

Lang, W. T. B., 1937, The Permian formation of the Pecos Valley of New Mexico and Texas: Am. Assoc. Petroleum Geologists Bull., v. 21, p. 833–898.

Laporte, L. F., 1971, Paleozoic carbonate facies of the central Appalachian shelf: Jour. Sed. Petrology, v. 41, no. 3, p. 724–740.

Lochman-Balk, Christina, 1971, The Cambrian of the craton of the United States, *in* Cambrian of the New World, V. 1: New York and London, Interscience Publishers, p. 79–167.

Logan, B. W., 1961, Cryptozoon and associate stromatolites from the Recent Shark Bay, Western Australia: Jour. Geology, v. 69, no. 5, p. 517–533.

Logan, B. W., and Cebulski, D. E., 1970, Sedimentary environments of Shark Bay, Western Australia, *in* Carbonate sedimentation and environments, Shark Bay, Western Australia: Am. Assoc. Petroleum Geologists Mem. 13, p. 1–37.

Logan, B. W., Rezak, Richard, and Ginsburg, R. N., 1964, Classification and environmental significance of algal stromatolites: Jour. Geology, v. 72, no. 3, p. 68–83.

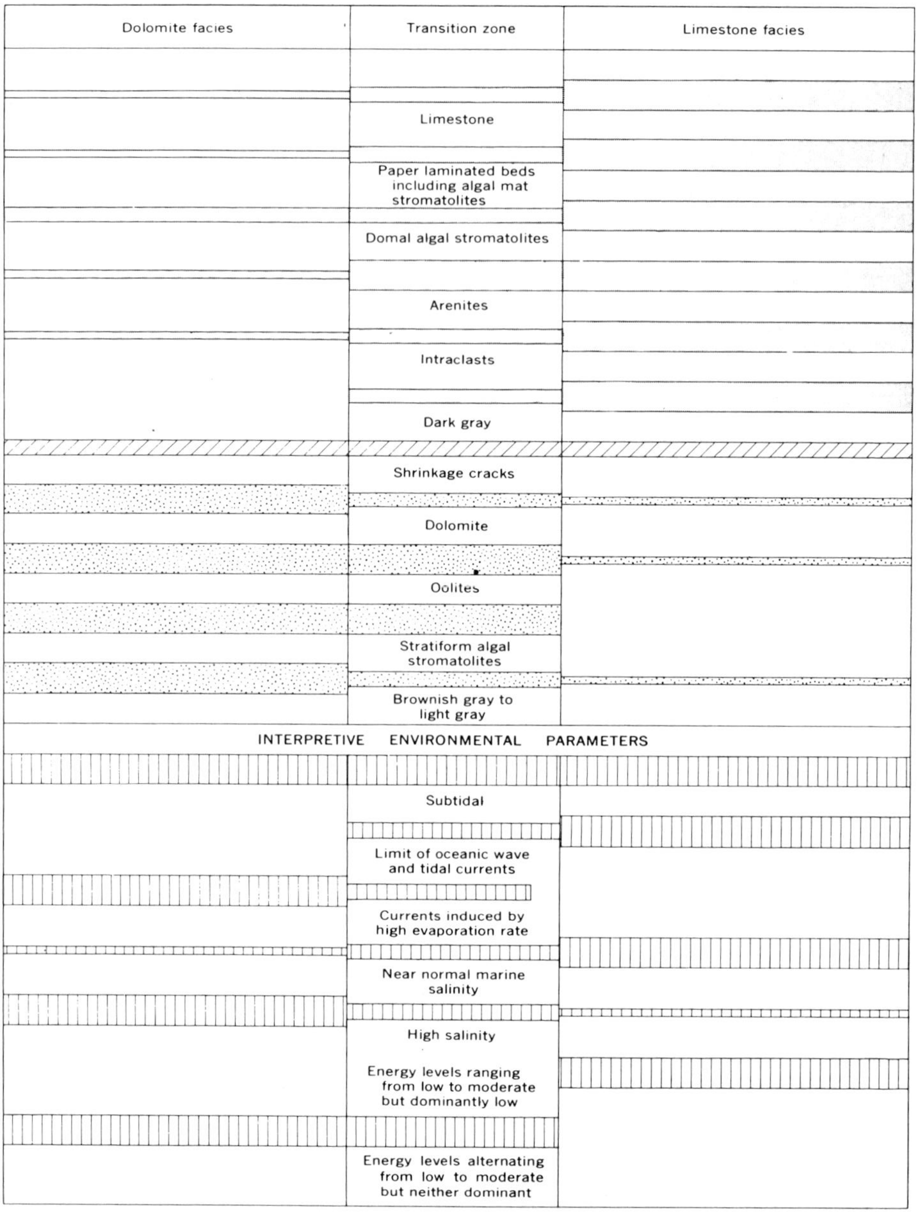

Figure 12.—The distribution of common elements of the dolomite, transition, and limestone facies of the Knox Group, and interpretive environmental parameters. Thickness of the bar emphasizes the relative importance of each element within each facies.

Lowenstam, H. A., 1954, Factors affecting the aragonite-calcite ratios in carbonate-secreting marine organisms: Jour. Geology v. 62, p. 284–322.

Monty, Claude, 1965, Recent algal stromatolites in the Windward Lagoon, Andros Island, Bahamas: Soc. Géol. Belgique Annales, v. 88, p. 269–276.

Newell, N. D., Imbrie, John, Purdy, E. G., and Thurber, D. L., 1959, Organism communities and bottom facies, Great Bahama Bank: Am. Mus. Nat. History Bull. v. 117, p. 177–228.

Newell, N. D., and Rigby, J. K., 1957, Geological studies on the Great Bahama Bank, in Regional aspects of carbonate deposition—A symposium: Soc. Econ. Paleontologists and Mineralogists Spec. Pub. 5, p. 15–72.

Ochsenius, Karl, 1877, Die Bildung der Steinsalzlager und ihrer Mutterlaugensalze: Halle, C. E. M. Pfeffer, 172 p.

Oder, C. R. L., and Miller, H. W., 1945, Stratigraphy of the Mascot-Jefferson City zinc district [Tenn.]: Am. Inst. Mining Metall. Engineers Tech. Pub. 1818, 9 p.; A.I.M.E. Trans., v. 178, p. 223–231 (1948).

Oder, C. R. L., and Ricketts, J. E., 1961, Geology of the Mascot-Jefferson City zinc district, Tennessee: Tennessee Div. Geology Rept. Inv. 12, 29 p.

Ostrom, M. E., 1970, Sedimentation cycles in the Lower Paleozoic rocks of western Wisconsin, in Field trip guidebook for Cambrian-Ordovician geology of western Wisconsin—Geol. Soc. America, Ann. Mtg., Milwaukee, Wis. 1970: Wisconsin Geol. and Nat. History Survey Inf. Circ. 11, p. 10–34.

Perry, W. J., Jr., 1964, Geology of Ray Sponaugle well, Pendleton County, West Virginia: Am. Assoc. Petroleum Geologists Bull., v. 48, no. 5, p. 659–669.

Prouty, C. E., 1947, Relationship of Cambro-Ordovician dolomites to facies barriers in Tennessee and Virginia [abs.]: Geol. Soc. America Bull., v. 58, no. 12, pt. 2, p. 1218–1219.

——— 1948, Trenton and sub-Trenton stratigraphy of northwest belts of Virginia and Tennessee, in Appalachian Basin Ordovician symposium: Am. Assoc. Petroleum Geologists Bull., v. 32, no. 8, p. 1596–1626.

Rodgers, John, 1968, The eastern edge of the North American Continent during the Cambrian and Early Ordovician, Chap. 10 in Zen, E-an, and others, eds., Studies of Appalachian geology, northern and maritime: New York and London, Interscience Publishers, p. 141–149.

Saxby, D. B., and Lamar, J. E., 1957, Gypsum and anhydrite in Illinois: Illinois State Geol. Survey Circ. 226, 26 p.

Scruton, P. C., 1953, Deposition of evaporites: Am. Assoc. Petroleum Geologists Bull., v. 37, no. 11, p. 2498–2512.

Shaw, A. B., 1964, Time in stratigraphy: New York, McGraw-Hill Book Co., 365 p.

Shinn, E. A., Ginsburg, R. N., and Lloyd, R. M., 1965, Recent supratidal dolomite from Andros Island, Bahamas, in Dolomitization and limestone diagenesis—A symposium: Soc. Econ. Paleontologists and Mineralogists Spec. Pub. 13, p. 112–123.

Smith, C. L., 1940, The Great Bahama Bank—[Pt.] 1, General hydrographical and chemical features: Jour. Marine Research, v. 3, no. 2, p. 147–170.

Stagg, A. K., and Fischer, F. T., 1970, Upper Knox stratigraphy of Middle Tennessee [abs.]: Geol. Soc. America Abstracts with Programs, v. 2, no. 7, p. 693.

Stewart, F. H., 1963, Marine evaporites, Chap. Y of Data of Geochemistry, 6th ed.: U.S. Geol. Survey Prof. Paper 440-Y, 52 p.

Swartz, F. M., 1948, Trenton and sub-Trenton of outcrop areas in New York, Pennsylvania, and Maryland, in Appalachian Basin Ordovician symposium: Am. Assoc. Petroleum Geologist Bull., v. 32, no. 8, p. 1493–1595.

Williams, C. H., Jr., 1969, Cross section from Mississippi-Tennessee State line to Horn Island in Gulf of Mexico: Jackson, Miss., Mississippi Geol. Survey, map.

Young, R. S., and Wedow, Helmuth, Jr., 1962, Geologic setting of the Myers zinc prospect, Smyth County, Virginia [abs.]: Geol. Soc. America Spec. Paper 68, p. 82.

24

Reprinted from *Geology* 3:283–285 (1975), courtesy of the Geological Society of America

Ground-water formation of dolomite in the Coorong region of South Australia

Chris C. von der Borch
David E. Lock
Douglas Schwebel
Discipline of Marine Geology and Geophysics
School of Earth Sciences, Flinders University
Bedford Park, South Australia 5042

ABSTRACT

The 90 X 200 km coastal plain of southeastern South Australia provides a setting which throughout Quaternary time was ideally suited for the sedimentation of dolomite and other carbonate minerals. Comparative studies between various present-day sedimentary environments and cores taken from regions of active dolomite formation near the Coorong Lagoon have defined a typical Holocene regressive sedimentary cycle for the area. A basal restricted marine and lagoonal aragonite and Mg-calcite unit, approximately 6,500 yr old, passes upwards into ephemeral lake units with carbonate mineralogies ranging from protodolomite +Mg-calcite to ordered dolomite. Hydrologic data show that the present coastal zone is the main area where seaward-flowing shallow ground waters emerge from unconfined carbonate aquifers beneath the coastal plain. This fact, and the observation that modern dolomite is also forming in isolated areas of ground-water discharge and evaporation remote from the coast, suggests that dolomite ideally forms from ions provided by ground waters, whereas aragonite, Mg-calcite, and protodolomite assemblages form from a marine or mixed sea-water–ground-water reservoir. The observed Holocene cycle is therefore explained by progressive seaward movement of a sea-water–ground-water interface, due to a slight relative fall of sea level about 6,500 yr ago combined with lateral sediment accretion. That the bulk of the dolomite formation is presently occurring near the coast is explained by the fact that this is the site, near base level, of most ground-water discharge and evaporation. Possible sources of Mg ions for the extensive formation of dolomite throughout Quaternary time are Mg depletion of high Mg-calcite allochems in the carbonate aquifers and leaching of ash deposits around Quaternary volcanic centers near ground-water source areas.

INTRODUCTION

The 90 X 200 km coastal plain of southeastern South Australia (Fig. 1) has been the locus of formation of a variety of fine-grained (1 μm) carbonate minerals throughout Quaternary time (Alderman and Skinner, 1957). These minerals include ordered and disordered dolomite, magnesite, hydromagnesite, high and low Mg-calcite, and aragonite (Skinner, 1963; von der Borch, 1965). These are presently forming in a modern coastal lagoon, the Coorong, and in associated shallow (0.5 to 1 m) embayments and ephemeral lakes. Dolomite is also in the process of forming away from the coast in ground-water–fed ephemeral lakes situated within "fossil" Pleistocene lagoons, which are now stranded by slow regional upwarp. Stratigraphic drilling in 1973 has verified the presence of dolomite in at least some of these inland flats. The dolomite occurs in extensive blankets several metres thick, separated by shoestringlike Pleistocene barrier islands that form conspicuous calcareous eolianite ridges in an otherwise flat terrane. Skinner (1963) also noted dolomite in auger samples taken from many of the inland interdune flats.

Most previous work has focused on the youngest part of the coastal plain near the Coorong Lagoon. Stratigraphic, mineralogic, and hydrologic research currently in progress is providing compelling evidence that the bulk of the dolomite in this region is forming not from a sea-water ionic reservoir but from shallow Ca and Mg-bicarbonate ground waters that actively discharge from carbonate aquifers beneath the coastal plain. The association of most areas of actively forming dolomite with the present coastline is due to the fact that it is mainly here, near base level, that the seaward-flowing shallow ground waters emerge, undergo summer evaporation, and are subjected to as yet undefined conditions that result in dolomite deposition.

Figure 1. Locality and physiography. Surface drainage after O'Driscoll, 1960. Ground-water flow lines plotted from isopotential map of M. Cobb, 1971. Core localities shown near northwestern corner. Most of flats northwest of volcanic centers are underlain by dolomite occurrences.

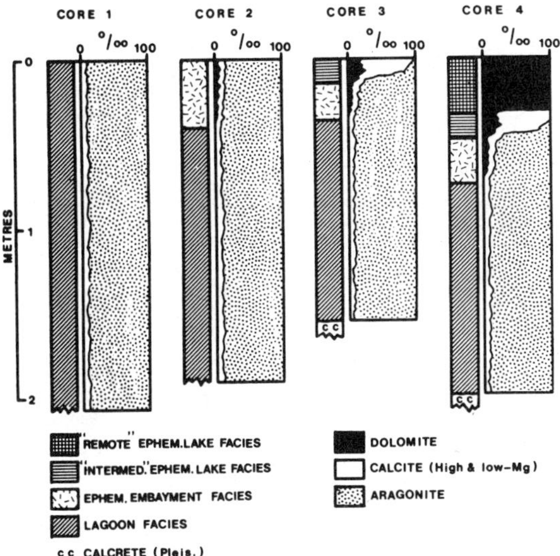

Figure 2. Generalized stratigraphic columns illustrating stratigraphy and mineralogy of typical Holocene cycle in coastal carbonato lakes and lagoon. High- and low-Mg calcite undifferentiated.

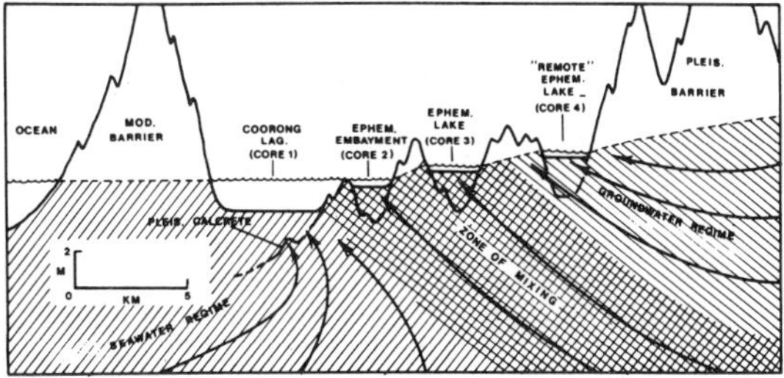

Figure 3. Generalized cross section illustrating lake and core localities with respect to the present ground-water–sea-water interface. Possible ground-water flow lines are indicated. Note large vertical exaggeration.

Thus, a microstratigraphic study of coastal dolomite areas will provide valuable clues to the origin of the widespread dolomite and other carbonate minerals beneath the Pleistocene coastal plain.

The stratigraphy of dolomites in the Coorong region bears some resemblance to that of supratidal dolomites of Persian Gulf sabkhas. Whether sabkha dolomites in general are forming from a sea-water or ground-water reservoir of ions is at present uncertain, although Hsü and Schneider (1973) tentatively suggested the latter, based on preliminary observations. It is the object of this preliminary paper to outline some of the regional controls bearing on this question.

STRATIGRAPHY

Semiquantitative x-ray diffractometer analyses of the mineralogies of about 20 3- to 4-m cores through upper Holocene sediments in ephemeral lakes associated with the Coorong Lagoon have documented the existence of a definite stratigraphic sequence. Figure 2 shows the stratigraphy and mineralogy of cores taken from four typical present-day environments that are sequentially removed in both distance and elevation from the present lagoon (see also Figs. 1 and 3). The first core is from the lagoon and the second from a seasonally drying (ephemeral) embayment of the lagoon. The third and fourth cores are from slightly higher-level (0.5 to 1.5 m) shallow ephemeral lakes that occur in hollows in the calcrete-veneered Pleistocene eolianites, which were originally connected with the lagoon.

The stratigraphy of the most remote lake (Fig. 2, core 4), which is typical of many coastal dolomite lakes, indicates that it has passed sequentially through each of the three stages represented by cores 1, 2, and 3 since inundation by the Holocene transgression about 6,000 yr ago. Basal sediments comprise a lagoonal unit of fetid dark gray pelletized aragonite and calcite skeletal mud, which contains a molluscan fauna typical of a protected marine environment. This grades upward into a lighter-colored pelletized aragonite and Mg-calcite skeletal mud with traces of dolomite, followed by a white protodolomite and Mg-calcite micrite. At this upper level, all traces of marine or lagoonal faunal elements disappear, suggesting that at the onset of protodolomite formation, between 3,000 and 5,000 yr ago, this lake had severed its lagoonal connection by a slight fall in relative sea level combined with lateral sediment accretion. The final event recorded in the core, representative of the present environment of "remote" ephemeral lakes, is the occurrence of a unit of white semi-indurated mud composed of moderately well-ordered dolomite. This has a surface radiocarbon age of 300 ± 250 yr (von der Borch and others, 1964).

HYDROLOGY

Ephemeral lakes of the types represented by cores 3 and 4 become filled during winter months to a depth of about 0.5 m by visible ground-water seepage. Piezometer measurements show a close correlation between the free-water surface of each lake and the regional ground-water table, verifying that the lakes are simply an "outcrop" of the unconfined ground water. The water-table and lake-water levels fall during the normally arid summers. Lake waters are concentrated by evaporation with sometimes complete desiccation during summer. Intimate connection between lake waters and regional ground waters is implied by the fact that, although high salinities are reached, only very minor quantities of halite crystals remain on dried lake floors.

Hydrologic studies in the area by O'Driscoll (1960) indicate that the unconfined ground waters in the upper Tertiary and Pleistocene carbonate aquifers flow sluggishly seaward and discharge into the lagoon and ocean. Main aquifer recharge is in the more pluvial and slightly higher-level areas in the eastern and southeastern sectors of the coastal plain. This is accompanied by more local recharge nearer the Coorong both by direct rainfall infiltration and from chains of freshwater swamps that occupy interdune corridors and flow slowly toward the northwest (Fig. 1).

As the ground waters approach the coast, hydrologic considerations suggest they override a zone saturated with higher-density ocean and lagoon waters in a manner suggested in Figure 3. A diffuse landward-dipping mixing zone of undefined width would occur at this interface.

Ground-water discharge zones are present in some inland localities in interdune corridors. Where these occur, seasonal evaporation takes place and dolomite is invariably noted. One such area is situated 80 km inland, and preliminary radiocarbon measurements indicate that the dolomite is modern. Physically and chemically these inland lakes are remarkably similar to many of their coastal counterparts. Regional isohalines (O'Driscoll, 1960) show shallow ground waters to be more saline than normal in a halo around the lakes. This may be due

TABLE 1. TYPICAL GROUND-WATER ANALYSES FROM BORES IN SOUTHEASTERN SOUTH AUSTRALIA, COMPARED WITH NORMAL SEA WATER

	Na^+ (%)	$CO_3^=$ (%)	Ca^{++} (%)	Mg^{++} (%)	Total salt (%)	Mg/Ca
Coastal dolomite region	31.5	5.4	1.5	3.9	0.64	2.52
	31.5	4.1	1.9	3.5	0.82	1.85
	21.1	4.7	7.7	5.2	0.33	0.67
	27.7	7.4	4.2	4.6	0.35	1.09
Inland dolomite regions	27.4	17.1	4.0	5.2	0.15	1.29
	23.8	17.3	8.1	4.7	0.16	0.58
	28.6	13.3	3.7	3.6	0.14	0.97
	28.4	13.1	3.1	5.1	0.19	1.62
	25.7	7.4	6.2	4.2	0.17	0.67
Surface interdune swamp (aquifer recharge zone)	27.3	6.1	4.0	4.6	0.27	1.15
	29.0	5.2	2.8	4.4	0.25	1.57
Sea water	30.61	0.41	1.16	3.69	3.5	3.18

to seasonal recharge of the aquifer by evaporatively concentrated water.

Table 1 illustrates selected analyses of regional waters from the shallow coastal-plain aquifer, compared with normal sea water. A relatively wide range of composition is evident, with coastal regions being more saline than inland areas. When compared with sea water, the ground waters are slightly depleted in Na^+, considerably enriched in total carbonate and Ca^{++}, and slightly enriched in Mg^{++}. Mg/Ca is invariably lower than normal sea water.

CONTROLS OF DOLOMITE FORMATION

The association described above of a modern, pure dolomite phase with ground-water discharge remote from a marine influence suggests strongly that this mineral can form anywhere on the coastal plain from evaporating ground waters. This fact, coupled with hydrologic observations, adequately explains the carbonate cycle observed in the coastal carbonate lakes.

It is proposed that the "outcrop" of the zone of mixing between ground water and sea water (Fig. 3) separates two distinct areas of carbonate mineral formation. Coastal ephemeral lakes, which are now remote from the parent lagoon (Fig. 2, core 4), are landward of the mixing zone and are now fed only by ground waters. Their floors are as much as 1.6 m above present mean sea level, and they invariably contain an uppermost unit of dolomite or dolomite + magnesite. Seaward of the mixing zone lies the lagoon and its restricted ephemeral embayments, where aragonite and Mg-calcite dominate the carbonate assemblages (Fig. 2, cores 1, 2). In an intermediate situation lie many ephemeral lakes in which are forming poorly crystallized proto-dolomite and Mg-calcite assemblages (Fig. 2, core 3). These lakes, with floors about 0.5 m above mean sea level, occur either adjacent to the lagoon or along the southeastern extension of its interdune corridor immediately landward of the modern barrier. They presumably lie within the mixing zone.

Approximately 6,500 yr ago, relative sea level in the area stood 1 m above its present level, and sediments had not accumulated in the lakes to any great degree. The ocean shoreline and the sea-water–ground-water interface would thus have been farther inland than at present. Coastal lakes now forming dolomite and classified as "remote" would then have been influenced by waters with the ionic balance of sea water in an environment comparable to that of the present lagoon. Since that time the slight fall of sea level, combined with sediment accretion, has caused shoreline regression accompanied by seaward migration of the ground-water interface. Remote lakes thus became increasingly under the influence of ground waters, with concomitant increases in dolomite formation.

Future events in the Coorong interdune area will almost certainly follow the pattern that occurred repeatedly in the Pleistocene areas. The lagoon will eventually fill with sediment and sea level will fall in response to the next glacial episode. This will cause the ground-water–influenced zone to migrate seaward and a "dolomite front" will thus move across the pres-

ent lagoon. If ground-water discharge into a seasonally evaporating lake chain were to continue, then a regressive blanket of dolomite would cap the sequence and this would continue to form until lowering of base level brought an end to ground-water discharge.

SOURCE OF IONS

Ground waters that discharge into the dolomite-forming areas have initial compositions illustrated by Table 1, although effects of chemical differentiation in lakes cause local variations (von der Borch, 1965).

The calcium, magnesium, and bicarbonate ions present in the regional ground waters may have been derived at least in part by dissolution of skeletal components in upper Tertiary limestones and Pleistocene calcareous sands that constitute the shallow aquifer, a possibility suggested by widespread karst topography that exists around inland margins of the coastal plain. Metastable Mg-calcite allochems, such as coralline algae and echinoids, are important constituents of these aquifers and could provide a source of Ca^{++} and Mg^{++}, whilst the Na^+ could come from wind-borne cyclic salt or from flushing of connate marine pore waters.

A more significant source of Mg ions in the ground waters could be from leaching of Tertiary and Quaternary basic volcanic rocks. Volcanism was active in eastern limits of the area (Fig. 1) until about 1,400 yr ago (Blackburn, 1966), and ash layers are intercalated with some of the coastal-plain sediments over a wide area. Much of the northwest-flowing surface water originates in or near this volcanic zone. Hutton (1974) has shown that ash in the area has lost 60 to 80 percent of its original calcium, magnesium, and sodium during weathering to clay minerals and that these ions have been leached into the ground waters. They also contribute to runoff waters that feed the natural northwest-flowing interdunal drains. It is thus possible that significant amounts of calcium, magnesium, and sodium ions from the ash would find their way northwest along the interdune corridors to less pluvial areas, which are climatically suited to the formation of dolomite. These drainage swamps significantly recharge the shallow aquifer (O'Driscoll, 1960), which provides ground water for the dolomite lakes.

REFERENCES CITED

Alderman, A. R., and Skinner, H.C.W., 1957, Dolomite sedimentation in the southeast of South Australia: Am. Jour. Sci., v. 255, p. 561–567.

Blackburn, G., 1966, Radiocarbon dates relating to soil development, coast-line changes, and volcanic ash deposition in southeast South Australia: Australian Jour. Sci., v. 29, p. 50–52.

Cobb, M., 1971, Unpublished records of Hydrogeology Section: South Australia Geol. Survey.

Hsü, K. J., and Schneider, J., 1973, Progress report on dolomitization-hydrology of Abu Dhabi sabkhas, Arabian Gulf, *in* Purser, B. H., ed., The Persian Gulf: New York, Springer-Verlag, p. 409–471.

Hutton, J. T., 1974, Chemical characterization and weathering changes in Holocene volcanic ash in soils near Mount Gambier, South Australia: Royal Soc. South Australia Trans., v. 98, p. 179–183.

O'Driscoll, E.P.D., 1960, The hydrology of the Murray Basin province in South Australia: Geol. Survey South Australia Bull. 35, 148 p.

Skinner, H.C.W., 1963, Precipitation of calcian dolomites and magnesian calcites in the southeast of South Australia: Am. Jour. Sci., v. 261, p. 449–472.

von der Borch, C. C., 1965, The distribution and preliminary geochemistry of modern carbonate sediments of the Coorong area, South Australia: Geochim. et Cosmochim. Acta, v. 29, p. 781–799.

von der Borch, C. C., Rubin, M., and Skinner, B. J., 1964, Modern dolomite from South Australia: Am. Jour. Sci., v. 263, p. 1116–1118.

ACKNOWLEDGMENTS

Reviewed by B. Hostetler and P. W. Choquette.

J. W. Holmes contributed valuable discussions on the hydrology of the area.

MANUSCRIPT RECEIVED DECEMBER 16, 1974

MANUSCRIPT ACCEPTED MARCH 5, 1975

25

Origin of Nonmarine Dolomite in Eocene Lake Gosiute, Green River Basin, Wyoming

CLAUDIA A. WOLFBAUER
RONALD C. SURDAM } *Department of Geology, University of Wyoming, Laramie, Wyoming 82071*

ABSTRACT

The distribution of dolomite in the Laney Shale Member of the Green River Formation in the western Green River Basin provides further documentation regarding the suitability of the playa-lake model for the Green River Formation in Wyoming. Four alternate mechanisms were considered in evaluating the origin of the dolomite: (1) detrital, (2) late diagenetic, (3) primary, and (4) penecontemporaneous. It is concluded that dolomite in the Laney Shale Member is the product of penecontemporaneous dolomite formation that results from the evaporative concentration of subsurface Mg-rich brines in the mud flats. Qualitative thermodynamic diagrams are constructed that depict carbonate equilibrium relations in the mud flats and adjacent environments. *Key words: dolomite, Green River Formation, Lake Gosiute, Eocene, Wyoming.*

INTRODUCTION

Recently, the playa-lake model, consisting of a vast alkaline-earth playa fringing an alkaline lake, has been proposed (Eugster and Surdam, 1973) to explain many of the chemical and sedimentologic aspects of the Green River Formation that previously were not understood. One of the most significant elements of the playa-lake model is that it accounts for the generation of large amounts of dolomite in the mud flats bordering Eocene Lake Gosiute. Our purpose is to describe in detail the processes leading to the formation of dolomite in the playa fringe.

Because of the vast amount of information now available concerning the climate, hydrology, paleontology, and geologic setting of the Green River Formation during Eocene time, it is possible to develop a model for the large-scale dolomitization of nonmarine sediments in a subhumid to semiarid environment. It is suggested that dolomite formed on broad, featureless mud flats adjacent to Lake Gosiute by the evaporative pumping of magnesium-rich subsurface brines through calcareous muds.

Specifically, this report discusses the rocks of a facies change between the Bridger Formation (Bridger "A" of McGrew and Sullivan, 1970) and the Laney Shale Member of the Green River Formation in a 400-mi² area along the western margin of Lake Gosiute (Figs. 1, 2). This is one of the few areas along the basin margin where positive stratigraphic correlations can be made. The complex relations that can be delineated here between fluviatile and lacustrine rocks are characteristic of the marginal facies of the entire Green River Formation. The abundance of vertebrate, invertebrate, and algal remains, together with the occurrence of a variety of sedimentary structures, enables rather exact determinations to be made regarding the types and geographic extent of environments associated with the lake.

STRATIGRAPHY

The Green River Basin is approximately 16,000 mi² in area and results from structural downwarping that began during late Paleozoic time (J. D. Love, oral commun.) and continued into middle Eocene time.

The basin fill (Fig. 3) consists of red and gray fluviatile deposits of the Wasatch Formation (Paleocene to early Eocene), lacustrine and playa-like deposits of the Green River Formation (early to middle Eocene), and gray and green fluviatile and paludal rocks of the Bridger Formation (middle Eocene; Wyoming Geological Association, 1973).

In Wyoming, the Green River Formation is subdivided into four members (Fig. 3): the widespread, fresh-water sediments of the Laney Shale Member; the restricted lacustrine and playa-like deposits of the Wilkins Peak Member; the oil-shale–rich sediments of the Tipton Shale Member; and the initial lacustrine deposits of the Ramsey Ranch Member. The Laney Shale Member consists of buff, light-brown, and light-gray limestone, dolomitic limestone, siltstone, mudstone, and shale. Low-grade oil shale occurs in small scattered lenses in the lower part of the sequence. Algal domes, mud cracks, and the nature of the associated fauna indicate that the carbonate units were deposited in a shallow nearshore environ-

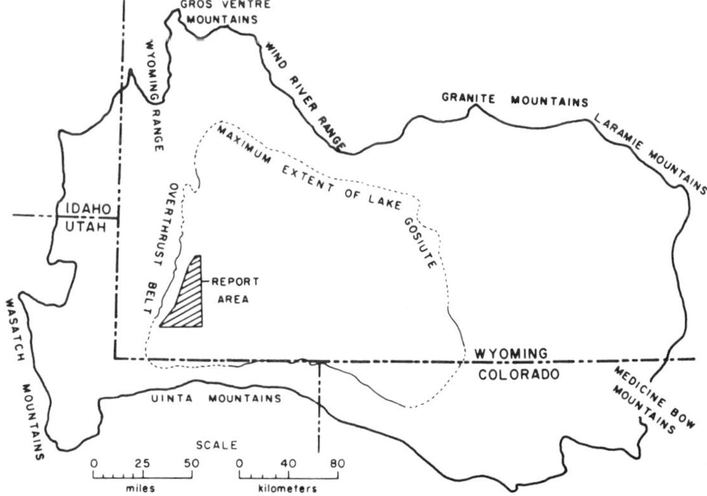

Figure 1. Hydrographic basin of Eocene Lake Gosiute showing maximum extent of lake during Laney stage (after Bradley, 1963).

331

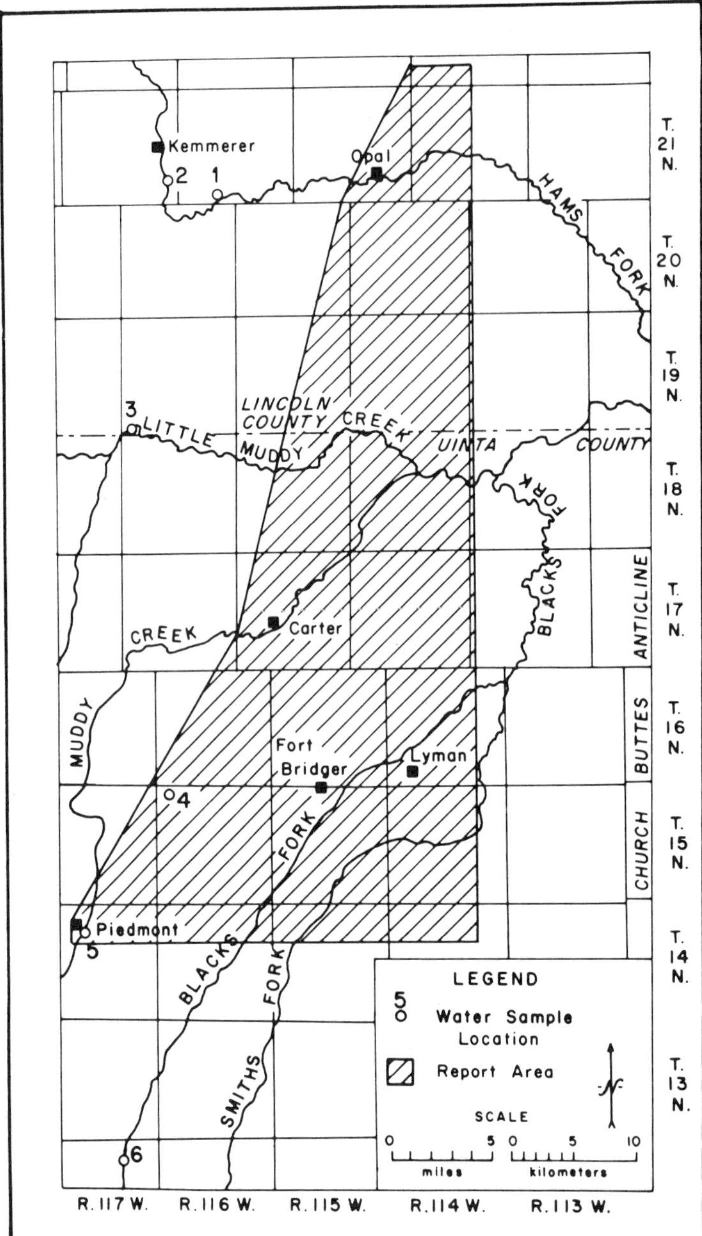

Figure 2. Index map of report area showing locations of water samples.

ment. It is possible that this area was partially isolated from the main body of Lake Gosiute by the Church Buttes uplift (Fig. 2). The Bridger Formation consists of gray and green sandstone, siltstone, and mudstone with minor limestone. Much of the formation contains altered volcanic ash of rhyolitic and andesitic composition (Koenig, 1949; Johannsen, 1914; Sinclair, 1906). Channel sandstone deposits are widespread and are characteristic of some stratigraphic intervals. Most limestones are thin (6 in. to 2 ft thick) and discontinuous and were deposited in small ponds and stream channel meanders (Pledge, 1969). A few, however, represent short-term widespread expansions of Lake Gosiute. The nature of the Bridger sediments and the associated vertebrate fossils indicate that the northern part of the report area was part of a large deltaic flood plain characterized by heavily forested swamps, marshes, and braided and meandering streams (Pledge, 1969). Figures 4 and 5 show the transition from the Bridger Formation in the north to the Laney Shale Member of the Green River Formation in the southern part of the report area. The line of cross section in Figure 5 is essentially parallel to the shoreline of Lake Gosiute, especially for sections 6 through 10. There is almost no outcrop in the dip direction. In general, marker units A through F (McGrew and Sullivan, 1970) and the *Goniobasis* marker unit (Wolfbauer, 1972b) are 1- to 5-ft-thick limestone, dolostone, and dolomitic limestone units that are traceable for distances of up to 40 mi or more. Locally, they contain mollusks, ostracodes, fish bones, algal structures, mud cracks, and clay blebs. In the report area, these units change in appearance and composition from north to south, and the identification of each unit was made by walking the outcrop from one section to the next.

ALGAE

Algal structures are associated with paludal deposits in the Bridger Formation and with the shoreline facies of the Laney Shale Member. The areal distribution of the various algal morphologies is valuable in reconstructing the paleoenvironment — especially in areas otherwise devoid of diagnostic faunas. Ancient shoreline positions can be delineated by mapping the occurrences of algal domes and mats in individual lacustrine units (Fig. 6). For example, the transition from the mud flats to the algal dome zone is exceptionally well illustrated near section 9 where marker unit A is exposed for approximately 0.5 mi in the dip direction. The unit is 6 ft thick and consists of limestone interbedded with thin layers of

Figure 3. Schematic diagram of Eocene stratigraphy in Green River Basin. Dashed line indicates approximate position of gradational contact between Wasatch and Bridger Formations.

Figure 4. Geologic map of report area.

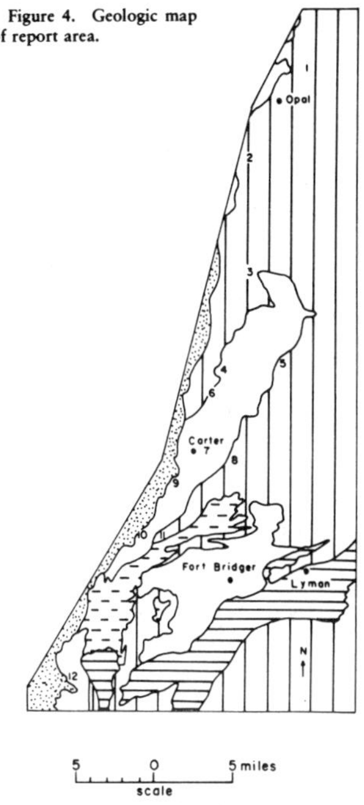

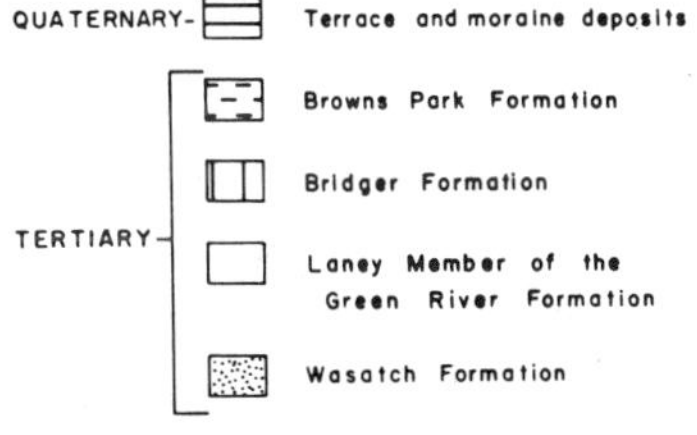

limy shale and mudstone. In its westernmost exposure, unit A contains small scattered algal mats and columns. Eastward toward Lake Gosiute, these features are replaced by larger, more continuous algal mats and encrusting algae and finally by algal domes. The uppermost part of the unit contains mud-crack polygons that are 1 to 2 in. wide in the western outcrop and up to 2 ft wide farther to the east. No mud cracks are associated with the algal domes.

Five different algal morphologies are recognized in the report area, each being a reflection of a particular biotope.

1. Encrusting Algae. Algae commonly encrust silicified logs and tree stumps. In some specimens, the original plant material decayed, leaving internal cavities which subsequently were filled with sediment and sparry calcite that formed geopetal features. This form of algae is abundant in the Bridger Formation.

2. Oncolites. Oncolites occur locally in the *Goniobasis* marker unit (Wolfbauer, 1972a). They consist of concentrically banded light and dark laminae of calcite micrite surrounding a recrystallized shell of *Goniobasis tenera*. No algal filaments have been preserved. The oncolites apparently formed as a result of seasonal algal growth on empty shells in the continual underwater agitation of a bicarbonate-rich stream (Minckley, 1963; Howe, 1932; Roddy, 1915; Weiss, 1969, 1970). The oncolites

were transported to the lake where they became incorporated in the lacustrine sediments.

3. Algal Mats. Algal mats occur locally on the shoreward side of the algal dome zone. By analogy with modern algal mats, the Eocene mats probably formed on the lower parts of the mud flats adjacent to Lake Gosiute where they were occasionally flooded for short periods of time.

4. Discoid Algae. Discoid algae occur in marker unit B near section 7. The algal deposit consists of masses of small, discrete, elongate discs, 0.5 to 1 in. in their longest dimensions. Bradley (1929) suggested that this particular morphology may have been due to a somewhat higher concentration of dissolved salts in the lake water because it is associated with mud cracks and salt crystal casts in the lower Wilkins Peak Member of the Green River Formation.

5. Algal Domes. "Puffball-shaped" algal domes are common, especially in association with marker units A, B, and C. They are usually simple in form and average 1 to 2 ft in diameter. The largest dome found (Fig. 7) is 3 ft in diameter and 4 ft high. The domes are analogous in form to those described by Logan (1961, Pl. 2, Fig. 1) from Shark Bay, Australia, and by Eardley (1938, Pls. 16 and 18) from Great Salt Lake, Utah, and probably formed in very shallow water along the shoreline of the lake.

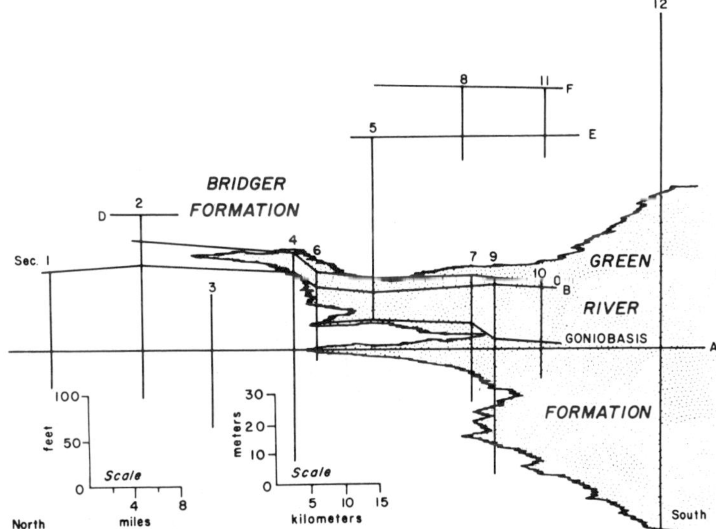

Figure 5. Correlation of measured stratigraphic sections 1 through 12, in western Green River Basin. Correlations based on marker units A through F and *Goniobasis* marker unit, shown as near-horizontal lines (Wolfbauer, 1972b).

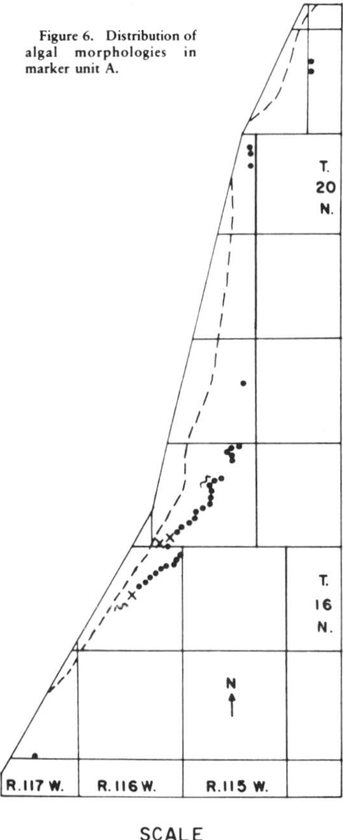

Figure 6. Distribution of algal morphologies in marker unit A.

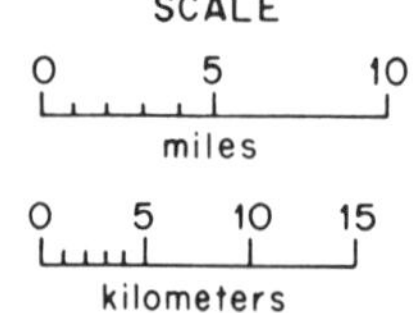

SCALE

0   5   10
miles

0   5   10   15
kilometers

ᛜ — Encrusting algae
x — Algal mats
• — Algal domes

- - - - - Approximate
shoreline position

MINERALOGY AND DOLOMITE GENESIS

The mineralogy of this sequence of rocks is surprisingly consistent. X-ray powder diffraction traces of more than 500 samples from 12 measured sections indicate that quartz and calcite are the most abundant minerals, with lesser amounts of montmorillonitic clay, feldspar, and dolomite also present. Barite, amphibole, gypsum, pyrite, biotite, clinoptilolite, illite, and kaolinite were also detected. The most striking mineralogic feature of these rocks is the distribution of dolomite. Dolomitic limestone and dolostone are found only in the Laney Shale Member, whereas dolomitic mudstone, siltstone, and sandstone occur in the Laney Shale Member and in the Bridger Formation. Dolomite does not occur in fluviatile and paludal rocks and is apparently restricted to rocks of the lacustrine and mud-flat environments.

The origin of the dolomite is extremely important to any reconstruction of the paleoenvironment of the Green River Formation. Four possible dolomite origins were considered — detrital, late diagenetic, primary, and penecontemporaneous.

Pre-Eocene Detrital Dolomite

In considering the first mechanism for the origin of the dolomite, we found the only probable source of older recycled dolomite to be the Overthrust belt, about 10 mi to the west. That part of the Overthrust belt bordering the report area, however, is composed primarily of shale and quartzose sandstone of the Cretaceous Hilliard and Frontier Formations. These same rock units were exposed during the Laney stage of Lake Gosiute (Lawrence, 1962).

In the vicinity of Opal, Wyoming, the Bridger Formation contains many channel sandstone deposits which together form a large deltaic complex (Wood, 1966; Pledge, 1969). The streams that built this complex drained the dolomite-bearing sequence of rocks in the northern part of the Overthrust belt. However, because dolomite is conspicuously absent not only in the channel sandstone but also the entire stratigraphic section at Opal, we do not believe that erosional processes in the hydrographic basin can account for the dolomite associated with the deposits of Lake Gosiute.

Late Diagenetic Dolomite

Considering the second mechanism, the paths of secondary solutions through these rocks must have been severely restricted by the abundance of relatively impermeable shale and mudstone units. The migration of dolomitizing solutions through the thin, permeable limestone might be expected either to (1) cause the more-or-less uniform dolomitization of these units, (2) cause increased dolomitization toward the source of the solutions, or (3) reflect a direct relation between dolomite abundance and some physical, mineralogic, or chemical property of the units (Murray and Lucia, 1967).

However, x-ray diffraction traces reveal large, unpredictable variations in the dolomite content of individual carbonate units from one outcrop to the next. In the dip direction, the occurrence of dolomite appears to be controlled by environmental factors rather than by the migration of late diagenetic solutions. The amount of dolomite in the rocks increases rapidly across the mud-flat zone and decreases abruptly when the "permanent" lacustrine zone is reached. The carbonate units are relatively homogeneous. Both dolomite and calcite are extremely fine grained (less than 10μ) and cannot be identified with a petrographic microscope. Staining with alizarin red-S solutions, however, shows that both minerals are evenly distributed throughout hand specimens. On a larger scale, however, the amount of dolomite commonly in-

Figure 7. Algal dome in Laney Shale Member of Green River Formation. Depocenter of Lake Gosiute is to right of photo.

creases toward the tops of dolomitic units. Regarding the second potential cause, there is no obvious correlation either between the grain size or the abundance and (or) type of noncarbonate minerals or fossil remains (Fig. 8) and the amount of dolomite in these units. Thus, late diagenetic alteration is an unlikely explanation for the Eocene dolomite.

Primary Lacustrine Dolomite

A third mechanism for the origin of the dolomite is that protodolomite was directly precipitated in the lake itself. The precipitation of protodolomite is dependent on concentrating the lake water sufficiently to produce a high molar ratio of Mg^{++}/Ca^{++}. Modern lacustrine protodolomite is forming in waters with Mg^{++}/Ca^{++} ratios ranging from 4.7 (Alderman and Skinner, 1957) to 22 (Nesbitt, 1974).

Hem (1959) reported that the Mg^{++}/Ca^{++} ratio for most fresh waters ranges from 0.2 to 1.0. Water samples were collected from four modern streams in the western Green River Basin that drain approximately the same assemblage of Paleozoic and Mesozoic clastic and carbonate rocks as their Eocene counterparts (Fig. 2). Chemical analyses (Table 1) show that the larger streams (Hams Fork and Blacks Fork) have an average Mg^{++}/Ca^{++} ratio of 0.6, identical to that of the Green River at Jensen, Utah (Bradley and Eugster, 1969). Smaller streams, however, reach values of as much as 1.6 (Little Muddy Creek). The two springs sampled had Mg^{++}/Ca^{++} ratios of 0.3 and 0.6, respectively (Fig. 2, locs. 1 and 4).[1] Obviously, waters of these compositions entering into Lake Gosiute would have to have been greatly concentrated by evaporation before protodolomite could have precipitated. Because the area under discussion represents the shoreline facies of a shallow lake and because of the very low topographic relief,[2] the shoreline would have moved far to the east of this area before the lake water could have been concentrated enough for protodolomite to precipitate.

In addition to the above physical evidence, there is also geochemical evidence, which suggests that dolomite and calcite did not precipitate from the same solution. Presumably, if protodolomite precipitated

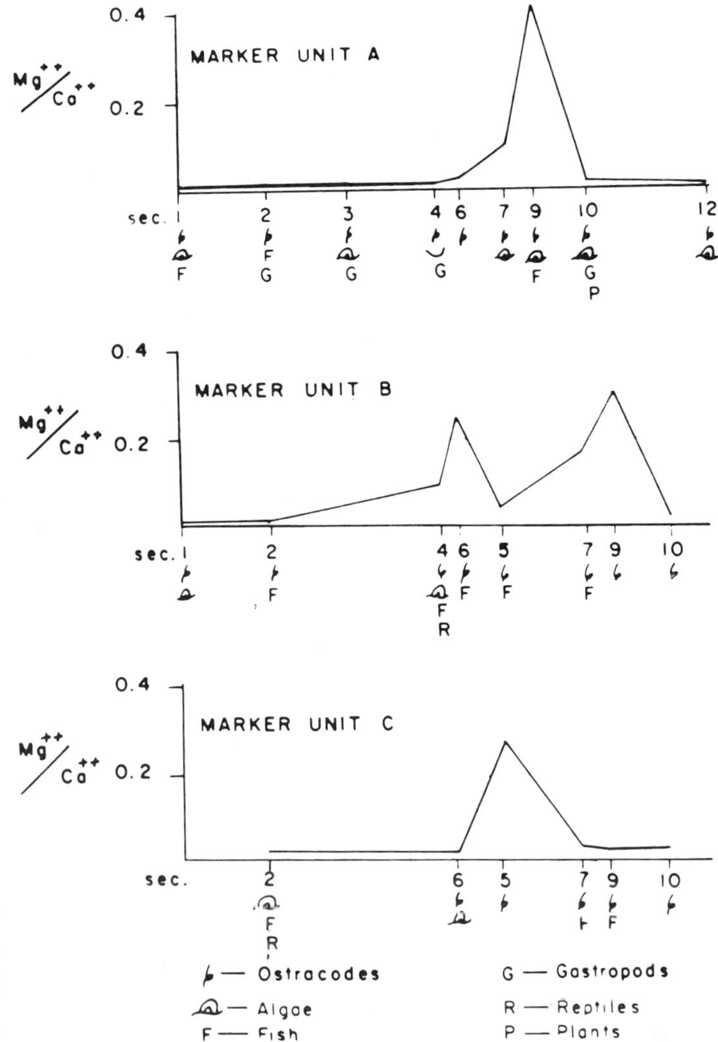

Figure 8. Mg/Ca ratios and faunal assemblages of marker units A, B, and C at sections 1 through 12.

[1] The climate of the Green River Basin today is semiarid. Under the somewhat more humid climatic conditions of the Laney stage of Lake Gosiute, it seems reasonable to assume that an even greater proportion of calcium would have been present in the streams, thereby lowering the ratios even further (see Livingstone, 1963, p. 4; Clarke, 1924, p. 494). The chemical composition of the lacustrine rocks also indicates that the Eocene streams were probably enriched in calcium and bicarbonate, with lesser concentrations of magnesium, sodium, and sulfate.

[2] Bradley (1964) stated that "the original dip probably did not exceed 1 or 2 feet per mile except in a rather narrow belt adjacent to the mountains."

directly from Lake Gosiute water, the chemistry of the lake water and of the precipitating phases might have been comparable to those of similar modern lakes in which protodolomite is now forming — specifically, the ephemeral lakes of the Coorong, Basque Lakes in British Columbia, Deep Springs Lake in California, and Lake Balaton in Hungary. At each of these locations, protodolomite is forming with or without Mg-calcite in waters with high Mg^{++}/Ca^{++} ratios.

Few reports on modern lakes are concerned with the mineralogy of the authigenic carbonates. Most workers group these sediments under the collective term "marl." It appears, however, that in many lacustrine environments, calcite precipitates from waters with relatively low Mg^{++}/Ca^{++} ratios. As these waters become concentrated by evaporation, the Mg^{++}/Ca^{++} ratio is increased and Mg-calcite becomes the dominant phase. With further concentration, protodolomite precipitates along with

TABLE 1. CHEMICAL ANALYSES OF WATER SAMPLES FROM MODERN SPRINGS AND STREAMS

	Samples (ppm)[*]						
	1	2	3	4	5	6	7
Si	25	0	5	17	9	4	10.5
Fe	0	0	0	0	0	0	0.1
Ca	64	33	52	92	65	7	41.1
Mg	12	11	50	31	40	3	13.8
Na	20	5	30	30	25	0	29.8
K	1	1	4	2	3	0.5	2.2
HCO_3	242	128	302	527	392	31	152.2
SO_4	33	25	99	41	12	0	80.6
Cl	4.2	2.6	20	7.9	16	1	12.0
NO_3	0	0.2	0.2	1.1	0.3	0.3	1.0
pH	8.4	8.1	8.0	8.1	8.2	7.1	
Mg^{++}/Ca^{++}	0.3	0.6	1.6	0.6	1.0	0.7	0.6

1, Opal Spring; 2, Hams Fork River; 3, Little Muddy Creek; 4, Cottonwood Spring; 5, Muddy Creek; 6, Blacks Fork River; 7, Green River at Jensen, Utah (U.S. Geological Survey, 1952). See Figure 2 for sampling locations. Analyses 1 through 6 performed in 1971 by K. Hilton and M. Purko.

Mg-calcite (Nesbitt, 1974; Jones, 1965; Skinner, 1963). Muller's (1970) data from Lake Balaton and Nesbitt's (1974) data from Basque Lakes indicate that when Mg-calcite is the only precipitating phase, the $MgCO_3$ content of the calcite increases with the increasing Mg^{++}/Ca^{++} ratio of the water. Similarly, Skinner (1963) found that when the waters of the Coorong are concentrated to the point at which protodolomite forms along with the Mg-calcite, the total amount of magnesium in both carbonate phases increases along with the Mg^{++}/Ca^{++} ratio of the waters.

Table 2 compares the mol percent of $MgCO_3$ found in the Mg-calcite and associated protodolomite of modern lacustrine environments with those associated with Eocene Lake Gosiute. These data suggest that, if the Eocene protodolomite was precipitated directly into the lake, it would be associated with Mg-calcite having a much broader and higher range in $MgCO_3$ content than was found in the Eocene sequence. It appears, then, that either the Green River calcite formed in fresher water than the protodolomite, or Mg-calcite coprecipitated with protodolomite, with magnesium subsequently being removed from the lattice during diagenesis. In view of the physical evidence presented earlier, we favor the former explanation.

Penecontemporaneous Dolomite

The fourth mechanism considered for the origin of the dolomite — the one envisioned for the sediments of Lake Gosiute — is that protodolomite formed in mud flats adjacent to the lake due to the evaporative concentration of ground water. During regressive phases of the lake, the waters receded to expose broad areas of sand, silt, and mud. The increased aridity that produced these climatically controlled drops in lake level resulted in intense evaporation on the newly exposed mud flats, thereby creating a hydraulic concentration gradient from the "fresh" water of the ground-water table below to concentrated brines with high Mg^{++}/Ca^{++} ratios near the surface of the mud flats. It was in these near-surface areas that protodolomite precipitated and (or) replaced pre-existing $CaCO_3$.

Paleobotanic work by MacGinitie (1969) indicates that the Eocene climate of this area was warm temperate to subtropical with an average annual temperature of 60 to 70°F. He determined that the precipitation (24 to 30 in./yr) was distinctly seasonal. Maximum rainfall occurred in late spring and early summer, followed by near-drought conditions in late summer and autumn.

Assuming an original sediment porosity of 40 percent and an annual evaporation rate of 146 cm/yr (Bradley, 1963), we calculate, using the method of Hsü and Siegenthaler (1969), that interstitial water containing 50 ppm magnesium[3] (Table 1, Little Muddy Creek, sample 3) could dolomitize 1 cm^3 of carbonate mud in 63 yr. This method assumes an annual evaporation rate of 73 cm^3 of water/cm^2 of sediment surface. The only feasible system for the continual supply and concentration of these brines combines the loss of water by evaporation from the sediment surface and its replacement by water drawn upward from the ground-water table.

This hypothesis explains the greater concentration of dolomite in the upper parts of some carbonate units, as well as its presence in small amounts in the detrital fraction of the lacustrine sediments. It also accounts for the commonly observed change in the composition of the carbonate marker units from limestone with small amounts of protodolomite in the distal parts of the mud-flat environment to dolostone composed entirely of protodolomite in the mud-crack zone adjacent to the lake, and finally to limestone in the lacustrine environment. Those rock units which were associated with areas having deeper ground-water circulation, with heavily vegetated areas having low rates of evaporation, or which were deposited during transgressive phases of the lake, were not exposed to concentrated brines and remained unaltered.

E. T. Degens (personal commun. in Ingerson, 1962) made a detailed study of fossiliferous and unfossiliferous beds in Mesozoic and Tertiary limestone. He concluded that the Mg/Ca ratio of these beds is a reflection of the salinity of the environment of deposition and that fossils occur only in formations in which the Mg/Ca ratio is less than 0.02. Figure 8 shows that there is no difference between the fossil content of dolomitic and nondolomitic out-

---

[3] This number is, at best, only a guess as to the minimum composition of the interstitial brines. Because the major rivers draining the hydrographic basin probably contained 10 to 15 ppm Mg (Table 1, samples 2 and 7), we hypothesize that the concentrated brines contained at least 50 ppm and perhaps considerably more.

TABLE 2. COMPARISON OF $MgCO_3$ IN CALCITE AND PROTODOLOMITE FROM MODERN ENVIRONMENTS WITH THAT OF ROCKS ASSOCIATED WITH EOCENE LAKE GOSIUTE

	$MgCO_3$ (mol %)	
	Calcite	Protodolomite
Lake Balaton (Muller, 1970)	6–11.5	42–44
Coorong (Skinner, 1963)	2–23	44–50
Basque Lakes (Nesbitt, 1974)	4–30	32–43
Eocene Lake Gosiute[*]	0– 4	39–50

Values for Lake Gosiute determined from x-ray diffraction patterns using the method of Goldsmith and Graf (1958).

crops of marker units A, B, and C. Considering the low-salinity tolerances of freshwater fish and ostracodes (Benson, 1961), these facts further support the conclusion that dolomite formed by the alteration of $CaCO_3$ sediments originally deposited in fresher water. Calcite high in Mg that formed prior to or along with protodolomite subsequently was altered to calcite low in Mg.

CARBONATE EQUILIBRIA

Using the data from modern environments and the basic principles of the phase rule, qualitative diagrams can be constructed which depict the equilibrium relations between Mg-calcite and protodolomite. The use, derivation, and construction of these diagrams has been described by Nesbitt (1974). In Figure 9 the curves for Mg-calcite, protodolomite, Ca-magnesite, and the concentration of the aqueous solution are plotted on a partial molar free energy versus composition diagram. By plotting the known (or assumed) Mg^{++}/Ca^{++} ratio of the solution on the aqueous solution curve, one can derive the possible range in $MgCO_3$ content of calcite and protodolomite forming from such solutions.

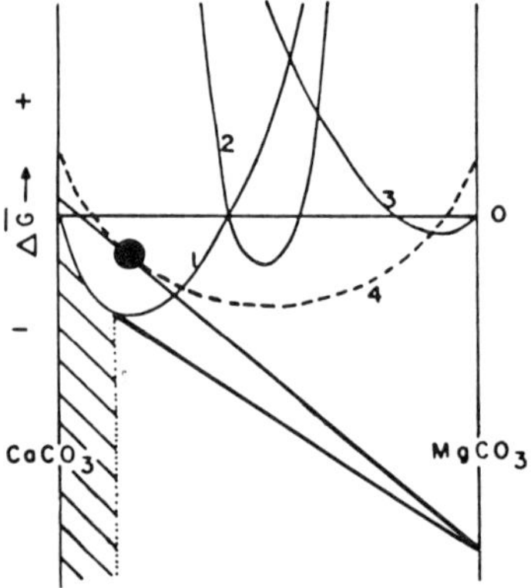

Figure 9. Example of qualitative thermodynamic diagram depicting equilibrium relations in $CaCO_3$-$MgCO_3$ system at 25°C and 1 atm. Partial molar free-energy curves are shown for Mg-calcite (1), protodolomite (2), Ca-magnesite (3), and aqueous solution (4). Dot = $CaCO_3/MgCO_3$ ratio of aqueous solution. Ruled area shows possible range in composition of phases forming from such a solution.

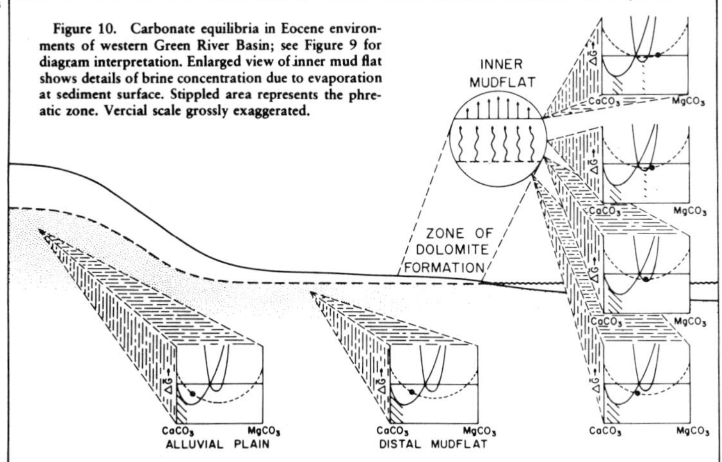

Figure 10. Carbonate equilibria in Eocene environments of western Green River Basin; see Figure 9 for diagram interpretation. Enlarged view of inner mud flat shows details of brine concentration due to evaporation at sediment surface. Stippled area represents the phreatic zone. Vercial scale grossly exaggerated.

A series of thermodynamic diagrams depicting carbonate equilibria conditions for the Eocene environments associated with the margins of Lake Gosiute is shown in Figure 10. The range in $MgCO_3$ content found in the Eocene rocks formed in these environments corresponds to the range predicted by the schematic diagrams. In the alluvial plain environment only a small amount of $MgCO_3$ would have been stable in the authigenic calcite. This particular diagram is based on the Mg^{++}/Ca^{++} ratio of present-day Hams Fork River water (Table 1, sample 2). In the distal parts of the mud flat, the waters were slightly more concentrated and, correspondingly, the $MgCO_3$ range in the calcites was slightly broader. The highest Mg^{++}/Ca^{++} ratio in the aqueous solutions occurred on the mud flats in the zone of most intense evaporation. There the solution concentrations were high enough for protodolomite to form along with Mg-calcite. In the lacustrine environment the water was fresher and in equilibrium with calcites having low $MgCO_3$ content.

CONCLUSIONS

Our interpretation of the dolomite-forming process which was operative in the Green River Basin along the margins of Lake Gosiute is shown schematically in Figure 10. Humid climatic periods resulted in occasional rises in the lake level and in flooding of adjacent swamps and alluvial plains. With the return to more arid conditions, the lake level dropped, exposing broad areas of mud flat. Because of the low topographic relief, several hundred square miles of mud flat were exposed. The increased aridity that produced this low stage of the lake also caused intense evaporation on the mud flat. Water was continually drawn upward by means of evaporative pumping from the ground-water table, and a hydraulic concentration gradient developed. The brines near the surface of the mud flat had high Mg^{++}/Ca^{++} ratios and caused the formation of protodolomite, probably as a penecontemporaneous replacement of pre-existing $CaCO_3$. The enormous amount of dolomite generated along the margins of Lake Gosiute was readily available for transportation to and redeposition in the adjacent lacustrine environment.

ACKNOWLEDGMENTS

We gratefully acknowledge the helpful discussions and suggestions made by P. O. McGrew, H. P. Eugster, and H. W. Nesbitt. We are particularly indebted to Nesbitt for suggesting that free energy data could be used to interpret the dolomitization patterns in the western Bridger basin and for suggesting to us the effectiveness of this type of diagram in understanding processes of dolomitization. An earlier draft of this manuscript was reviewed by Nesbitt and Richard Sheppard. Support was given by Petroleum Research Fund Grant 3809–A2 and National Science Foundation Grant GA–33789.

REFERENCES CITED

Alderman, A. R., and Skinner, H. C., 1957, Dolomite sedimentation in the southeast of South Australia: Am. Jour. Sci., v. 225, p. 561–567.

Benson, R. H., 1961, Ecology of ostracode assemblages, in Moore, R. C., ed., Treatise on invertebrate paleontology, Pt. Q, Arthropoda 3: Boulder, Colo., Geol. Soc. America, p. Q56–Q63.

Bradley, W. H., 1929, Algae reefs and oolites of the Green River Formation: U.S. Survey Prof. Paper 154–G, p. 203–223.

——1963, Paleolimnology, in Frey, D. G., ed., Limnology in North America: Madison, Univ. Wisconsin Press, p. 621–652.

——1964, Geology of the Green River Formation and associated Eocene rocks in southwestern Wyoming and adjacent parts of Colorado and Utah: U.S. Geol. Survey Prof. Paper 496–A, 86 p.

Bradley, W. H., and Eugster, H. P., 1969, Geochemistry and paleolimnology of the trona deposits and associated authigenic minerals of the Green River Formation of Wyoming: U.S. Geol. Survey Prof. Paper 496–B, 71 p.

Clarke, F. W., 1924, The data of geochemistry: U.S. Geol. Survey Bull. 770, 841 p.

Eardley, A. J., 1938, Sediments of Great Salt Lake, Utah: Am. Assoc. Petroleum Geologists Bull., v. 22, p. 1305–1411.

Eugster, H. P., and Surdam, R. C., 1973, Depositional environment of the Green River Formation of Wyoming: A preliminary report: Geol. Soc. America Bull., v. 84, p. 1115–1120.

Goldsmith, J. R., and Graf, D. L., 1958, Relation between lattice constants and composition of the Ca-Mg carbonates: Am. Mineralogist, v. 42, p. 84–101.

Hem, J. D., 1959, Study and interpretation of the chemical characteristics of natural water: U.S. Geol. Survey Water-Supply Paper 1473, 269 p.

Howe, M. A., 1932, The geologic importance of the lime secreting algae: U.S. Geol. Survey Prof. Paper 170–E, p. 57–64.

Hsü, K. J., and Siegenthaler, C., 1969, Preliminary experiments on hydrodynamic movement induced by evaporation and their bearing on the dolomite problem: Sedimentology, v. 12, p. 11–25.

Ingerson, E., 1962, Problems of the geochemistry of sedimentary carbonate rocks: Geochim. et Cosmochim. Acta, v. 26, p. 815–847.

Johannsen, A., 1914, Petrographic analysis of the Bridger, Washakie, and other Eocene Formations of the Rocky Mountains: Am. Mus. Nat. History Bull., v. 33, p. 209–222.

Jones, B. F., 1965, The hydrology and mineralogy of Deep Springs Lake, Inyo County, California: U.S. Geol. Survey Prof. Paper 502–A, 56 p.

Koenig, K. J., 1949, Bridger Formation in the Bridger basin, Wyoming [Ph.D. dissert.]: Urbana, Univ. Illinois, 119 p.

Lawrence, J. C., 1962, Wasatch and Green River Formations of the Cumberland Gap area, Lincoln and Uinta Counties, Wyoming [M.A. thesis]: Laramie, Univ. Wyoming, 102 p.

Livingstone, D. A., 1963, Chemical composition of rivers and lakes, in Fleischer, M., ed., Data of geochemistry: U.S. Geol. Survey Prof. Paper 440, chap. G, p. G1–G64.

Logan, B., 1961, Cryptozoon and associate stromatolites from the Recent, Shark Bay, Western Australia: Jour. Geology, v. 69, p. 517–533.

MacGinitie, H. D., 1969, Eocene Green River flora of northwestern Colorado and northeastern Utah: California Univ. Pubs. Geol. Sci., v. 83, 203 p.

McGrew, P. O., and Sullivan, R., 1970, Stratigraphy and paleontology of Bridger A: Contr. Geology, v. 9, p. 66–85.

Minckley, W. L., 1963, Studies of the ecology of a spring stream: Doe Run, Meade County, Kentucky: Wildlife Monographs, no. 11, p. 1–124.

Muller, G., 1970, High-magnesian calcite and protodolomite in Lake Balaton (Hungary) sediments: Nature, v. 226, p. 749–750.

Murray, R. C., and Lucia, F. J., 1967, Cause and control of dolomite distribution by rock selectivity: Geol. Soc. America Bull., v. 78, p. 21–35.

Nesbitt, H. W., 1974, Mineralogy and geochemistry of the Basque Lakes, British Columbia [Ph.D. dissert.]: Baltimore, Johns Hopkins Univ.

Pledge, N. S., 1969, Paleoenvironments and paleoecology of part of the lower Bridger Formation, south of Opal, Wyoming [M.S. thesis]: Laramie, Univ. Wyoming, 74 p.

Roddy, H. J., 1915, Concretions in streams formed by the agency of blue-green algae and related plants: Am. Philos. Soc. Proc., v. 54, p. 246–258.

Sinclair, W. J., 1906, Volcanic ash in the Bridger beds of Wyoming: Am. Mus. Nat. History Bull., v. 22, p. 273–280.

Skinner, H.C.W., 1963, Precipitation of calcian dolomites and magnesian calcites in the southeast of South Australia: Am. Jour. Sci., v. 261, p. 449–472.

U.S. Geological Survey, 1952, Quality of surface waters of the United States, 1947: U.S. Geol. Survey Water-Supply Paper 1102, 651 p.

Weiss, M. P., 1969, Oncolites, paleoecology, and Laramide tectonics, central Utah: Am. Assoc. Petroleum Geologists Bull., v. 53, p. 1105–1120.

——1970, Oncolites forming on snails (*Goniobasis*): Jour. Paleontology, v. 44, p. 765–769.

Wolfbauer, C. A., 1972a, Oncolites from the Green River Formation of western Wyoming: Contr. Geology, v. 11, p. 61–62.

——1972b, Chemical petrology of nonmarine carbonates in the western Bridger Basin, Wyoming [Ph.D. dissert.]: Laramie, Univ. Wyoming, 80 p.

Wood, C. B., 1966, Stratigraphy and paleontology of the Bridger Formation northeast of Opal, Lincoln County, Wyoming [M.S. thesis]: Laramie, Univ. Wyoming, 112 p.

Wyoming Geological Association, 1973, Wyoming stratigraphic nomenclature chart, 25th Ann. Field Conf., 1973: Casper, Wyo., p. 9.

Manuscript Received by the Society June 28, 1974

Revised Manuscript Received March 22, 1974

[*Editors' Note:* Claudia Wolfbauer Frahme may be contacted at Mobil Oil Corporation, P.O. Box 5444, Denver, Colorado 80217.]

26

ORGANIC DOLOMITE FROM POINT FERMIN, CALIFORNIA

J. H. Spotts and S. R. Silverman, *Chevron Research Company, Box 446, La Habra, California.*

Abstract

Petrography and stable carbon isotope data indicate that small dolomite rhombohedra that occur with tar as cement in some California Miocene sands were partially derived from the alteration of the tar or other organic matter. The dolomite cement occurs in blue-schist turbidite sandstones that are exposed in sea cliffs at Point Fermin, Los Angeles County, California. The euhedral crystals are attached to sand grains and are also apparently embedded in the tar cement. The dolomite crystals have several types of nuclei: (1) rhombohedral to spherical tar globules similar to the tar surrounding the euhedra, (2) gaseous bubbles, and (3) combinations of these. Carbon isotope values of the rhombohedra indicate that the carbonate in the dolomite was partially derived from oxidation of organic matter. The dolomite probably crystallized around oil or tar droplets and carbon dioxide bubbles that were produced as discontinuous phases during the oxidation of the organic matter. The rhombohedra exhibit unusual growth habits related to the symmetry. Complex apical grooves rather than simple dihedral edges are developed along the six rhombohedral edges that intersect the *c*-axis. In addition to tar and dolomite, the cement complex in these sandstones contains sparry calcite and a zeolite mineral, clinoptilolite.

Introduction

Recent geochemical and mineralogical evidence indicates that organic and inorganic geological materials are not necessarily independent of each other and thus cannot be considered as separate and unrelated entities. Isotope geochemistry and clay mineralogy have shown in the past few years that organic and inorganic substances can be intimately and systematically intermixed. In an increasing number of occurrences, co-existing "so-called" organic and inorganic materials have proven to be genetically related and we might anticipate that detailed isotope studies will reveal many more such hybrid organic-inorganic substances. Organosilicate complexes and several types of carbonate derived directly from organic material are examples of different types of mergence of organic and inorganic substances. Dolomite rhombohedra formed partially from carbon dioxide derived from the oxidation of bitumen were revealed in a petrographic study of Miocene sandstones of the Los Angeles Basin. The rocks are well exposed in the sea cliffs at Pt. Fermin in San Pedro, California, near the Los Angeles-Long Beach Harbor area.

This occurrence of organic carbonate is described in detail here because microscopy of the dolomite in tar sands provides evidence on the physico-chemical mechanisms of the formation of carbonates from oil. Another unique characteristic of the Pt. Fermin dolomite is the unusual crystal habit. It is possible that there is a genetic relationship between

339

the modified rhombohedral habit and the organic influence in carbonate deposition but this is impossible to determine on the basis of one occurrence.

A similar occurrence of dolomite enclosing globular hydrocarbon nuclei is described by Osipova (1964) from Paleogene carbonate rocks in the northeastern part of the Fergana Depression in the Kara-Dar'i area of Central Asia. Osipova observed that dolomite grains that form the linings of cavities grew around petroleum microglobules and thus were formed later than the petroleum. Diagenesis in these carbonate rocks is enhanced by the presence of petroleum and the associated waters which are rich in carbon dioxide.

MINERALOGY AND PETROGRAPHY

The dolomite occurs as cement embedded in tar in the pores of an unusual sandstone or schist arenite in the Altamira sand member of the Monterey Formation. The sandstone (Fig. 1) consists predominantly of low grade metamorphic rock fragments derived from the Catalina schist facies of the Franciscan Series. These rock and mineral fragments include glaucophane, epidote, lawsonite, fibrous glaucophane schist, quartz-glaucophane schist, chlorite schist, quartzite, lawsonite schist, epidote-albite schist, and various combinations or gradations among these rock types. Quartz and feldspar are the other principal grain types in the sandstone and these minerals account for approximately 10–20 per cent of the rock (Fig. 1).

The dolomite crystals are rhombohedral and occur as disseminated individual rhombs and aggregates. The crystals range up to 0.10 mm on the rhombohedral edge with an average edge length of approximately 0.045 mm. In thin section, most of the dolomite rhombs are attached to the sand grains although there are many grains that appear to be unattached and "floating" in the tar that fills many of the pores. Textural evidence indicates that the dolomite has crystallized in fluid-filled pore space exclusively and has not replaced any of the preexisting mineral constituents of the sandstone. Some pores contain several rhombohedra that project inward from the pore walls and have been subsequently filled with a zeolite, clinoptilolite. Portions of the sandstone are also cemented by sparry calcite and veins transverse to bedding are also calcite-filled. Clinoptilolite occurs with both types of carbonate cement; however, dolomite has not been observed adjacent to calcite.

The dolomite is colorless to very light brown; the color is apparently due to hydrocarbon inclusions. The refractive indices are $\omega = 1.678 \pm 0.002$ and $\epsilon = 1.506 \pm 0.002$ ($\omega-\epsilon = 0.172$). No zonation was observed, and there are no twin lamellae. The dolomite does not show wavy ex-

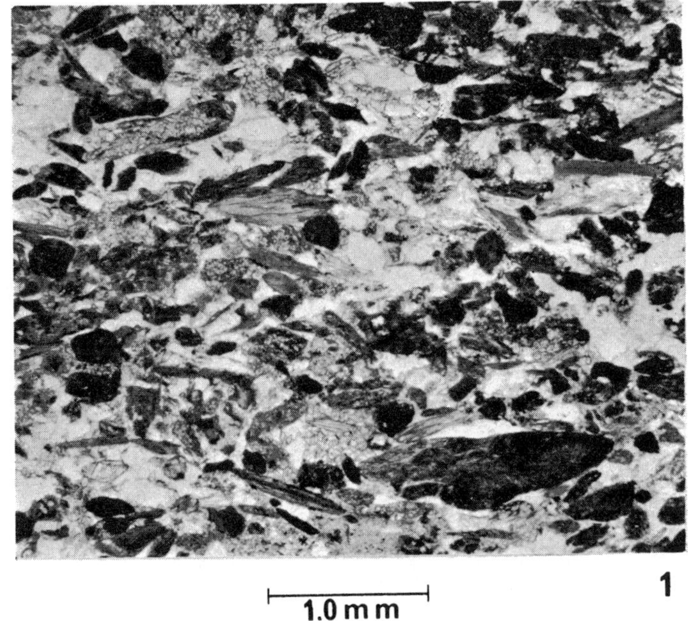

1.0 mm

1

0.1 mm

2

tinction or curvilinear crystal faces that are characteristic of many types of dolomite.

The most unique petrographic property of the dolomite is the presence of hydrocarbon and gaseous nuclei that occur in most of the grains. The discovery of the abundant nuclei first led us to suspect a genetic relationship between the carbonate and the hydrocarbons. There are three types of nuclei; (1) small globular to rhombohedral masses of tar and (2) gaseous inclusions or nuclei and (3) combinations of these (Fig. 2). The tar nuclei are the larger of the two types and range up to 0.03 mm in diameter. The tar masses usually represent approximately one tenth of the volume of the crystal and the masses are commonly centrally located in the grains. The shape ranges from perfectly spherical to irregularly globular to rhombohedral. The rhombohedral tar masses could be described as negative crystals of dolomite filled with tar. The gaseous inclusions which are interpreted as carbon dioxide bubbles derived from the oxidation of organic material also range in shape from spherical to rhombohedral. These gaseous nuclei or inclusions are also located near the centers of the crystals but they tend to be smaller in size and more predominantly and perfectly spherical than the tar nuclei. A small per cent of the rhombohedra contain compound nuclei consisting of a tar globule and a gas bubble that is attached or enclosed in the tar (Fig. 2, Nos. 3 and 6). Dolomite rhombohedra with the presumed gaseous nuclei were dissolved under the microscope. The nuclei were observed to rise to the top of the acid and disappear just as the carbon dioxide produced by acidization, thus indicating that the nuclei are gaseous.

In thin section, approximately half the crystals contain nuclei; however, the actual percentage is considerably higher because a random thin section could not intersect nuclei in every case. Grain mounts of dolomite separated by disaggregation and heavy liquids indicate that approximately 75 per cent of the dolomite rhombohedra contain observable tar or gaseous nuclei. It seems likely that submicroscopic globules and bubbles may also have served as nucleation centers, thus possibly accounting for the small per cent of rhombohedra without observable nuclei.

FIG. 1. Photomicrographs of Altamira Sandstone, a schist sandstone cemented with tar, organically derived dolomite and locally with calcite and clinoptilolite. No. 1 Sand grains consist of several types of schist and minor quartz and feldspar. Matrix consists of tar and dolomite. Plane polarized light. No. 2 Tar-filled pores of schist sandstone with fine grained rhombohedral organic dolomite and clinoptilolite (C).

The crystal habit is modified rhombohedral r {100}. The modifications of the regular rhombohedron are such that the six normal dihedral edges leading to or intersecting the c-axis are actually grooves or trenches or a series of trenches bounded by r {100} planes or crystal faces. This habit is illustrated in Fig. 3. There are therefore two types of unique views in thin section and grain mounts as a result of this crystal habit (Fig. 4). One consists of a lensoid section with the apparent rounding due to the grooves or trenches on the apical edges of the rhombs. The other unique cross-section is the view down the c-axis which is a hexagon modified by three to six notches related to the apical grooves or trenches.

X-ray diffraction spectrometry indicates that the organic dolomite is ideal dolomite as defined by Goldsmith and Graf (1958). Ideal dolomite approximates 1:1 molar $CaCO_3$:$MgCO_3$ in composition and Goldsmith and Graf have shown that non-ideal dolomite with an excess of a few mol per cent $CaCO_3$ has weakened order reflections relative to the ideal or stoichiometric dolomite. The x-ray diffraction pattern for Pt. Fermin dolomite is shown in Table 1. The spacings and relative intensities were determined with an x-ray diffractometer, using Ni-filtered Cu radiation and hand ground packed-pellet samples prepared from essentially pure dolomite separations obtained through heavy liquid and magnetic separator techniques.

Diffractometer traces for Pt. Fermin dolomite compare closely with those for ideal dolomite from Gabbs, Nevada and Binnental, Switzerland reported by Goldsmith and Graf (1958). The data in Table 2 indicate this comparison for selected diffraction lines. The {10$\bar{1}$} and {11$\bar{2}$} are sets of planes parallel to the c-axis; {222} and {444} are sets of planes perpendicular to c. The reflection labelled {01·5} is an order reflection. The two reflections from {444} and {11$\bar{2}$} are both sharp and of equal intensity which is also characteristic of ideal dolomites.

ISOTOPE GEOCHEMISTRY

The mode of occurrence of the dolomite rhombs, especially their close association with organic matter in the rock suggested that the dolomite

---

FIG. 2. Photomicrographs of dolomite rhombohedra and tar cement (brown) in schist sandstone, illustrating several types of nuclei and dolomite. No. 1. Several dolomite euhedra with spherical to irregular globular nuclei. No. 2. Subhedral rhombohedral nucleus of tar. Crystal is surrounded by tar. No. 3. Bubble nucleus which is interpreted as carbon dioxide bubble. Extremely thin "membrane" of tar surrounds bubble; nucleus is two-phase. No. 4. Two negative crystal nuclei (rhombohedral), one is tar (brown) and one is gaseous, possibly carbon dioxide (black, actually it is colorless but with extreme negative relief). No. 5. Cluster of rhombohedra with gaseous nuclei (dark bubbles) with extreme negative relief. No. 6. Two-phase nucleus consisting of relatively large tar mass and small bubble inclusion. Note glaucophane sand grain at bottom of photomicrograph.

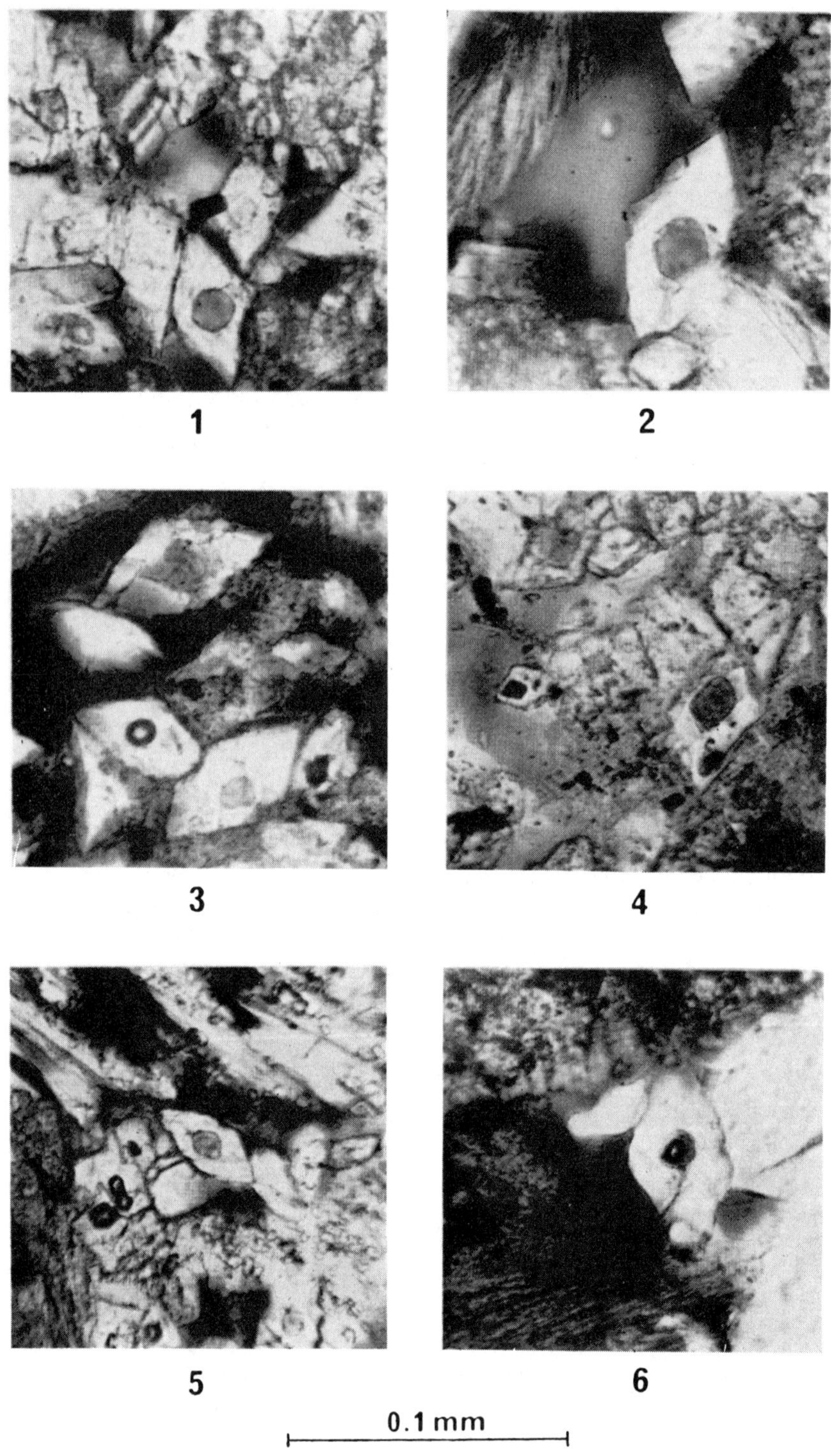

1   2

3   4

5   6

0.1 mm

[*Editors' Note:* Organic dolomite from Point Fermin, California, photographed by J. H. Spotts and S. R. Silverman. In the original, Fig. 2 appears in color. In this reproduction, the tar appears dark. In No. 4, the gaseous nucleus is on the left.]

may have been wholly or partly derived from oxidized organic matter. Evidence for this type of carbonate formation mechanism has been demonstrated by isotopic studies of carbonates from a wide range of localities (Silverman *et al*; in press). C^{13}/C^{12} δ-values of organically derived carbonates range from −10 to −55 per mil. Primary carbonates deposited under typical marine conditions, as well as secondary carbonates formed by recrystallization of primary inorganic carbonate minerals, have a distinctly higher range of carbon isotope ratios (+4 to −4 per mil). This range of differences is illustrated and compared with carbon isotope ratios of organic materials in Fig. 5.

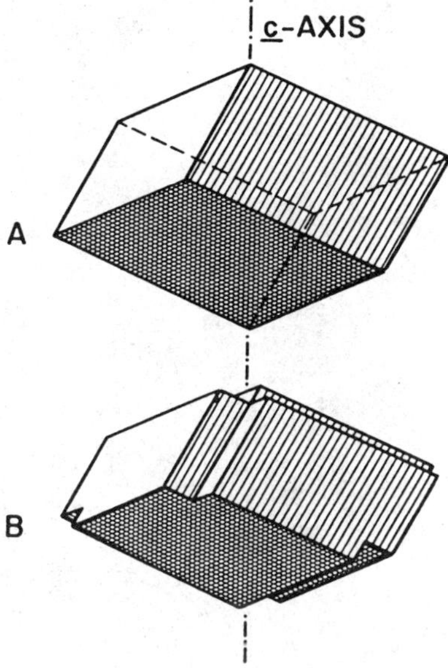

Fig. 3. Crystal habits of organic dolomite from Point Fermin, California. A. Common dolomite rhombohedral habit r {100} which is rare at Point Fermin. B. Characteristic modified rhombohedral habit for Point Fermin dolomite with rhombohedral "grooves" along apical rhomb edges, i.e. rhombohedral edges that intersect the *C*-axis.

Two Pt. Fermin samples were selected for isotope analysis. One sample (Sample A) contained organic matter which was not associated with dolomite rhombs, although this specimen contained calcite cement. In a second sample (Sample B), the organic matter was associated with aggregates of rhombohedral dolomite crystals. The calcite cement and dolomite rhombs were isolated from Samples A and B, respectively. Both isolates were purified and treated in the manner described by Epstein *et al.* (1953). Carbon dioxide liberated from the purified carbonate by acid treatment was analyzed for C^{13}/C^{12} ratio by mass spectrometry. These measurements are reported as deviations (δ-values), in parts per

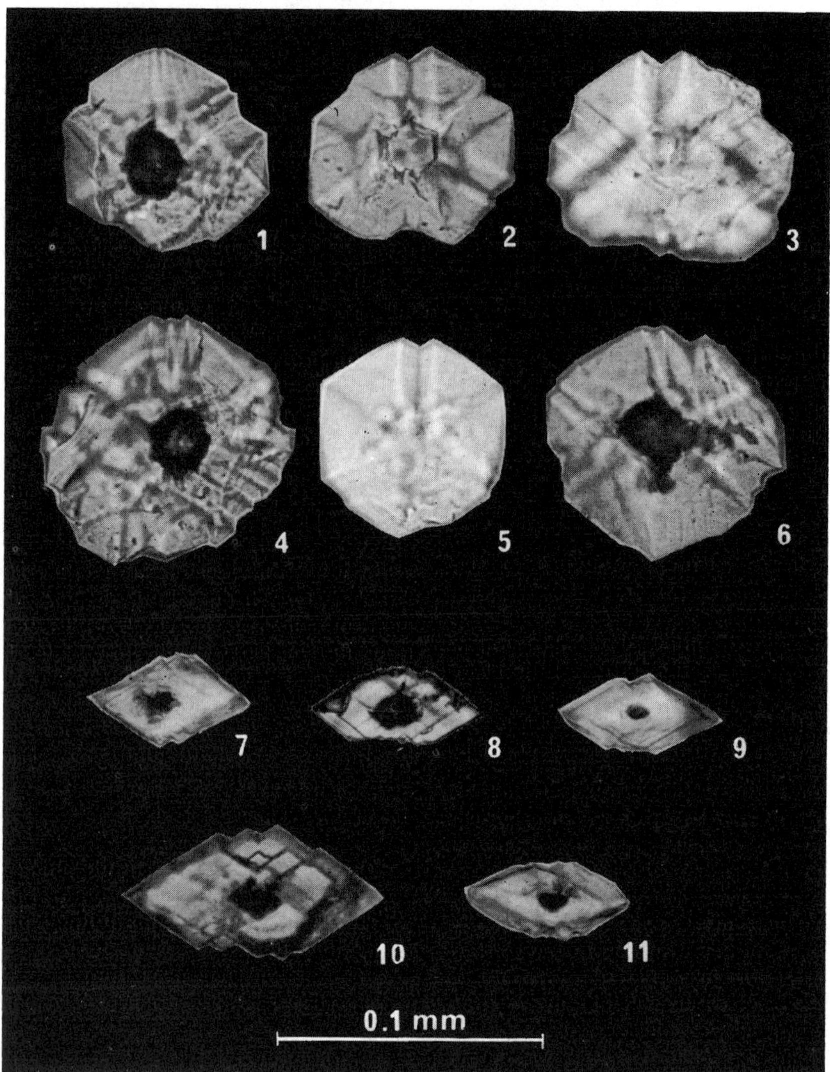

Fig. 4. Photomicrographs of organic dolomite grains illustrating unusual crystal habit.
Grain mount photographs in Aroclor 4465, Nos. 1–6. Dolomite crystals oriented with c-
axis normal to plane of photograph. Note pseudo-twin appearances and re-entrants at
periphery of figures which are actually due to sets of apical edge grooves. See Fig. 3. Three
crystals (Nos. 1, 4 and 6) have gaseous nuclei with extreme negative relief. Two rhombohedra (Nos. 2 and 3) have pale hydrocarbon nuclei. No. 5 shows no apparent nucleus. Nos.
7–11. Dolomite crystals oriented with c-axis approximately parallel to plane of photograph.
"Abraded" lensoid outline is again due to prevalence of sets of apical edge grooves which
have not affected non-apical edges and thus these appear with sharp points in sections with
this orientation.

TABLE 1. X-RAY DIFFRACTION DATA FOR ORGANIC DOLOMITE, PT. FERMIN, CALIFORNIA

2θ (CuKα Radiation)	d	I
10.9	4.46	VW
24.1	3.69	VW
31.1	2.87	VS
33.6	2.67	M
35.4	2.53	W
37.4	2.40	W
41.2	2.19	W
43.9	2.06	VW
45.0	2.01	W
49.3	1.85	VW
50.5	1.80	M
51.1	1.78	M
59.9	1.54	VW
63.3	1.46	VW
64.3	1.44	VW
65.0	1.43	W
67.7	1.38	W
70.3	1.33	W
72.7	1.29	VW
79.8	1.20	W

thousand (per mil), of the C^{13}/C^{12} ratio of sample from that of PDB-1 standard. Bitumen content of both samples was removed by extraction with carbon disulfide. Isotope ratios of the bitumen carbon were determined by analysis of the carbon dioxide prepared from the bitumen by

TABLE 2. COMPARISON OF SEVERAL DOLOMITE REFLECTIONS

(2θ, CuKα Radiation)

	Sets of Planes				
Hexagonal Indices	{00·6}	{01·5}	{111·0}	{03·0}	{00·12}
Rhombohedral Indices	{222}	{221}	{10$\bar{1}$}	{112}	{444}
Sample Locality					
Gabbs, Nevada (Goldsmith and Graf, G-424)	33.45	35.25	37.30	67.35	70.50
Binnental, Switzerland (Goldsmith and Graf, G-1199)	33.40	35.25	37.25	67.35·	70.45
Pt. Fermin, California	33.45	35.35	37.40	67.65	70.25
Levy County, Florida (Goldsmith and Graf G-371) 54.6 mol per cent CaCO$_3$	33.15	35.05	37.10	67.10	69.90

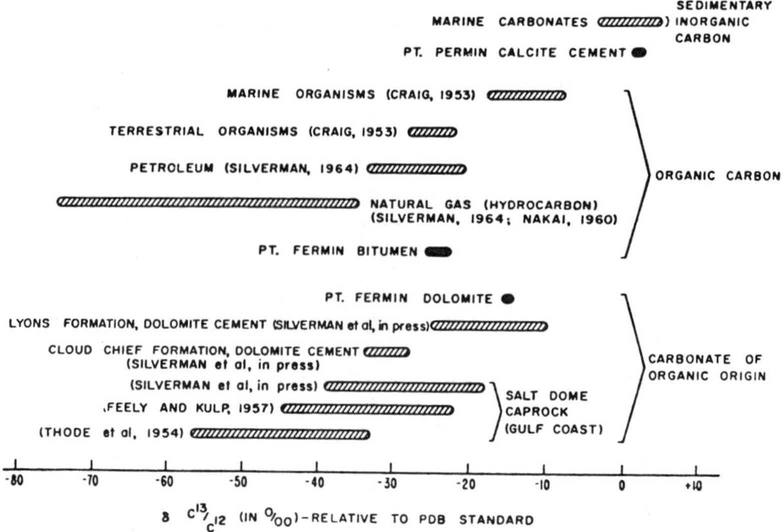

Fig. 5. Carbon isotope ratios of carbonates and organic materials.

the combustion technique described by Silverman and Epstein (1958). Results of these determinations, shown in Table 3, indicate that the δ-values of the bitumen fractions fall within the range of values noted for petroleums. The calcite fraction of Sample A is typical of inorganic marine carbonate, but the C^{13}/C^{12} ratio of the dolomite rhombs in Sample B is sufficiently lower than that of inorganic carbonate to imply that its carbon is derived, at least in part, from oxidized organic matter. Carbonate cements and vein materials from several other boulders from the same outcrop but essentially devoid of tarry material were analyzed and the isotopic compositions are typical of inorganic marine carbonate, similar to the calcite cement in Sample A.

The intimate association of organic dolomite, gaseous and tarry nuclei

TABLE 3. C^{13}/C^{12} δ-VALUES OF CARBONATE AND BITUMEN FRACTIONS OF PT. FERMIN SAMPLES A AND B

Sample	Fraction	C^{13}/C^{12} (in °/oo)[1]
A	Dolomite-free Calcite Cement	+1.8
	Bitumen Fraction	−25.0
B	Dolomite Crystals	−16.4
	Bitumen Fraction	−22.9

[1] Relative to PDB-1 Standard.

in the dolomite, and the tar in which the rhombohedra are embedded suggests but does not necessarily prove *in situ* oxidation of the tar to produce carbon dioxide that became incorporated in the dolomite. Oxidation of the organic matter, regardless of the *in situ* possibility, can occur by several means:

(1) reaction with oxygenated formations waters in the subsurface or by aerial exposure in the outcrop, (2) oxidation-reduction reactions involving the organic matter and oxygen-bearing ions in the water such as sulfate, and (3) oxidation-reduction reactions involving other oxygen-bearing minerals.

Organic carbonates in the Lyons Formation (Permian, Colorado) were formed by this latter mechanism, in which iron oxide was reduced simultaneously with oxidation of bitumen and with subsequent precipitation of organic dolomite cement (Silverman, *et al*, in press). The abundance of gaseous and tarry nuclei centrally located in the rhombohedra suggests (1) that the carbonate formed in a multi-phase medium, possibly water-tar-CO_2 and (2) that the crystallization of the dolomite began selectively on the fluid phase boundaries or bubble (globule) surfaces.

ACKNOWLEDGMENT

The authors wish to thank Dr. R. W. Rex for assiastnce in obtaining the x-ray diffraction data.

REFERENCES

CRAIG, H. (1953) The geochemistry of the stable carbon isotopes. *Geochim. Cosmochim. Acta*, **3**, 53–92.

EPSTEIN, S., R. BUCHSBAUM, H. A. LOWENSTAM AND H. C. UREY (1953) Revised carbonate-water isotopic temperature scale. *Bull. Geol. Soc Am.* **64**, 1315–1326.

FEELY, H. W. AND J. L. KULP (1957) Origin of Gulf Coast salt-dome sulfur deposits. *Bull. Am. Assoc. Petroleum Geol.* **41**, 1802–1853.

GOLDSMITH, J. R. AND D. L. GRAF (1958) Structural and compositional variations in some natural dolomites. *Jour. Geol.* **66**, 678–693.

NAKAI, N. (1960) Carbon isotope fractionation of natural gas in Japan. *Jour. Earth Sci. Nagoya Univ.* **8**, 174–180.

OSIPOVA, A. I. (1964) Catagenic changes in petroleum-bearing carbonate rocks. In, *Chemistry of the Earth's Crust* (Khimiya Zemnoi Kory) Vol II, ed. A. P. Vinogradov. Izdatel'stvo "Nauka," Moscow, 415–427.

SILVERMAN, S. R. (1964) Investigations of petroleum origin and evolution mechanisms by carbon isotope studies. In, *Isotopic and Cosmic Chemistry*, ed. H. Craig, S. L. Miller and G. J. Wasserburg. North-Holland Publishing Co., Amsterdam, 92–102.

SILVERMAN, S. R. AND S. EPSTEIN (1958) Carbon isotopic composition of petroleums and other sedimentary organic materials. *Bull. Am. Assoc. Petroleum, Geol.* **42**, 998–1012.

———, D. W. LEVANDOWSKI, L. C. BONHAM AND W. G. TOLAND (in press) Carbonates from oil *Bull. Am. Assoc. Petroleum Geol.*

THODE, H. G., R. K. WANLESS AND R. WALLOUCH (1954) The origin of native sulfur deposits from isotope fractionation studies. *Geochim. Cosmochim. Acta* **5**, 286–298.

Manuscript received, December 3, 1965; accepted for publication, February 12, 1966.

27

Reprinted from *Jour. Sed. Petrology* **35**:448–453 (1965)

POSSIBLE ROLES OF CLAY MINERALS IN THE FORMATION OF DOLOMITE[1]

CHARLES F. KAHLE
University of Toledo, Toledo, Ohio

ABSTRACT

Evidence is accumulating which strongly suggests that clay minerals effect the formation of some dolomites. Clays may behave somewhat as catalysts by providing a source of magnesium ions or by serving as membranes which effect ionic migration. Alternately, clays may assume a more active role in dolomitization by serving as centers of nucleation, or by entering into chemical reactions which involve dolomite as one of the products.

INTRODUCTION

The possible ways that clay minerals may effect the hydrodynamics and chemistry of subsurface water are known very roughly. Less is known concerning the possible cause and effect relations between clay minerals and the nature, distribution, and abundance of dolomite. Evidence that clay minerals may be important in some cases of dolomitization has resulted primarily from petrographic and mineralogic investigations of carbonates (Bisque and Lemish, 1960; Dunbar and Rogers, 1957, p. 223–224; Kahle, 1962, p. 99; and Murray, 1960, p. 75). Together these writers delt with samples ranging from Cambrian to Recent and representing many different lithologic types of carbonates. A definite tendency was present in all groups of samples for a positive correlation between the percentages of insoluble residue and dolomite or magnesium content. This tendency may be indicative of diagenetic changes in which clay minerals played an important role.

In this paper dolomites are classified genetically as follows: primary dolomites are those formed by direct precipitation from sea water above the depositional interface; secondary includes all other modes of occurrence. Included in the latter catagory are detrital dolomites which are not discussed in this paper. Penecontemporaneous dolomites result from a reaction or exchange between overlying or interstitial water with an initial aragonitic or calcitic sediment. Metasomatic dolomites result from a reaction between pore solutions and a rock mass.

The purpose of this paper is to call attention to some salient data and discuss some hypotheses, and perhaps point out a few directions of work which could prove helpful in solving the problem.

[1] Manuscript received June 23, 1964.

MAGNESIUM AND CLAY MINERALS

Normal sea water contains about three times as much magnesium as calcium on a weight basis and thus provides an obvious source of magnesium necessary for the formation of many dolomites. Although this idea has been used repeatedly in explaining the formation of many ancient dolomites, it involves a materials balance problem. Analysis of hypersaline interstitial waters from Bahaman sediments for example, has shown that even if Mg^{++} from the water as well as Mg^{++} from the $CaCO_3$ fraction in the sediment could be withdrawn, only about two percent of the resulting rock could be made into dolomite without the addition of Mg^{++} from other sources (Cloud, 1962, p. 30, 106). A practical answer to this problem may be related to the length of time that the original sediment remains in contact with sea water or to repeated reflux of sea water into underlying sediments. However, many carbonate sediments have been in contact with sea water for long periods of time and yet display no evidence of dolomitization. In addition, the reflux mechanism does not provide a unique solution which seemingly can be applied to the formation of all ancient dolomites. Clay minerals could provide a source for at least some, and possibly all of the magnesium needed for the dolomitization of a particular sediment or rock.

Clay minerals may contain appreciable quantities of Mg^{++} either in interlattice sites or in exchange positions resulting from broken edges or unbalanced charges. Magnesium is especially common in montmorillonite, chlorites, and vermiculites which may contain 25 percent or more of Mg O. Along with magnesium, other common cations found in exchange positions in clay minerals are Ca^{++}, Na^{++}, and K^+. The general order of replaceability of these cations in clays is: Na < K < Mg < Ca (Carroll, 1959).

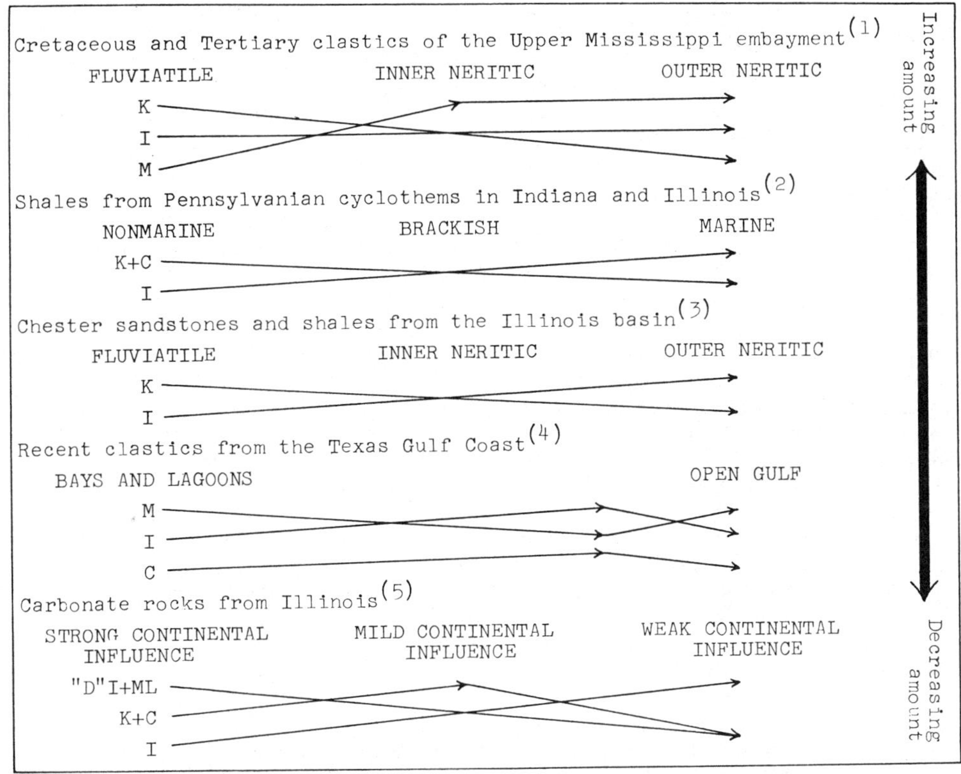

Fig. 1.—Clay mineral variations encountered within and adjacent to the near shore marine environment. Abbreviations: K—kaolinite; I—illite; M—montmorillonite; C—chlorite; ML—mixed layer clay; "D"—degraded. (1)—Pryor and Glass, 1961, p. 48; (2)—Murray, 1953; (3)—Smoot, 1960; (4)—Grim and Johns, 1953; (5) Ostrom, 1959, p. 142.

ENVIRONMENTAL FACTORS

A critical question at this point concerns the environment of deposition of clay minerals. Clay minerals introduced into the marine environment may be subjected to a number of possible influences including winnowing, selective sorting, and diagenetic modifications. A detailed consideration of all of these variables is beyond the scope of this paper. These topics have been discussed in detail previously (Weaver, 1958; Grim, 1958).

Clay minerals ordinarily tend to be winnowed out of agitated shallow water environments and deposited in deeper water. For example, the insoluble residue content of sediments on the bank portion of the Bahaman platform is extremely low (Cloud, 1962). Clay minerals are relatively common, however, in the Sigsbee deep in the Gulf of Mexico (Murray and Harrison, 1956). Oölitic limestones are generally considered to represent a product of a shallow water, agitated environment. A study of nearly 200 sample of oölitic limestones ranging in age from Cambrian to Recent indicated that the average insoluble residue content was about 5 percent by weight (Kahle, 1962); these same samples average only about 2 percent by weight of clay minerals.

A number of geologists have recognized variations in the distribution and abundance of clay mineral types within and adjacent to marine environments (fig. 1). Whether or not such variations are produced mainly by diagenesis or are merely functions of provenance, sorting, and other variables is debatable. Figure 1 tends to substantiate the idea that magnesium bearing clays such as montmorillonite generally increase seaward. Montmorillonite generally occurs in smaller particle sizes than does most other common clay minerals and could, therefore, be expected to settle out of suspension much more slowly. The salinity of sea water is also known to cause fairly rapid flocculation of clays such as kaolinite and illite whereas montmorillonite flocculates more slowly and at higher salinites (Whitehouse and Jeffrey, 1955). In the Bahaman area Cloud (1962, p. 59), found that montmorillonite was conspicuous in every clay fraction from the straits whereas no montmorillonite at all was found in sediments from the bank areas.

If clays, especially Mg^{++}-bearing varieties, are generally more abundant in a basinward direction, then it might be expected that in some instances, there would be a relation between localization of clay minerals and of dolomites. Evidence bearing on this possibility is not conclusive. Siever and Glass (1957, p. 56) in their study of some Pennsylvanian carbonate rocks in the Illinois Basin found no apparent correlation of dolomite content with basin or shelf occurrence. It has been argued that near shore carbonate sediments are more susceptable to dolomitization (Cloud and Barnes, 1957; Fairbridge, 1957). However, studies of shallow water, oölitic sediments in the Bahaman area by Cloud, (1962), and by Stehli and Hower (1961, p. 360, 369), have failed to disclose any evidence of dolomitization. Ancient oölitic rocks also are generally not extensively dolomitized (Kahle, 1962). Sabins (1962) found that primary dolomites are more abundant in a basinward direction in certain Cretaceous units in the Western Interior of the United States. Dolomites have also been found to be more abundant toward the central parts of the Upper Paleozoic basin of the Russian Platform (Ronov, 1956). A recent study of the Brassfield limestone (Silurian), in southern Indiana indicates that dolomites become more abundant toward the center of the basin (Craig Hatfield, personal communication).

Some of the observations noted above may be related to recent work by Bredehoeft and others (1963), who have investigated the concentration of brines in theoretical and actual basins. Their work points out that the concentration of dissolved ions in brines can be expected to increase with depth so that the highest concentrations will be found toward the central portion of a basin. Supporting evidence is in the form of chemical analyses of natural brines plus theoretical consideration of models. Both theoretical considerations and laboratory evidence indicate that clay minerals may act as membranes which impede the migration of cations and anions, but from the standpoint of ionic radii would normally be pervious to cations. In some cases the anions may hold the cations in order to obtain electrical and chemical neutrality. As a consequence, increasing concentrations of all or selected ions will tend to occur in units adjacent to shale membranes. Static or mobile brines which have Mg^{++} concentrated in this manner, might then precipitate as a high-Mg^{++} calcite or calcian dolomite.

ION EXCHANGE

Additional ways in which magnesium concentrations could be increased within sediments includes various aspects of ion exchange. The idea of ion exchange as a mechanism for producing Mg^{++} which could contribute to dolomitization would seem to have maximum application during the earlier stages of diagenesis. The movement of pore fluids, especially in the early stages of diagenesis, is generally in an upward direction, often toward the margins of the basin. Upward migrating pore fluids should in some cases, have ample opportunity to take part in ion exchange processes and gain increasing concentrations of magnesium. If this process were operative, it could result in effects opposite to those expected from brine concentration, namely, that dolomitization should increase toward the basin margins.

Divalent cations such as magnesium are usually bound more tightly than monovalent cations. However, if the material with bound divalent cations is placed in a solution for a long enough period of time, chemical equilibrium can be obtained. Establishment of equilibrium would involve a reversable reaction which obeys the law of mass action. As an example of this, the following reaction may be considered: $2Na^+(X) + Mg^{++} = Mg^{++}(X)_2 + 2Na^+$ where (X) represents a unit of exchange capacity in the solid material. In fresh water solutions this reaction should tend to move to the right due to magnesium ions from solution replacing absorbed sodium on the exchange material. Alternately, the reaction would tend to move to the left in situations where the exchange material was in contact with a highly mineralized sodium chloride brine. A reaction such as this might occur during the diagenesis of carbonate sediments where meteoric water encroachs into sediments containing connate water (Piper, 1953, p. 182).

An additional ion-exchange mechanism might be activated by dissolution of carbonate rocks and the release of appreciable quantities of Ca^{++} ions to pore waters. According to Chave (1960, p. 367), the addition of such ions to interstitial waters would increase the amount of exchangeable Mg^{++} that could be removed from clay minerals and replaced by Ca^{++}. Other ions besides Ca^{++} may of course, substitute for Mg^{++}. However, the relative amount of Ca^{++} compared to other ions in subsurface waters, along with the ease of replacement between various cations, indicates that fairly large amounts of Mg^{++} could be released to interstitial waters by ion exchange with Ca^{++}. A variation of this idea has been presented by Powers (1957), and involves his concept of an equivalence level. This level is the depth where potassium and magnesium ions have the same likelihood of being absorbed by clay minerals. According to this concept, progressively larger amounts of Mg^{++} would become available with increasing depth due to preferential exchange of K^+ over Mg^{++} at exchange sites in clays. It is possible that this mechanism or the exchange of Ca^{++} for Mg^{++} is responsible for the increasing magnesium content with depth noted in several dolomitized units in western Ohio (Stout, 1941).

Both the clay membrane effect, ion exchange, or both provide mechanisms whereby the concentration of magnesium in pore water could be progressively increased relative to that of calcium. Time and temperature would be major factors in either mechanism. Eventually, the magnesium content could be built up to the "threshold" level necessary for the following reaction:

$$2CaCO_3 + Mg^{++} \rightarrow CaMg(CO_3)_2 + Ca^{++}$$

Total porosity and permeability as well as localization of these variables at a specific stage of diagenesis, coupled with an adequate amount of magnesium bearing clays, might thus give rise to either widespread or localized dolomitization. Probably much of the magnesium added to the original connate water would be derived from shale units within the basin. In some cases, however, chemical evidence suggest that waters may be squeezed out of shale units external to a basin and migrate for hundreds of miles to their present location (Chave, 1960). Even the minor amounts of clays which generally occur within carbonates are a potential source of magnesium. According to Chilingar (1960, p. 183), the clay fraction of calcareous sediments in the Bahamas is high in magnesium content and the percentage of Mg^{++} increases with increasing distance from shore.

ADDITIONAL ROLES OF CLAYS

In general, clay minerals are associated with the finer grained fractions of carbonate rocks. The finer grained fractions are in turn, more susceptible to dolomitization than are the coarser grained fractions (Fairbridge, 1957, p. 156, 168; Hobbs, 1957, p. 40; Hofmann, 1962, p. 284; and Peterson, 1962, p. 21, 26). This observation, noted independently by several geologists, could be related to the known relation between small grain size and chemical reactivity due to the increase in surface area. It is also possible that clay minerals associated with the finer fraction have contributed in some way to dolomitization. Clay minerals may serve as some sort of catalyst or as centers of nucleation for crystal of dolomite. Similar suggestions have been made with reference to the possible effect of magnesium bearing algae on dolomitization. Randomly distributed clays within an initial carbonate sediment could serve as impurity centers which would affect nucleation. The splotchy or cloudy appearance of some rhombs in partially dolomitized limestones, contrasted to clearer seemingly impurity free surrounding calcite grains, may be related to this idea. One of the most recent examples of the effects of clays on dolomitization has been presented by Schmidt (1964, p. 345–346), in a study of the Upper Jurassic Gigas Beds in Germany. He found that dolomitization affected the high-clay portions first, even prior to replacement of aragonitic calciclastics. It was also observed that dolomitization increased as the clay content of the rocks increased.

Possible ways that clay minerals within sediments may take part in reactions which involve the formation of dolomite have been investigated by several workers but especially by Zen (1959). The more important clay mineral-carbonate phase assemblages developed tentatively by Zen are shown in figure 2. In every case the assumption is made that quartz is present in excess, and that H_2O and CO_2 are mobile constituents. Possible reactions involving assemblages shown in figure 2 may be illustrated schematically as follows:

Figure 2A–2B
   calcite + chlorite + carbon dioxide = dolomite + kaolinite + quartz + water
Figure 2C–2D
   montmorillonite + calcite + carbon dioxide = dolomite + kaolinite + quartz + water

Limited petrographic and mineralogic work supports the essential validity of these reactions although numerous problems remain concerning the chemistry involved and the quantitative importance of dolomites which may have been derived by such reactions.

Other layer-silicate minerals beside clays have also been suggested as being reactants involved in the production of dolomite. For example, Calvert (1964, p. 184, 185), has proposed that the pre-Trenton sediments of the Cincinnati Arch region were eventually dolomitized by "diagenetic absorption of feldspar and biotite" especially the latter. He proposes that magnesium-rich biotite mica could decompose in a

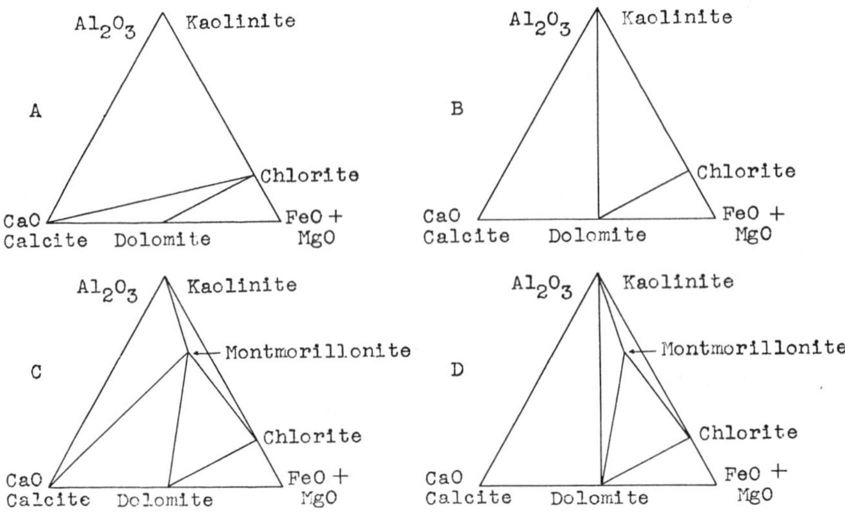

FIG. 2.—Alternative phase relations in carbonate rocks. After Zen (1959).

calcium carbonate environment to yield $MgCO_3$ which would then recrystallize in the presence of $CaCO_3$ to produce dolomite.

Conceivably, the colloidal nature of clays could influence the formation of some dolomites. Colloidal material material can have a considerable effect on the rate and type of nucleation of any material crystallizing from a solution. Colloidal suspensions of clay minerals could serve as a catalyst to slow down reaction rates Siegel (1960, p. 152–153), has shown that a finely divided suspension of activated charcoal induces increased ordering in the lattice of artifically prepared protodolomite. The effect of the suspension is to produce a material which more closely resembles natural dolomite. Clay minerals in nature might have a similar effect. Material crystallizing from a solution containing impurities which promote the formation of a colloidal suspension will normally either utilize the impurities as centers of nucleation or will digest considerable impurity material during crystal growth. In an environment in which primary dolomite was precipitating, much of the potential impurity material in suspension would probably be in the form of clay size material, especially clay minerals. Similar reasoning might also apply to clay minerals trapped in saturated pore solutions in which nucleation of dolomite was taking place.

SUMMARY

The more important ways in which clay minerals could contribute to dolomitization include various aspects of ion exchange, nucleation and crystal growth processes, and participation in certain clay mineral-carbonate reactions. At present, only limited information is available concerning these topics. With the possible exception of dolomite-evaporate assemblages, evidence is not compelling that dolomitization is invariably a phenomenon associated with toward-shore or shoaling conditions.

Future work should involve attempts to elucidate any trends or patterns which would indicate a definitive relationship between either types, abundance, or distribution of clay minerals with respect to dolomite in individual units, or better still, within entire basins. Additional research on whether or not clay minerals have any effect on artificially prepared protodolomites would also be of interest.

ACKNOWLEDGMENTS

Appreciation is expressed to Dr. Hobart E. Stocking of Oklahoma State University for his encouragement and helpful suggestions. The writer is indebted to Dr. Frederic Siegel of the Kansas Geological Survey for his criticism of the manuscript.

REFERENCES

BISQUE, R. E., AND LEMISH, JOHN, 1959, Insoluble residue-magnesium content relationship of carbonate rocks from the Devonian Cedar Valley Formation: Jour. Sedimentary Petrology, v. 29, p. 73–76.

BREDEHOEFT, J. D., AND OTHERS, 1963, Possible mechanism for concentration of brines in subsurface formations: Am. Assoc. Petroleum Geologists Bull., v. 47, p. 257–269.

CALVERT, W. L., 1964, Pre-Trenton sedimentation and dolomitization, Cincinnati arch province; theoretical considerations: Am. Assoc. Petroleum Geologists Bull., v. 48, p. 166–190.

CARROLL, DOROTHY, 1959, Ion exchange in clays and other minerals: Geol. Soc. America Bull., v. 70, p. 749–780.
CHAVE, K. E., 1960, Evidence on history of sea water from chemistry of deeper subsurface waters of ancient basins: Geol. Soc. America Bull., v. 44, p. 357–370.
CHILINGAR, G. V., 1960, Ca/Mg ratios of calcareous sediments as a function of depth and distance from shore: The Compass, v. 37, no. 3, p. 182–186.
CLOUD, P. E., JR., AND BARNES, V. E., 1957, Early Ordovician seas in central Texas: Geol. Soc. America Mem. 67, p. 163–214.
CLOUD, P. E., JR., AND OTHERS, 1962, Environment of calcium carbonate deposition west of Andros Island, Bahamas: U. S. Geol. Survey Prof. Paper 350, 138 p.
DUNBAR, C. O., AND RODGERS, J., 1957, Principles of stratigraphy. John Wiley and Sons, Inc., New York, p. 223–224.
FAIRBRIDGE, R. W., 1957, The dolomite question, *in* Leblanc, R. J., and Breeding, J. G., eds., Regional Aspects of Carbonate Deposition: Soc. Econ. Paleontologists and Mineralogists, Spec. Publ. no. 5, p. 125–178.
GRIM, R. E., 1958, Concept of diagenesis in argillaceous sediments: Am. Assoc. Petroleum Geologists Bull., v. 42, p. 246–253.
GRIM, R. E., AND JOHNS, W. D., 1954, Clay mineral investigations of sediments in the northern Gulf of Mexico: Clays and Clay Min. Bull., Nat. Acad. Science-Nat. Research Council, Pub. 327, p. 81–103.
HOBBS, C. R. B., JR., 1957, Petrography and origin of dolomite-bearing carbonate rocks of Ordovician age in Virginia: Bull. Virginia Polytech. Inst., Series 116, v. 50, no. 5, 128 p.
HOFMANN, H. J., 1963, Ordovician Chazy group in southern Quebec: Am. Assoc. Petroleum Geologists Bull., v. 47, p. 270–301.
KAHLE, C. F., 1962, Diagenesis in oolitic and pellet-type limestones: Unpublished Ph.D. Dissertation, Univ. of Kansas, 138 p.
MURRAY, H. H., 1954, Genesis of clay minerals in some Pennsylvanian shales of Indiana and Illinois: Clays and Clay Minerals Bull., Nat. Acad. Science-Nat. Research Council, Pub. 327, p. 47–67.
MURRAY, H. H., AND HARRISON, J. L., 1956, Clay mineral composition of recent sediments from Sigsbee Deep: Jour. Sedimentary Petrology, v. 26, p. 363–368.
MURRAY, R. C., 1960, Origin of porosity in carbonate rocks: Jour. Sedimentary Petrology, v. 30, p. 59–84.
OSTROM, M. E., 1959, Clay mineralogy of some carbonate rocks of Illinois: Unpublished Ph.D. Thesis, Univ. of Illinois.
PETERSON, M. N. A., 1962, The mineralogy and petrology of Upper Mississippian carbonate rocks of the Cumberland plateau in Tennessee: Jour. Geology, v. 70, p. 1–31.
PIPER, A. M., GARRET, A. A., AND OTHERS, 1953, Native and contaminated waters in the Long Beach-Santa Ana Area, California: U. S. Geol. Survey Water-Supply Paper 1136.
POWERS, M. C., 1957a, Adjustment of clays to chemical change and the concept of the equivalence level: Clays and Clay Minerals, Sixth Nat. Conf., Pergamon Press, N. Y., p. 309–326.
PRYOR, W. A., AND GLASS, H. D., 1961, Cretaceous-Tertiary clay mineralogy of the Upper Mississippian embayment: Jour. Sedimentary Petrology, v. 31, p. 38–51.
RONOV, A. B., 1956, Chemical composition and conditions of formation of Paleozoic carbonate deposits of the Russian platform, *in* Type of dolomite rocks and their genesis: Trudy Geol. Inst., Izd. Akak, Nauk SSSR, v. 4, p. 256–343.
SCHMIDT, VOLKMAR, 1964, Facies, diagenesis and related reservoir properties in the Gigas Beds (Upper Jurassic), Northwestern Germany (abs.): Am. Assoc. Petroleum Geologists Bull., v. 48, p. 545–546.
SIEGEL, F. R., 1961, Factors influencing the precipitation of dolomitic carbonates: Kansas Geol. Survey Bull. 152, pt. 5, p. 127–158.
SIEVER, R., AND GLASS, H. D., 1957, Mineralogy of some Pennsylvanian carbonate rocks of Illinois: Jour. Sedimentary Petrology, v. 27, p. 56–63.
SMOOT, T. W., 1960, Clay mineralogy of pre-Pennsylvanian sandstones and shales of the Illinois Basin, Part 3: Illinois Geol Survey, Circ. 293, 19 p.
STEHLI, F. G., AND HOWER, J., 1961, Mineralogy and early diagenesis of carbonate sediments: Jour. Sedimentary Petrology, v. 31, p. 358–371.
STOUT, WILBER, 1941, Dolomites and limestones of western Ohio: Ohio Geol. Survey, 4th Ser., Bull. 42, 468 p.
WEAVER, C. E., 1958, A discussion of the origin of clay minerals in sedimentary rocks: Clays and Clay Min. Bull., Nat. Acad. Science-Nat. Research Council, Pub. 566, p. 159–173.
WHITEHOUSE, U. G., AND JEFFREY, L. M., 1955, Peptization resistance of selected samples of kaolinitic, montmorillonitic, and illitic clay materials: Clays and Clay Mins. Bull., Nat. Acad. Science-Nat. Research Council, Pub. 395, p. 260–281.
ZEN, E-AN, 1959, Clay mineral-carbonate relations in sedimentary rocks: Am. Jour Sci., v. 257, p. 29–43.

28

Reprinted from *Jour. Sed. Petrology* **43**:603–613 (1973)

ALGAL ORIGIN OF DOLOMITE LAMINATIONS
IN STROMATOLITIC LIMESTONE[1,2]

CONRAD D. GEBELEIN AND PAUL HOFFMAN
Bermuda Biological Station for Research, St. George's West, Bermuda
Geological Survey of Canada, Ottawa, Canada

ABSTRACT: Many ancient carbonate rocks consist of dolomite interlayered with limestone on a centimeter or millimeter scale. Many such rocks contain cryptalgal laminites or stromatolites. Comparable Recent stromatolitic and flat laminated algal sediments are composed of alternating layers of particulate carbonate sediment and algal mats. The characteristics of algal-rich and sediment-rich laminae are identical to those of the dolomite and calcite laminae, respectively, of ancient forms. On this basis, dolomite laminae in ancient interlaminated sediments are considered equivalent to, and hence derived from, algal-rich lamina in Recent stromatolitic sediments; calcite laminae are derived from sediment-rich laminae. Modern algal sediments of this type, however, are not known to contain thin layers of dolomite.

It is suggested that the dolomite layers are secondary. During deposition, magnesium ions are complexed organically in the algal mat layers. The algal sheath material, in which the magnesium is concentrated, is very stable and does not decompose until long after deposition, or even lithification. Only when the organic matter is decomposed is the magnesium released to form dolomite in the micro-environment of the relict algal mat layers. Thus the dolomite, although secondary, conforms to the primary algal mat layers of the sediment.

Laboratory experiments support the hypothesis. Sheath material of the major stromatolite-forming blue-green alga *Schizothrix calcicola* grown in sea water shows a three to four fold increase in Mg/Ca ratio relative to that in the sea water medium. Sufficient magnesium is complexed in a single 2 mm thick algal mat layer to produce a layer of dolomite 1 mm thick. No dolomite was precipitated in the algal mat layers experimentally, but crystallization of 17–20 mole percent-Mg calcite in contact with the sheath material was achieved, showing the influence of the organic layer micro-environment on the composition of the carbonate mineral produced. The actual formation of dolomite may be a long term process, probably achieved by the partial replacement of the carbonate matrix in contact with the relict algal mat layers, and/or by selective replacement of high-Mg calcite within the algal mat layers.

The hypothesis explains why thinly interlayered dolomite is not found abundantly in Recent sediments, and yet why the dolomite layers so faithfully conform to the primary sedimentary structures in ancient rocks. The hypothesis refers specifically to laminated cryptalgal rocks, but should be considered for other problematical types of ancient dolomite where the distribution of dolomite may be related to the original distribution of organic matter in the sediment.

INTRODUCTION

The vast thicknesses of lower Paleozoic and Proterozoic carbonate rocks in North America are characterized by beds in which dolomite and limestone are interlayered on a centimeter, or even millimeter scale. The alternating layers of

dolomite and limestone commonly conform perfectly to the primary sedimentary structures. Many of the beds contain mudcracks, truncated ripple marks, cryptalgal laminites (Aitken, 1967) and stromatolites indicating a very shallow water or intertidal depositional environment.

Several generations of geologists have been puzzled as to the origin of the dolomite. Most have concluded that it must be either: (1) primary, precipitated directly from sea water; or (2) penecontemporaneous, formed by replace-

---

[1] Manuscript received November 9, 1972; revised January 30, 1973. A preliminary abstract of this paper appears in Bricker, O. E. (*ed.*), 1970, Carbonate Cements, Johns Hopkins University Press.

[2] Contribution number 565 from Bermuda Biological Station for Research.

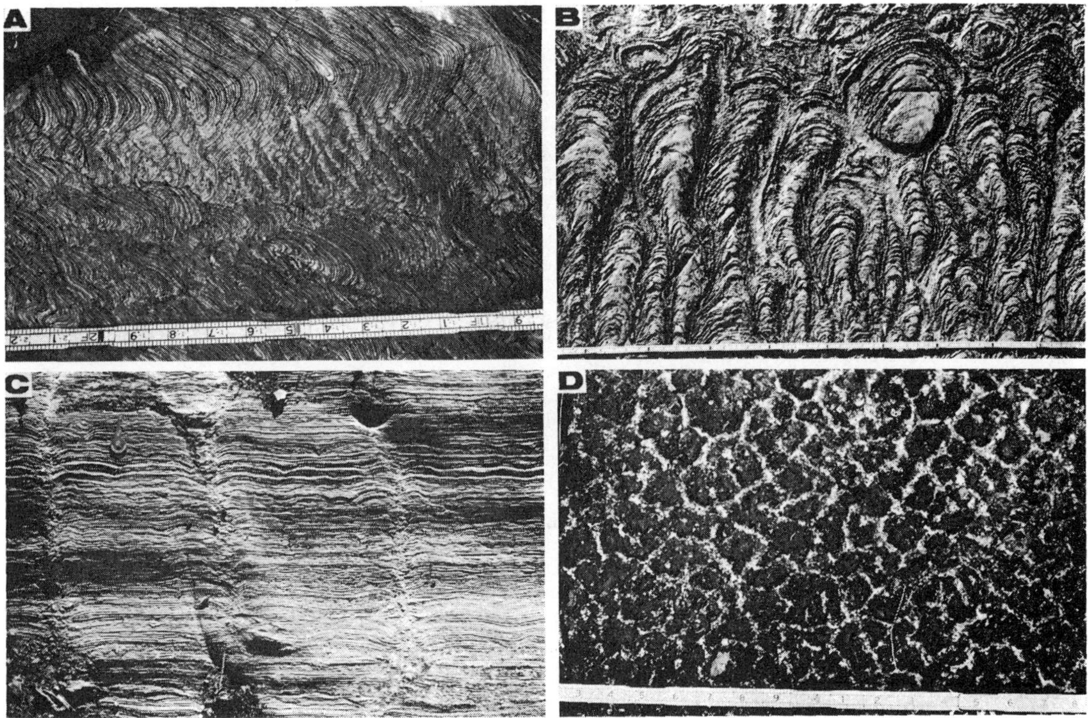

Fig. 1.—Ancient stromatolites with dolomite laminations (scale in tenths of feet).
    1A. Compound domal stromatolites on outcrop surface showing dolomite laminations (dark) and limestone (light). Wildbread Formation (Proterozoic), Great Slave Lake.
    1B. Columnar stromatolites composed of interlaminated dolomite (resistant) and limestone (recessive). Taltheilei Fm. (Proterozoic), Great Slave Lake.
    1C. Undulatory stromatolites with interlaminated dolomite (resistant) and limestone (recessive). The vertical disruptions are polygonal prism cracks. Conococheague Fm. (Cambrian), near Clear Spring, Md.
    1D. Bedding surface with mudcracked, dolomitized algal mats (dark) and limestone (light). Utsingi Fm. (Proterozoic), Great Slave Lake.

ment of aragonite or calcite by magnesium-rich brines soon after deposition. According to either theory, analogous Recent sediments should also contain alternating layers of dolomite and calcium carbonate.

Most modern carbonate sediments, however, contain little or no dolomite, even those which otherwise are exact replicas of the ancient dolomite-limestone beds. Primary dolomite (Alderman and Skinner, 1957; Von der Borsch, 1965) is exceedingly rare, if it occurs at all. Detrital dolomite is a minor constituent of many Recent sediments (Friedman and Sanders, 1967) and truly Recent (i.e. non-detrital) dolomite is reported from the Persian Gulf (Illing *et al.* 1965, Bonaire (Deffeyes *et al.* 1965), the Bahamas (Shinn *et al.* 1965) and Florida (Shinn, 1968).

Recent dolomites are forming by penecontemporaneous replacement of some other calcium carbonate mineral in hypersaline brines.

Mechanisms for the production of these brines have been proposed for each of the two major types of dolomite occurrences. "Capillary concentration" of interstitial waters by intense evaporation on supratidal and intertidal banks is proposed to explain the formation of lithified dolomite crusts on carbonate mudflats. "Refluxion" of dense hypersaline brines downward through porous carbonate sediments is proposed for the dolomites found on Bonaire. Both mechanisms postulate the presence of high salinity waters in which the Mg/Ca ratio has been markedly increased from that of sea water. Increase in the Mg/Ca ratio is presumed to occur in response to the precipitation of calcium salts, usually gypsum.

Many massive or thick-bedded ancient dolomites may have formed by the "capillary concentration" or "refluxion" mechanisms. However, thinly bedded or interlaminated dolomites present another problem. None of the Recent dolomitic sediments are known to consist of such an intimate interlamination of dolomite and calcite (or aragonite). Moreover, many of the ancient interlayered dolomite-limestone beds are not associated with gypsum or anhydrite, nor do they contain crystal cast pseudomorphs of these minerals. In fact, it is not clear how either of the postulated mechanisms of penecontemporaneous dolomite formation could account for the intimate, regular association of dolomite and limestone observed in ancient rocks. Concerning this important type of dolomite occurrence, Friedman and Sanders (1967) state that "no satisfactory explanation can be offered why some limestone interbeds are preferentially dolomitized whereas others are not."

These facts leave but two alternatives. Either sediments analogous to the ancient interlayered dolomite-limestone beds are not forming in the world today (or haven't been found yet), or the dolomite is of secondary origin. Pursuing the second alternative (while nevertheless encouraging others to pursue the first), if the dolomite results from some secondary process long after sedimentation, then the analogous Recent deposits would not be expected to contain dolomite layers.

What kind of secondary process could possibly yield dolomite with such a highly selective distribution? There are three problems. First, conditions favoring dolomite formation must exist within certain layers, but not in others. Second, because the layers conform with primary structures, the contrasting conditions in alternate layers must be inherited from the primary sedimentary environment. Third, sufficient Mg must be deposited in or transported to the sites of dolomite formation.

This paper presents a speculative solution to these problems, based on studies of the Cambro-Ordovician carbonates of the central Appalachians, Proterozoic carbonates of the Northwest Territories in Canada, Recent carbonate sediments in Florida, the Bahamas, Bermuda, western Australia and the Persian Gulf, and on experimental studies with modern blue-green algae.

EVIDENCE FROM ANCIENT ROCKS

In ancient rocks, the interlayering of dolomite and limestone is very conspicuous. On weathered outcrops the dolomite layers are light yellow-brown and stand in relief, whereas the limestone layers are blue-gray and recessive. In sawed slabs etched with dilute acid the dolomite is chalky white and the limestone is dark grey. In thin sections the limestone can be preferentially stained with Alizarin Red-S.

Fig. 2.—Undulatory stromatolites and discoidal oncolites on natural outcrop surface showing dolomite laminations (dark) and limestone (light). Scale in tenths of feet.

Formations with thinly interlayered dolomite and limestone examined by us are in the Cambrian Conococheague Group and the Ordovician Beckmantown Group (described by Sando, 1957; Sarin, 1962) in Maryland and Pennsylvania, the Devonian Manlius Formation (Laporte, 1964, 1967) in New York, the Cambrian Lyell Formation (Aitken, 1966) in Alberta and the Proterozoic Pethei Group (Hoffman, in press) in the Northwest Territories of Canada. The classic example is the so-called Lofer facies in the Upper Triassic of the Austrian Alps, made famous by the work of Sander (1936, 1951) and Fischer (1964).

The most important observation is that the laminations in the interlayered dolomite-limestone beds are "cryptalgal" (Aitken, 1967) in origin. The distinctive texture and morphology of the laminations are the result of mats of blue-green algae which intermittently covered the sediment surface during deposition. Some of the beds contain domal (Fig. 1A), columnar (Fig. 1B) or polygonal (Figs. 1C–D) stromatolites (classification of Aitken, 1967). Other beds are made up of detached stromatolites, or oncolites (Fig. 2). More commonly, the laminations are flat or gently undulatory (Figs. 1C and 1D). Such beds have been called undulatory stromatolites (Donaldson, 1963), planar stromatolites (Howe, 1968) and cryptalgal laminites (Aitken, 1967). Photomicrographs of cryptalgal laminites with alternating layers of dolomite and limestone are well illustrated by Sander (1951, Figs. 1, 4 and 8), Fischer (1964, Figs. 14 and 19), and Laporte (1967, Figs. 5, 6 and 13). The criteria for establishing the cryptalgal origin of such laminites are outlined in detail by Ginsburg *et al.* (1954) and Aitken (1967).

The dolomite layers are mostly less than 1

358

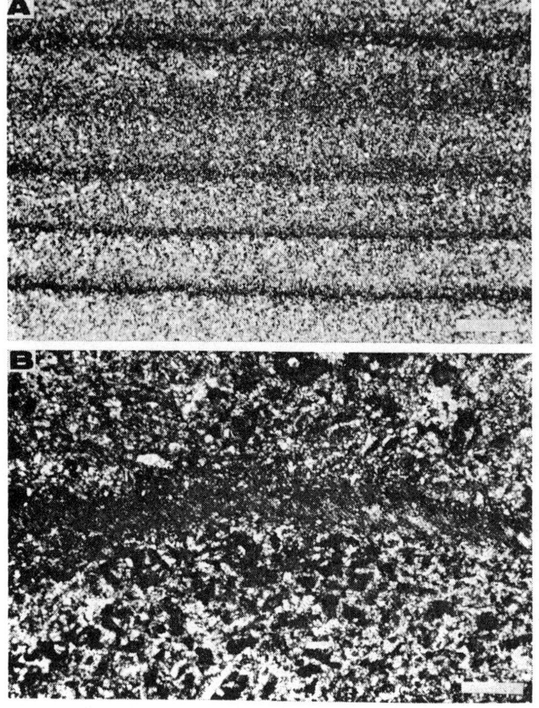

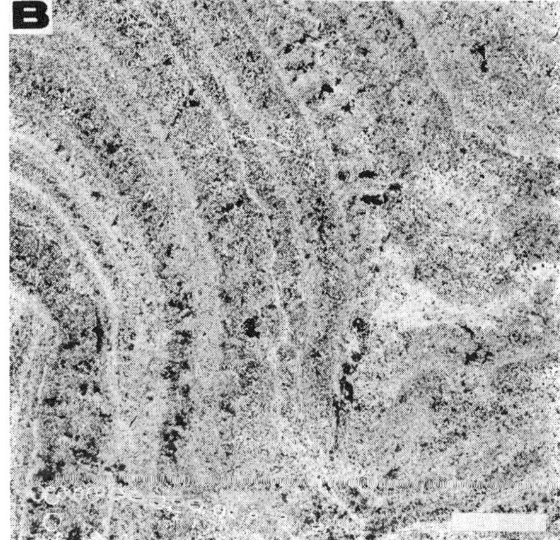

Fig. 3.—Petrographic character of ancient dolomite and limestone laminations.

3A. Thin section of a domal stromatolite (Taltheilei Fm.). Note the smooth, dark (bituminous) thin laminae-dolomite-alternating with coarse, thick laminae of calcite. Bar — 2 mm.

3B. Thin section of a domal stromatolite (Manlius Fm., Devonian of New York). Note the fine-grained layer with bituminous inclusions (dolomite) between coarse biopelsparite layers (calcite). Bar = 1 mm.

Fig. 4.—Ancient stromatolites with dolomite tufts.

4A. Etched slab of a Proterozoic domal stromatolite (Wildbread Fm., Great Slave Lake) showing dark dolomite laminations with carbonaceous residues and light calcite laminations. Note the radial dolomite pattern. Area in photo is 1.5 cm high.

4B. Thin section of slab shown above. Note radial tufts (dolomite) traversing stromatolitic laminae. Bar = 1 mm.

mm thick and tend to be uniform along their length. They are composed of tightly interlocking rhombs, 1–10 microns in diameter, in some cases transected by tiny tubes, believed to be molds of algal filaments (Donaldson, 1963, plate III; Fischer, 1964, Figs. 14 and 16). The dolomite layers contain, or are capped by, dark, presumably bituminous residues (Fig. 3A; Laporte, 1967, Fig. 7; Donaldson, 1963). They normally lack epiclastic textures (for an exception to this see Schenk, 1967, Figs. 8, 9, and 10). They are commonly smooth (Fig. 3B), but may be tightly crenulated (Sander, 1951, Fig. 8) or have tufts which protrude into the overlying limestone layer. In domal stromatolites, the tufts have a radial pattern (Fig. 4; Komar, 1966, plate XVII-1) which resembles some algal tufa (Irion and Müller, 1968, Figs.

4 and 5). All evidence of algal growth is found within the dolomite layers.

The limestone layers consist of a coarse-grained (10–50 μ) mosaic of blocky calcite and are highly variable in thickness (Fig. 3A; Laporte, 1967, Figs. 6 and 13). Thickness variation along the length of limestone layers in stromatolites is related to the prevailing direction of sediment transport (Hoffman, 1967, Fig. 5). The limestone layers contain scattered intraclasts, oolites, pellets, quartz silt and single dolomite rhombs (Fig. 5). Thus, in contrast to the dolomite layers, the limestone layers appear to have been originally a particulate sediment.

EVIDENCE FROM RECENT SEDIMENTS

Recent cryptalgal laminites and stromatolites examined by us occur in Florida (Gebelein and Hoffman, 1968; Gebelein, 1971), Bermuda (Gebelein, 1969), Bahamas (Shinn *et al.* 1969; Neumann *et al.* 1970), Western Australia (Logan, 1961; Logan *et al.* in press) and the Persian Gulf (Kendall and Skipwith, 1968). The lamination in all of the laminites and most of the stromatolites is composed of alternations of algal filaments and particulate carbonate sediment.

The layers of algal filaments form as surficial mats during periods of non-deposition of particulate sediment, or periods of abnormally rapid growth of the algae. These layers are mostly between 0.1 and 5.0 mm in thickness, approaching the minimum figure in the subtidal environment and the maximum figure in intertidal areas where ponding of seawater occurs (Gebelein, 1971). The thickness of the algal layers is normally uniform along their length. Algal layers are commonly smooth (Fig. 6A; Kendall and Skipwith, 1968, Fig. 8), but they may be crenulated tightly (Figs. 6B and C; Kendall and Skipwith, 1968, Fig. 2F) or have vertical tufts of algal filaments which project upward into the overlying sediment layers. In Recent domal stromatolites the algal tufts have a radial pattern (Fig. 6D; Black, 1933, Fig. 28; Monty, 1965, Fig. 4; Hardie, 1969). The algal layers are commonly composed of almost 100% organic matter.

The sediment layers range from 1.0 to 20 mm in thickness, and are not uniform along their length. Lateral thickness variation is related to the direction of sediment transport (Gebelein and Hoffman, 1968; Gebelein, 1969; Hoffman *et al.* 1970). The particulate nature of the sediment layers is easily seen in this section (Fig. 7; Logan, 1961). The sediment layers normally have less than 5% organic matter (by weight).

The association of carbonate lithification with algal mats and stromatolites has been reported by several workers. Dalrymple (1964) and Hardie (1969) have reported precipitation of fine grained carbonate in algal mats from intertidal sediments of Baffin Bay, Texas, and Andros Island, Bahamas, respectively. In these localities, precipitation of cryptocrystalline aragonite occurs on or within the mucilaginous sheaths surrounding individual algal threads.

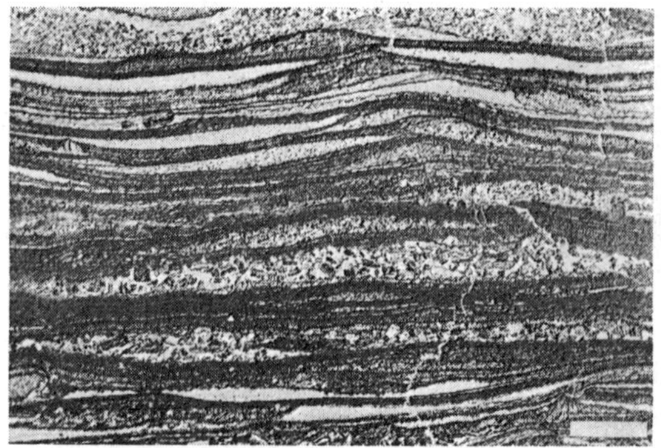

FIG. 5.—Etched slab of a Proterozoic cryptalgalaminite (Wildbread Fm.). Thin, dark dolomite layers alternate with coarse, clastic (lithoclasts, oolites) calcite layers. Note ripple forms in clastic calcitic layers. Bar = 4 mm.

A similar association of carbonate crystals and sheath material was described by Monty (1967) for algal stromatolites and mats growing in fresh to brackish supratidal and intertidal settings on the east coast of Andros Island, Bahamas. Crystals were either aragonite or 5–10 mole percent Mg calcite. An association between blue-green algae and carbonate precipitation also exists at Cape Sable, Florida (Gebelein, 1971). In the intertidal zone of large, shallow ponds (salinity varies seasonally from 5 to 65‰), thick (2–5 mm) organic-rich algal mats are underlain by laminae composed exclusively of photosynthetic and non-photosynthetic bacteria. Couplets of algal-bacterial laminae extend to about two feet below the present depositional surface. Rates of decomposition of the algal mucilage are much greater in these mats than in mats formed in other Recent environments, due to the intimate association of algae with extremely high bacterial cell concentrations. (Golubic describes a similar mat in analogous ponded areas on the Persian Gulf tidal flats; pers. comm.). At Cape Sable, partially decomposed algal laminae one centimeter or deeper below the surface are permeated by tremendous concentrations of minute (1–4 μ) carbonate crystals. Crystals are 14–19 mole percent Mg calcite. All crystals grow attached to the surface of mucilaginous sheaths of the blue-green algae; thus, a clear-cut genetic relationship between algal mucilage, bacterial decomposition and Mg carbonate nucleation and crystal growth can be demonstrated in these mats.

Table 1 summarizes the similarities between the algal layers in Recent sediments and the dolomite layers in ancient rocks, and between

TABLE 1.—*Comparison of Ancient Dolomite-Limestone Interlaminations with Recent Stromatolitic Laminations*

	Dolomite laminae (ancient) Algal-rich laminae (Recent)	Limestone laminae (ancient) Sediment-rich laminae (Recent)
Bituminous residue or organic material	present	absent
Thickness	uniform usually less than 1 mm	variable 1 mm or greater
Relation to direction of sediment supply	none	thickest on the side facing the sediment source
Epiclastic texture	absent	present
Tufts and radial filaments	present	absent

the sediment layers in Recent sediments and the limestone layers in ancient rocks. This comparison suggests that the production of interlayered limestone and dolomite is related, not to changes in water chemistry leading to alternate precipitation of dolomite and calcite, nor to alternations in the chemistry of interstitial waters, but to the production of algal-sediment laminae.

Acting upon this apparent relationship between algae and dolomite, one of us (Gebelein) performed a series of experiments to elucidate the chemical properties of the algal material relevant to the problem of dolomite formation.

EVIDENCE FROM EXPERIMENTAL STUDIES

Mg/Ca Ratios

Laboratory experiments have been performed using the blue-green alga *Schizothrix calcicola*, which is the major stromatolite forming alga in Florida, Bermuda and Western Australia. Cultures of *S. calcicola* have been isolated from these areas and grown in artificial sea water media. The cultures were allowed to grow and, finally, to deplete the limiting nutrients. All cells died, leaving only the mass of gelatinous sheath material secreted around the cells during growth. This sheath material was dried, ashed and analyzed for Mg and Ca by atomic absorption. When the cultures were grown in a sea water media with a Mg/Ca ratio of 4.5/1, the same ratio in the sheath material was 15/1. When the ratio of the media was 8/1 or greater, the ratio in the sheath material was 25–30/1. Thus magnesium is concentrated 3–4 times in the algal sheath material relative to the sea water media.

The mechanism for this preferential Mg uptake onto the sheath is not known. It may be biological, providing a source of Mg for the cells. Alternatively, the process may be purely physicochemical; many organic compounds are known to preferentially chelate divalent cations.

Having demonstrated the ability of algal layers to concentrate magnesium ions from sea water, two further questions must be answered: (1) Is sufficient magnesium present in a single algal layer to produce a layer of dolomite of the same thickness? and (2) Does a layer of algal sheath material provide a suitable substrate for crystallization of dolomite?

Absolute Amount of Mg

Density calculations on algal mats from Florida, Western Australia and the Persian Gulf have been combined with the experimental data on magnesium concentrations in algal sheath material to provide an approximate value for the absolute amount of magnesium present in a single algal layer. These calculations are outlined below:

(1) density of the algal layer (no included sediment), based on 32 measurements from 6 localities, averages 1.27 gm/cc;
(2) the weight of a dried and ashed algal mat sample is 25% of its *in situ* weight;
(3) 60% by weight, of the ashed sample of algal mats grown under laboratory conditions is Mg; mats found in nature contain 25 to 65% Mg;
(4) therefore, 1 cc of *in situ* algal mat can yield 0.19 gm of Mg;
(5) 1 cc of dolomite contains 0.37 gm of Mg;
(6) therefore, each algal layer 2 mm thick provides sufficient Mg for the potential production of a dolomite layer 1 mm thick.

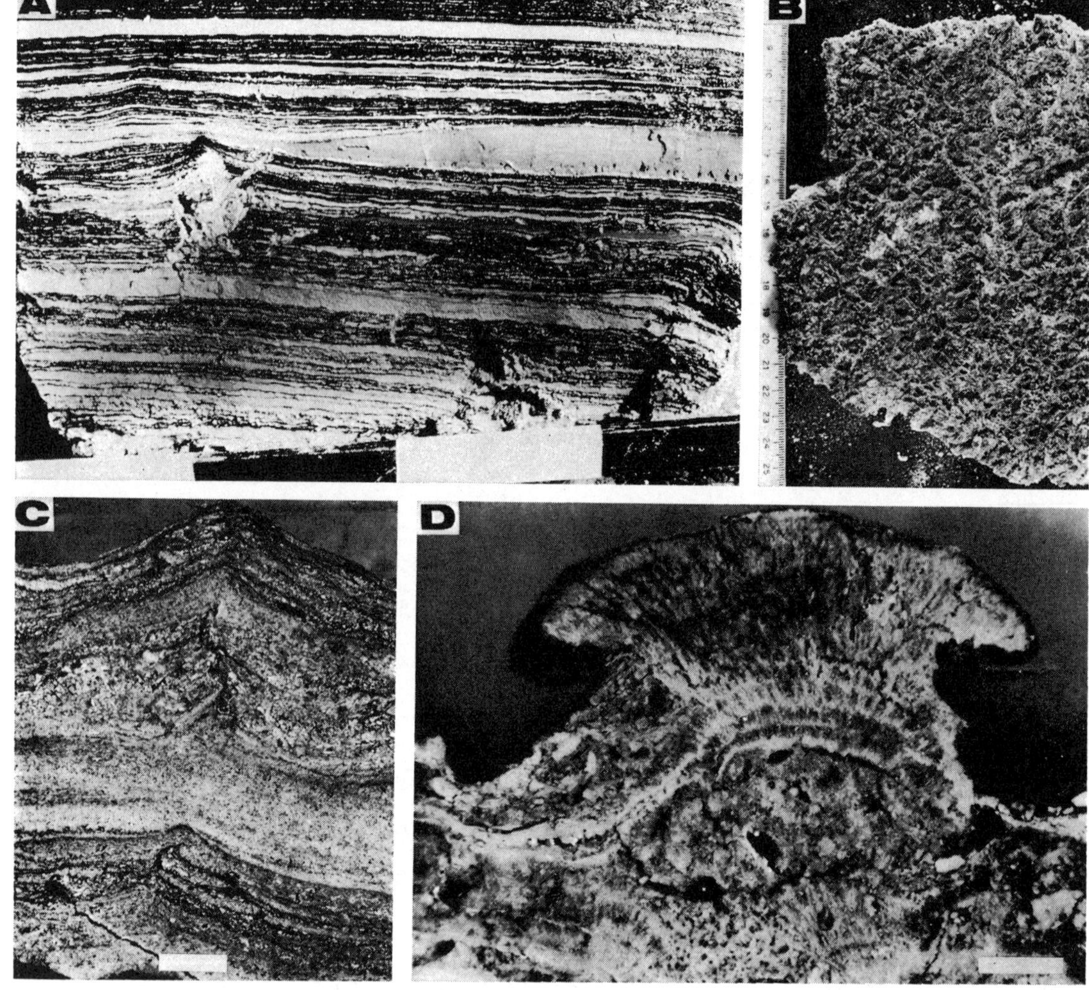

Fɪɢ. 6.—Recent algal stromatolites and flat-laminated algal sediments.
6A. Vertical section through the algal flat, Abu Dhabi, Trucial Oman. Note thin, dark continuous algal layers and thick, discontinuous sediment layers. Each bar = 10 cm.
6B. Surface of a crenulate algal mat, intertidal pond, Duck Key, Florida Bay. Growth habit of the algae gives this distinctive texture. Scale in centimeters.
6C. Vertical section through a crenulate mat, Hamelin Pool, Shark Bay, Australia. Note the superposition of mat structures and continuity of dark algal layers. Bar = 1 cm.
6D. Vertical section through a lithified domal stromatolite, intertidal zone, western Andros Island, Bahamas. Note the radial algal tufts in the interior of the dome. Bar = 4 mm.

It is evident from these estimates that the algal layers in many modern tidal flat sediments may concentrate sufficient magnesium to produce dolomite layers of the thickness found in the ancient interlayered dolomite-limestone rocks.

Crystallization of Mg Carbonates

Dolomite could not be crystallized in the algal layers experimentally. This is to be expected, for if the kinetics of dolomite formation were such that it could be produced under sedimentary conditions in the laboratory in a matter of weeks, then surely it would be abundant as a primary and penecontemporaneous mineral in modern sediments. However, it was possible to demonstrate that the concentration of magnesium ions in the algal layers influences the subsequent precipitation and composition of carbonate minerals.

Small quantities of algal sheath material, with Mg/Ca ratios of 20–30/1, were transferred to flasks containing natural or artificial sea water

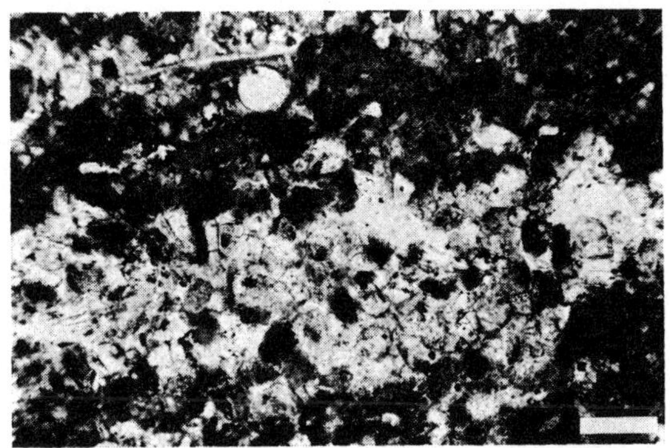

FIG. 7.—Thin section of an algal laminated tidal flat sediment, Cape Sable, Florida. Note the alternate thin, dark algal layers and the thick, light, coarse clastic layers. Bar = 0.6 mm.

of varying salinity and Mg/Ca ratio. Controls were run in which no sheath material was added. Both natural and artificial sea waters were filtered through a 0.45 μ filter and sterilized before the addition of the sheath material to minimize any possible effects of particulate organic matter and bacteria. Minute amounts of ammonium carbonate (0.004 or 0.006 M)

were then added to each flask in order to simulate breakdown of the organic material and to accelerate precipitation. The experimental variables and results are summarized in Table 2. Two hundred combinations of salinity, Mg/Ca ratio, ammonium carbonate and light were run. Light and dark experiments were compared to insure that no photosynthesis was occurring, i.e. that no intact algal cells remained. No difference was noted.

The formation of small (0.5–5.0 μ, average 2.0 μ) rhombic crystals occurred within four days in flasks to which sheath material plus ammonium carbonate had been added. The crystals grew firmly attached to the algal sheaths (Fig. 8). Further crystal growth occurred parallel to the filaments and in optical continuity with the original crystals. X-ray analyses showed that the crystals were all 17–20 mole percent Mg calcite.

Trace amounts of aragonite and 10 mole percent Mg calcite formed in the control vessels. The precipitation of 10 mole percent Mg calcite as an initial precipitate from sea water at 68°F is predicted by Winland (1969). The precipitation of 20 mole percent Mg calcite in contact with the sheath material demonstrates

TABLE 2.—*Variables and Results of Precipitation Experiments*

	0.0 (NH$_4$)$_2$CO$_3$		0.004 M, 0.006 M (NH$_4$)$_2$CO$_3$	
	Sheath material added	No sheath material	Sheath material added	No sheath material
S‰ 30, 70, 35, 80, 60, 90				
Mg/Ca in solution 2.5/1, 9.0/1 4.5/1, 12.0/1 6.0/1, 17.0/1	No precipitation		Abundance of crystals of 17–20 M% Mg calcite. Most precipitation at 35–70‰	Trace of crystals of aragonite and 10 M% Mg calcite
Illumination Continuous light, 12 hours light/ 12 hours dark, Continuous dark				

Constants:
      pH = 8.2
      temperature = 68°F
Algal Material:
      Schizothrix calcicola used for all experiments
      Cells grown in artificial sea water
      Salinity of growth medium = 35‰
      Mg/Ca in growth medium:
         4.5/1, 6.0/1, 8.0/1, 15.0/1

the ability of the organic matter (via complexed Mg) to influence the composition of the final precipitate.

DISCUSSION

Concentration of magnesium in algal sheath material provides a mechanism for producing alternating layers with high and low Mg/Ca ratios in modern tidal flat sediments. Although the precise mechanism for magnesium concentration is not known, the uptake of both magnesium and calcium ions onto organic substrates is well documented. Both Greenfield (1963) and Berner (1968) have shown that Ca and Mg ions are complexed and concentrated from sea water onto natural organic materials. The relative concentration of these two ions appears to vary with the type of organic matter employed in the study: Kitano and Kanomori (1966) produced high Mg calcites experimentally at low temperature and pressure in the presence of certain organic compounds capable of complexing Ca ion. Moreover, Mg ion is a necessary component of the photosynthetic pigment, chlorophyll, found in blue-green algae. High Mg concentration in the algal sheath may represent a reserve for the production of chlorophylls by the living algae. Some of the Mg found in the sheaths may have been derived from preexisting chlorophyll molecules. However, it is unlikely that chlorophyll itself was present at the time of analysis, as this pigment is rapidly photooxidized, losing its Mg, when cells die and lyse.

Initial precipitation of high Mg calcite in the sea water experiments (and in natural algal sediments; Gebelein, 1971) reflects the Mg/Ca ratio of the microenvironment of crystallization within the sheath, and not the Mg/Ca ratio in the surrounding waters. Thus, a large portion of the Mg in the sheath is readily available at or near surface sites of nucleation on the sheath. Under natural field conditions, mobilization of the remaining Mg complexed in the sheath material may be a long term process, dependent on the decomposition of the organic material. This material is very stable to decomposition and organically complexed Mg in the sediments may persist long after deposition and even cementation.

High Mg calcite is not, however, a necessary precursor to dolomite in this scheme. The experimental production of Mg calcite merely indicates: (1) that Mg complexed in the sheath *is* available for encorporation into a carbonate precipitate; and (2) that, under certain conditions, the organic matrix *can* serve as a suitable substrate for nucleation of Mg carbonates.

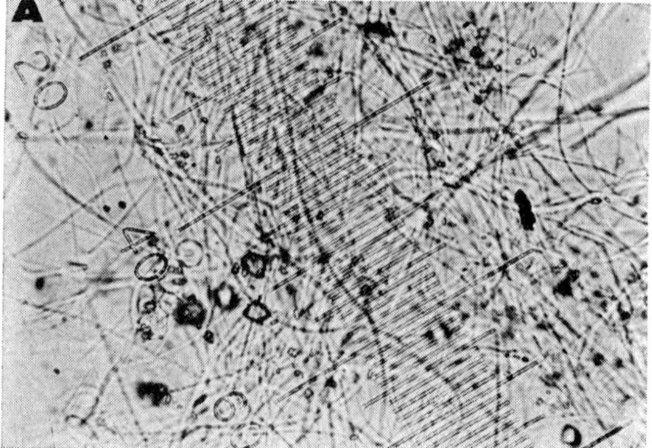

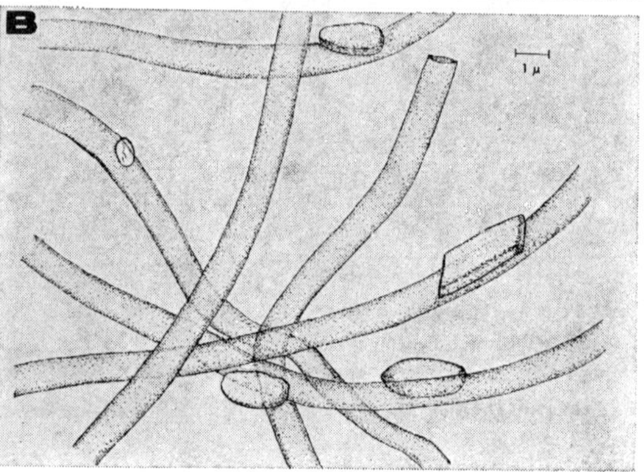

FIG. 8.—Experimental crystallization onto algal sheaths.

8A. Photomicrograph of sheathes of *S. calcicola* coated with rhombs of 17–20 mole % Mg calcite. Note attachment of crystals to sheaths only. Each line = 1 micron.

8B. Drawing illustrating the nature of attachment of crystals to algal sheaths. Note elongation of crystals along the sheaths.

When complete decomposition finally does occur, and the excess Mg liberated, dolomitization may take place. Dolomite may either crystallize directly within the organic layers, or, more probably, may partially replace the carbonate matrix in contact with the organic layers. High Mg calcite initially precipitated in the organic layers may be selectively dolomitized (as suggested in a different context by Schlanger, 1957).

It is important to note that this mechanism is not proposed to explain all types of bedded or massive dolomitic rocks. Careful petrographic examination of dolomitic laminae and interlaminated limestone-dolomite rocks must be made to determine the stromatolitic or cryptalgal nature of the rocks. Nor does this mechanism preclude the possibility that other chemical or sedimentological attributes of an algal-laminated deposit may enhance dolomitization. Selective dolomitization of micrite (for example, Illing *et al.*, 1965; Asquith, 1967) may occur in conjunction with algal-influenced dolomitization in laminae where fine sediment grains are preferentially trapped within the algal laminae (cf. Gebelein, 1969). Occurrence of clay minerals within predominately algal laminae also may augment dolomitization (for example, Kahle, 1965; Kepper, 1966).

The key point, however, is that the organic layers as described in this paper provide "micro-reservoirs" of excess magnesium in layers conformable with the primary structures of the sediment. These "reservoirs" may persist long after deposition and penecontemporaneous diagenesis. Thus the dolomite layers in the ancient rocks may be secondary in spite of their highly selective distribution. Both petrographic and experimental evidence indicate that this mechanism alone is capable of producing interlaminated limestone-dolomite fabrics. Further petrographic study is urgently needed to determine the extent of this dolomitization in rocks of all ages. Moreover, it may be profitable to consider this mechanism of dolomite formation when studying limestones with dolomite mottling, preferentially dolomitized burrow structures, and grains with dolomite coatings or envelopes, as well as the common interlayered dolomite-limestone rocks.

CONCLUSIONS

Many ancient shallow water carbonate sequences, particularly in the lower Paleozoic and Proterozoic, contain beds in which dolomite is intimately interlayered with limestone on a centimeter or millimeter scale. Most of these beds are stromatolitic (cryptalgal), having been composed originally of alternating layers of carbonate sediment and blue-green algal mats. Structurally, the dolomite layers correspond to the original algal mat layers, and the limestone layers to the sediment layers between the mats. Modern cryptalgal sediments of this type, however, do not contain dolomite layers. This fact makes it difficult to believe that the dolomite is primary or penecontemporaneous in origin. We suggest that the dolomite layers are secondary, and form by the following mechanism:

1) the original deposit consists of alternating algal-rich and sediment-rich layers;
2) magnesium ions are complexed organically and thus concentrated in the algal layers (at this stage the magnesium is not in the form of dolomite);
3) after deposition, and perhaps long after lithification, decomposition of the organic residue releases the excess magnesium in the micro-environment of the algal layer;
4) dolomite is formed, perhaps over a period of thousands or even millions of years, probably by replacement of the surrounding calcium carbonate matrix in contact with the algal layers (the extent of dolomitization, and hence the ultimate thickness of the dolomite layers, depends on the amount of magnesium originally complexed in the algal layers).

Therefore, the dolomitization process is secondary, explaining its absence from Recent stromatolitic sediments, but controlled by primary organic-rich layers, explaining its selective distribution.

ACKNOWLEDGMENTS

Laboratory studies were supported in part by API Research Project 97 to R. K. Matthews, Brown University. Atomic absorption analyses were performed by Oiva Joensuu, Institute of Marine Sciences, University of Miami, and Robert Fifer, Brown University.

REFERENCES

AITKEN, J. D., 1966, Middle Cambrian to Middle Ordovician cyclic sedimentation, southern Rocky Mountains of Alberta: Bull. Canadian Petroleum Geology, v. 14, p. 405–441.

———, 1967, Classification and environmental significance of cryptalgal limestones and dolomites, with illustrations from the Cambrian and Ordovician of southwestern Alberta: Jour. Sed. Petrology, v. 37, p. 1163–1178.

ALDERMAN, A. R., AND H. C. W. SKINNER, 1957, Dolomite sedimentation in the southeast of South Australia: Am. Jour. Sci., v. 255, p. 561–567.

ASQUITH, G. B., 1967, Marine dolomitization of the Mifflin member, Platteville Limestone in southwest Wisconsin: Jour. Sed. Petrology, v. 37, p. 311–326.

BERNER, R. A., 1968, Calcium carbonate concretions formed by the decomposition of organic matter: Science, v. 159, p. 195–197.

BLACK, MAURICE, 1933, The algal sediments of Andros Island, Bahamas: Royal Soc. London Philos. Trans., ser. B, v. 222, p. 165–192.

DALRYMPLE, D. W., 1964, Recent sedimentary facies of Baffin Bay, Texas, Unpubl. Ph.D. dissertation, Rice University, Houston, Texas, 192 p.

DEFFEYES, K. S., F. J. LUCIA, AND P. K. WEYL, 1965, Dolomitization of Recent and Plio-Pleistocene sediments by marine evaporite waters on Bonaire, Netherlands Antilles: Soc. Econ. Paleontologists and Mineralogists Spec. Pub. no. 13, p. 71–88.

DONALDSON, J. A., 1963, Stromatolites in the Denault Formation, Marion Lake, Coast of Labrador, Newfoundland: Canada Geol. Survey Bull. 102, p. 1–33.

FISCHER, A. G., 1964, The Lofer cyclothems of the Alpine Triassic: *in* Merriam, D. F. (*ed.*),

365

Symposium of Cyclic Sedimentation, State Kansas Geol. Survey Bull. 169, v. 1, p. 107–148.

FRIEDMAN, G. M., AND J. E. SANDERS, 1967, Origin and occurrence of dolostones: in Chillingar, G. V., Bissell, H. J., and Fairbridge, R. W. (*eds.*), Carbonate Rocks, Part A, Elsevier, p. 267–348.

GEBELEIN, C. D., 1969, Distribution, morphology, and accretion rate of Recent subtidal algal stromatolites, Bermuda: Jour. Sed. Petrology, v. 39, p. 49–69.

———, 1971, Sedimentology and ecology of a Recent carbonate facies mosaic, Cape Sable, Florida; Unpubl. Ph.D. dissertation, Brown University, Providence, Rhode Island, 225 p.

———, AND PAUL HOFFMAN, 1968, Intertidal stromatolites and associated facies from Cape Sable, Florida (abs.): Geol. Soc. America Spec. Paper 121, p. 109.

GINSBURG, R. N., L. B. ISHAM, S. J. BEIN, AND JOEL KUPERBERG, 1954, Laminated algal sediments of south Florida and their recognition in the fossil record: Unpubl. rept. no. 54–21, Marine Laboratory, Coral Gables, Florida, Univ. Miami, p. 1–33.

GREENFIELD, L. J., 1963, Metabolism and concentration of calcium and magnesium and precipitation of calcium carbonate by a marine bacterium: New York Acad. Sci. Annals, v. 109, p. 23–45.

HARDIE, L. A., 1969, Algal crusts from Bahamas (abs.): Am. Assoc. Petroleum Geologists Annual Meeting Program, Dallas, Texas, p. 69–70.

HOFFMAN, P. F., 1967, Algal stromatolites: use in stratigraphic correlation and paleocurrent determination: Science, v. 157, p. 1043–1045.

HOFFMAN, P. F., 1970, Precambrian shallow to deep water carbonate facies transition in the Pethei Group, Great Slave Lake, District of Mackenzie: Canada Geol. Survey Paper (in press).

———, B. W. LOGAN, AND C. D. GEBELEIN, 1969, Biological versus environmental factors governing the morphology and internal structure of Recent algal stromatolites in Hamelin Pool, Shark Bay, Western Australia (abs.): Geol. Soc. America Northeastern Sect., 4th Ann. Mtg., Pt. 1, p. 28–29.

HOWE, W. B., 1968, Planar stromatolite and burrowed carbonate mud facies in Cambrian strata of the St. Francois Mountain area: Missouri Div. Geol. Survey and Water Resources, Rept. Inv. no. 41, 113 p.

ILLING, L. V., A. J. WELLS, AND J. C. M. TAYLOR, 1965, Penecontemporaneous dolomite in the Persian Gulf: Soc. Econ. Paleontologists and Mineralogists Spec. Pub. no. 13, p. 89–111.

IRION, G., AND G. MÜLLER, 1968, Mineralogy, petrology and chemical composition of some calcareous tufa from the Schwäbische Alb, Germany: in Müller, G., and Friedman, G. M. (*eds.*), Recent Developments in Carbonate Sedimentology in Central Europe, Springer-Verlag, New York, p. 157–171.

KAHLE, C. F., 1965, Possible roles of clay minerals in the formation of dolomite: Jour. Sed. Petrology, v. 35, p. 448–453.

KENDALL, C. G. ST. C., AND SIR P. A.D'E SKIPWITH BT., 1968, Recent algal mats of a Persian Gulf lagoon: Jour. Sed. Petrology, v. 38, p. 1040–1058.

KEPPER, J. C., 1966, Primary dolostone patterns in the Utah-Nevada Middle Cambrian: Jour. Sed. Petrology, v. 36, p. 548–562.

KOMAR, V. A., 1966, Stromatolity verkhnedokembriiskikh otlozhenii severa sibirskoi platformy i ikh stratigraficheskoe znachenie (Upper Precambrian stromatolites in the northern part of the Siberian Platform and their stratigraphic significance): Akad. nauk SSR, Geol. Inst., Trans., v. 154, 122 p.

LAPORTE, L. F., 1964, Supratidal dolomitic horizons within the Manlius Formation (Devonian) of New York: in Purdy, E. G., and Imbrie, J. (*eds.*), Carbonate Sediments, Great Bahama Bank, Geol. Soc. America Guidebook for Field Trip no. 2, Miami, p. 59–66.

———, 1967, Carbonate deposition near mean sea level and resultant facies mosaic, Manlius Formation (Lower Devonian) of New York State: Am. Assoc. Petroleum Geologists Bull. v. 51, p. 73–101.

LOGAN, B. W., 1961, *Cryptozoon* and associated stromatolites from the Recent, Shark Bay, Western Australia: Jour. Geology, v. 69, p. 517–533.

MONTY, CLAUDE, 1965, Recent algal stromatolites in the Windward Lagoon, Andros Island, Bahamas: Soc. Géol. Belgique Annals, v. 88, p. 269–276.

———, 1967, Distribution and structure of Recent stromatolitic algal mats, eastern Andros Island, Bahamas: Soc. Géol. Belgique Annals, v. 90, p. 57–93.

NEUMANN, A. C., C. D. GEBELEIN, AND T. P. SCOFFIN, 1970, Composition, structure, and erodibility of subtidal mats, Abaco, Bahamas: Jour. Sed. Petrology, v. 40, p. 274–297.

SANDER, BRUNO, 1936, Beiträge Zur Kenntnis der Anlargerungsgefüge (Rhythmische Kalke und Dolomite aus Tirol): Tschermaks Mineralog. u. Petrogr. Mitt., v. 46, p. 27–209.

———, 1951, Contributions to the Study of Depositional Fabrics (Rhythmically deposited Triassic limestones and dolomites): Am. Assoc. Petroleum Geologists, Tulsa, Oklahoma, 207 p.

SANDO, W. J., 1957, Beekmantown Group (Lower Ordovician) of Maryland: Geol. Soc. America Mem. 68, 161 p.

SARIN, D. D., 1962, Cyclic sedimentation of primary dolomite and limestone: Jour. Sed. Petrology, v. 32, p. 451–471.

SCHENK, P. E., 1967, The Macumber Formation of the Maritime Provinces, Canada—a Mississippian analogue to Recent strand-line carbonates of the Persian Gulf: Jour. Sed. Petrology, v. 37, p. 365–376.

SCHLANGER, S. O., 1957, Dolomite growth in coralline algae: Jour. Sed. Petrology, v. 27, p. 181–186.

SHINN, E. A., 1968, Selective dolomitization of Recent sedimentary structures: Jour. Sed. Petrology, v. 38, p. 612–616.

SHINN, E. A., R. N. GINSBURG, AND R. M. LLOYD, 1965, Recent supratidal dolomite from Andros Island, Bahamas: Soc. Econ. Paleontologists and Mineralogists Spec. Pub. no. 13, p. 112–123.

SHINN, E. A., R. M. LLOYD, AND R. N. GINSBURG, 1969, Anatomy of a modern carbonate tidal flat, Andros Island, Bahamas: Jour. Sed. Petrology, v. 39, p. 1202–1208.

VON DER BORCH, C., 1965, The distribution and preliminary geochemistry of modern carbonate sediments of the Coorong area, South Australia: Geochim. et Cosmochim. Acta, v. 29, p. 781–799.

WINLAND, H. D., 1969, Stability of calcium carbonate polymorphs in warm, shallow sea water: Jour. Sed. Petrology, v. 39, p. 1579–1587.

29

Detrital Dolomite in Onondaga Limestone (Middle Devonian) of New York: Its Implications to the "Dolomite Question"[1]

R. C. LINDHOLM[2]

Washington, D. C. 20006

Abstract Dolomite occurs in the matrix of the Onondaga Limestone (Middle Devonian) in New York as scattered grains ranging in size from 4 to 150 μ. Detrital quartz is associated with the dolomite. Study of etched and stained thin sections shows a correlation in grain size between the dolomite and quartz. Limited data show a correlation in grain size among dolomite, quartz, and detrital calcite (silt to fine sand) matrix. In addition, there is a correlation in abundance between dolomite and quartz, where high dolomite values are present with high quartz values.

These data suggest that dolomite in the Onondaga is detrital. Source of the dolomite is uncertain, but reworked penecontemporaneous supratidal sediments and older (e.g., Silurian) dolomite are suggested possibilities. Wind is a likely mechanism for transport of the detritus.

Deposition of detrital dolomite followed by later diagenetic overgrowth on the detrital nuclei is suggested as a mechanism for "dolomitization." This process is compatible with three phenomena observed in dolomitic rocks: (1) association of insoluble detritus with dolomite, (2) presence of dolomite in fine-grained limestone, (3) fine-grained texture of dolomite interpreted as "primary" and coarse-grained texture of dolomite interpreted as "replacement." Two models for the origin of dolomitic rocks are proposed.

Introduction

This paper is part of a general petrologic study of the Onondaga Limestone (Lindholm, 1967). The Onondaga included in the study crops out for 270 mi between Buffalo and Albany, New York, with the best exposures in quarries. During the summers of 1964 and 1965, 29 outcrops were studied.

Stratigraphy of Onondaga Limestone

Before 1964, the most comprehensive study of Onondaga stratigraphy was that done by

[1] Manuscript received, April 3, 1968; accepted, June 12, 1968.

Read at the SEPM Ancient Carbonates Session, AAPG-SEPM annual meeting, Dallas, Texas, April 14–16, 1969.

[2] Assistant professor of geology, The George Washington University. Field work for this study was made possible by a Sigma Xi Grant-in-Aid of Research and by funds supplied by the New York State Museum and Science Service. The writer is indebted to many persons for suggestions and helpful criticism during the course of work and preparation of the manuscript, especially W. A. Oliver, who suggested the problem, F. J. Pettijohn (supervisor), R. N. Ginsburg, and J. W. Pierce.

Oliver (1954, 1956a,b, 1960). He divided the formation into four members, from oldest to youngest, the Edgecliff, Nedrow, Moorehouse, and Seneca. Members are subdivided into 12 zones (A-L, from oldest to youngest). The subdivisions are based mainly on paleontology, although bedding and gross lithology also are considered.

Lithofacies are distinguished by the proportion of allochems (almost entirely fossil debris) and fine-grained carbonate matrix (plus sparry calcite cement; Lindholm, 1967). Carbonate rock classification follows the scheme proposed by Folk (1959, 1962). Rocks with the highest allochem content are most abundant in eastern (Albany area) and western (Buffalo area) New York, as well as in the lowermost beds of the Onondaga throughout the area (Fig. 1). The Onondaga in central New York (Syracuse) is characterized by less abundant fossil debris.

Dolomite

Techniques.—One quarter of each thin section was stained and one quarter was etched to facilitate study of noncalcareous materials. Dolomite and quartz contents were determined by point counting on the etched part of each slide. To check the results, x-ray analyses were made on nine samples. Dolomite percentages were obtained by comparison of x-ray intensities for the 3.30 Å calcite peak and 2.88 Å dolomite peak (Tennant and Berger, 1957). These data agreed with dolomite content determined by point counting.

Grain-size distribution of fine-grained calcite, as well as dolomite and quartz, was determined by point counting stained and etched thin sections under the highest power objective available. The longest dimension of each grain was used in measuring grain size.

Morphology.—Dolomite grains are generally subhedral to euhedral rhombohedrons; anhedral grains constitute less than 20 percent of the total dolomite. A few grains are polycrystalline. Many rhombs contain dark, cloudy interiors; less commonly, the interior contains many opaque inclusions (pyrite?), which range

367

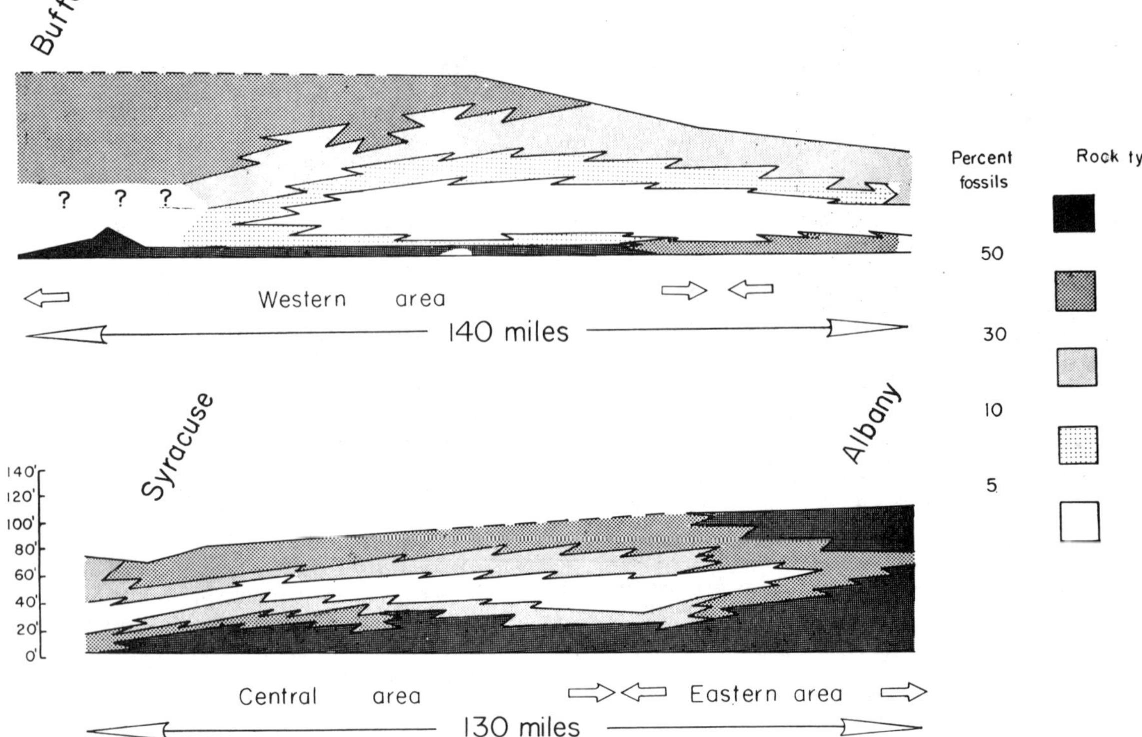

Fig. 1.—West-east cross section showing distribution of carbonate rock type in Onondaga Limestone. Fossil content is indicated as less than 5 percent, 5–10 percent, 10–30 percent, 30–50 percent, and greater than 50 percent.

in size from 1 to 5μ. In most grains the shape of the interior shows no crystal form, suggesting overgrowth on a round to subround core.

Composition.—X-ray analyses of the dolomite from the Onondaga show the principal peak at 2.90 Å. This value indicates that the dolomite has a "non-ideal" crystal structure, and contains between 3 and 6 percent excess $CaCO_3$ (Goldsmith and Graf, 1958).

Distribution.—Dolomite is present in most samples as small grains scattered throughout the limestone matrix. Dolomite replacement of fossils was not observed, although dolomite grains were found in spar-filled cavities of skeletal debris. The cavities are commonly primary features of the fragment or shell, *e.g.*, the cavity between brachiopod valves, or the axial canal in crinoid columnals. Less commonly, the cavities are the result of postmortem boring.

Dolomite is most abundant (5–20 percent of total rock) in limestone where allochems are subordinate to fine-grained calcite matrix. Such rocks generally are present in central New York (Fig. 2). Dolomite and detrital quartz are absent in biosparites.

Grain size and abundance.—There are significant relations between size and abundance of dolomite grains and size and abundance of detrital quartz grains. Maximum size of both dolomite and quartz grains was determined in 70 thin sections (Fig. 3). The data suggest that dolomite and quartz grains present together are approximately the same size (correlation coefficient = +0.76).

Comparable data were obtained from grain-size distribution in two calcisiltites (fine-grained carbonate, 4–62 μ), one coarse grained and the other very fine grained (Fig. 4). All three constituents (calcite, dolomite, and quartz) are much finer in the very fine-grained calcisiltite (Fig. 4B) than in the coarse-grained calcisiltite (Fig. 4A). Correspondence between calcite grain size and grain size of the other two components is significant because the calcite matrix in the Onondaga is considered to be calcareous silt, formed by comminution of skeletal material (Lindholm, 1967, p. 54–59). The samples described in Figure 4 were chosen to show the maximum contrast present in the Onondaga.

In addition to the grain-size relation, there is a correlation in relative abundance between dolomite and detrital quartz (correlation coefficient = +0.66). High dolomite values are present with high quartz values, as shown in Figure 5.

ORIGIN OF DOLOMITE

Scattered dolomite rhombs in other formations are attributed to preferential growth of dolomite in $CaCO_3$ mud (Bergenback and Terriere, 1953; Bluck, 1965; Lucia, 1962; Murray, 1960; Murray and Lucia, 1967). Correlation in abundance between quartz silt and dolomite (Fig. 5), and the association of dolomite, quartz silt, and calcareous silt of approximately the same size (Figs. 3, 4) suggest that dolomite rhombs in the Onondaga are detrital. Detrital dolomite also is known from other ancient (Amsbury, 1962; Bluck, 1965; Sabins, 1962) and recent (Illing et al., 1965; Sugden, 1963) carbonates.

Source of the dolomite in the Onondaga is unknown. It might have been penecontemporaneous supratidal sediments which subsequently were eroded. The supratidal area is the only normal marine environment in which Holocene dolomite is known to form.

The source also might have been older dolomite exposed to erosion during Onondaga deposition. Silurian dolomite underlies the Onondaga in western New York and may have been exposed in land areas adjacent to the Onondaga sea.

The mechanism of transport of detrital dolomite found in the Onondaga is purely speculative. Transport by water currents is one obvious possibility and cannot be ruled out.

Windborne dust is an alternative, both for the dolomite and the quartz. Studies of continental deposits (Arrhenius, 1963; Scheidig, 1934) known to be of eolian origin show that the material is composed largely of grains in the 20–40 μ range. Sediments of that size are blown at least 50 mi into the Pacific Ocean from the deserts of northern Mexico (Bonatti and Arrhenius, 1965). Eolian sediment from the Sahara is present 1,000 mi off the west coast of Africa; at approximately 200 mi, nearly 60 percent of the quartz in some deep-sea deposits has been identified as of desert origin (Barth et al., 1939). Samples collected by ships in the same area demonstrate eolian transport in ex-

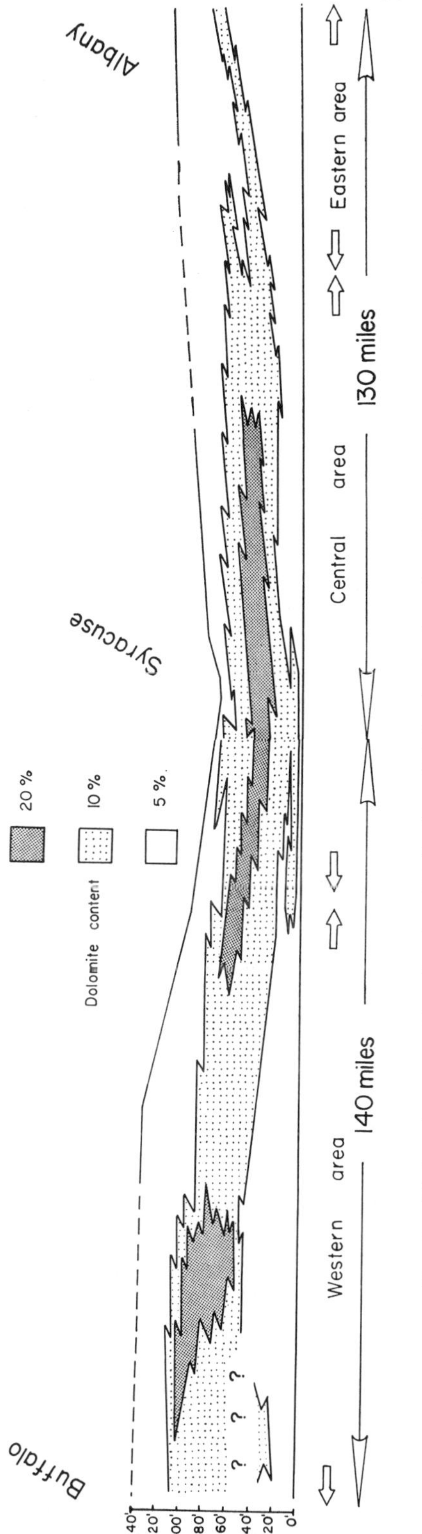

Fig. 2.—West-east cross section showing distribution of dolomite in Onondaga Limestone. Content indicated as less than 5 percent, 5–10 percent, 10–20 percent.

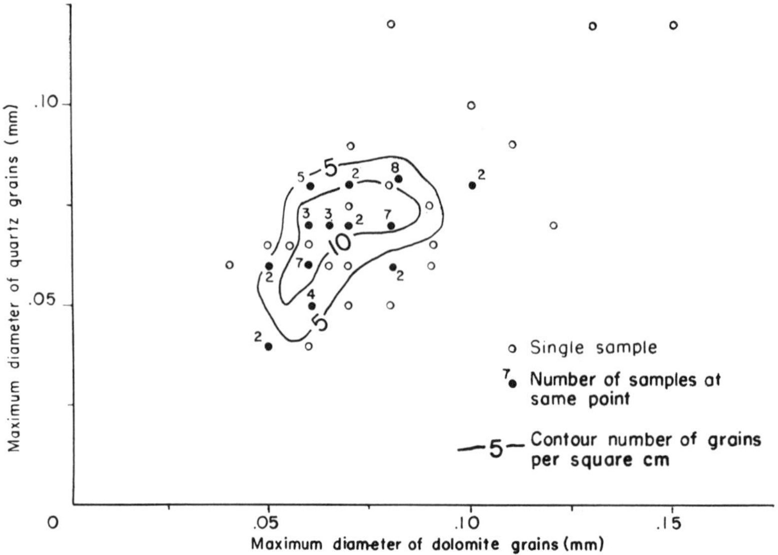

FIG. 3.—Maximum size of quartz grains plotted against maximum size of dolomite grains. Based on 70 thin sections of Onondaga Limestone. Correlation coefficient = +0.76; for N = 70, correlation is significant at the 1-percent level (Dixon and Massey, 1957, p. 468, Table A-30a).

cess of 1,000 mi (Kuenen, 1950, p. 214; Schott, 1942). Airborne sediment derived from the Sahara and containing dolomite silt has been collected at Barbados (Delany *et al.*, 1967). Eolian dust collected on ships in the Red Sea and the Persian Gulf has a grain-size range from 5 to 100 μ, with an average size of 24 μ (Emery, 1956, p. 2367).

Thus, sediment of approximately the same size as the dolomite and quartz in the Onondaga can be transported great distances as wind-blown dust. The argument for this means of transport is strengthened by the association with the Tioga Bentonite, a very widespread unit in the Onondaga which is obviously a wind-transported deposit. In addition, volcanic biotite is present in numerous samples of the fine-grained part of the Onondaga.

IMPLICATIONS TO DOLOMITIZATION PROBLEM

Deposition of detrital[3] dolomite, followed by diagenetic overgrowth on the detrital nuclei, is proposed as a possible mechanism of "dolomiti-

zation." The process is compatible with several geologic phenomena.

First, large amounts of insoluble residues of detrital origin are associated with abundant dolomite in limestone (Lesley, 1879; Fairbridge, 1957, p. 154–156) and recent carbonate sediments (Taft and Harbaugh, 1964, p. 126–127). This relation would be expected if detrital dolomite was introduced concurrently with detrital quartz into the calcareous sediment.

Second is the association of dolomite with fine-grained limestone and its scarcity in current-deposited carbonate sand (Murray and Lucia, 1967). Detrital dolomite silt would be washed out of current-deposited sand, and concentrated in the $CaCO_3$ mud.

Third, the kinetics needed to initiate crystallization of dolomite are more rigorous than those required to continue growth on previously formed nuclei. Supratidal flats (Deffeyes *et al.*, 1965; Illing *et al.*, 1965; Shinn, 1967; Shinn *et al.*, 1965) and saline lakes (Alderman, 1959; Alderman and Skinner, 1957; Alderman and Von Der Borch, 1960, 1961; Skinner, 1963) are the only modern environments where conditions favor initiation of dolomite crystallization. If supratidal dolomites were reworked into intertidal and subtidal carbonate sediments, the stage would be set for later dolomitization of the entire rock. Later dolomitiza-

---

[3] Detrital, in this case, means that the dolomite was eroded and transported. Erosion from older dolomitic rocks is not implied, although it is possible. Penecontemporaneous diagenetic dolomite, subjected to erosion and transportation, is also considered to be detrital.

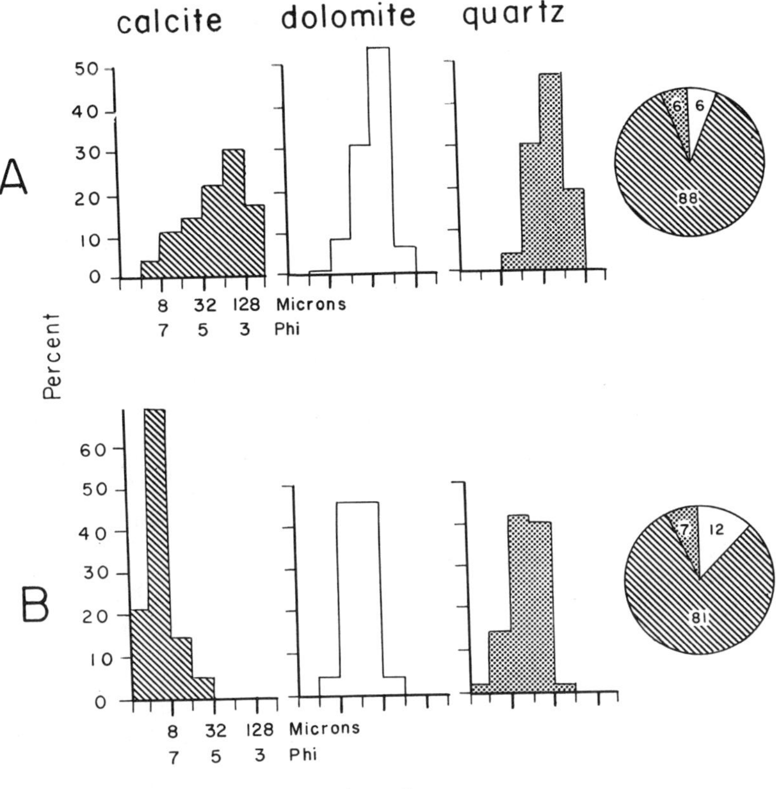

Fig. 4.—Grain-size distribution of calcite, dolomite, and quartz in coarse-grained calcisiltite (A) and very fine-grained dolomite and quartz are present in fine-grained calcisiltite, and coarse-grained dolomite and quartz are present in coarse-grained calcisiltite. Percentage of calcite, dolomite, and quartz is shown by pie diagram.

A. Coarse-grained calcisiltite, Moorehouse Member, coral facies (western area).

	M_z	S_I	S_G
calcite	4.4 ϕ (0.048 mm)	1.44 ϕ	1.22 ϕ
dolomite	4.9 ϕ (0.032 mm)	0.73 ϕ	0.75 ϕ
quartz	4.7 ϕ (0.038 mm)	0.70 ϕ	0.75 ϕ

B. Very fine-grained calcisiltite, Moorehouse Member (central area).

	M_z	S_I	S_G
calcite	7.6 ϕ (0.005 mm)	0.75 ϕ	0.76 ϕ
dolomite	6.0 ϕ (0.016 mm)	0.58 ϕ	0.58 ϕ
quartz	6.2 ϕ (0.014 mm)	0.70 ϕ	0.64 ϕ

tion (at any time) could proceed under conditions much less restrictive than those needed to nucleate dolomite. If, in a limestone containing 4 percent detrital dolomite, the grain size of the dolomite were increased by a factor of 3 (*e.g.*, from 20 to 60μ) by diagenetic growth, the final rock would be 100 percent dolomite.

Ancient dolomite interpreted as being "primary" or detrital is fine grained, as is recent dolomite; ancient dolomite showing clear indications of replacement is coarse grained (Krynine, 1957). The proposed mechanism of dolomitization by initial deposition of detrital dolomite followed by later diagenetic growth, fits these observations. If detrital dolomite (silt size) were not modified by later growth, the evidence of detrital origin would be easily recognized. If, given a primary dolomite content of

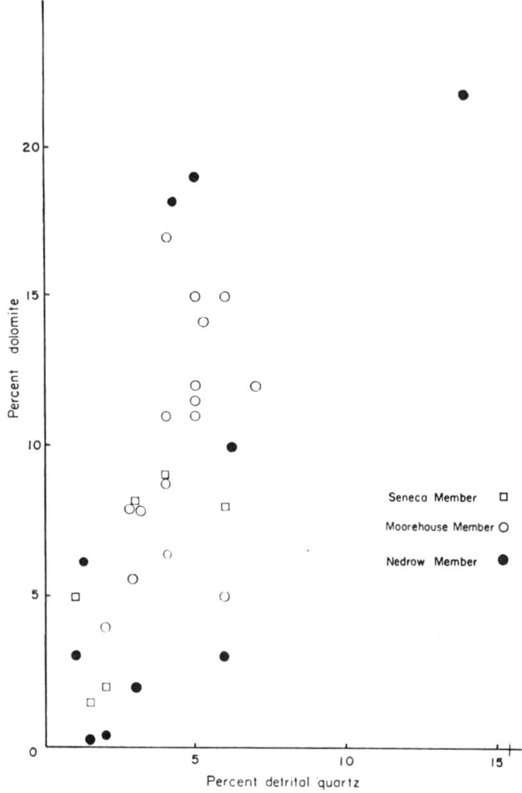

FIG. 5.—Percentage of dolomite plotted against percentage of detrital quartz. Based on point counts of 32 thin sections from Onondaga Limestone, central area. Correlation coefficient = +0.66; for N = 32, correlation is significant at 1 percent level (Dixon and Massey, 1957, p. 468, Table A-30a).

4 percent, dolomite overgrowths enlarged the original grain dimensions by a factor of 3 (volume increased by a factor of 27) and formed a rock composed entirely of dolomite, evidence of replacement would mask any evidence of a "primary" or detrital origin. In the latter example a combination of mechanisms, commonly considered to be mutually exclusive, is operative.

This mechanism of "dolomitization" is presented as one hypothesis that should be considered in determining the origin of a particular dolomite and is not intended to explain all dolomites. Certainly, some are formed by reflux of hypersaline brines through carbonate sediments that probably contained no detrital dolomite (Adams and Rhodes, 1960; Deffeyes *et al.*, 1965). Others may be produced by late diagenetic replacement of dolomite-free (detri-

tal) limestone. Before a detrital origin is proposed for any dolomite, the evidence should be well established. Evidence observed in the Onondaga includes correlation in abundance between dolomite and insoluble detritus and correlation in grain size among dolomite, detrital calcite, and insoluble detritus.

The role of detrital dolomite in carbonate sedimentation and diagenesis warrants attention in future work. Two models might be used in analyzing dolomitic rocks.

The first model involves introduction of detrital dolomite and quartz into calcareous sediments, without later diagenetic increase in dolomite grain size (Fig. 6A). The grain size of the detritus increases toward the source (to the right). If the source of detrital dolomite were penecontemporaneous supratidal flats, the size range would be small, from 1 to 5 μ (based on studies of Holocene dolomite; *e.g.*, Deffeyes *et al.*, 1965; Illing *et al.*, 1965). A much broader range of grain sizes would be available if the source were older dolomite exposed on the land. In addition, the percentage of detritus (dolomite and quartz) in the sediment increases toward the source.[4] Values for size and content of detritus are only estimates, given to suggest order of magnitude. The abscissa (percentage values) indicates distance from the source area and may be in tens or hundreds of miles. Trends, not absolute values, are intended in these models.

The second model (Fig. 6B) involves introduction of detrital dolomite and quartz into calcareous sediment, followed by diagenetic growth of the dolomite, so that the calcareous sediment is totally "replaced" by dolomite. The grain size of the detrital dolomite is kept constant (10μ) to simplify the illustration (in fact it probably would be varied as in the first model). The grain size and amount of detrital quartz increase toward the source (to the right), but the grain size of the dolomite (diagenetic) increases away from the source. It is assumed that diagenetic dolomite crystallizes only on the detrital nuclei, and that no later diagenetic modifications occur. Grain size therefore is controlled by mutual interference of dolomite crystals as the rock nears total replacement, and hence is a function of the spacing of the detrital nuclei (*i.e.*, percentage of nuclei in rock).

[4] Deep-sea samples from the Atlantic show an increase in dolomite toward the North African coast. This dolomite is derived from the desert regions of North Africa and transported westward by the trade winds (Delany *et al.*, 1967).

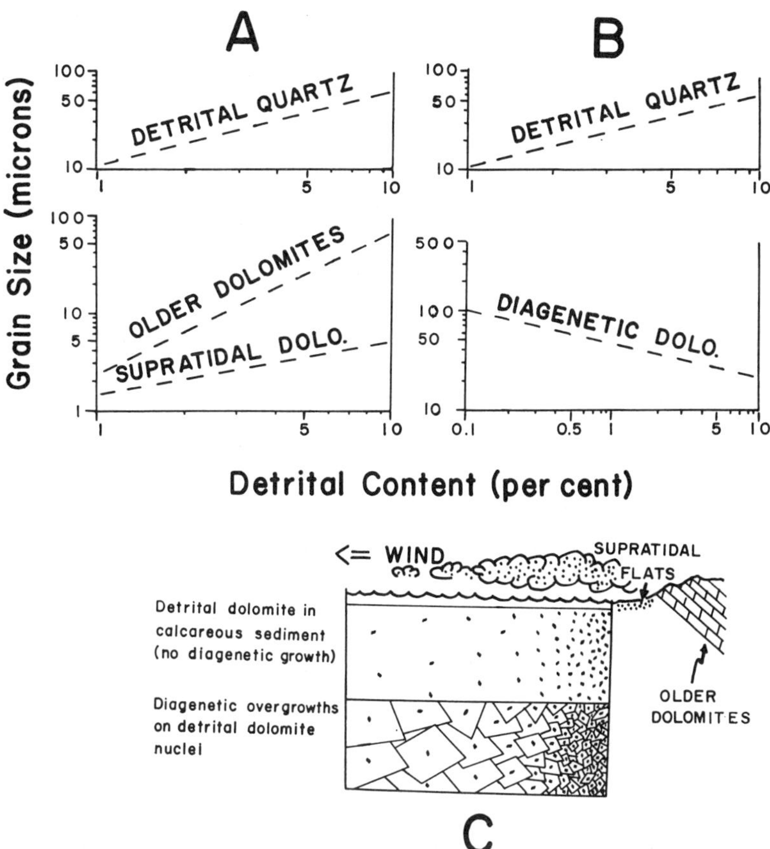

FIG. 6.—Models for dolomitic rocks.

A. Grain size and abundance of detrital dolomite and quartz in calcareous sediment. Grain size of dolomite would depend on nature of source. If source were supratidal, variation in grain size would be small $(1-5 \mu)$; if source were older dolomite, exposed on land, expected variation would be much wider. Assumed source to right.

B. Grain size and abundance of detrital quartz and dolomite after diagenetic crystallization on detrital nuclei has completely "dolomitized" calcareous sediment (or rock). Grain size of diagenetic dolomite plotted against percentage of detrital dolomite (nuclei for diagenetic dolomite) in sediment. Grain size of detrital dolomite kept constant at 10μ. Assumed source to right.

C. Schematic summary of *A* and *B*, showing abundance of detrital dolomite and grain size of "diagenetic dolomite" resulting from crystallization on detrital nuclei.

The models are presented as an aid to stratigraphic interpretation of dolomite and dolomitic limestone. Many variables would affect the models; for example, a source area with no silt-size quartz. Further, the models are entirely conceptual, based on observations of detrital dolomite in the Onondaga Limestone, as well as general characteristics of dolomitic rocks. The ideas warrant attention in future work on dolomites.

REFERENCES CITED

Adams, J. E., and M. L. Rhodes, 1960, Dolomitization by seepage refluxion: Am. Assoc. Petroleum Geologists Bull., v. 44, p. 1912–1920.

Alderman, A. R., 1959, Aspects of carbonate sedimentation: Geol. Soc. Australia Jour., v. 6, p. 1–10.

——— and H. C. W. Skinner, 1957, Dolomite sedimentation in the southeast of South Australia: Am. Jour. Sci., v. 255, p. 561–567.

——— and C. C. Von Der Borch, 1960, Occurrence of hydromagnesite in sediments in South Australia: Nature, v. 188, p. 931.

——— and ——— 1961, Occurrence of magnesite-dolomite sediments in South Australia: Nature, v. 192, p. 861.

Amsbury, D. L., 1962, Detrital dolomite in central Texas: Jour. Sed. Petrology, v. 32, p. 5–14.

Arrhenius, G., 1963, Pelagic sediments, *in* The sea, ideas and observations on progress in the study of the seas—V. 3, The earth beneath the sea: New York, Interscience, p. 655–727.

Barth, T. F. W., C. W. Correns, and P. Eskola, 1939,

Die Entstehung der Gestein: Berlin, Springer, 422 p.

Bergenback, R. E., and R. T. Terriere, 1953, Petrography and petrology of Scurry reef, Scurry County, Texas: Am. Assoc. Petroleum Geologists Bull., v. 37, p. 1014–1029.

Bluck, B. J., 1965, Sedimentation of Middle Devonian carbonates, southeastern Indiana: Jour. Sed. Petrology, v. 35, p. 656–682.

Bonatti, E., and G. Arrhenius, 1965, Eolian sedimentation in the Pacific off northern Mexico: Marine Geology, v. 3, p. 337–348.

Deffeyes, K. S., F. J. Lucia, and P. K. Weyl, 1965, Dolomitization of recent and Plio-Pleistocene sediments by marine evaporite waters on Bonaire, Netherlands Antilles, *in* Dolomitization and limestone diagenesis—a symposium: Soc. Econ. Paleontologists and Mineralogists Spec. Pub. No. 13, p. 71–88.

Delaney, A. C., *et al.*, 1967, Airborne dust collected at Barbados: Geochim. et Cosmochim. Acta, v. 31, p. 885–909.

Dixon, W. J., and F. J. Massey, 1957, Introduction to statistical analysis: New York, McGraw-Hill, 488 p.

Emery, K. O., 1956, Sediments and water of Persian Gulf: Am. Assoc. Petroleum Geologists Bull., v. 40, p. 2354–2383.

Fairbridge, R. W., 1957, The dolomite question, *in* R. J. LeBlanc and J. G. Breeding, eds., Regional aspects of carbonate deposition—a symposium: Soc. Econ. Paleontologists and Mineralogists Spec. Pub. No. 5, p. 125–178.

Folk, R. L., 1959, Practical petrographic classification of limestones: Am. Assoc. Petroleum Geologists Bull., v. 43, p. 1–38.

———— 1962, Spectral subdivision of limestone types, *in* Classification of carbonate rocks: Am. Assoc. Petroleum Geologists Mem. 1, p. 62–84.

Goldsmith, J. R., and D. L. Graf, 1958, Structural and compositional variations in some natural dolomites: Jour. Geology, v. 66, p. 678–693.

Illing, L. V., A. J. Wells, and J. C. M. Taylor, 1965, Penecontemporary dolomite, Persian Gulf, *in* Dolomitization and limestone diagenesis: Soc. Econ. Paleontologists and Mineralogists Spec. Pub. No. 13, p. 89–111.

Krynine, P. D., 1957, Dolomites (abs.): Geol. Soc. America Bull., v. 68, p. 1757.

Kuenen, P. H., 1950, Marine geology: New York, Wiley and Sons, 568 p.

Lesley, J. P., 1879, Notes on a series of analyses of dolomitic limestone rocks of Cumberland County, Pa., . . .: 2d Geol. Survey Pennsylvania (1876–78), MM, p. 311–362.

Lindholm, R. C., 1967, Petrology of the Onondaga Limestone (Middle Devonian), New York: Unpub. Ph.D. thesis, Johns Hopkins Univ., 188 p.

Lucia, F. J., 1962, Diagenesis of crinoidal sediment: Jour. Sed. Petrology, v. 32, p. 848–866.

Murray, R. C., 1960, Origin of porosity in carbonate rocks: Jour. Sed. Petrology, v. 30, p. 59–84.

———— and F. J. Lucia, 1967, Cause and control of dolomite distribution by rock selectivity: Geol. Soc. America Bull., v. 78, p. 21–36.

Oliver, W. A., 1954, Stratigraphy of the Onondaga Limestone (Devonian) in central New York: Geol. Soc. America Bull., v. 65, p. 621–652.

———— 1956a, Stratigraphy of the Onondaga Limestone in eastern New York: Geol. Soc. America Bull., v. 67, p. 1441–1474.

———— 1956b, Biostromes and bioherms of the Onondaga Limestone in eastern New York: New York State Mus. and Sci. Service Circ. 45, 23 p.

———— 1960, Coral faunas in the Onondaga Limestone of New York: U.S. Geol. Survey Prof. Paper 400-B, p. 172–174.

Sabins, F. F., 1962, Grains of detrital, secondary and primary dolomite from Cretaceous strata of the western interior: Geol. Soc. America Bull., v. 73, p. 1183–1196.

Scheidig, A., 1934, Der Löss und seine gastechnischen Eigenschaften: Dresden-Leipzig, Steinkoph, 223 p.

Schott, W., 1942, Geographie des Atlantischen Ozeans: Hamburg, C. Boysen, 438 p.

Shinn, E. A., 1964, Recent dolomite, Sugarloaf Key, *in* R. N. Ginsburg, comp., South Florida carbonate sediments: Geol. Soc. America Fld. Trip 1 Guidebook, p. 62–67.

———— R. N. Ginsburg, and R. M. Lloyd, 1965, Recent supratidal dolomite from Andros Island, Bahamas, *in* Dolomitization and limestone diagenesis—a symposium: Soc. Econ. Paleontologists and Mineralogists Spec. Pub. No. 13, p. 71–88.

Skinner, H. C. W., 1963, Precipitation of calcian dolomites and magnesian calcites in the southeast of South Australia: Am. Jour. Sci., v. 261, p. 449–472.

Sugden, W., 1963, Some aspects of sedimentation in the Persian Gulf: Jour. Sed. Petrology, v. 33, p. 355–364.

Taft, W. H., and J. W. Harbaugh, 1964, Modern carbonate sediments of southern Florida, Bahamas, and Espiritu Santo Island, Baja California: a comparison of their mineralogy and chemistry: Stanford Univ. Pubs. Geol. Sci., v. 8, no. 2, 133 p.

Tennant, C. B., and R. W. Berger, 1957, X-ray determination of dolomite-calcite ratio of a carbonate rock: Am. Mineralogist, v. 42, nos. 1–2, p. 23–29.

30

SELECTIVE DOLOMITIZATION OF RECENT SEDIMENTARY STRUCTURES[1]

EUGENE A. SHINN

Shell Development Co., Coral Gables, Florida

ABSTRACT

Relatively non-dolomitic limestone nodules which float in lithified dolomitic sediment are forming on a supratidal mud flat in the lower Florida Keys. Storm tides periodically deposit layers of lime mud above normal high tide level which dry and crack to form typical mud-crack polygons. These polygons erode into flattened nodules that subsequently become buried in relatively more porous and permeable sediment. Magnesium-enriched brines concentrated through evaporation are more readily transmitted through the more permeable sediment than are the less permeable nodules. Selective dolomitization of the more permeable sediment produces features bearing a strong similarity to sedimentary structures commonly found in ancient dolomitic rocks.

INTRODUCTION

Relatively non-dolomitic, horizontally bedded limestone nodules interlaminated with dolomite or dolomitic limestone are common features in ancient, especially Palaeozoic, carbonates (Van Tuyl, 1914; Matter, 1967). Such features generally occur within 1 to 3 foot thick beds often grading into pure dolomite beds. Often, similar structures have been completely dolomitized (Matter, 1967).

Several discoveries of Recent dolomite have shown that supratidal sedimentation and penecontemporaneous dolomitization can form beds and layers (Shinn, Ginsburg, and Lloyd, 1965; Deffeyes, Lucia, and Weyl, 1965; Illing, Wells, and Taylor, 1965). The purpose of this note is to demonstrate that Recent dolomitization, in addition to forming beds and layers, can be selective and produce relatively nondolomitic planar nodules similar to those observed in ancient rocks.

DISTRIBUTION

Figure 1 is an aerial photograph of the muddy southwestern shore at Sugar Loaf key in the lower Florida Keys. Figure 2, a schematic cross section of the sedimentary zones, shows that there is less than 2 feet of Recent carbonate sediment overlying Pleistocene oolitic limestone. This pelleted muddy sediment is similar in the marine, intertidal and supratidal zone, but in the supratidal zone much of the surficial sediment is cemented to form a 1 to 4 inch thick crust. (The dolomitic crust is not to be confused with a laminated soilstone crust which coats the underlying Pleistocene rock at Sugarloaf Key and other Florida Keys. The non-dolomitic laminated crust, which is up to 3 inches thick, has recently been described and attributed to soil

processes by Multer and Hoffmeister, 1967.) The crust, similar to those in the Bahamas (Shinn, Ginsburg, and Lloyd 1965), is dolomitic, contains mud cracks, algal mats, scattered marine fossils, and birdseye structures (Shinn, 1968). The dolomite, which does not exceed 25 percent of the crust, occurs as calcium rich (60 mole percent Ca, 40 mole percent Mg), poorly ordered 1 to 3 micron crystals, the distribution of which, in thin section, is difficult to determine because of their small size. Distribution of dolomite was determined by X-ray diffraction analysis of minute samples taken from selected points on the face of vertically sawn crusts (fig. 3).

DISTRIBUTION OF DOLOMITE AND CEMENT IN CRUSTS

Dolomite is not evenly distributed and is restricted, mainly, to the more tightly cemented light tan part of crusts. Floating in the light tan cemented dolomitic matrix are white to grey compacted, but not cemented, lime mud nodules (fig. 3). The nodules are plate-like, form discontinuous layers, and attain a maximum thickness of 1 inch. The maximum dolomite content in the nodules is 10 percent (see fig. 3), but the intervening cemented layers contain up to 25 percent. Sediment infilling burrows which penetrate the lime mud nodules also contain up to 25 percent dolomite (fig. 3).

CARBON 14 AGE DETERMINATIONS

Carbon 14 age determinations of dolomite separated from crusts by acid leaching were perfomed by E. Martin of Shell Development Co., Houston, Texas. A sample from the uppermost quarter inch of a crust gave a C^{14} age of 300 years, whereas one from 4 inches below the surface gave an age of 900 years. The dolomite is

[1] Manuscript received September 20, 1967.

375

FIG. 1.—Oblique aerial photograph of sedimentary zones near south western shore of Sugar Loaf Key. "X" near Overseas highway indicates location of sample shown in figure 3.

clearly of Recent origin, and sedimentary relationships indicate that the sediment which became only partially dolomitized was also deposited in Recent times, at approximately present sea level.

SEDIMENTATION AND CONDITIONS FOR DOLOMITIZATION

Numerous observations over a period of several years showed that the area is flooded only by high spring and storm tides. During spring tide flooding, the saline water is relatively clear and apparently little, if any, sediment is deposited. During storms, however, marine sediment in Florida Bay is stirred into suspension and deposited in the intertidal and supratidal areas, as observed after the passage of Hurricane Donna (Ball, Shinn, and Stockman, 1967). During hurricane Donna as much as 4 inches of lime mud was deposited on several islands in Florida Bay. After a few months this dried and compacted to form a stiff mud cracked layer ranging from 1/16 of an inch to several

inches in thickness. During the years that followed, the characteristic "V" shaped mudcracks were modified by weathering and became rounded to such an extent that original cracks became almost unrecognizable. Many of the mud cracks weathered and increased in width until they were as wide as adjacent poly-

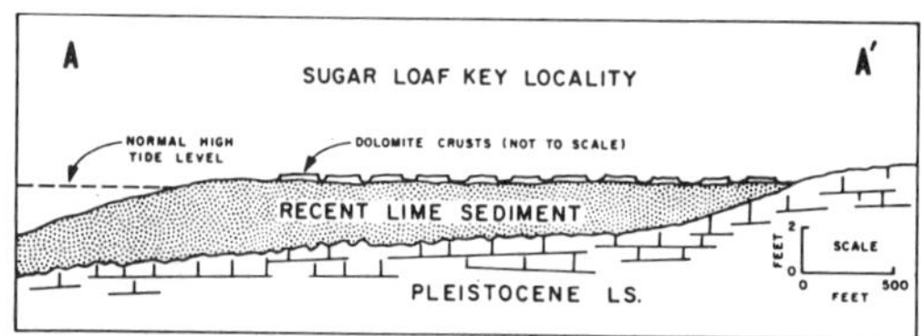

FIG. 2.—Diagrammatic section through sedimentary zones shown in figure 1. Cemented crust containing up to 25 percent dolomite overlies unconsolidated sediment. Numerous pinacles of underlying weathered limestone protrude through supratidal sediment (not shown in diagram). The underlying limestone is coated with a non-dolomitic laminated dark brown "caliche-like" layer not related to the present environment.

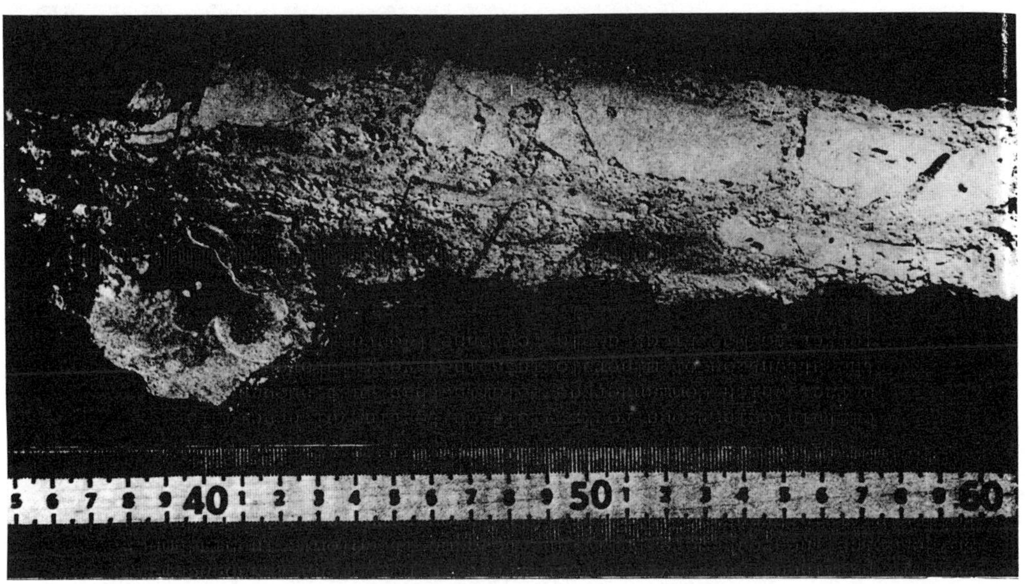

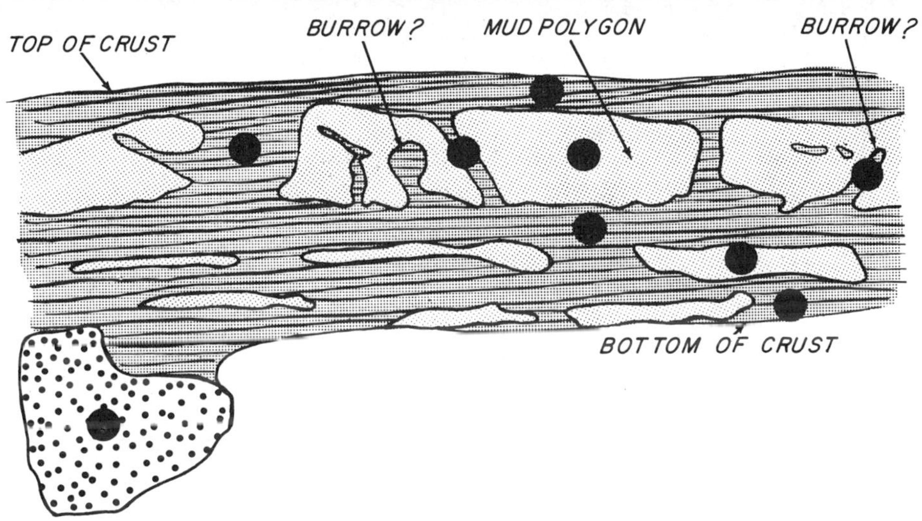

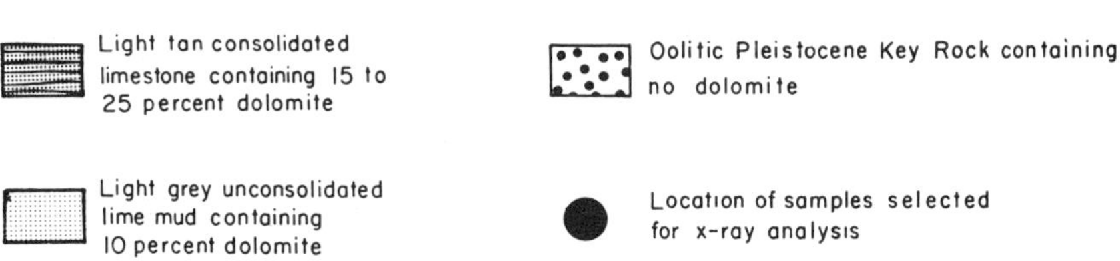

Fig. 3.—Dolomitic crust from Sugar Loaf Key. Photograph shows slabbed face; drawing below shows distribution of X-ray samples.

gons, and sand-sized chips and flakes, which "spalled-off" weathered polygons, filled these cracks. In addition, marine and reworked inter- and supratidal sediment was deposited in the cracks during spring and mild storm tide flooding. Often both the filled cracks and adjacent polygons were blanketed with $\frac{1}{2}$ inch layer of similar sediment.

The loosely packed, sandy filling of cracks and the blanketing sediment was observed to be more porous and permeable than the enclosed dried and compacted mud polygons. In many cases blue-green algae, which normally form continuous mats, grew preferentially on, and accentuated, the outlines of filled mud cracks (fig. 4). The algally covered sediment filling inter-polygon areas was generally wet (to the touch), whereas adjacent polygons were dry.

During periods of subaerial exposure, evaporation produced hot (up to 50°C) interstitial brines in the top inch of sediment. Water, squeezed from surface sediment during dry periods, was 5 to 6 times more concentrated than normal sea water and had Mg/Ca ratios as high as 40:1. Intertidal and underlying sediment contained water of lower salinity with a Mg/Ca ratio at about 5:1. The evaporating brine at the surface is, apparently, continually replenished by upward capillary movement of less saline water. The underlying water is seemingly replenished during spring and storm flooding. Chemical conditions in the supratidal zone are, thus, being constantly modified by rain, dew, and tidal floodings.

The origin of the high Mg/Ca ratios in the interstitial brines is not understood, especially since no gypsum is present in the sediment. Although the chemistry of the process of dolomitization is not understood, the pertinent factor determining the distribution of dolomite here is permeability. More water is able to pass through the porous and permeable crack filling and blanketing sediment than through the relatively tight and impermeable mud polygons. Dolomitization is thus selective and results in the preser-

Fig. 4.—Hurricane Donna storm layer after 2 years exposure at Sugar Loaf key. Algal mat growing preferentially on porous and permeable crack-filling sediment. Crust in figure 3 shows that dolomitization is relatively more rapid in the sediment filling mud cracks. Circular objects include a 1 gallon paint can "core", and its lid, (dia. approx. 15 cm). Sedimentary structures in this core can be found in figure 2 in reference 9. (See reference list.)

vation of sedimentary textures and structures which are characteristic of the supratidal environment in this climate.

If the sediments described above remain in the supratidal environment, it is likely that they will eventually become completely dolomitized. However, if relatively rapid subsidence or sea level rise results in a change to a less Mg-enriched intertidal or marine environment, then these sedimentary structures may be preserved undolomitized. Sander (1936) and, more recently, Matter (1967) have documented analogous completely dolomitized structures in Triassic and Cambro-Ordovician rocks, respectively. Sarin (1962) documented the latter type in Cambro-Ordovician rocks. Selectively dolomitized supratidal features of Devonian age have recently been documented by Laporte (1967), and numerous examples of various ages are discussed by Friedman and Sanders (1967).

REFERENCES

BALL, M. M., SHINN, E. A., AND STOCKMAN, K. W., 1967, The geologic effects of Hurricane Donna: Jour. Geology, v. 75, p. 583–597.

DEFFEYES, K. S., LUCIA, J. F., AND WEYL, P. K., 1965, Dolomitization of Recent and Plio-Pleistocene sediments by marine evaporite waters on Bonaire, Netherlands Antilles: SEPM spec. publ. no. 13, p. 71–88.

FRIEDMAN, G. M., AND SANDERS, J. E., 1967, Origin and occurrence of dolostones, p. 267–348 *in* Chilingar, G. V., Bissel, H. J., and Fairbridge, R. W., eds. Carbonate Rocks. Elsevier Publishing Co., New York 417 p.

ILLING, L. V., WELLS, A. J., AND TAYLOR, J. C. M., 1965, Penecontemporary dolomite in the Persian Gulf: SEPM spec. publ. no. 13, p. 89–111.

LAPORTE, L. F., 1967, Carbonate deposition near mean sea level and resultant facies mosaic, Manlius formation (Lower Devonian) of New York State: Am. Assoc. Petroleum Geol. Bull, v. 51, p. 73–101.

MATTER, ALBERT, 1967, Tidal flat deposits in the Ordovician of western Maryland: Jour. Sedimentary Petrology, v. 37, p. 601–609.

MULTER, H. G., AND HOFFMEISTER, J. E., 1967, Subaerial laminated crust in the Florida Keys: Geol. Soc. America Bull., v. 78, p. 183–192.

SANDER, BRUNO, 1936, Contributions to the study of depositional fabrics: published in English (1951) by American Association of Petroleum Geologists, translated by Knopf, E. B., 207 p.
SARIN, D. D., 1962, Cyclic sedimentation of primary dolomite and limestone: Jour. Sedimentary Petrology, v. 32, p. 451–471.
SHINN, E. A., GINSBURG, R. N., AND LLOYD, R. M., 1965, Recent supratidal dolomite from Andros Island, Bahamas: SEPM spec. publ. no. 13, p. 112–123.
——— 1967, Practical significance of birdseye structures in carbonate rocks: Jour. Sedimentary Petrology v. 38, p. 215–223.
VAN TUYL, F. M., 1914, The origin of dolomite: Iowa Geol. Surv. Ann. Proc., v. 25, p. 251–422.

Part VIII

SUMMARIES OF IMPORTANT CONTRIBUTIONS ON DOLOMITIZATION FROM THE FRENCH, GERMAN, AND SOVIET LITERATURE

Editors' Comments
on Papers 31, 32, and 33

31 BOURROUILH-LE JAN
Historical Résumé of French Contributions to the Study of Dolomitization

32 CHILINGAR
Summary of Some Important Soviet Contributions to the Field of Dolomitization

33 FÜCHTBAUER
Significant Contributions on Dolomitization from the German Literature

Part VIII represents a departure from the format of previous Benchmark volumes. In this part we include brief summaries of significant papers on dolomitization by foreign researchers written by F. G. Bourrouilh-Le Jan (French), G. V. Chilingar (Soviet), and H. Füchtbauer (German). (Space limitations prevented our considering contributions in other languages.) Based on the philosophy of the Benchmark Series, the contributors were asked to convey highlights of the important foreign research as they see it and according to their own organization. Most of the papers summarized have been previously published only in the language of that particular country. In addition to the partial translation of de Saussure's early analysis of dolomite (1792), Dr. Bourrouilh-Le Jan has elected to discuss the major developments in the French study of dolomitization beginning with Dolomieu (1791) and de Saussure (1792) up to the present. We believe this is the first such history presented in the English language. Dr. Chilingar presents his choices within the following headings: primary dolomite, classification of dolomites, porosity and permeability in dolomites, and dolomite reservoir rocks and their properties. Here, there is a strong complement to the rest of the book. Dr. Füchtbauer begins by providing a summary of Bruno Sander's classic article (1936) on interpretation of the fabric of Triassic carbonates. He follows this with synopses of five other papers, which primarily emphasize ancient dolomites, ranging from early to late diagenetic and which include petrology and geochemistry as well as field studies.

31

Original article written expressly for this Benchmark volume

HISTORICAL RÉSUMÉ OF FRENCH CONTRIBUTIONS TO THE STUDY OF DOLOMITIZATION

F. G. Bourrouilh-Le Jan
Université de Pau et des Pays de l'Adour

It is appropriate that Part VIII begins with a consideration of French contributions to our knowledge of dolomitization because of the initial description of the rock by the naturalist Déodat de Dolomieu (1791; Fig. 1). De Dolomieu's paper (*Sur un genre de Pierres calcaires très-peu effervescentes avec les Acides & phosphorescentes par la collision*), a letter to Mr. Picot de la Peyrouse, has been recently translated by Carozzi and Zenger (1981) who included some interesting footnotes as well as a brief sketch of de Dolomieu's life. Their translation of the beginning of de Dolomieu's paper is presented on the first page of the introduction to this book.

Based on his observations in the Tyrolean Alps, de Dolomieu reported the occurrence of a marble, known to the sculptors as *marmo graeco duro* to distinguish it from other softer Greek marbles that possess the same scaly texture or large, saltlike crystals. He noted its resistance to acid as its chief characteristic. By calcination, he obtained lime nearly as readily as through the use of ordinary limestones. After examination of many other examples in the "primitive mountains," he concluded that such rocks may differ from each other in texture and color. De Dolomieu compared and contrasted these marbles with younger horizontal and fossiliferous rocks in the "secondary mountains" of the Italian Alps, which were later named after him. De Dolomieu also discovered that the marbles in the "primitive mountains" possessed, to varying degrees, the property of phosphorescence when struck or scraped with an iron nail or another piece of the rock. He further noted the great resistance to weathering of these rocks, which cap many of the Tyrolean peaks.

383

Figure 1   Sketch of Déodat de Dolomieu based
            on photograph taken by F. G. Bour-
            rouilh-Le Jan of portrait in the
            Archives de l'Académie des Sciences,
            Institut de France. Sketch by Ann
            Zenger. Courtesy of Messieurs les
            Secrétaires Perpétuels de l'Académie
            des Sciences, Institut de France.

It remained for de Saussure (1792), in his paper *Analyse de la dolomie* published just over a year after de Dolomieu's announcement, to describe the rock more fully and to name it *dolomie* in his honor. Following is a translation of the first few paragraphs of de Saussure's paper:

> The slow and nearly imperceptible effervescence which
> certain calcareous rocks yield on testing with acid is
> a fact which Commander de Dolomieu has just brought to
> the attention of mineralogists in a very interesting
> letter to Mr. Picot de la Peyrouse (Journal de Physique,
> 1791).

> Mr. Fleuriau de Bellevue, on returning from his trip
> in the Tirol, was kind enough to give me a few pieces
> of this material, until now little known, and Mr. de
> Dolomieu, to whom I then wrote, sent me some superb
> specimens of its principal types.
>
> In all respects, this rock deserves a special name,
> but that of *slightly effervescent calcareous rock* is
> vague and inappropriate. It is befitting to derive its
> name from that of the famous naturalist who introduced
> it to us.

De Saussure insisted that we need not confuse dolomie with pearl
spar, rhodocrosite, or some ores of siderite despite the fact that all
are weakly effervescent in acid. He recognized that dolomie was harder
and its specific gravity higher than ordinary limestone, the average
of five different varieties being 2.852. He judged the various colors
to be related to iron content, and he noted a disposition toward a
conchoidal fracture. The little-known property of phosphorescence,
originally discovered by de Dolomieu, was further tested by de Saussure
who also reported that it is not characteristic of all dolomites; in
addition, some limestones possess this property. The first chemical
analysis of dolomite was presented by de Saussure: lime, 44.29; clay,
5.86; magnesia, 1.40; iron, 0.74; and carbon dioxide, 46.10; for a
total of 98.39, with 1.61 lost. He stated that the clay was not "acci-
dental;" he had found it in all his specimens. De Saussure concluded
(p. 162):

> We have seen that dolomite differs in some character-
> istics from the nature of ordinary calcareous rocks and
> that it approximates them in all respects after its cal-
> cination and its saturation with carbon dioxide and wa-
> ter; we did recognize that the gas which saturates it
> prior to calcination is really carbon dioxide and the
> dominant "earth" (element) is calcareous.

SOME NINETEENTH-CENTURY HIGHLIGHTS

Systematic geological mapping in France began early in the 1800s.
For example, Elie de Beaumont (1832), d'Archiac (1843), and others
reported the discovery of dolomitic units near Paris and elsewhere in
France.

The Plutonist-Neptunist controversy involved dolomitization. Von
Buch (1823) ascribed the formation of dolomite to volcanic vapors of
deep-seated origin. In 1831, Boué argued that the "shelly nature" of
dolomite attested to a Neptunian origin. In a letter to Arago, Virlet
(1835) distinguished two kinds of dolomite: primitive dolomitic rocks,
of any age, formed by the simultaneous deposition of calcium carbonate
and magnesium carbonate, and dolomitic rocks of "transmutation." The
Plutonist theory for dolomite formation was strenuously attacked by
Arago (1835), Virlet (1835), and Fournet (1845). On the other hand,
it was defended by such famous geologists as von Buch and Elie de
Beaumont (1837, 1854). Coquand (1841) verified that the Muschelkalk
(Triassic) in southern France contained crystalline limestone and
dolomite only around basaltic eruptions. In 1846, Durocher introduced

the word *magnésification* for the transformation of limestones to dolomite, which occurs at the contact of various igneous rocks.

Cordier (1862) wrote that there exists a great number of mixed rock masses, which indicated a transition between sedimentary limestone and nearly pure dolomite.

From an early date, numerous research efforts concentrated on the chemistry of waters depositing magnesium carbonate, including dolomite. Girardin (1837) noted calcium and magnesium carbonates in Saint Allyres spring near Clermont-Ferrand. Moitessier (1863) noticed carbonate material precipitating from the mineral waters of Lamalou in the south of France, the crystals having the chemical composition of dolomite (22.16 percent MgO and 30.27 percent CaO). In 1871, Delesse reported that the mud from Platen Lake in Hungary contains more than 15 percent magnesium carbonate, but he was not able to say whether it formed in situ or was reworked.

Attempts to synthesize dolomite began in the mid-1800s, paralleling those in Germany and England, and include the work of Marignac (1849), Durocher (1851), and Sainte-Claire Deville (1858). Durocher seems to have produced dolomite artificially by heating a mixture of magnesium chloride and fragments of porous limestone for three hours in a sealed gun barrel. In another geochemical study, Damour (1851) published his analyses of the magnesium content of certain coralline algae including *Lithophyllum*, *Melobesia*, and *Amphiroa* (ranging from 12.76 to 19.29 percent) as contrasted with green algae, such as *Halimeda* and *Galaxaura* (0.64 to 1.11 percent).

As long ago as 1849 Coquand initiated the concept that magnesium-enriched springs could supplement the magnesium in seawater, thus providing the explanation for the dolomitization of the overlying sediments "per ascensum." Daubrée in 1887 was the first to invoke the role of underground waters, enriched in sodium carbonate, in dolomitization. The Belgian Klement (1895) attempted to explain dolomitization by the action of heated and concentrated seawater on the aragonite of organisms.

TWENTIETH-CENTURY WORK

The Plutonist-Neptunist controversy over the origin of dolomite had largely ended and the sedimentary origin of most dolomites was accepted. This debate was superseded at about the beginning of the twentieth century by the primary versus replacement issue; direct precipitation was espoused by Suess in *La Face de la Terre* (1900, p. 435). Delépine (1911) demonstrated a facies relationship between dolomite and limestone of Dinantian (Lower Carboniferous) age in the Namur Basin of Belgium. Longchambon (1914a) treated fetid dolomite with ether, and on evaporation of the solution, he obtained an organic, syrupy, malodorous extract. He also (1914b) associated Pyrenean sedimentary dolomite with postorogenic (Caledonian and Hercynian) erosion, which he theorized contributed magnesium to seawater. Following the work of Damour (1851) and that of Lemoine (1911) on red algae, Jourdy (1914) attempted to demonstrate the importance of melobesioids as a source of magnesium for nearby or distant dolomitization.

The term *dedolomitization* was introduced in 1848 in a letter from de Morlot to de Beaumont as a result of the former's experiments involving dolomite powder, a gypsiferous solution, and the production of calcium carbonate and magnesium sulfate.

Certainly Lucien Cayeux ranks as one of history's great sedimentologists; his contributions are many. He clearly dissociated the problems of silicification from those of dolomitization (1916, 1925a, 1925b, 1929, 1935). He believed that silicification always precedes dolomitization and that silicification is selective, replacing limestone and chalk rather than dolomite. His monograph *Les Roches Sédimentaires de France: Roches Carbonatées* (1935) is a classic on carbonate rocks in general, including a wealth of field and petrographic observations, which are well illustrated. More specifically, Cayeux presented a classification of dolomites and emphasized the significance of *épigenèse* ("replacement") in dolomitization. He demonstrated the existence of several phases, with an accompanying enrichment of dolomite, following an initial stage. He recognized the complexity of the environmental control on dolomitization and various timings of the process, ranging from penecontemporaneous to late. He criticized the theory of dolomitizing "per ascensum" springs; his field and microscopic investigations convinced him of the opposite, a dolomitization "per descensum," or from above, from a hard-ground or emergent surface. He observed a relationship among sharp upper limits of dolomite, lack of sedimentation, and emergent surfaces.

Rivière (1939a, 1940) designed experiments to determine the effect of calcium carbonate on seawater and found a lowering of pH and the alkaline reserve. Appreciating that the phenomena are complex, he concluded that magnesium could be fixed in the calcium carbonate and thus in the sediments (1939b). He (1940) believed that the "trapping" of magnesium by calcareous sediments would occur in shallow water, such as in reefs, as contrasted with the quiet, continuously submerged deeper sea floor.

Several notable geochemical studies were published between about 1950 and 1980. For example, Lucas (1948a, 1948b) experimented with sodium carbonate added to seawater. After several months, aragonite formed, followed by nesquehonite. Lalou (1957) failed to obtain true dolomite, as determined by X-ray diffraction (Ricour, 1960), in carbonates within a mat at the surface of an aquarium culture with mud, glucose, and bacteria. Ricour (1960), however, insisted on an important role of bacteria in the process of dolomitization. Work on dolomite synthesis was published by Baron and Favre (1959) and then by Baron (1960) who described and illustrated directly precipitated dolomite obtained from solutions of $CaCl_2$ and $MgCl_2$ at high temperature (150°C) and high partial pressure of CO_2 and with the addition of Na_2CO_3. Sureau (1974) succeeded in experimentally dolomitizing calcite crystals at 150°C and various CO_2 pressures in the system $CaCO_3$-$MgCl_2$-H_2O-CO_2. He found the dolomitization to be a three-stage process involving successive dissolutions and recrystallizations of the different minerals: Ca-magnesite, protodolomite, and Ca-dolomite. Cailleau et al. (1979) experimentally showed the influence of Mg^{2+} and organic matter on the recrystallization of the various carbonates. Isotope studies of dolomitizing fronts (Javoy and Fayard, 1966) revealed very small-scale changes (scale of microns) in Mg^{2+} content. Although isotope results were consistent within each dolomite, there were differences between the two studied zones. Crystallographic measurements revealed a disordered dolomite located at the diagenetic front. Several other noteworthy publications on stable isotope analyses of dolomitic rocks appeared subsequently. Fritz (1967) reported on the isotope geochemistry of dolomitic nodules at certain horizons in the Paris Basin. Fontes, Kulbicki, and Létolle

(1969) described the isotope analyses of carbonates, including dolomites, on Mururoa Atoll. Fontes, Fritz, and Létolle (1970) determined the genesis of dolomites in the Paris Basin largely on the basis of stable isotope studies.

Many field-based studies of dolomites were reported in the mid-1900s. Canal (1947) began a systematic project on Aquitan Basin dolomites using a staining technique on polished sections. He was able to distinguish various kinds of calcite and dolomite veins and their chronology. Gèze (1949) demonstrated the influence of magnesium-enriched solutions migrating along faults in both Paleozoic and Mesozoic strata in southern France. He introduced the phrase *dolomite diapirs* for tower-like structures in certain layers. Contrary to Gèze, Fondeur et al. (1952) could discern no relation between tectonism and dolomitization in the Causses area. There, secondary dolomitization in oolitic limestones is localized in lenses of various size at the top of the sequence. The replacement seems to be very early diagenetic and related to evaporite formation. However, Garreau et al. (1959) showed a relationship between faults and fissures and epigenetic dolomitization. At about the same time, Pèlissonnier (1959) emphasized a tectonic-dolomitization association in Paleozoic units in the Pyrenees. The well-known paleontologist Lecompte (1958), studying Devonian reefs in the Belgian Ardennes, believed that dolomitization was localized in shallow-water, high-energy stromatoporoid limestones of biostromes in a back-reef position relative to bioherms.

Michard (1960, 1969) published two review papers on dolomitization and distinguished various kinds of dolomite: chemically precipitated, early diagenetic, late diagenetic, and late structural or epigenetic.

The association of emergent surfaces, including those representing paleokarst, and late diagenetic dolomitization in the Dolomites and Carnic Alps were described by Cros and Lagny (1969), Lagny (1974), and Cros (1977). Early dolomitization related to synsedimentary tectonism and emergence in the Middle Lias (Jurassic) of the High Atlas Mountains (Morocco) was reported by Dagallier (1977).

Aware of the lack of detailed study of dolomicrite, Delmas (1975), using ultrathin sections, found that some dolomicrites are composed of interlocking crystalline amoeboid grains (as are the calcite micrites), and he concluded that these early or "quasi-primary" dolomites resulted from the action of an excess of CO_3^{2-} ions on Mg-calcite or aragonite with an attendant reduction in porosity. Conversely, if Mg^{2+} is in excess, Delmas claimed that porosity would increase by the formation of euhedral dolomicrite.

Pelagic dolomite crusts in the Mediterranean were described by Froget (1972) and Desforges and Cros (1975). Anglada, Froget, and Récy (1975) invoked submarine diagenesis to explain enrichment of the sediments in magnesium (dolomite), iron, and phosphate southeast of New Caledonia.

French workers have been particularly active in studying the dolomitization in uplifted atolls. For example, Bourrouilh-Le Jan (1972) detailed for the first time the horizontal distribution of dolomite and its lateral equivalents on such a raised Pacific atoll (Lifu, New Caledonia). She concluded that dolomitization of the subtidal reef and calcitization in the laterally adjacent carbonates were synchronous diagenetic events related to a phreatic lens with variable ionic concentrations. Later, Répellin and Trichet (1975) and Répellin (1977) attributed the massive dolomitization of the middle zone of the uplifted atoll of Mururoa (French Polynesia) to the "Dorag" model, as did Coudray (1976) for Miocene to Holocene deposits in New Caledonia.

French carbonate sedimentologists have been involved with modern dolomite sedimentation. Bourrouilh-Le Jan (1973) linked the dolomite crusts on Andros Island (Bahamas) tidal flats with low-salinity ponds, tidal channels, and tidal estuaries. Later (1978, 1980), in a more detailed way, she related the early dolomitization to the climate and hydrology. Early dissolution of the carbonate minerals, particularly aragonite, in the freshwater ponds and lakes of western Andros, produces a hyperalkaline environment. Sediments are exposed to the fluctuations of the water table, and dolomitization occurs along the periphery of these lakes during evaporation. Elsewhere, Perthuisot (1975) attributed the dolomite at Sebka El Melah near Zarzis, Tunisia, to magnesium underflow with huntite as a precursor. Baud (1975) envisioned similar hypersaline conditions to explain the sedimentation and diagenesis of Triassic carbonates in the Prealps of Switzerland where dolomite occurs with quartzine, celestite, and fluorite in a subtidal to supratidal environment.

Outstanding field-based investigations by Purser (1975), Bouroullec and Deloffre (1977), and Bouroullec, Delfaud, and Deloffre (1979) permitted them to relate oil-field occurrences to early diagenesis and commonly early diagenetic dolomitization.

ACKNOWLEDGMENTS

I am grateful to Donald Zenger for editorial assistance. Zenger, with Virginia Crosby (Pomona College) and Albert Carozzi (University of Illinois at Urbana-Champaign), assisted with the partial translation of de Saussure. Ann Zenger prepared the sketch of de Dolomieu (Fig. 1).

REFERENCES

Anglada, R., C. Froget, et J. Récy, 1975, Sédimentation ralentie et diagenèse sous-marine au SE de la Nouvelle Calédonie (Dolomitisation, Ferruginisation, Phosphatisation), *Sed. Geology* 14:301-317.

Arago, D. F., 1835, Formation des dolomies, *Acad. Sci. Comptes Rendus* 1:192-193.

Archiac, A. d', 1843, Description géologique du département de l'Aisne, *Soc. Géol. France Mém.* 5:317-319.

Baron, G. A., 1960, Sur la synthèse de la dolomite. Application au phénomène de la dolomitisation, *Inst. Français Pétrole Rev.* 15:3-68.

Baron, G. A., et J. H. Favre, 1959, Contribution française à l'étude de la dolomitisation. III. Recherches expérimentales sur le rôle des facteurs physico-chimiques dans la synthèse de la dolomie, *Fifth World Petroleum Cong. Proc.* (sec. 1), pp. 53-81.

Baud, A., 1975, Diagenèse de sédiments carbonatés sous des conditions "hypersalines:" quartzine, célestine, fluorine dans les calcaires du Trias moyen des Préalpes Médianes (Domaine Briançonnais, Suisse occidentale), *IXème Congrès Intern. Sédimentol., Nice,* pp. 18-24.

Beaumont, E. de, 1832, Découverte d'une couche de dolomie à Beynes, près de Grignon, *Soc. Géol. France Bull.* 2:419.

Beaumont, E. de, 1837, Application du calcul à l'hypothèse de la formation par l'épigénie des anhydrites, des gypses et des dolomies, *Soc. Géol. France Bull.* (sér. 1) 8:174-177.

Beaumont, E. de, 1854, Note sur l'origine présumée des dolomies, *Acad. Sci. Comptes Rendus* 39:526.

Boué, A., 1831, Compte rendu des progrès de la Géologie, *Soc. Géol. France Bull.* 1:115.

Bouroullec, J., et R. Deloffre, 1977, Support sédimentologique à l'interprétation diagraphique. Exemple d'un paléorivage Jurassique aquitain (France SW), in *Essai de caractérisation sédimentologique des dépôts carbonatées. 2. Eléments d'interprétation.*, Centre de Recherches de Boussens et de Pau, Elf-Aquitaine, pp. 151-157.

Bouroullec, J., J. Delfaud, et R. Deloffre, 1979, Quelques aspects sédimento-diagénétiques de la plate-forme carbonatée du Dogger-Oxfordien aquitain, in *La Sédimentation du Jurassique W-Européen,* Assoc. Sédimentol. Français Pub. Spéc. 1, pp. 271-291.

Bourrouilh-Le Jan, F. G., 1972, Diagenèse récifale: calcitisation et dolomitisation. Leur répartition dans un atoll soulevé. Ile Lifou, archipel des Loyauté, Territoire de la Nouvelle Calédonie, *Cahiers ORSTOM Sér. Géol.* 4:121-148.

Bourrouilh-Le Jan, F. G., 1973, Les dolomies et leurs genèses, *Centre Recherches Pau Bull.* 7:111-135.

Bourrouilh-Le Jan, F. G., 1978, Rôle des ouragans et des cyclones tropicaux sur la sedimentation carbonatée: la plaine d'estran de l'ouest d'Andros, Bahama. Interférence de la climatologie, de l'hydrologie et de la diagenèse, *Acad. Sci. Comptes Rendus* 287:907-910.

Bourrouilh-Le Jan, F. G., 1980, Hydrologie des nappes d'eau superficielles de l'île Andros, Bahama. Dolomitisation et diagenèse de plaine d'estran en climat tropical humide, *Centre Recherches Explor.-Prod. Elf-Aquitaine Bull.* 4:661-707.

Buch, L. von, 1823, Sur la dolomie du Tyrol, *Annales Chimie* 23:396-409.

Cailleau, P., C. Jacquin, D. Dragone, A. Girou, H. Roques, et L. Humbert, 1979, Influence des ions étrangers et de la matière organique sur la cristallisation des carbonates de calcium, *Inst. Français Pétrole Rev.* 34:83-112.

Canal, P., 1947, Observations sur les caractères pétrographiques des calcaires dolomitiques et des dolomies, *Soc. Géol. France Compte Rendu* 17:161-162.

Carozzi, A. V., and D. H. Zenger, 1981, On a Type of Calcareous Rock That Reacts Very Slightly with Acid and That Phosphorescences on Being Struck (Sur un genre de Pierres calcaires très-peu effervescentes avec les Acides & phosphorescentes par la collision), translation with notes on de Dolomieu's paper reporting the discovery of dolomite, *Jour. Geol. Education* 29:4-10.

Cayeux, L., 1916, *Introduction a l'étude pétrographique des roches sédimentaires,* France Carte Géol. Mém., Imprimerie Nationale, Paris, 524p.

Cayeux, L., 1925a, L'âge relatif des phtanites et des dolomies du calcaire carbonifère du Nord de la France et de la Belgique, *Acad. Sci. Comptes Rendus* 180:843-846.

Cayeux, L., 1925b, L'âge relatif des silex et des dolomies de la craie du Bassin de Paris, *Acad. Sci. Comptes Rendus* 180:1354-1356.

Cayeux, L., 1929, *Les roches sédimentaires de France. Roches siliceuses,* France Carte Géol. Mém., Imprimerie Nationale, Paris, 774p.

Cayeux, L., 1935, *Les Roches Sédimentaires de France. Roches Carbonatées (Calcaires et Dolomies),* Masson et Cie, Paris, 463p.

Coquand, H., 1841, Modifications éprouvées par les calcaires au voisinage des roches ignées, *Soc. Géol. France Bull.* 12:314-352.

Coquand, H., 1849, Observations sur l'origine de la dolomie du Tyrol, *Soc. Géol. France Bull.* (sér. 2) 6:322.

Cordier, P. L. A., 1862, De l'origine des roches calcaires qui n'appartiennent pas au sol primordial, *Acad. Sci. Comptes Rendus* 54:293-299.

Coudray, J., 1976, Recherches sur le Néogène et le Quaternaire marins de la Nouvelle Calédonie. Contribution de l'étude sédimentologique à la connaissance de l'histoire géologique post-éocène, *Expédition Français sur les Récifs coralliens de la Nouvelle Calédonie,* Fondation Singer-Polignac, Paris, 8:1-275.

Cros, P., 1977, Données nouvelles sur la dolomitisation des carbonates triasiques des Dolomites italiennes, *Sci. Terre* 21:307-355.

Cros, P., et P. Lagny, 1969, Paléokarsts dans le Trias moyen et supérieur des Dolomites et des Alpes Carniques Occidentales, *Sci. Terre* 14: 139-195.

Dagallier, G., 1977, Une série carbonatée littorale: le Lias moyen (à Pb-Ba) de Mibladen (Maroc). Dolomitisation et dissolutions polyphasées en environnement tectonique semi-mobile, *Sci. Terre* 21:53-101.

Damour, A., 1851, Note sur la composition des Nullipores et de quelques Corallinées, *Acad. Sci. Comptes Rendus* 32:253-255.

Daubrée, A., 1887, *Les eaux souterraines aux époques anciennes,* Dunod, Paris, 443p.

Delépine, G., 1911, Recherches sur le Calcaire Carbonifère de Belgique, *Lille Facultés Catholiques Mém. Travaux* 8:1-419.

Delesse, A., 1871, *Lithologie des mers de France et des mers principales du globe. Lithologie du fond des mers,* éd. E. Lacroix, Librairie Scientifique Industrielle et Agricole, Paris, 136p.

Delmas, M. R., 1975, La formation et l'évolution des micrites et dolomicrites, *Centre Recherches Pau Bull.* 9:77-97.

Desforges, G., et P. Cros, 1975, Diagenèse calcaréo-dolomitique précoce en milieu pélagique profond (Quaternaire récent, Méditerranée occidentale), *IXème Congrès Intern. Sédimentol., Nice, Thème 8,* pp. 19-24.

Dolomieu, D. de, Sur un genre de Pierres calcaires très-peu effervescentes avec les Acides & phosphorescentes par la collision, *Jour. Physique* 39:3-10.

Durocher, J., 1846, Etude sur le métamorphisme des roches, *Soc. Géol. France Bull.* (sér. 2) 3:546-647.

Durocher, J., 1851, Production artificielle de la dolomie sous l'influence de vapeurs magnésifères, *Acad. Sci. Comptes Rendus* 33:64-66.

Fondeur, C., M. Gottis, J. Rouire, et A. Vatan, 1952, Quelques aspects de la dolomitisation au Jurassique en France, *XIXème Congrès Géol. Intern., Alger* 15:471-491.

Fontes, J.-C., G. Kulbicki, et R. Létolle, 1969, Les sondages de l'atoll de Mururoa: aperçu géochimique et isotopique de la série carbonatée, *Cahiers Pacifique* 13:69-74.

Fontes, J.-C., P. Fritz, et R. Létolle, 1970, Composition isotopique, minéralogique et genèse des dolomies du Bassin de Paris, *Geochim. et Cosmochim. Acta* 34:279-294.

Fournet, J., 1845, Note sur les résultats sommaires d'une expédition géologique du Tyrol méridional et de quelques parties des régions subalpines de l'Italie, *Soc. Géol. France Bull.* (sér. 2) 3:30.

Fritz, P., 1967, Les rognons dolomitiques du Bassin de Paris; pétrographie et analyse isotopique, *Rev. Géographie Phys. et Géologie Dynam.* (sér. 2) 9:329-334.

Froget, C., 1972, Exemples de diagenèse sous-marine dans les sédiments pliocènes et pléistocenes: dolomitisation, ferruginisation (Méditerranée nord-occidentale, sud de Marseille), *Sedimentology* 19:59-83.

Garreau, B., O. L. de Charpal, L. Montadert, Y. G. Gubler, P. E. Rouge, G. A. Baron, et J. H. Favre, 1959, Contribution française a l'étude de la dolomitisation, *Fifth World Petroleum Cong. Proc.* (sec. 1), pp. 53-81.

Gèze, B., 1949, La dolomitisation des calcaires de la Montagne Noire et des Causses, *Soc. Hist. Nat. Toulouse Bull.* 84:113-128.

Girardin, J., 1837, Analyse de l'eau minérale de Sainte-Allyre près Clermont, département du Puy-de-Dôme, *Annales Mines* (sér. 3) 11:457-462.

Javoy, M., et M. Fayard, 1966, Etude de "fronts de dolomitisation" dans le Lias des Causses et le Dévonien de la Montagne Noire, *Geol. Rundschau* 55:78-85.

Jourdy, E., 1914, Coraux, Mélobésiées, Bactéries, Dolomies. Origine et genèse des dolomies sédimentaires, *Soc. Géol. France Bull.* (sér. 4) 14:279-309.

Klement, C., 1895, Sur l'origine de la dolomie dans les formations sédimentaires, *Soc. Belge Géologie, Paléontologie et Hydrologie Bull. Proc.-Verb.* 9:27; *Mém.* 9:3-23.

Lagny, Ph., 1974, Emersions successives, karstification et sédimentation continentale au Trias moyen dans la region de Sappada, *Sci. Terre* 19:195-233.

Lalou, C., 1957, Etude expérimentale de la production de carbonates par les bactéries des vases de la baie de Villefranche-sur-Mer, *Inst. Océanog. Annales* 33:201-267.

Lecompte, M., 1958, Les récifs paléozoïques en Belgique, *Geol. Rundschau* 47:384-401.

Lemoine, Mme. P., 1911, Structure anatomique des Mélobésiées, *Inst. Océanog. Annales* 2:1-213.

Longchambon, M., 1914a, Sur le rôle de la magnésie dans les cycles sédimentaires, *Acad. Sci. Comptes Rendus* 158:267-269.

Longchambon, M., 1914b, Sur structure primitive des dolomies pyrénéennes, *Acad. Sci. Comptes Rendus* 158:953-955.

Lucas, G., 1948a, La sédimentation calcaire. Action du carbonate sur l'eau de mer, *Acad. Sci. Comptes Rendus* 226:937-939.

Lucas, G., 1948b, La sédimentation calcaire. Quelques modalités de l'alcalinisation de l'eau de mer en présence d'ions carboniques, *Acad. Sci. Comptes Rendus* 226:1023-1025.

Marignac, Ch. de, 1849, Note sur l'origine de la dolomie, *Acad. Sci. Comptes Rendus* 28:364-366.

Michard, A., 1960, Sur la variétés des processus de genèse dolomitique, *Acad. Sci. Comptes Rendus* 251:3025-3027.

Michard, A., 1969, Les dolomies. Une revue, *Alsace-Lorraine Service Carte Géol. Bull.* 22:1-92.

Moitessier, M. A., 1863, Formation de la dolomie dans les eaux minérales, *Acad. Sci. et Let. Montpellier Sec. Sci. Proc.-Verb.* (4), p. 18.

Morlot, M. A. de, 1848, Sur l'origine de la dolomie, *Acad. Sci. Comptes Rendus* 26:311-315. (Extr. d'une lettre de M. A. de Morlot à M. Elie de Beaumont.)

Pèlissonnier, H., 1959, Dolomitisation de la série dévono-dinantienne dans les Pyrénées de la Haute-Garonne et de l'Ariège, *Soc. Géol. France Bull.* 7:625-631.

Perthuisot, J.-P., 1975, La Sebkha el Melah de Zarzis. Genèse et évolution d'un bassin salin paralique, *Trav. Lab. Géol. Ec. Norm. Super. Paris* 9:1-252.

Purser, B. H., 1975, *Sédimentation et diagenèse précoce des séries carbonatées du Jurassique moyen de Bourgogne,* Thèse Etat, Univers. Paris-Sud "Centre-Orsay," Paris, 383p.

Répellin, P., 1977, Contribution à l'étude d'un récif corallien: le sondage "Colette," atoll de Mururoa (Polynésie française), *Cahiers Pacifique* 20:1-210.

Répellin, P., et J. Trichet, 1975, Un exemple de diagenèse de carbonates récifaux (atoll de Mururoa-Polynésie française), *IXème Congrès Intern. Sédimentol., Nice, Thème 7,* pp. 179-186.

Ricour, J., 1960, De la genèse de certaines dolomies, *Acad. Sci. Comptes Rendus* 251:1798-1800.

Rivière, A., 1939a, Sur la dolomitisation des sédiments calcaires, *Acad. Sci. Comptes Rendus* 209:597-599.

Rivière, A., 1939b, Observations nouvelles sur le mécanisme de dolomitisation des sédiments calcaires, *Acad. Sci. Comptes Rendus* 209:691-692.

Rivière, A., 1940, L'eau de mer et les sédiments calcaires, *Soc. Géol. France Compte Rendu* 5:40-42.

Sainte-Claire Deville, Ch., 1858, Sur l'action des chlorures et des sulfates alcalins et terreux dans le métamorphisme des roches sédimentaires, *Acad. Sci. Comptes Rendus* 47:89-98.

Saussure, Th. de, 1792, Analyse de la dolomie, *Jour. Physique* 40:161-173.

Suess, Ed., 1900, *La face de la terre,* (tome II), Armand Colin, Paris, 878p.

Sureau, J. F., 1974, Etude expérimentale de la dolomitisation de la calcite, *Soc. Française Mineralogie Cristallographie Bull.* 97:300-312.

Virlet, T., 1835, Lettre de M. Théodore Virlet à M. Arago sur le phénomène de la dolomitisation et la transformation des roches en général, *Acad. Sci. Comptes Rendus* 1:268-272.

32

Original article written expressly for this Benchmark volume

SUMMARY OF SOME IMPORTANT SOVIET CONTRIBUTIONS TO THE FIELD OF DOLOMITIZATION

G. V. Chilingar
University of Southern California

The progress of lithologic investigations (including dolomites) in the Soviet Union from 1918 through 1968 was reviewed by Yablokov (1970). Indeed impressive are the immense contributions to the field of carbonate sediments and rocks, including dolomites, of such giants of Soviet geology as G. I. Teodorovich and N. M. Strakhov. Their deaths during recent years were a great loss to the scientific world. However, the studies of such Soviet workers as V. N. Kholodov, K. I. Bagrintseva, Yu. I. Mar'enko, A. M. Ivanov, A. P. Lisitsin, and Yu. P. Kazanskiy, among others, are adding to our growing knowledge of dolomites and dolomitization.

PRIMARY DOLOMITES

The possibility of direct precipitation of (primary) dolomite from seawater, which is a continuing controversy (see, for example, Chilingar et al., 1979), has been studied by Kazanskiy (1972, 1975), Kazanskiy, Kataeva, and Shugurova (1969), and Beloborodova et al. (1972), all of whom have indicated the possibility of protodolomite precipitation from seawater during the Precambrian. The major factors that appear to affect the precipitation of dolomite are alkalinity (HCO_3^- + CO_3^{2-} content; Table 1); pH; the relative proportions of cations, particularly the Mg/Ca ratio; sulfate ion concentration in seawater; and organic matter influences. According to these authors, the favorable factors for the precipitation of protodolomite include Mg/Ca ratio close to unity, higher concentrations of Ca^{2+} and Mg^{2+} relative to Na^+, and high alkalinity and pH, both of which insure high concentrations of CO_3^{2-} anions in the solution. High partial pressures of CO_2 in the Precambrian atmosphere are believed to have played a critical role in increasing seawater alkalinity. According to the studies of Kazanskiy, Kataeva, and Shugurova (1969) on inclusions in Precambrian siliceous rocks, Precambrian seawater apparently contained high concentrations of Ca^{2+} and Mg^{2+}. During the late Precambrian and early Paleozoic, the Mg/Ca ratio of seawater may have been shifted toward higher values because of the withdrawal of considerable amounts of Ca^{2+} by sulfate and organic carbonate deposits. As a result, dolomite accumulation during

394

the early Paleozoic conceivably shifted toward basins of relatively
higher salinity. More pronounced shifts toward such basins also may
have occurred during the Devonian, when there may have been a considerable decrease in atmospheric CO_2 because of the introduction of land
plants. Many Soviet investigators do, in fact, believe in the direct
precipitation of dolomite from seawater.

pH

Form	4	5	6	7	8	9	10
$H_2CO_3 + CO_2$	99.7	96.2	71.5	20.0	2.4	0.2	--
HCO_3^-	0.3	3.8	28.5	80.0	97.2	95.7	70.4
CO_3^{2-}	--	--	--	--	0.4	4.1	29.6

Table 1   Relative proportions of various forms of carbonate
complexes in waters of different pH, activity not
taken into consideration (after Alekin, 1970).

As noted by Chilingar et al. (1979), however, it is difficult to
distinguish early penecontemporaneous-replacement dolomites and the
primary dolomites that may precipitate directly from seawater. Soviet
scientists are attempting such a distinction using, among other criteria, $^{18}O/^{16}O$ and $^{13}C/^{12}C$ ratios. Definite conclusions, however, have
not been reached regarding such distinctions. Vinogradov, Ronov, and
Ratynskiy (cited by Strakhov and Bushinskiy, 1956, p. 258) made an
extensive study of the temporal variation of Ca^{2+} and Mg^{2+} contents of
carbonate rocks of the Russian Platform (Fig. 1). The general decrease
in the average Mg/Ca ratio corresponds to a decrease in the abundance
of dolomites in younger rocks (see also Chilingar, Paper 4).

CLASSIFICATION OF DOLOMITIC ROCKS

In an excellent discussion on the origin of dolomites, Teodorovich
(1958, pp. 302-311) recognized four groups of dolomites: normal-marine,
slightly calcitic dolomites, and dolomitic limestones (replacement
types); primary pelitomorphic dolomites and sulfate-dolomite rocks;
calcitic dolomite deposits in saline (increasing salinity) seas, lagoon-
like bays on reefs, and so forth; and primary, pelitomorphic, chemical,
calcitic dolomitic muds forming in continental lakes in dry and hot
(even if not periodical) climates. Strakhov (1956) attributed the origin
of an additional type of dolomite to mixing of fresh river waters and
saline waters of bays. Teodorovich (1958, pp. 271-301) also proposed a
textural classification of dolomites (Fig. 2) based primarily on whether
the "relics" are positive or negative (that is, precursor fossils and
minerals are identifiable or discernible but not readily recognizable,
respectively). Recognizing that there were more than thirty different

395

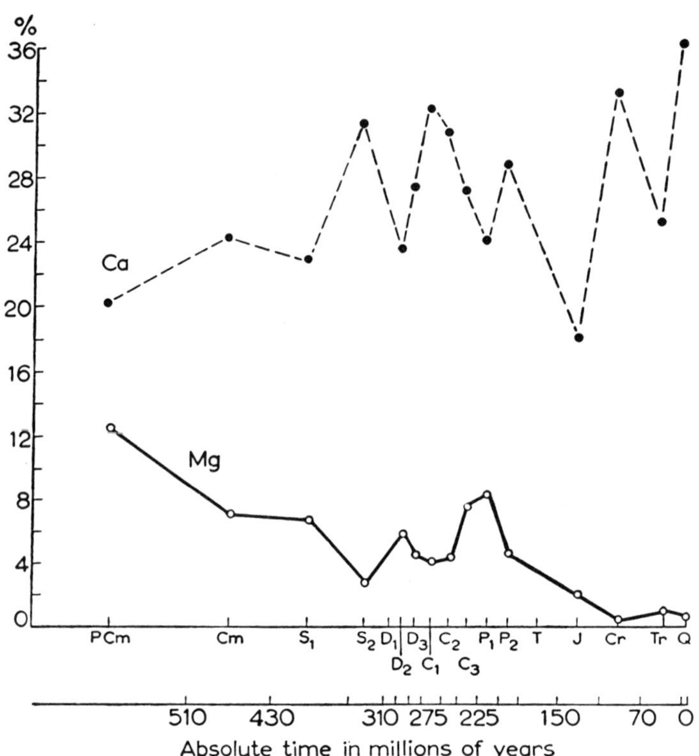

Figure 1   Changes with time in the average Ca
            and Mg contents (in weight %) of car-
            bonate rocks of the Russian Platform
            (after Vinogradov, Ronov, and Ratynskiy,
            cited in Strakhov and Bushinskiy, 1956,
            p. 258). The curves are based on: 97
            analyses of average assays (weighted for
            individual formation thicknesses) from
            3659 samples and literature review (2826
            analyses for Ca and 2583 analyses for Mg).

structural and genetic classifications of carbonate rocks, Mar'enko
(1978) presented two elaborate textural-genetic classifications of
both limestones and dolomites. These widely used classifications are
modified, expanded, and improved versions of Teodorovich's schemes.
The results of many basic sedimentological studies have had significant
application to exploration for and exploitation of oil and gas from
carbonate reservoir rocks in the Soviet Union.

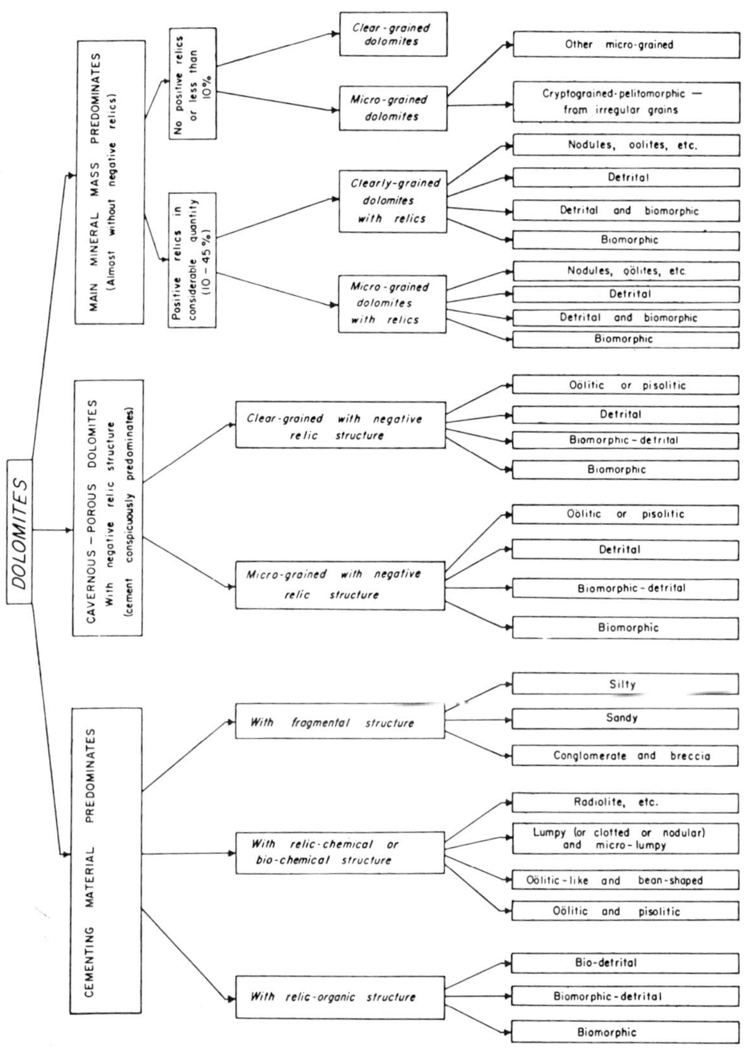

Figure 2   Textural classification of dolomites
(after Teodorovich, 1958, p. 295).

397

DOLOMITE RESERVOIR ROCKS

Although carbonate rocks constitute only 15 to 30 percent of the **total volume of sedimentary rocks, about 60 percent of the total hydro-**carbon reserves in the world are in carbonate reservoirs, particularly in dolomites. However, the percentage of the total in situ oil recovered often is very low (10 to 20 percent) because of the lithologic complexity of such rocks. Thus, I stressed a great need for cooperation between petroleum engineers and geologists on numerous occasions (for example, Chilingar, Mannon, and Rieke, 1972). The origin, composition, and diagenetic and catagenetic processes (mainly dolomitization) in large measure determine the petrophysical properties of carbonate rocks and, consequently, the behavior of carbonate reservoirs. It, therefore, was encouraging to discover that three excellent texts on carbonate reservoir rocks were published recently in the Soviet Union: Bagrintseva (1977), Ivanov (1976), and Mar'enko (1978). These treatises, as well as many other Soviet references relating to reservoir rocks (for example, Bagrintseva, 1972; Bagrintseva and Per'kova, 1971), may be obtained through interlibrary loan from the Seaver Science Library at the University of Southern California.

In the Soviet Union, studies are continuing on the effects of postsedimentation diagenetic processes on the properties of carbonate reservoir rocks (for example, Sarkisyan, Politykina, and Chilingarian, 1973). Many bioclastic carbonate reservoirs of high porosity and permeability were deposited in shallow, high-energy areas of sedimentary basins. Such processes as stylolitization commonly improve reservoir rock characteristics; in contrast, sulfatization, calcitization, and silicification adversely affect reservoir rock properties. However, secondary mineralization indirectly improves the flow capacity (permeability) of such rocks by enhancing heterogeneity, which favors the subsequent formation of fractures and solution cavities. Dolomitization generally creates and/or enhances porosity and, therefore, most carbonate reservoir rocks are either dolomites or dolomitic limestones.

POROSITY AND PERMEABILITY IN DOLOMITES

According to Maksimova and Entsov (1968), the replacement of calcite by dolomite involves a physical contraction of 12 to 13 percent (porosity increase) if (1) the reaction proceeds according to

$$2CaCO_3 + Mg^{2+} \rightarrow CaMg(CO_3)_2 + Ca^{2+};$$

(2) the replacement is on a mole-for-mole basis; and (3) Ca^{2+} ions are removed. Obviously, if replacement is on a volume-for-volume basis and Ca^{2+} cations are not removed, then dolomitization would not result in porosity formation (Sarkisyan, Politykina, and Chilingarian, 1973, p. 1307). There is no general agreement among Soviet workers as to whether dolomitization creates porosity. Frolova (1962, 1963) believed that dolomitization results in volume reduction and porosity increase regardless of the lithogenetic stage. According to Gmid and Zvonitskaya (1965) and Per'kova (1966), however, a complex relation, which is not always direct, exists between porosity and dolomitization. According to Sarkisyan, Politykina, and Chilingarian (1973), dolomitization causes porosity unless it is of early diagenetic (for example, penecontemporaneous) origin, and porosity is destroyed by subsequent compaction.

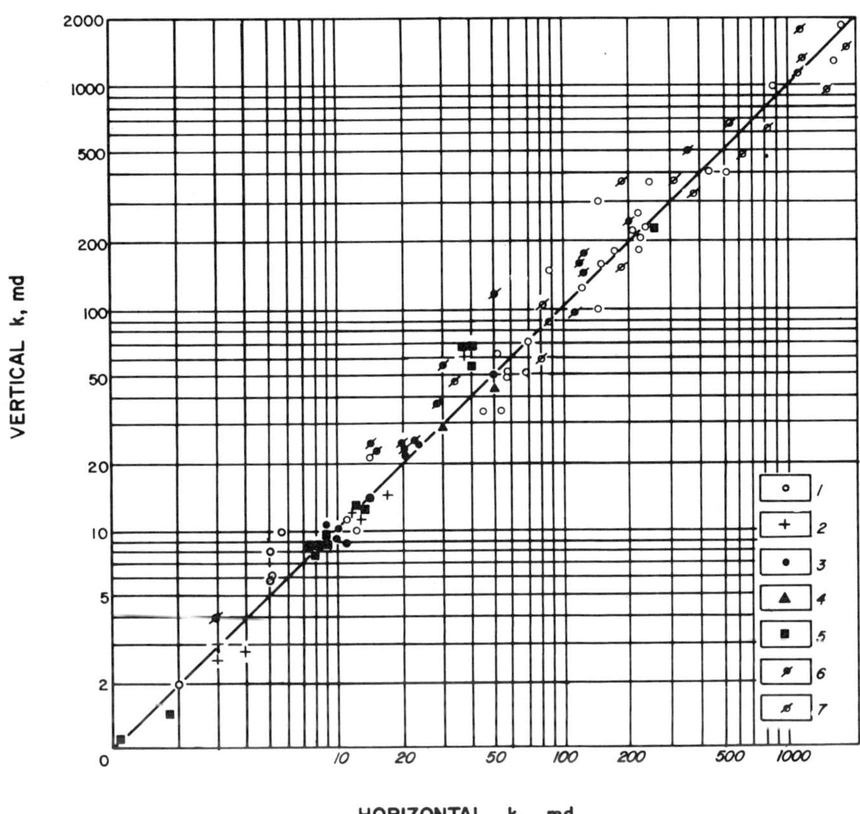

Figure 3   Relationship between horizontal and
vertical permeability in dolomites
with intergranular porosity. 1: Kuybyshev,
Along-Volga Region; 2: Saman-Tepe; 3:
Uchkyr; 4: Adam-Tash; 5: Orenburg; 6:
Vyktyl; 7: Urta-Bulak (after Bagrintseva,
1977, p. 45).

Geologists and petroleum engineers are well aware that vertical
permeabilities in carbonate reservoirs commonly exceed horizontal
permeabilities, especially in the case of reefs. Such relationships
possibly are due to the dissolving effects of compaction-derived fluids
moving vertically (upward), creating solution channels, vugs, and

399

caverns, and enlarging preexisting fractures. In contrast, horizontal permeabilities generally exceed vertical permeabilities in sandstone reservoirs. In dolomites with intergranular porosity, vertical permeabilities commonly are practically equal to horizontal permeabilities (Fig. 3).

Geologists and petroleum engineers in the Soviet Union recognize three types of porosity: total porosity, also referred to as physical or absolute porosity, equivalent to sum volume of all pores, caverns, vugs, and fractures regardless of whether they are isolated or inter-communicative; open or saturation porosity, which includes only inter-communicating pores; and effective ("useful" or "dynamic") porosity, which includes open porosity minus irreducible fluid saturation (for example, immobile fluid retained by capillary forces in minute pores, crevices, and cracks). In the United States "open" porosity is synony-mous with "effective" porosity (see Bagrintseva, 1977, pp. 18-19).

Chilingar (1964) demonstrated a reasonable correlation between the porosity and permeability of cores having irreducible (immobile) fluid saturation in the minute pores, crevices, and so forth, which do not have a major effect on the flow of fluids through the rock. Thus, in my opinion, it is advisable to adopt the Soviet definitions of porosity types in dolomitic reservoirs.

Bagrintseva (1977, p. 134) developed the following relationship between the permeability (k) in millidarcies, and the ratio of effec-tive porosity (ϕ_e) to open porosity (ϕ_o) for some dolomites and dolo-mitic limestones of the Kuybyshev (Along-Volga) region and Central Asia:

$$\ln\,(k+1) = 0.372 + 7.328\left(\frac{\phi_e}{\phi_o}\right)^2 .$$

The coefficient of correlation is 0.90, whereas the standard devi-ation was found to be 0.746. Bagrintseva (1977) also established an interrelationship among irreducible fluid saturations, median diameter of pores, porosity (both effective and open), permeability (both effec-tive and absolute), specific surface area, and coefficient of hetero-geneity for various carbonate rocks. The results obtained by Bagrintseva are indeed of great practical value.

ACKNOWLEDGMENTS

The writer is greatly indebted to Drs. S. J. Mazzullo and D. H. Zenger for their very valuable suggestions concerning improvement of this manuscript.

REFERENCES

Alekin, O. A., 1970, *Principles of Geochemistry,* Gidrometeoizdat, Leningrad, 444p.

Bagrintseva, K. I., 1972, About Porosity of Carbonate Rocks of Various Composition and Genesis, *Lithology and Mineral Resources* 4:99-106.

Bagrintseva, K. I., 1977, *Carbonate Rocks--Oil and Gas Reservoirs,* Izd. Nedra, Moscow, 231p.

Bagrintseva, K. I., and Ya. N. Per'kova, 1971, About Influence of Genetic Characteristics on Formation of Reservoir Properties of Carbonate Rocks, in *Materials of IV All-Union Conference on Reservoirs,* Izd. Nedra, Moscow, pp. 297-301.

Beloborodova, G. V., Yu. V. Davydov, Yu. P. Kazanskiy, V. N. Kataeva, and N. T. Mandrikova, 1972, Possibility of Chemical Precipitation of Dolomite from Sea Waters of Riphaean Era, *Akad. Nauk SSSR Izv. Ser. Geol.* 4:152-155.

Chilingar, G. V., 1964, Relationship Between Porosity, Permeability and Grain-Size Distribution of Sands and Sandstones, in *Deltaic and Shallow Marine Deposits,* ed. J. U. van Straaten, Elsevier, Amsterdam, pp. 71-75.

Chilingar, G. V., R. W. Mannon, and H. H. Rieke, 1972, *Oil and Gas Production from Carbonate Rocks,* American Elsevier, New York, 408p.

Chilingar, G. V., D. H. Zenger, H. J. Bissell, and K. H. Wolf, 1979, Dolomites and Dolomitization, in *Diagenesis in Sediments and Sedimentary Rocks,* ed. G. Larsen and G. V. Chilingar, Elsevier, Amsterdam, pp. 423-536.

Frolova, E. K., 1962, Lithologic Characteristics of Productive Carbonate Formations of Kuleshov Oil Field, in *Geology, Geochemistry, and Geophysics Symposium,* Kuybyshev Sci. Res. Inst. Petroleum Industry USSR, vol. 11, pp. 62-84.

Frolova, E. K., 1963, Lithologic Characteristics of Productive Horizon of Bashkir Suite (A_4 Stratum) of Kuybyshev Area, in *Geology, Geochemistry, and Geophysics Symposium,* Kuybyshev Sci. Res. Inst. Petroleum Industry USSR, vol. 20, pp. 22-63.

Gmid, L. P., and I. V. Zvonitskaya, 1965, Lithological and Petrographical Investigations of Carbonate Rocks and Formation of Their Porosity, *Vses. Neft. Nauchno-Issled. Geol.-Razved. Inst. Trudy Proc.* 242:49-73.

Ivanov, A. M., 1976, *Complex Study of Carbonate Rocks as Oil and Gas Reservoirs,* Izd. Nedra, Moscow, 295p.

Kazanskiy, Yu. P., 1972, About Relation of Composition of Atmosphere with Development of Life During Precambrian and Paleozoic Time, in *Environment and Life During Geologic Past,* Izd. Nauka, Novosibirsk, Vypusk 1.

Kazanskiy, Yu. P., 1975, Experimental Investigations on Synthesis of Carbonates from Carbonate-Chloride Solutions and Their Application Towards Development of Theory of Dolomitization, in *Crystallography of Minerals and Geologic Problems,* ed. A. G. Kossouskaya, Izd. Nauka, Moscow, pp. 242-244.

Kazanskiy, Yu. P., V. N. Kataeva, and N. A. Shugurova, 1969, Experiment on Studying Composition of Gaseous and Liquid Phases as Relics of Ancient Atmospheres and Hydrospheres, *Geology and Geophysics, No. 11.*

Maksimova, G. A., and I. I. Entsov, 1968, Formation of Secondary Porosity and Caverns as a Result of Metasomatic Dolomitization of the Limestones of the Perm'-Kama Region, *Geologiya Nefti i Gaza* 11:35-38.

Mar'enko, Yu. I., 1978, *Oil and Gas-Bearing Carbonate Rocks,* Izd. Nedra, Moscow, 240p.

Per'kova, Ya. N., 1966, Influence of Dolomitization Process on Porosity of Carbonate Rocks in Section of Productive Horizons of Kuybyshev **Along-Volga Region,** *Geologiya Nefti i Gaza* 3:36-49.

Sarkisyan, S. G., M. A. Politykina, and G. V. Chilingarian, 1973, Effect of Postsedimentation Processes on Carbonate Reservoir Rocks in Volga-Urals Region, USSR, *Am. Assoc. Petroleum Geologists Bull.* 57:1305-1313.

Strakhov, N. M., 1956, *Toward Knowledge of Diagenesis. Problems of Mineralogy of Sedimentary Formations,* vols. 3 and 4, L'vov Government University.
Strakhov, N. M., and G. I. Bushinskiy, eds., 1956, Types of Dolomite Rocks and Their Genesis, *Akad. Nauk SSSR Izd. 4, Moscow,* 378p.
Teodorovich, G. I., 1958, *Study of Sedimentary Rocks,* Gostoptekhizdat, Leningrad, 572p.
Yablokov, V. S., 1970, Development of Lithologic Investigations in USSR During the Last Fifty Years (Organization, Cadres, Literature), in *Status and Problems of Soviet Lithology,* ed. A. V. Sidorenko, Repts. Plen. Meetings VIII All-Union Lithologic Conf. May 28, 1968–June 2, 1968, vol. 1, Izd. Nauka, Moscow, pp. 18–42.

33

Original article written expressly for this Benchmark volume

SIGNIFICANT CONTRIBUTIONS ON DOLOMITIZATION FROM THE GERMAN LITERATURE

H. Füchtbauer
Ruhr-Universität Bochum

Six articles have been selected that were published in the German literature and have contributed to our knowledge of dolomites. To begin with, Sander's classic and frequently cited work (1936) is a presentation of an early microscopic examination of calcitic and dolomitic loferites including discussions of the sedimentary and diagenetic environments, which, in fact, interfinger in these rocks. Two more general papers on dolomite follow, including an experimental investigation of equilibria by Usdowski (1967) and an X-ray diffraction study based on many samples of different formations from oil wells by Füchtbauer and Goldschmidt (1965). The next two papers by Lang (1964) and Bausch (1963) deal with Upper Jurassic dolomites in southern Germany. Lang did fieldwork and thin-section petrology in the Swabian Alb, and Bausch presented an example of descendant early dolomitization in the Franklin Alb. The last paper, by Richter (1974), presents a modern investigation of one of the classic dolomite areas in Germany, the Devonian of the Eifel Mountains. I am aware that this selection is rather subjective and that these papers cannot cover all subjects under dolomitization. By selecting these, however, I tried to cover at least some of the different aspects of dolomite formation.

Sander (1936) applied his method of quantifying and carefully describing microscopic and megascopic textures in the Triassic limestones and dolomites of the Alps including the Dachsteinkalk (now known as loferites), the Wettersteinkalk (-dolomit) and, in Part II, the loferites and the Main Dolomite of the southern Alps. He documented his findings by excellent photographs and emphasized external and internal sedimentation and precipitation of carbonates. The internal precipitation (cementation) in large voids, for example in fenestrae, is in many cases *belteropor*; in other words, the shape and/or orientation of the cement mosaic is governed by the shape of the voids and their *wegsamkeit* ("permeability").

Sander was the first to observe that internal dolomite sediment is more coarsely crystalline than the surrounding laminated dolomite (approximately 6 µm and < 3 µm, respectively; p. 63), without giving an explanation for such relationships. Many of his rock sections show "paradiagenetic folding" (wavelength about 2 cm), with separation of the laminae and infilling of the space created by internal sediment and cement (calcite or dolomite); such curled layers may give rise to microbreccias (p. 88). Other laminations (such as shown in his Fig. 47) would today be interpreted as stromatolitic algal mats; Sander explained them as surface phenomena (p. 89).

Rhythmicity was considered by Sander as a fundamental property of sediments caused by one or more controls with identical or different time periodicity (p. 40). Although Sander considered a tidal origin of the laminated units of the loferites (his "mm-Rhythmite," pp. 196, 198), he admitted lack of experience with such sediments and favored an interpretation as annual varves deposited at a depth of about 100 meters (p. 195). As for the massive *Megalodon* beds (his "m-Rhythmite"), he suggested agitated shallower water (*frischeres Wasser*) which would destroy the laminae, compared with the stagnant conditions (*unfrischeres Wasser*) of the laminated units (pp. 197, 199). Glacially induced eustatic fluctuations were believed to be responsible for these cycles.

The most characteristic phenomenon of the Wettersteinkalk (Middle Triassic) are the *Grossoolithe* (multiple-layer cement in cavities). Such cements commonly are composed of alternating layers of calcite and dolomite and may even be interrupted by laminae of internal sediment. These features were interpreted by Sander as cements formed in shallow-marine environments (p. 125). Sander distinguished between primary, in general cryptocrystalline dolomites, and secondary, metasomatic, microcrystalline or macrocrystalline dolomites, which are commonly partially dolomitized limestones.

Usdowski (1967), in his monograph on dolomite formation, dealt with equilibria in the system $CaCl_2$-$CaCO_3$-$CaSO_4$-$MgCl_2$-$MgCO_3$-$MgSO_4$ at temperatures of 50°C to 180°C. An English summary of this monograph has been published in *Recent Developments in Carbonate Sedimentology in Central Europe* (G. Müller and G. M. Friedman, eds., 1968, pp. 21-32, Springer-Verlag). Although these equilibria are not reached in nature during sedimentation and early diagenesis, they can be attained during late diagenesis. The interstitial solutions in sediments and sedimentary rocks cover a wide range of concentrations although solutions with 20 to 50 mole % Mg^{2+} and 50 to 80 mole % Ca^{2+} are most common, a relation that differs strongly from sea water (83 mole % Mg). The anion distribution is such that most analyses of interstitial water project near the corners for Cl_2^{2-}, CO_3^{2-}, and SO_4^{2-}. Replacement reactions between calcite, dolomite, Mg-calcite, Ca-dolomite, and anhydrite are also considered. Usdowski's diagrams permit determinations of the composition of the solutions that lead, under equilibrium conditions, to such replacements.

According to experiments, dolomitization is favored by increasing temperature and concentration of the solutions. This suggests that early diagenetic (that is, low-temperature) dolomitization occurs at high concentration, whereas lower concentrations are sufficient for late diagenetic (that is, higher temperature) dolomitization.

Dolomite is stable in solutions with 18 mole % Mg^{2+} and 82 mole % Ca^{2+}, above 80°C, whereas calcite is stable below 80°C under these same ionic compositions. Considering porewater composition and the magnesium contributed by Mg calcite, it was possible by a mass-balance calculation to explain generally the large quantities of late diagenetic dolomites, although chlorite contributes to the decrease of Mg/Ca in the interstitial water compared with seawater. The increasing Mg/Ca ratio in carbonate rocks with increasing geological age is explained by an increasing probability for dolomitization.

Füchtbauer and Goldschmidt (1965), in their X-ray investigation of dolomites, were concerned mainly with Ca excess and the degree of lattice order as indicated by the intensity ratio of (01.5) (at about $2\theta = 35.3°$, CuK_α) and (11.0) (at about $2\theta = 37.3°$), the first being present only in ordered dolomites (Goldsmith and Graf, 1958). This ratio--that is, the ordering degree--decreases with increasing Ca excess and with decreasing crystal size of certain early diagenetic Tertiary dolomites in Libya.

The latter occurrence is in a cyclic sequence of beds 0.5 to 3 m in thickness composed in their upper portion of dolomite containing dasyclads and miliolids and merging downward into limestone with normal-marine foraminifera and echinoderms. Within the dolomite layers; the crystal size increases downward from two to greater than twenty microns. This relationship is believed to indicate a tendency toward late diagenesis downward in each cycle. In the same direction, the Ca excess decreases and the lattice order increases.

These observations suggest that a higher Ca excess may develop under conditions of early diagenesis with its attendant high nucleation rate, whereas in slowly forming late diagenetic dolomites, under otherwise comparable conditions, the Ca excess is lower. Two more relationships of the Ca excess have been observed in dolomitic rocks. First, with increasing admixtures of calcite, the Ca excess also increases in the dolomite lattice. This is true for early but not for late diagenetic dolomites (see Füchtbauer, 1964, p. 514) as demonstrated for early diagenetic dolomites by Fritz (1966, p. 82) and Marschner (1966, Upper Triassic), and for late diagenetic dolomites by Richter (1974; Fig. 4 of this report). The reason is possibly a decreasing supply of Mg^{++} and, in late diagenetic dolomites, the tendency toward well-ordered stoichiometric crystals. Second, with decreasing "salinity" (Upper Permian, Germany; Paleocene, Libya; Upper Jurassic, Germany; Füchtbauer and Goldschmidt, 1965, Fig. 4) or decreasing aridity (Persian Gulf-Bonaire-Florida), the Ca excess of dolomite increases, presumably because of a decreasing Mg^{++}/Ca^{++} ratio.

Dolomites with a high Ca excess occur mainly in Tertiary to Recent dolomites, though $Ca_{.59}$ dolomites are reported from Upper Jurassic and $Ca_{.56}$ dolomites from Devonian rocks. In general, however, the dolomites adjust their lattice toward stoichiometric composition during diagenesis. This adjustment is enhanced in highly porous dolomites but is impeded in dense dolomites as demonstrated in Füchtbauer and Goldschmidt's Figs. 7 and 8. (Certain figures from this publication are used in Füchtbauer, 1974. Fig. 4 corresponds to Fig. 5-77; Figs. 7 and 8 are merged into Fig. 5-78.)

Lang (1964) investigated dolomitization and dedolomitization in the Upper Jurassic "Malm delta and epsilon" limestones southeast of Stuttgart, Germany. This sequence consists in part of limestones and in part of quiet-water bioherms of calcitized siliceous sponges overgrown by blue-green algae. Fragments of sponges and other skeletons encrusted by algae ("tuberoids") occur in biostrome-like bedded limestones. See Table 1 for compositional rock types.

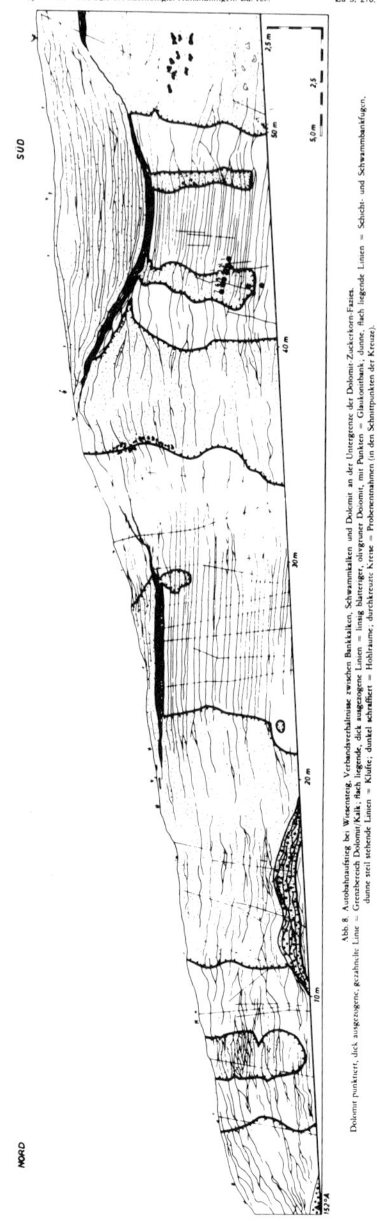

Figure 1. Outcrop along Autobahn Wiesensteig southeast of Stuttgart. Lower surface of dolomitization (dots = dolomite; dentate lines = dolomite-limestone boundaries; thick dots = glauconitic; heavy lines = foliated green dolomite; thin lines = bedding planes that are narrowly spaced in bedded limestones and widely spaced in bio- herms; vertical lines = partings; dark hatched = cavities) (from Lang, 1964, Fig. 8).

Table 1. Compositional Rock Types

	Wt. Percent MgCO$_3$	Crystal Size	Occurrence
Limestone	0.5-2.5*	(Cryptocrystalline)	Bedded limestones and bioherms
Dolomite	42+	0.05-0.5 mm	Bioherms
Dedolomite	0.05	Coarse-crystalline	Bioherms

*Traces of dolomite cannot be ruled out, though not detected in thin sections.
+In the dolomite lattice there is 2 to 3 percent excess CaCO$_3$.

In many cases the dolomitizing solutions migrated downward and were stopped or deflected by marly layers as shown in Fig. 1. Commonly, the dolomite is restricted to the central parts of the bioherms, which are overlain and underlain by limestone. The matrix is preferentially dolomitized in contrast to the skeletal particles, which leads to the conclusion of a relatively early prelithification dolomitization. Mg-calcitic skeletons as well as downward percolation of seawater were considered as sources of Mg^{2+}.

Sucrosic limestones, which occur in the areas of dolomitization and which are frequently connected with joints and karst fissures, were interpreted as dedolomites that formed under the influence of meteoric water (Fig. 2). This was corroborated by isotope analyses (Fritz, 1966, 1967).

The first stage is a disintegration of the dolomite rock into dolomite "sand," which then was easily accessible by solutions and became calcitized.

Bausch (1963) described an excellent example of descendant dolomitization, also in the Malm (Malm epsilon-zeta) but farther to the east, between Nürnberg and Regensburg, where dolomitization of bioherms is more common than in the area discussed by Lang. He noted that the upper limit of dolomitization above the bioherms starts at the upper surfaces of successive limestone beds about 2 m in thickness and decreases downward within these beds. These dolomitization cycles are similar to those described from Libya (Füchtbauer and Goldschmidt, 1965). Bausch's interpretation was that dolomitization occurred after deposition of each bed and before deposition of the overlying one. These dolomites are coarsely crystalline and barren of fossils, in contrast to the associated limestones. Dolomite pebbles embedded in limestone also point to an Upper Jurassic age of the dolomitization.

In my opinion, the dolomites Bausch described are of early diagenetic origin, which, during subsequent recrystallization, received properties, such as coarse crystallinity, of late diagenetic dolomites. **The general paleogeographic situation at the end of the Late Jurassic** supports the assumption of an environment favoring early diagenetic dolomitization: regression and shallowing of the sea combined with increasing clay content (*Zementmergel* in the western area), as well as lagoonal and even evaporitic deposits near Munich (Meyer, 1977).

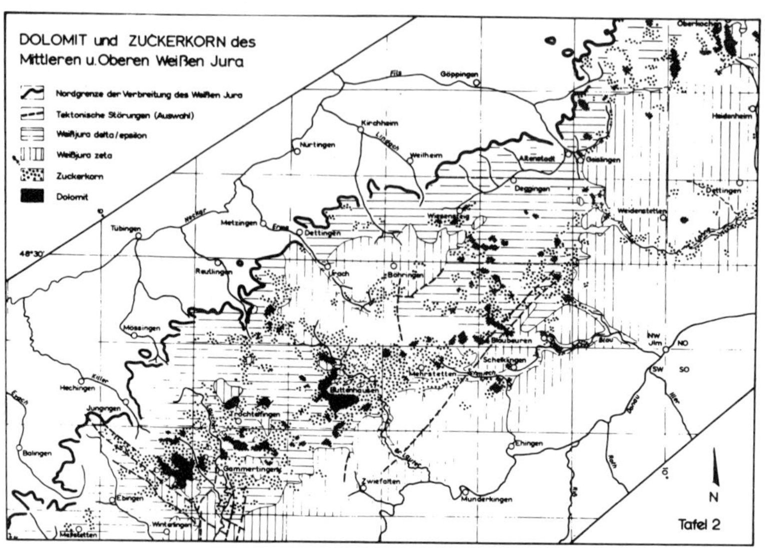

Figure 2. Map of the Upper Jurassic in the Swabian Alb area showing limestone, dolomite (dark, crosshatched), and calcitized dolomite (= "dedolomite," dotted) (from Fritz, 1967, Fig. 1).

The dolomitization of the bioherms (Lang, 1964; Fritz, 1967) was, according to Bausch and Hoefs (1972), late diagenetic. These authors found in subsurface samples oxygen isotope figures at variance with the figures by Fritz (1967): $\delta^{18}O$ = -4.5 in the dolomites and -3.3 in the massive limestones from which the dolomites formed (mean values). This difference clearly supports a dissolution-reprecipitation process instead of solid diffusion as a mechanism of dolomitization. The latter mechanism is also ruled out because of the requirement of too much time even for small crystals to be dolomitized (Brätter, Möller, and Rösick, 1972). Taking into account the theoretical increase of $\delta^{18}O$ during dolomitization by 5-7°/$_{\circ\circ}$ (Bausch and Hoefs, 1972), I believe that the observed decrease points to a Dorag-type model of dolomitization (infiltration of meteoric water into the bioherms).

Examples of ascendant dolomitization in the northern Rheinisches Schiefergebirge were investigated by Gotthardt (1962). Coarse-crystalline dolomite "dykes" associated with silicification and PbS cross the folds of the Devonian limestones. They occur also in the Lower Carboniferous and are now considered hydrothermal, as indicated by our recent work showing a quickly fading greenish cathodoluminescence of enclosed quartz crystals.

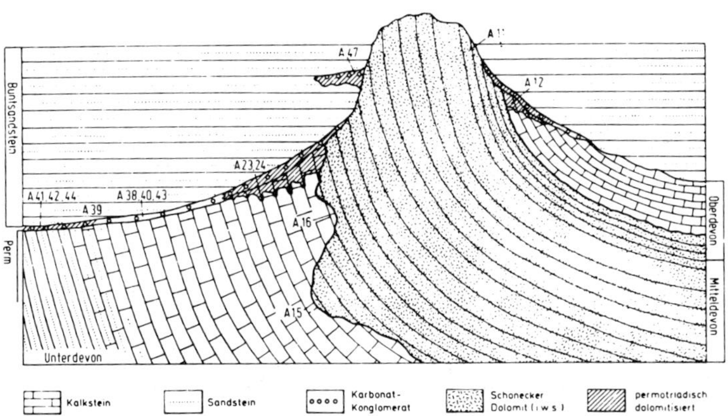

Figure 3. Dolomite types in the Eifel Mountains, idealized and exaggerated section. A 11, 12 = Devonian peritidal and biostromal dolomites, early to late diagenetic. A 23-44 = conglomerates, breccias, and underlying Devonian limestones, dolomitized in Permian time in a continental environment. A 47 = as in A 23-44 but dolomitization in Triassic time (from Richter, 1974, Fig. 47).

Richter (1974) reexamined the dolomites of the Eifel Mountains (Variscian fold belt), which have been investigated numerous times since Leopold von Buch (1824), who interpreted them to be related to volcanic processes. Twenty different interpretations were suggested by twenty-six authors. On the basis of fieldwork, thin section study, and X-ray and chemical analyses, Richter was able to characterize two main groups of dolomite occurrences (Figs. 3, 4). First are stratiform peritidal, nearly stoichiometric ($Ca_{.48}$ - $Ca_{.51}$) early diagenetic dolomites in the Middle Devonian and near the Middle-Upper Devonian boundary. These rocks are laminated, microcrystalline, and overlie biostromes dolomitized by downward percolation during late diagenesis as schematically shown in Fig. 4 ($Ca_{.48}$ - $Ca_{.51}$ passing downward into $Ca_{.50}$ - $Ca_{.54}$, both macrocrystalline). Ascendant dolomitization by compaction occurred in a 0.5 to 5 m thick layer of limestone deposited on top of the peritidal dolomites, in deeper water ($Ca_{.51}$ - $Ca_{.56}$). The second group represents continental dolomitization (**$Ca_{.50}$ - $Ca_{.57}$**) of Permian conglomerates composed of Devonian limestone fragments, and of the underlying Devonian limestone themselves, as well as of calcite cement in Lower Triassic sandstones. This was interpreted as dolomitization under arid conditions, in depressions near Devonian dolomite hills, though Dr. Richter would now consider the Dorag model as a more appropriate explanation (personal communication, 1980).

409

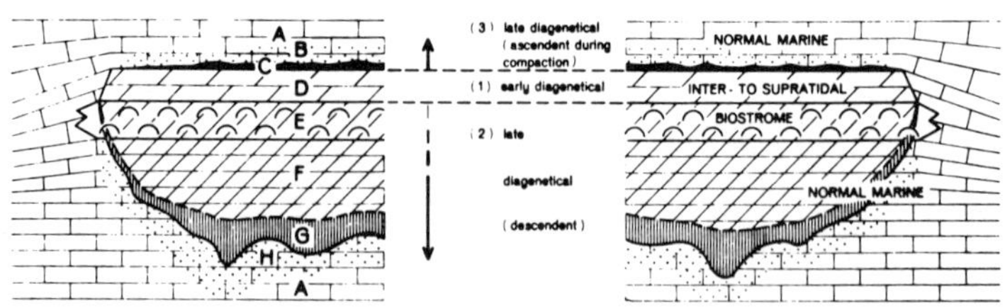

	Carbonates		Within dolomitized skeletons : structures + = preserved − = not preserved	Crystal size dolomite ∗	FeO weight - % mean			Sr²⁺ ppm mean			Inclusions in authigenic quartz
					min		max	min		max	
A	calcite							60	**200**	600	calcite
B	Ca₅₁.₅₆-dolomite	+ calcite	−		0,1	**0,5**	1,2	25	**140**	300	calcite
C		no calcite	(+)								calcite
D+E	Ca₄₈.₅₁-dolomite		+		0,01	**0,15**	0,6	0	**15**	40	dolomite / dolomite + calcite
F	Ca₄₈.₅₁-dolomite		+ / −								dolomite + calcite / calcite
G	Ca₅₀.₅₄-dolomite	no calcite	(+) / −		0,01	**0,4**	1,9	15	**30**	45	calcite
H		+ calcite	−		0,01	**0,3**	0,9	15	**85**	230	calcite
A	calcite							60	**200**	600	calcite

∗ = crystal size depends additionally on clay content
and on secondary crystal enlargement

Figure 4. Devonian peritidal and biostromal dolomites in the Eifel Mountains (from Richter, 1974, Fig. 48).

Richter (1974) includes the following general observations:
1. During late diagenesis, marly layers were preferentially dolomitized.
2. Dolomite crystal size decreases with increasing insoluble residue (clay and silt): 1 percent residue, 0.6 mm (mainly skeletons); 2 percent, 0.13 mm; 7 percent, 0.04 mm. The same relationship was reported from limestones by Bausch (1968) and interpreted as resulting from clay particles hampering the crystal enlargement.
3. In general, the crystal size is larger in late than in early diagenetic dolomite; such relationships are noted by numerous others workers.
4. Matrix is preferentially dolomitized compared with skeletons and, to lesser degrees, cement.
5. Fibrous cement A is frequently dolomitized prior to isopachous cement B mosaics, possibly because of an originally higher Mg content of cement A.
6. The occurrence of Mg dolomite (Ca.₄₈ - Ca.₅₀) in the early diagenetic dolomites and in the rocks immediately below is remarkable.

7. Zoned dolomite crystals occur above and below the early diagenetic dolomite. Their central parts grew faster and are richer in inclusions and higher in Ca and Fe than the clear rims.
8. X-ray peak width at quarter height and asymmetry helped considerably in the identification of the composition of zoned dolomites.
9. Sr, Fe, and Mn contents are higher in the late than in the early diagenetic dolomites.
10. Spotty isolated dolomite crystals syntaxially replacing echinoderm fragments in optical continuity testify to a closed-system dolomitization using the skeletal magnesian calcite as an Mg source (Richter's Fig. 12).
11. The dedolomites are higher in Mg than the primary limestones (Richter's Fig. 39).
12. Authigenic quartz and potassium feldspar were formed after the early and before the late diagenetic dolomitization of Devonian limestones. This interpretation is based on the following observations: quartz and feldspar crystals in early diagenetic dolomites contain dolomite inclusions and therefore were formed after the dolomitization. On the other hand, such crystals in late diagenetic dolomite contain inclusions of calcite containing 1 to 4 mole % $MgCO_3$ and therefore were formed before late dolomitization. Moreover, this calcite represents an earlier stage of calcite diagenesis compared with that represented by the present composition of adjacent Devonian limestones, (that is, 0 to 2 mole % $MgCO_3$).
13. Fast-growing diagenetic (and hydrothermal) quartz crystals are strongly elongated. They are rich in inclusions, whereas their diagenetic, slower-growing overgrowths are less elongate and are poor in inclusions.

REFERENCES

Bausch, W. M., 1963, Geologisches Erscheinungsbild eines Dolomitisierungsprozesses, *Geol. Blätt. No-Bayern Erlangen* 13:89-92.

Bausch, W. M., 1968, Clay Content and Calcite Crystal Size of Limestones, *Sedimentology* 10:71-75.

Bausch, W. M., & J. Hoets, 1972, Die Isotopenzusammensetzung von Dolomiten und Kalken aus dem süddeutschen Malm., *Contr. Mineralogy and Petrology* 37:121-130.

Brätter, P., P. Möller, & U. Rösick, 1972, On the Equilibrium of Coexisting Sedimentary Carbonates, *Earth and Planetary Sci. Letters* 14:50-54.

Buch, L. von, 1824, Über das Vorkommen des Dolomits in der Nähe der vulkanischen Gebilde der Eifel, *Nöggerath, Das Gebirge in Rheinland-Westfalen nach Mineralogischem und Chemischem Bezuge* 3:280-283.

Fritz, P., 1966, *Zur Genese von Dolomit und zuckerkörnigem Kalk im Weissen Jura der Schwäbischen Alb (Württemberg): Mikroskopische Untersuchungen und Isotopenanalysen,* Diss., Univ. Stuttgart, 104p.

Fritz, P., 1967, Oxygen and Carbon Isotopic Composition of Carbonates from the Jura of Southern Germany, *Canadian Jour. Earth Sci.* 4:1247-1267.

Füchtbauer, H., 1964, Fazies, Porosität und Gasinhalt der Karbonatgesteine des norddeutschen Zechsteins, *Deutsch. Geol. Gesell. Zeitschr.* 114:484-531.

411

Füchtbauer, H., 1974, *Sediments and Sedimentary Rocks I,* Halsted Press Div., John Wiley, New York, 464p.

Füchtbauer, H., & H. Goldschmidt, 1965, Beziehungen zwischen Calciumgehalt und Bildungsbedingungen der Dolomite, *Geol. Rundschau* 55:29-40.

Goldsmith, J. R., and D. L. Graf, 1958, Structural and Compositional Variations in Some Natural Dolomites, *Jour. Geology* 66:678-693.

Gotthardt, R., 1962, *Geologie des Dornaper Massenkalkes,* Diss., T. H. Aachen, 107p.

Lang, H. B., 1964, Dolomit und zuckerkörniger Kalk im Weissen Jura der mittleren Schwäbischen Alb (Württemberg), *Neues Jahrb. Geologie u. Paläontologie Abh.* 120:253-299.

Marschner, H., 1966, *Mineralogisch-petrographische Untersuchungen an karbonatreichen Gesteinen aus dem Unteren Keuper des Weserberglandes,* Diss., Univ. Hamburg, 137p.

Meyer, R. K. F., 1977, Stratigraphie und Fazies des Frankendolomits und der Massenkalke (Malm), *Erlanger Geol. Abh.* 104:1-40.

Richter, D. K., 1974, *Entstehung und Diagenese der devonischen und permotriassischen Dolomite in der Eifel: Contributions to Sedimentology II,* E. Schweizerbart'sche Verlagsbuchhandlung, Stuttgart, 101p.

Sander, B., 1936, Beiträge zur Kenntnis der Anlagerungsgefüge (Rhythmische Kalke und Dolomite aus der Trias), *Tschermaks Mineralog. u. Petrog. Mitt.* 48:27-139, 141-209.

Usdowski, H.-E., 1967, Die Genese von Dolomit in Sedimenten, *Mineralogie und Petrographie in Einzeldarstellungen 4,* Springer-Verlag, Berlin, 95p.

AUTHOR CITATION INDEX

Adams, J. E., 124, 164, 179, 204, 225, 324, 373
Agnew, A. F., 243, 245
Agueda Villar, J. A., 287
Aitken, J. D., 365
Alderman, A. R., 104, 124, 204, 225, 243, 298, 330, 337, 365, 373
Alekin, O. A., 400
Alexandersson, T., 179
Allen, H. A., 245
Alway, F. J., 255
Amsbury, D. L., 309, 373
Anglada, R., 389
Arago, D. F., 389
Archiac, A. d', 389
Arrhenius, G., 373, 374
Arrington, F., 287
Asquith, G. B., 204, 243, 309, 365
Atwood, D. K., 7, 129, 204, 210, 225

Baas-Becking, L. G. M., 138
Back, W., 179, 225, 243, 244, 254, 259, 298
Badiozamani, K., 243, 254, 259
Baer, O., 114
Bagrintseva, K. I., 400, 401
Bajor, M., 104
Ball, M. M., 152, 378
Banks, N. G., 298
Barnes, I., 225, 298, 309
Barnes, V. E., 7, 16, 105, 124, 183, 194, 244, 324, 355
Baron, G. A., 124, 389, 392
Barrell, J., 245
Barth, T. F. W., 373
Bates, R. L., 266
Bathurst, R. G. C., 16, 129
Baud, A., 389
Bausch, W. M., 411
Bays, C. A., 244
Beales, F. W., 114, 266
Beaumont, E. de, 7, 44, 389, 390

Beck, K. C., 226
Behre, C. H., Jr., 243, 245
Behrens, E. W., 16, 183, 204, 210, 244, 254
Bein, S. J., 184, 194, 366
Beloborodova, G. V., 401
Benson, R. H., 337
Bergenback, R. E., 374
Berger, R. W., 245, 374
Berner, R. A., 7, 16, 17, 155, 179, 225, 226, 244, 254, 311, 365
Berry, W. B. N., 7, 259
Bertrand-Geslin, 44
Bevan, A. C., 244
Bien, G. S., 17, 226, 311
Bird, K. J., 7
Bischoff, J. L., 225, 244, 254
Bissell, H. J., 7, 16, 104, 124, 156, 266, 401
Bisque, R. E., 354
Black, A. P., 225
Black, M., 124, 324, 365
Blackburn, G., 330
Blackwelder, E., 7, 44, 124
Blatt, H., 254
Blix, R., 115
Bluck, B. J., 374
Bonatti, E., 374
Bonham, L. C., 349
Bonython, C. W., 138
Boucot, A. J., 7, 259
Boué, A., 7, 44, 390
Bouroullec, J., 390
Bourrouilh-Le Jan, F. G., 254, 390
Bowen, V. T., 255
Boyd, D. W., 69
Bradley, J. S., 156
Bradley, W. F., 83
Bradley, W. H., 309, 337
Bramlette, M. N., 106
Brandon, A., 9, 184, 205, 287
Brätter, P., 411
Braun, M., 183, 204

Bray, E. E., 106
Bredehoeft, J. D., 354
Brinkley, S. R., Jr., 125
Brown, C. W., 124, 266
Brown, E., 225
Bubb, J. N., 7, 129, 204, 210, 225
Buch, L. von, 43, 390, 411
Buchsbaum, R., 105, 244, 349
Bundy, W. M., 165
Burst, J. F., 83, 194, 324
Burt, D. M., 299
Bushinskiy, G. I., 402
Butler, G. P., 129, 130

Cailleau, P., 390
Calvert, W. L., 354
Calvin, S., 47, 244
Canal, P., 390
Carozzi, A. V., 7, 390
Carpenter, A. B., 7
Carroll, D., 355
Carver, R. E., 225
Cayeux, L., 114, 390
Cebulski, D. E., 325
Charpal, O. L. de, 392
Chave, K. E., 83, 105, 114, 225, 244, 324, 355
Chazen, P. O., 16
Cheetham, A. H., 225
Chen, C. S., 225
Chenoweth, P. A., 324
Cherry, R. N., 179
Chilingar, G. V., 7, 16, 69, 104, 105, 114, 124, 156, 266, 324, 355, 401
Choquette, P. W., 246, 266, 309, 311
Cifelli, R., 255
Clarke, F. W., 46, 69, 165, 337
Claypool, G. E., 246
Clayton, R. N., 105, 225, 244, 245, 266, 299
Cloud, P. E., Jr., 7, 16, 105, 124, 194, 244, 324, 355
Cobb, M., 330
Cockbain, A. E., 205
Cohee, G. V., 244
Collegno, P. de, 44, 210
Conley, R. F., 165
Cooper, B. N., 114
Coquand, H., 44, 390, 391
Cordier, P. L. A., 391
Correns, C. W., 373
Cotecchia, V., 254
Cotter, E., 179
Coudray, J., 391
Craemer, M., 267, 311
Craig, H., 105, 114, 244, 349
Crawford, J., 324

Crickmay, G. W., 16
Cros, P., 391
Culkin, F., 298
Cullis, C. G., 52
Curtis, R., 225

Dagallier, G., 391
Dalrymple, D. W., 365
Daly, R. A., 7, 16, 44, 69, 105, 114, 124
Damour, A., 391
Dana, J. D., 44, 298
Daubrée, A., 391
Davies, G. R., 204, 324
Davies, P. J., 309
Davis, R. A., Jr., 325
Davydov, Yu. V., 401
Deffeyes, K. S., 8, 105, 130, 152, 156, 179, 204, 225, 244, 254, 299, 309, 365, 374, 378
Degens, E. T., 105, 106, 244
De Groot, K., 179
Deike, R. G., 244, 254, 259
Deininger, R. W., 244
Delaney, A. C., 374
Delfaud, J., 390
Dellwig, L. F., 165
Delmas, M. R., 391
Delépine, G., 391
Delesse, A., 391
Deloffre, R., 390
Desforges, G., 391
De Sitter, L. U., 254
Dickey, P. A., 254
Dixon, E. E. L., 47
Dixon, W. J., 374
Dolomieu, D. de, 8, 391
Donaldson, A. C., 194, 325
Donaldson, J. A., 365
Dott, R. H., Sr., 16
Dragone, D., 390
Dreyer, R. M., 299
Dunbar, C. O., 266, 355
Dunham, J. B., 10, 17, 259, 267, 311
Dunham, R. J., 287
Durocher, J., 43, 391
Dwornik, E. J., 311

Eardley, A. J., 105, 124, 225, 309, 337
Ehrlich, R., 16
Emery, K. O., 106, 254, 325, 374
Engel, A. E. J., 105, 244, 266, 299
Entsov, I. I., 401
Epstein, S., 105, 115, 165, 210, 244, 245, 254, 266, 299, 349
Eskola, P., 373

España Instituto Geológico y Minero, 287
Ethington, R. L., 10, 17
Eugster, H. P., 309, 337
Evans, G., 194, 225

Fadeyev, F. I., 204
Fairbridge, R. W., 105, 124, 165, 204, 355, 374
Faust, G. T., 124
Favre, A., 43
Favre, J. H., 389, 392
Fay, R. C., 106
Fayard, M., 392
Feely, H. W., 349
Ferguson, J., 309
Fischbeck, R., 254
Fischer, A. G., 156, 179, 184, 194, 287, 365
Fischer, F. T., 327
Fisher, W. L., 156, 184, 225
Fleming, R. H., 138
Folk, R. L., 204, 225, 254, 259, 287, 325, 374
Fondeur, C., 391
Fontes, J. C., 244, 391
Forchhammer, G., 44
Förstner, U., 255, 310
Fournet, J., 8, 44, 391
Frapolli, 43
Freeman, T., 8, 204, 266
Friedman, G. M., 8, 16, 156, 183, 184, 204, 225, 244, 254, 266, 287, 310, 366, 378
Friedman, I., 244, 267, 299, 311
Fritz, P., 244, 266, 391, 411
Froget, C., 389, 391
Frolova, E. K., 401
Füchtbauer, H., 411, 412
Fujimoto, C. K., 255
Fyfe, W. S., 225, 244, 254

Garman, R. K., 130, 210, 225, 244, 299, 310
Garreau, B., 392
Garrels, R. M., 105, 244, 299
Garret, A. A., 355
Garrett, P., 204
Garrison, R. E., 287
Gastner, M., 245
Gebelein, C. D., 204, 325, 366
Gèze, B., 392
Ginsburg, R. N., 9, 130, 180, 184, 194, 205, 225, 226, 245, 255, 287, 325, 327, 366, 374, 379
Girardin, J., 392
Girou, A., 390
Glass, H. D., 355
Gmid, L. P., 401
Goldberg, E. D., 244

Goldschmidt, H., 412
Goldsmith, J. R., 83, 140, 152, 204, 225, 244, 245, 254, 337, 349, 374, 412
Goode, H. D., 299
Goodell, H. G., 130, 210, 225, 244, 299, 310
Goreau, T. F., 287
Gotthardt, R., 412
Gottis, M., 391
Grabau, A. W., 44
Grady, J. R., 106
Graf, D. L., 83, 105, 124, 140, 152, 204, 225, 244, 310, 337, 349, 374, 412
Grandjean, B.-V., 44
Green, W. L., 44
Greenfield, L. J., 366
Grim, R. E., 355
Gross, M. G., 225, 244
Gubler, Y. G., 392
Gutstadt, A. M., 204

Hadding, A., 194
Haimovitz, A., 287
Hall, C. W., 44
Hall, W. E., 244
Halla, F., 83, 114
Hallam, A., 287
Halliday, W. R., 254
Hanshaw, B. B., 179, 225, 243, 244, 254, 259
Harbaugh, J. W., 374
Hardie, L. A., 366
Hardman, E. T., 44, 47
Hardy, J. L., 266
Harris, L. D., 325
Harris, W. H., 225
Harrison, J. L., 355
Harriss, R. C., 225, 244
Hatch, F. H., 124
Hatfield, C. B., 204
Healy, H. G., 225
Heard, H. C., 225, 244
Helgeson, H. C., 244
Hem, J. D., 226, 299, 337
Herbert, P., Jr., 325
Hermite, H., 287
Hershey, O. H., 244
Hewett, D. F., 266, 299
Heyl, A. V., Jr., 243, 245
Hickox, J. E., 156
Hite, R. J., 325
Hobbs, C. R. B., Jr., 355
Hoefs, J., 411
Hoffman, P., 366
Hofmann, H. J., 355
Hoffmeister, J. E., 378
Högbom, A. G., 44

Holland, C. H., 325
Holland, H. D., 8, 225, 245, 254, 299
Hollister, J. S., 287
Hood, D. W., 254
Horne, J., 49
Horr, C. A., 245
Hossfeld, P. S., 138
Howe, M. A., 337
Howe, W. B., 366
Hower, J., 355
Hsü, K. J., 8, 17, 130, 156, 179, 204, 210,
    225, 245, 254, 299, 330, 337
Hubbert, M. K., 179
Hudson, J. D., 204
Huebner, J. S., 225, 254, 299
Hülsemann, J., 106
Humbert, L., 390
Hunt, T. S., 44, 114
Hutton, J. T., 330
Hydrographic Office Pub., 299

Illing, L. V., 8, 124, 130, 152, 165, 179, 180,
    194, 204, 225, 245, 266, 299, 310,
    325, 366, 374, 378
Imbrie, J., 194, 325, 327
Inden, R. F., 310
Ingerson, E., 83, 105, 114, 204, 337
Irion, G., 255, 310, 366
Irwin, M. L., 310, 325
Isham, L. B., 184, 194, 366
Ivanov, A. M., 401

Jackson, A. W., 44
Jackson, J. A., 266
Jacquin, C., 390
Jaeger, J. C., 299
Jamieson, J. C., 165, 245
Javoy, M., 392
Jeffrey, L. M., 355
Jodry, R. L., 266
Joensuu, O. I., 83
Johannsen, A., 337
Johns, W. D., 355
Johnson, J. G., 259
Johnson, M. W., 138
Jones, B. F., 225, 226, 244, 310, 337
Jordan, C. F., 7
Jourdy, E., 392
Judd, J. W., 44

Kahle, C. F., 355, 366
Kanehiro, Y., 255
Kataeva, V. N., 401
Kay, M., 245
Kazanskiy, Yu. P., 401
Keith, M. L., 245

Kendall, C. G., 325, 366
Kendall, G. W., 259
Kent, P. E., 287
Kepper, J. C., 366
Khodak, Yu. A., 124
King, R. H., 156, 165, 325
Kinsman, D. J. J., 130, 179, 204, 225, 245,
    254, 299
Kirsipu, T. V., 225, 254, 299
Kirwan, R., 8
Kitano, Y., 254
Klähn, H., 254
Klement, C., 114, 392
Knetsch, G., 105
Koehler, E., 114
Koehn, H., 310
Koenig, K. J., 337
Kohout, F. A., 226, 254, 259
Komar, V. A., 366
Kopp, O. C., 310
Korzhinskii, D. S., 226
Kramer, J. R., 124
Krauskopf, K., 299
Krey, F., 69
Krynine, P. D., 124, 374
Kuenen, P. H., 69, 374
Kulbicki, G., 391
Kulp, J. L., 69, 349
Kuperberg, J., 184, 194, 366

Ladd, H. S., 106, 254
Lagny, P., 391, 392
Lalou, C., 124, 392
Lamar, J. E., 69, 184, 327
Land, L. S., 8, 16, 17, 183, 204, 210, 226,
    244, 245, 254, 259, 287, 299, 310
Landes, K. K., 8, 245
Lang, H. B., 412
Lang, W. B., 165, 325
Langmuir, D., 156, 179, 226, 254
Laporte, L. F., 204, 287, 325, 366, 378
Lawrence, J. C., 337
Lecompte, M., 392
Leitmeier, H., 254
Lemberg, J., 19
Lemish, J., 354
Lemoine, Mme. P., 392
Lerman, A., 226
Lesley, J. P., 44, 374
Létolle, R., 391
Levandowski, D. W., 349
Liebermann, O., 17, 130
Linck, G., 44, 114
Lindgren, W., 8, 44
Lindholm, R. C., 204, 374
Lippmann, F., 204, 310

Livingstone, D. A., 337
Lloyd, R. M., 180, 205, 226, 245, 255, 287, 327, 366, 374, 379
Lochman-Balk, C., 325
Lock, D., 131, 310, 311
Logan, B. W., 194, 205, 325, 337, 366
Long, M. B., 194
Longchambon, M., 392
Lovering, T. S., 205, 255, 299
Lowenstam, H. A., 105, 115, 165, 244, 245, 327, 349
Lucas, G., 392
Lucia, F. J., 8, 130, 152, 156, 179, 204, 225, 244, 254, 299, 311, 338, 365, 374, 378
Lyons, E. J., 243, 245

McCarthy, J. H., 299
McCrea, J. M., 105, 245
MacDonald, C., 287
McDonald, G., 165
MacGinitie, H. D., 337
McGrew, P. O., 337
McIntire, W. L., 245
McKee, E. H., 259
Mackenzie, F. T., 244, 255
McKenzie, J. A., 130, 156
McKinney, C. R., 115, 245
Mägdefrau, K., 114
Maiklem, W. R., 255
Maksimova, G. A., 401
Mandrikova, N. T., 401
Manheim, F. T., 259
Mannon, R. W., 401
Mansfield, C. F., 310
Mar'enko, Yu. I., 401
Martil Pérez, R., 287
Marignac, Ch. de, 392
Marschner, H., 412
Martin, E. L., 309
Massey, F. J., 374
Master, I. M., 17
Matter, A., 205, 378
Mattes, B. W., 8, 267, 310
Matthews, R. K., 225, 245
Matti, J. C., 259
Mawson, D., 130, 138
Mazzullo, S. J., 184
Meigen, W., 18
Melson, W. G., 255
Mennig, J. J., 125
Mensink, H., 105
Meyer, R., 179
Meyer, R. K. F., 412
Michard, A., 392
Middleton, G., 194, 254
Miller, D. N., 105, 267

Miller, H. W., 327
Miller, R. H., 259
Milliman, J. D., 245
Minckley, W. L., 338
Moitessier, M. A., 392
Möller, P., 411
Montadert, L., 392
Monty, C., 327, 366
Moore, C. H., 254
Moore, G. W., 255
Morey, G. W., 83
Morlot, A. de, 8, 44, 115, 392
Morris, H. T., 299
Morrow, D. W., 8, 210, 259
Mossler, J. H., 184, 205, 255, 310
Mottl, M. J., 8
Mountjoy, E. W., 8, 267, 310
Muir, M., 310
Müller, G., 156, 179, 205, 254, 255, 310, 338, 366
Muller, J., 245
Multer, H. G., 378
Murata, K. J., 267, 311
Murray, A. N., 9
Murray, H. H., 355
Murray, J. W., 287
Murray, R. C., 9, 17, 156, 179, 245, 254, 311, 338, 355, 374

Nakai, N., 349
Narkiewicz, M., 267
Neher, J., 125
Nesbitt, H. W., 338
Neuman, R. B., 194
Neumann, A. C., 366
Newell, N. D., 125, 156, 165, 226, 327
Nier, A. O., 245
Nolan, T. B., 125
Norrish, K., 138

Ochsenius, K., 327
Oder, C. R. L., 327
O'Driscoll, E. P. D., 330
Odum, H. T., 115
Oliver, W. A., 374
Olson, E. R., 259
O'Neil, J. R., 309
Oppenheimer, C. H., 17
Osipova, A. I., 349
Ostrom, M. E., 245, 327, 355
Owen, B. B., 125
Owen, D. D., 245
Oxburgh, U. M., 225, 254, 299

Page, R., 325
Papike, J. J., 311

Parry, W. T., 226
Parsons, L. M., 9, 52
Patterson, R. J., 130
Patton, J. B., 69
Peach, B., 49
Pearson, E. F., 205
Pèlissonnier, H., 392
Perconig, E., 287
Per'kova, Ya. N., 401
Perry, W. J., Jr., 327
Perthuisot, J.-P., 392
Peterson, M. N. A., 17, 226, 311, 355
Pettijohn, F. J., 194
Pfaff, F. W., 44, 115
Phillipi, E., 44
Pilkey, O. H., 225, 226, 244
Piper, A. M., 355
Pittman, J. S., 287
Playford, P. E., 205
Pledge, N. S., 338
Politykina, M. A., 401
Poole, F. G., 259
Potter, E. C., 259
Potter, P. E., 194
Powers, M. C., 355
Pray, L. C., 17, 266, 311
Prouty, C. E., 245, 327
Pryor, W. A., 355
Purdy, E. G., 194, 325, 327
Purser, B. H., 393
Pytkowicz, R. M., 255

Querol, R., 287
Quirk, J. P., 138

Raasch, G. O., 244
Rankama, K., 115
Rastall, R. H., 124
Ratynskii, V. M., 17
Récy, J., 389
Reeves, C. C., Jr., 226
Reineck, H.-E., 194
Répellin, P., 393
Reuling, H. T., 17, 184
Reuter, H., 105, 106
Reuter, J. H., 106
Rezak, R., 194, 205, 325
Rhodes, M. L., 124, 179, 204, 225, 324, 373
Riba, O. A., 287
Rich, J. L., 165
Richardson, S. H., 106
Richter, D. K., 412
Ricketts, J. E., 327
Ricour, J., 393
Riedel, W. R., 106

Rieke, H. H., 401
Rigby, J. K., 125, 156, 327
Riley, J. P., 299
Ristvet, B. L., 255
Rittenberg, S. C., 106
Ritter, F., 83
Rivière, A., 17, 115, 393
Roberts, H. H., 254
Rodda, P. U., 156, 184, 225
Roddy, H. J., 338
Rodgers, J., 125, 266, 327, 355
Roedder, E., 267, 299
Roehl, P. O., 9, 179, 184
Rogers, L. E. R., 138
Rohrbacher, T. J., 204
Rohrer, E., 125
Ronov, A. B., 17, 355
Roques, H., 390
Rosenburg, P. E., 299
Rösick, U., 411
Ross, R. J., 259
Rouge, P. E., 392
Rouire, J., 391
Rubin, M., 106, 130, 131, 225, 226, 244,
    299, 330
Rubinson, M., 225
Rucker, J. B., 225
Runnells, D. D., 9, 210, 226, 245
Rutte, E., 106
Rutten, M. G., 287

Sabins, F. F., Jr., 106, 311, 374
Sainte-Claire Deville, Ch., 393
Salem, M. R., 8, 210, 259
Sánchez de la Torre, L., 287
Sander, B., 9, 106, 125, 287, 366, 379, 412
Sanders, J. E., 8, 16, 156, 204, 244, 254, 266,
    287, 310, 366, 378
Sando, W. J., 194, 366
Sardeson, F. W., 44
Sarin, D. D., 194, 366, 379
Sarkisyan, S. G., 401
Saunders, D. F., 125
Saussure, N. T. de, 9, 393
Saxby, D. B., 184, 327
Scheerer, T., 44
Scheidig, A., 374
Schenck, P. E., 9, 184, 205, 366
Schlanger, S. O., 17, 179, 311, 366
Schloemer, H., 83
Schmidt, A., 48, 299
Schmidt, R. A., 287
Schmidt, V., 205, 311, 355
Schneider, J. F., 130, 156, 330
Schneider, R., 226

Scholle, P. A., 184, 255
Schott, W., 374
Shrock, R. R., 9
Schroeder, J. H., 311
Schuchert, C., 245
Schultz, F., 255
Scoffin, T. P., 366
Scruton, P. C., 165, 327
Seibold, E., 106
Sevier, R., 299
Sharma, T., 245
Shaw, A. B., 311, 327
Shearman, D. J., 225, 287
Shell Development Company, 141
Sherman, G. D., 255
Shimp, N. F., 105, 124, 225
Shinn, E. A., 9, 130, 152, 180, 194, 205, 226, 245, 255, 287, 327, 366, 374, 378, 379
Shrock, R. R., 194
Shugurova, N. A., 401
Sibley, D. F., 311
Siebenthal, C. E., 299
Siedlecka, A., 254, 255
Siegel, F. R., 9, 125, 355
Siegenthaler, C., 204, 254, 337
Siever, R., 226, 244, 355
Silverman, S. R., 349
Simancas, R., 287
Sinclair, W. J., 338
Skeats, E. W., 9, 44, 47, 55, 125
Skinner, B. J., 106, 130, 131, 226, 299, 330
Skinner, H. C. W., 104, 106, 124, 130, 204, 205, 225, 226, 243, 299, 330, 337, 338, 365, 373, 374
Skipwith, P. A., 325, 366
Skirrow, G., 299
Skougstad, M. W., 245
Sledz, J. J., 310
Sloss, L. L., 226, 245
Smith, C. L., 327
Smith, M., 9, 184, 205, 287
Smoot, T. W., 355
Sonnenfeld, P., 9, 17, 205, 245, 311
Sorby, H. C., 19, 44, 255
Spörli, B., 267
Sprigg, R. C., 138
Stagg, A. K., 327
Stehli, F. G., 355
Steidtmann, E., 9, 17, 44, 69, 125, 210, 245
Steinen, R. P., 309
Stevenson, R. E., 325
Stewart, F. H., 327
Stewart, J. H., 259
Stockman, K. W., 152, 378
Stout, W., 355

Strakhov, N. M., 17, 69, 115, 125, 402
Stringfield, V. T., 226
Suess, Ed., 46, 393
Sugden, W., 374
Sullivan, R., 337
Surdam, R. C., 309, 337
Sureau, J. F., 393
Sverdrup, H. U., 138
Swartz, F. M., 327

Taft, W. H., 106, 287, 311, 374
Taylor, J. C. M., 8, 130, 152, 179, 180, 194, 204, 225, 245, 299, 325, 366, 374, 378
Teis, R. V., 115
Templeton, J. S., 245
Tennant, C. B., 245, 374
Teodorovich, G. I., 125, 156, 402
Terriere, R. T., 374
Terry, R. D., 69
Textoris, D. A., 9, 184, 205
Thiel, G. A., 255
Thode, H. G., 349
Thomas, H. H., 48
Thompson, A. M., 9, 205
Thompson, G., 255
Thompson, J. B., Jr., 226
Thompson, M. E., 105, 244, 299
Thorstenson, R. C., 255
Thrailkill, J., 255
Thurber, D. L., 327
Tietz, G., 156, 179, 205
Tikhomirov, V. V., 125
Toland, W. G., 349
Tracey, J. I., Jr., 106, 254
Trichet, J., 393
Trusheim, F., 106
Turekian, K., 69
Twenhofel, W. H., 125

Ubisch, H. von, 115
Udluft, H., 115
U.S. Geological Survey, 338
Urey, H. C., 105, 115, 244, 245, 349
Usdowski, H.-E., 156, 180, 255, 267, 412

Vacher, H. L., 245
Van Hise, C. R., 44, 115
Van Straaten, L. M. J. U., 194
Van Tuyl, F. M., 9, 17, 33, 47, 48, 125, 245, 299, 379
Vatan, A., 9, 125, 391
Vernon, R. O., 226, 245
Vinogradov, A. P., 17, 69, 125
Virgili, C., 287

Virlet, T., 393
von der Borch, C. C., 17, 104, 124, 131, 225,
        226, 298, 299, 310, 311, 330, 366, 373

Walch, C. A., 259
Wallouch, R., 349
Walters, L. J., Jr., 246
Walther, J., 44
Wanless, H. R., 267
Wanless, R. K., 349
Waring, G. A., 226, 299
Weaver, C. E., 355
Weber, J. N., 245
Wedow, H., Jr., 327
Weeks, L. G., 69
Weigelin, M., 46
Weiss, M. P., 338
Weller, S., 9, 44
Wells, A. J., 8, 9, 106, 130, 131, 140, 152,
        179, 194, 204, 225, 226, 245, 299,
        325, 366, 374, 378
Welte, D., 105, 106
West, I. M., 9, 184, 205, 287
Weyl, P. K., 8, 9, 105, 130, 152, 156, 179,
        204, 225, 244, 246, 254, 299, 365,
        374, 378

Weynschenk, R., 115
White, D. E., 226, 255, 299
Whitehouse, U. G., 355
Whiteman, A. J., 156
Wickman, F. E., 115
Williams, C. H., Jr., 327
Willmann, H. B., 245
Wilson, J. L., 194, 287
Winland, H. D., 366
Wolf, K. H., 7, 16, 266, 401
Wolfbauer, C. A., 338
Wood, C. B., 338
Woolheater, C., 287
Wyoming Geological Association, 338

Yablokov, V. S., 402
Yanat'eva, O. K., 299
Young, R. S., 325, 327

Zeller, E. J., 125
Zen, E-an, 226, 355
Zenger, D. H., 7, 9, 10, 16, 17, 205, 255,
        259, 266, 267, 311, 390, 401
Zirkel, F., 10, 44
Zvonitskaya, I. V., 401

SUBJECT INDEX

Abu Dhabi, 128-129, 139, 155
Afghanistan, 303
Africa, 303
Alberta Basin, Canada, 265, 308
Algae
    Mg^{2+} source, 302-303, 306-307, 361-365
    stromatolitic limestones, 182-183, 356-366
Alpine Triassic carbonates, 4
Alps. *See* Swiss Alps
Andros Island, 141
Appalachians, 186, 306
Arkansas, 5, 264
Ascendant dolomitization, 407-409
Asia, 266
Atolls
    Funafuti, 12, 19, 182
    Pacific Ocean, 3
Australia, 4, 128, 132, 303, 305, 328-330
Authigenic dolomites, 307

Baffin Bay, 14
Bahamas, 4, 129, 141, 155, 182, 202 203,
    209, 306, 308
Bahama syndrome, 4
Barred basins, 160, 317
Base metals, 263
Bighorn dolomite, 3
Biotic dolomitization, 6
Bitumen, 339
Bonaire, 4, 129, 154-155, 308
Brazil, 265
Burial dolomitization. *See* Deep-burial
    dolomitization

Calcite bus, 247
California, 183, 197, 303, 305, 309, 339
Cambrian dolomites
    California, 183, 197
    eastern United States, 303, 312
    Mississippi Valley, 37
    Montana, 265

New York, 183, 197
    regional, 263
    Wyoming, 265
Ca-Mg ratio, 3, 14, 59, 112, 120-122, 209, 395
Canary Islands, 154-155
Capillary concentration, 4, 155
Carboniferous limestone, 4, 13, 45, 52
Caribbean area, 182-183
Celestite, 389
Cement, dolomite as, 13, 29, 281, 348, 404
Chemistry. *See* Synthesis and chemistry
Clay minerals
    formation of dolomite, 306, 350, 352-354,
        408
    Mg^{2+} source, 306, 350
    replacement of, 309
Compaction fluids, 265, 305
Connate waters, 265, 305
Continental dolomites, 15, 95, 112, 331
Continental drift
    Jurassic dolomites, 6
    seafloor spreading, 6
Convective circulation, 288
Coorong Lagoon, 4, 14, 116, 128, 132, 303,
    328
Coral reefs, 18
    Tertiary, in the Pacific Ocean, 3
Cretaceous deposits
    Texas, 154-155, 307
    Uinta Basin, 307

Dachsteinkalk, 403
Dalmatians. *See* Uroliths
Dead Sea, 303
Dedolomitization, 249, 270, 408
Deep-burial dolomitization, 15, 262,
    264-266, 288-298
Deep Spring Lake, 303
Deep-water dolomitization, 15, 95, 107,
    116, 249-250, 308-309
Descendant dolomitization, 408

Subject Index

Detrital dolomite, 3, 6, 13, 35, 47, 201-202,
279, 307-308, 334
crystal morphology, 369-370
wind-derived, 201-202, 367, 369-370
with overgrowths, 201-202, 307, 367,
369-370
Devonian dolomites
Alberta Basin, 265, 308
Alberta Rockies, 265
California, 183, 197
Germany, 403
Iowa, 40
Manilus Formation, 182, 185
New York, 183, 185, 197, 367
Dolomie, 1
Dolomite. *See also* Laboratory synthesis;
Synthesis and chemistry; *specific
variety of form*
affected by bacterial influence, 15, 34,
304-305
affected by climate, 304, 308
affected by structural influences, 209, 262
anticlines, 228, 252, 263
catagenetic, 262
dolomite dikes, 263
domes, 263
faults, fractures, 47, 263, 268, 288
joint-controlled, 47, 264, 280, 408
karst, 408
paleogeography, 112, 258
postorogenic, 263-264, 268
subunconformity, 263
thrusts, 263
unloading, 264
classification of, 6, 13, 45, 229, 279,
288-289, 350, 382, 395-397
origin
contact with Mg-bicarbonate, 2
direct precipitation, 2
effects of Mg-rich vapors, 2
effects of volcanic vapors, 2
primary versus replacement, 2-3
spatial distribution
regionally extensive, 2, 3, 5-6, 129, 157,
211, 228, 258-259, 288, 303, 312
stratiform bodies, 263, 304, 409
"Dolomite bus," 247
"Dolomite dikes," 263, 407
Dolomitization, with depth, 20-21, 119,
140-146. *See also* Environment;
Models; Sites; Timing
Dolostone, 1
Dorag model, 5, 231, 409

Ellenburger Group, 4
England, 4, 13, 263

Environment, of dolomitization
caliche, 249-250
caves (spelean), 250, 309
continental, 15, 409
deep-burial, 15, 262, 264-266, 288-298
deep-water, 15, 95, 107, 116, 249-250, 309
fluvial, 307, 309
hydrothermal, 262, 265, 288
intertidal, 108, 182-183, 185, 192-193,
303, 308
lacustrine, 4, 6, 128, 249-250, 302-304, 309
lagoonal, 4, 14, 128, 154, 166, 303, 328
marginal marine, 4, 263-264, 303
marine, 3, 15, 155-156, 209
metamorphic, 34, 262, 288, 309
metasomatic replacement, 109
playa, 304, 331
sabkha, 4, 13, 128-129, 155, 249-250
shallow-burial, 2, 5, 262-263, 305
shallow-marine, 15, 56, 182-183, 312,
317-324
shelf seas, 6, 312, 317-324
subtidal, 6, 182-183, 208-209, 256-258,
303, 312, 317-324
supratidal, 155, 182, 195, 259, 279, 303,
308, 376-378
uroliths, 305
volcaniclastic, 309
Eocene time period, 304, 331
Epeiric seas, 303
groundwaters, 5, 155, 169, 171-172,
328-329
vadose versus phreatic, 155

Geologic record, distribution in, 4-6, 14, 59,
120, 183, 195, 262-263, 265-266,
307-308
uniformitarianism, 3, 183, 195, 395
Green River Formation, 304

Hydrothermal dolomite, 72, 262, 265, 288,
295, 297, 407

Igneous activity, intrusives, 288
Incongruent dissolution, 306
Insoluble residues, 306
Internal dolomite sediment, 403
Intertidal dolomitization, 109, 182-183, 185,
303, 308
Inyo Mountains, 263
Isotope studies, 15, 85, 263-264, 305

Joachim Formation, 264
Jurassic dolomites, 6, 403

Keuper time period, 268
Knox Group, 312

Laboratory synthesis
    effect of organic matter, 3
    effect of pressure, 2
    effect of salts, 3
    effect of temperature, 2-3
    precipitation of dolomite, 2
    replacement of dolomite, 2
Lacustrine dolomites, 4, 6, 95, 128, 132,
        302-304, 309, 328
    playa lake, 304, 331
Lagoonal dolomites, 4, 14, 128, 154, 166,
        303, 328
Lake Balkash, USSR, 116, 303
Lake Gosiute, 331
Laney Shale Member. *See* Green River
        Formation
Leduc reefs, 265
Length-slow chalcedony, 273
Limpid dolomite, 249, 252
Lithified crusts, 129, 141, 178, 308, 375-376
Lithophyllum, 19
Lithothamnion, 19
Little Falls dolomite, 197, 264
Lockport Formation, 197
Loferites, 403
Lost Burro Formation, 197

Manilus Formation, in New York, 182, 185
Marbre, 1n
Marine dolomites, 3, 15, 155-156, 209,
        302
    marginal marine, 4, 263-264, 308-309
Maryland, 182, 186
Megalodon beds, 404
Metamorphic dolomites, 34, 262, 288, 309
Mg-purging, 252
Mg^{2+}, source of, 1, 2, 6, 33, 66, 95, 108,
        112, 118, 128, 155, 171, 209, 218,
        262-265, 283, 288, 302-307, 330, 365
Michigan, 265
Microfabrics and petrography, 12, 31-32,
        40-42, 48, 128-129, 146, 154, 183, 199,
        209-210, 249, 258, 266, 339, 403-404
    fabric obliteration, 13, 215, 249, 266,
        270
    fabric preservation, 13, 264
Miette reefs, 265, 308
Mifflin Member, 232, 233
Mineralogy
    hydromagnesite, 14-15, 107, 116
    nesquehonite, 14-15, 107
    ordered dolomite, 2, 14, 71, 128, 133,
        199, 209, 217, 303, 305, 328, 330

protodolomite, 3, 14, 15, 70, 128-129,
        133, 199, 209, 247, 303, 305, 328,
        335, 375
    stoichiometry, 2, 14, 409
Miocene Monterey Formation, 305
Mississippian dolomites, 4, 37, 263, 308
Models, of dolomitization
    brackish waters, 222-224, 228, 240-242,
        248
    brines, 129, 154, 159, 213, 248, 305
    capillary concentration, 4, 155, 202
    compaction of shales, 265, 306
    compaction waters, 265
    connate waters, 248, 265, 305
    deep subsurface, 263, 265
    Dorag model, 5, 231, 409
    evaporative pumping, 5, 6, 155, 166, 168,
        177-178, 183, 202, 208, 302, 304, 331,
        336
    evaporitic, 202-203, 249-250
    expulsion of formation waters, 263, 265
    flooding, 4, 150-152, 155
    freshwater, 5, 155, 209, 248, 251, 262-264
    groundwater, 5, 155, 208-209, 218, 224,
        229, 239, 248, 251, 256, 288, 303,
        328, 355
    hypersaline waters, 13-14, 135, 150-152,
        154, 160, 182-183, 208-209, 264
    late diagenetic, 2-6, 15, 262-265, 303,
        307, 334, 382
    meteoric waters, 208-209, 248, 250-251,
        264
    mixing-zone, 5-6, 208-209, 222-224, 228,
        240-242, 248, 252, 257-259, 330
    normal salinity, 5, 14, 208-209, 248, 305
    penecontemporaneous, 2, 4-5, 13, 16,
        128-129, 154-155, 182-183, 209, 264,
        303-304, 336
    peritidal, 409. *See also* Peritidal
        depositional features
    postlithification, 5, 268
    schizoid, 5, 13, 209, 251
    seawater interactions, 4, 208-209, 248
    seepage reflux, 4, 129, 154-155, 159-163,
        168, 208, 249-250, 265, 280, 306
    subtidal, 199, 312, 317
    volcanic, 409
Montana, 265
Monterey Formation. *See* Miocene
        Monterey Formation
Muschelkalk, 5, 264, 268

Nevada, 256
New Market limestone, 186
New Mexico, 154
Northwest Territories, Canada, 306

Onondaga limestone, in New York, 202, 308, 367
Ordovician dolomites
  Arkansas, 5, 264
  Bighorn dolomite, 3-4
  eastern United States, 303, 312
  Ellenburger Group, 4
  Maryland, 182-183, 186
  Wisconsin, 209, 228
Ore deposits, economic significance of, 6, 263-264, 288
Organic dolomite, 339, 348
Organic matter, 3, 6, 112, 128, 339
  inorganic origin, 13, 304-305, 356

Pacific settings
  atolls, 12, 13
  Funafuti, 12-13, 182
  Tertiary coral reefs, 3
Paleozoic dolomites, 3, 6, 14-15, 209, 256, 304, 306
  in New York, 183
Paradiagenetic folding, 403
Penecontemporaneous dolomitization, 2, 4-5, 13, 16, 128-129, 141, 154-155, 182-183, 185, 209, 264, 303-304
Peritidal depositional features
  algal laminae, 146, 182, 188
  association with evaporites, 2, 128, 139, 160, 198, 264, 273
  birdseyes, 182, 185, 188, 199, 275
  desiccation (mud) cracks, 5, 146, 182, 185, 199, 275, 315
  intraclasts, 188, 199
  laminated dolomites, 182, 185, 188, 273
  lumpy carbonates, 182, 188
  paucity of fossils, 51, 182, 185, 188, 199
  ribboned limestone, 182-183, 188
  stromatolites, 183, 188, 199, 315-317
  thin beds, 182, 188
Permian dolomites
  New Mexico, 154, 158
  Texas, 154, 158
Persian Gulf, 4, 139, 155, 182-183
Petroleum, economic significance of, 6, 398, 403
Photosynthesis, 305
Pierre, 1n
Plants, 117, 304
  Ruppia, 132
Plastic dolomites, 133, 303
Platteville Formation, 231-233
Plattin Formation, 264
Porosity-permeability controls, 6, 16, 21-23, 154, 308, 382, 398-399, 403

Precambrian dolomites, 3, 6, 14-15, 263, 304, 394-395
Primary dolomite, 2, 13-15, 46, 63, 112, 116, 120, 128-129, 183, 303, 328
Progressive dolomitization, 21, 39
Proterozoic dolomites, 306
Protodolomite. *See* Mineralogy
Pseudo-breccia, 13, 41, 47

Quartzine, 389

Radiocarbon-dated dolomite, 146-147, 307, 375-376
Reefs
  coral, 18
  foram-algal, 266
  Leduc, 265
  Miette, 265, 308
  Silurian, Michigan, 265
Regressive sedimentation, 256, 303-304
Replacement dolomite, 2, 4-6, 13, 15-16, 128-129, 154-155, 183, 208-209, 364-365
Reservoir rocks, economic significance of, 382, 398
Reservoirs, 398-399
Rhythmicity, 404
Rocky Mountain region, 263, 304

Sabkha dolomite, 4, 13, 128-129, 155
Sabkha syndrome, 4
Salinity gradients, 159, 303, 317
Schizothrix calciola, 356, 361, 364
Seafloor spreading, 6
Seepage refluxion, 4, 129, 154, 159, 168, 202, 208, 249, 265, 306
Selective dolomitization, 307
  of aragonite, 129
  of cements, 13, 308
  of cryptalgal laminae, 356, 364-365
  fabric specificity, 6, 13, 38, 49, 111, 303, 308
  of mud matrix, 13, 155, 308, 408
  precursor mineralogies, 2, 15, 129, 304-305
  of sedimentary structures, 6, 13, 303, 307, 375-378
  of skeletal fragments (allochems), 13, 308
Shallow-burial dolomitization, 2, 5, 262-263, 305
Shallow-marine dolomite, 15, 182
Sites, of dolomitization
  catagenetic, 262
  epigenetic, 2, 262-263

mesogenetic, 262
telogenetic, 262
Sodium, in dolomite, 239, 253, 394
Solution-cannibalism, 208-209, 218-219,
    228, 305
Spain, 5, 264, 268
Spelean dolomite, 309
Stoichiometry. *See* Mineralogy
Strontium, in dolomite, 237-239
Subsidence, 119, 159, 252
Subtidal dolomite, 6, 182-183, 208-209, 303
Supratidal dolomite, 129, 141, 155, 182,
    195, 199, 303, 308
Swiss Alps, 263
Synthesis and chemistry, 2, 14
    chemistry, 78, 147-150, 213, 230-231, 248,
        291-295, 404
    effect of
        alkalinity, 394
        depth, 16
        eH, 16
        Mg/Ca ratio, 128-129, 202-203, 209,
            212-215, 247-248, 264, 285, 288, 303,
            305, 330, 361-362, 394-395
        pCO_2, 14, 64, 109, 119, 394-395
        permeability, porosity, 16, 111, 154,
            288, 308, 383
        pH, 16, 117, 120, 128, 137, 394
        pressure, 2, 14-16, 36-37, 71, 108
        salts, 2, 109
        sulfates, 394
        temperature, 2, 3, 6, 14-16, 36-37, 71,
            108, 119, 264, 288, 404
        time (age of the rocks), 16, 120, 222,
            356, 364-365
    equilibria, 213, 248, 264 265, 291-295,
        336-337, 403
    fluid inclusion studies, 264, 297, 394, 411
    isotopes, fractionation, 15, 85, 233-235,
        264, 305, 343, 345, 347-349, 395,
        407
    isovolumetric replacement, 6
    leaching, 6
    mechanisms, 352-353
        solid state diffusion, ion exchange
            (migration), 14-16, 128, 306, 407
    molecular replacement, 6
    nuclei, 202, 305
    ordering, 14, 71, 146, 150, 203, 213, 247
    precipitation, 2, 14-15, 247, 394
    precursor mineralogies, 2, 15, 129, 289,
        304-305
    process effectiveness, 155, 177
    rates, 14, 120, 155, 172, 209, 247
    salinity, 118, 128, 248, 395, 405

stability, 3, 14, 107, 146, 150, 170, 230,
    248, 264-265, 288, 404
timing, 2, 128, 262-264
volume reduction, 6

Tar, 305, 339-340
Tertiary time period
    Asia, 265-266
    Brazil, 265-266
    Florida, 208, 215
    Pacific Ocean, 3
    Yucatan, 208, 215
Texas
    Baffin Bay, 14
    Cretaceous, 154-155, 307
    Ellenburger Group, 4
    Permian, 154, 159
Timing, of dolomitization
    catagenetic, 261
    early diagenetic, 2, 199, 228, 240-242,
        262-263, 403
    eogenetic, 262-263
    epidiagenetic (epigenetic), 2, 101, 262,
        263
    late diagenetic, 2, 4, 15, 93, 101, 262-264,
        279, 303, 307, 334, 382
    mesogenetic, 262
    penecontemporaneous, 2, 4, 13, 15, 54,
        91, 128-129, 141, 154-155, 182-183,
        185, 192-193, 195-199, 209, 264, 279,
        303-304, 336
    postdepositional, 55, 199
    postlithification, 4
    postorogenic, 111, 263
    preorogenic, 111
    relevance, 128, 199, 262-263
    secondary dolomite, 118, 356, 364-365
    subaerial exposure, 228, 240-242, 280
    telogenetic, 262
Triassic dolomites, 382
    Alpine, 4, 403
    Spain, 4, 268
Tribes Hill Formation, New York, 41, 197
Turkey, 303
Turner Valley Formation, 308

Uinta Basin, 307
Uniformitarianism
    Ca/Mg ratio in time, in carbonate rocks, 3
    distribution of dolomite in time, 3
Uroliths, 305
USSR, 303, 394-402

Venezuela, 4
Vermiculite. *See* Clay minerals

Volcaniclastic dolomites, 309
Volcanic vapors, 2, 409

Weathering
   of clays, 306
   of ultrabasic rocks, 304-305
Wettersteinkalk, 403

Wisconsin, 209, 227
Wyoming, 3, 265

X-ray analyses, diffraction, 78, 216, 236-237, 403

Zoned dolomite, 23-24
Zoning, in dolomitic crusts, 25-28

About the Editors

DONALD H. ZENGER received the B.S. degree in geology from Union College (New York) in 1954, the M.A. degree from Dartmouth College in 1959, and the Ph.D. degree from Cornell University in 1962. Since 1962 he has taught at Pomona College where he is professor of geology. For several field seasons Dr. Zenger has been temporary geologist for the New York State Geological Survey. Most of his research work has been involved with the dolomitization of lower and middle Paleozoic carbonates in New York, California, and Wyoming, and he has published numerous research papers and review articles on dolomitization. From 1966 to 1974 he conducted a project entitled "Petrology of Dolomitic Rocks" supported by the Petroleum Research Fund (American Chemical Society). In 1968-1969 he spent a sabbatical as a Science Faculty Fellow (National Science Foundation) at the University of Liverpool. He is a member of several professional societies, as well as the honorary societies of Phi Beta Kappa and Sigma XI. He served as president of the Pacific Section, Society of Economic Paleontologists and Mineralogists, in 1973-1974.

S. J. MAZZULLO is presently a petroleum geological consultant in Midland, Texas. From 1978 to 1981 he was senior staff stratigrapher and manager of the Stratigraphic Exploration Group of Union Texas Petroleum Corporation in Midland. From 1975 to 1978 he was professor of earth sciences at the University of Texas-Permian Basin, Odessa, Texas, and consultant to the petroleum industry. Dr. Mazzullo received the bachelor's and master's degrees in geology at Brooklyn College, and the Ph.D. in sedimentology at Rensselaer Polytechnic Institute in 1974. His doctoral dissertation concerned stratigraphy and paleodepositional environments of Lower Ordovician limestones and dolomites in eastern New York State and western Vermont. His research interests include carbonate sedimentology and diagenesis and their relation to petroleum exploration. He is an active member of the American Association of Petroleum Geologists, the Society of Economic Paleontologists and Mineralogists, the International Association of Sedimentologists, the American Association for the Advancement of Science, Sigma Xi, and other professional societies.

Benchmark Papers
in Geology

Series Editor: Rhodes W. Fairbridge
Columbia University

Volume

 1 ENVIRONMENTAL GEOMORPHOLOGY AND LANDSCAPE CONSERVATION, Volume I: Prior to 1900 / *Donald R. Coates*
 2 RIVER MORPHOLOGY / *Stanley A. Schumm*
 3 SPITS AND BARS / *Maurice L. Schwartz*
 4 TEKTITES / *Virgil E. Barnes and Mildred A. Barnes*
 5 GEOCHRONOLOGY: Radiometric Dating of Rocks and Minerals / *C. T. Harper*
 6 SLOPE MORPHOLOGY / *Stanley A. Schumm and M. Paul Mosely*
 7 MARINE EVAPORITES: Origin, Diagenesis, and Geochemistry / *Douglas W. Kirkland and Robert Evans*
 8 ENVIRONMENTAL GEOMORPHOLOGY AND LANDSCAPE CONSERVATION, Volume III: Non-Urban / *Donald R. Coates*
 9 BARRIER ISLANDS / *Maurice L. Schwartz*
10 GLACIAL ISOSTASY / *John T. Andrews*
11 GEOCHEMISTRY OF GERMANIUM / *Jon N. Weber*
12 ENVIRONMENTAL GEOMORPHOLOGY AND LANDSCAPE CONSERVATION, Volume II: Urban Areas / *Donald R. Coates*
13 PHILOSOPHY OF GEOHISTORY: 1785–1970 / *Claude C. Albritton, Jr.*
14 GEOCHEMISTRY AND THE ORIGIN OF LIFE / *Keith A. Kvenvolden*
15 SEDIMENTARY ROCKS: Concepts and History / *Albert V. Carozzi*
16 GEOCHEMISTRY OF WATER / *Yasushi Kitano*
17 METAMORPHISM AND PLATE TECTONIC REGIMES / *W. G. Ernst*
18 GEOCHEMISTRY OF IRON / *Henry Lepp*
19 SUBDUCTION ZONE METAMORPHISM / *W. G. Ernst*
20 PLAYAS AND DRIED LAKES: Occurrence and Development / *James T. Neal*
21 GLACIAL DEPOSITS / *Richard P. Goldthwait*
22 PLANATION SURFACES: Peneplains, Pediplains, and Etchplains / *George F. Adams*
23 GEOCHEMISTRY OF BORON / *C. T. Walker*
24 SUBMARINE CANYONS AND DEEP-SEA FANS: Modern and Ancient / *J. H. McD. Whitaker*
25 ENVIRONMENTAL GEOLOGY / *Frederick Betz, Jr.*
26 LOESS: Lithology and Genesis / *Ian J. Smalley*

27 PERIGLACIAL PROCESSES / *Cuchlaine A. M. King*

28 LANDFORMS AND GEOMORPHOLOGY: Concepts and History / *Cuchlaine A. M. King*

29 METALLOGENY AND GLOBAL TECTONICS / *Wilfred Walker*

30 HOLOCENE TIDAL SEDIMENTATION / *George deVries Klein*

31 PALEOBIOGEOGRAPHY / *Charles A. Ross*

32 MECHANICS OF THRUST FAULTS AND DÉCOLLEMENT / *Barry Voight*

33 WEST INDIES ISLAND ARCS / *Peter H. Mattson*

34 CRYSTAL FORM AND STRUCTURE / *Cecil J. Schneer*

35 OCEANOGRAPHY: Concepts and History / *Margaret B. Deacon*

36 METEORITE CRATERS / *G. J. H. McCall*

37 STATISTICAL ANALYSIS IN GEOLOGY / *John M. Cubitt and Stephen Henley*

38 AIR PHOTOGRAPHY AND COASTAL PROBLEMS / *Mohamed T. El-Ashry*

39 BEACH PROCESSES AND COASTAL HYDRODYNAMICS / *John S. Fisher and Robert Dolan*

40 DIAGENESIS OF DEEP-SEA BIOGENIC SEDIMENTS / *Gerrit J. van der Lingen*

41 DRAINAGE BASIN MORPHOLOGY / *Stanley A. Schumm*

42 COASTAL SEDIMENTATION / *Donald J. P. Swift and Harold D. Palmer*

43 ANCIENT CONTINENTAL DEPOSITS / *Franklyn B. Van Houten*

44 MINERAL DEPOSITS, CONTINENTAL DRIFT AND PLATE TECTONICS / *J. B. Wright*

45 SEA WATER: Cycles of the Major Elements / *James I. Drever*

46 PALYNOLOGY, PART I: Spores and Pollen / *Marjorie D. Muir and William A. S. Sarjeant*

47 PALYNOLOGY, PART II: Dinoflagellates, Acritarchs, and Other Microfossils / *Marjorie D. Muir and William A. S. Sarjeant*

48 GEOLOGY OF THE PLANET MARS / *Vivien Gornitz*

49 GEOCHEMISTRY OF BISMUTH / *Ernest E. Angino and David T. Long*

50 ASTROBLEMES—CRYPTOEXPLOSION STRUCTURES / *G. J. H. McCall*

51 NORTH AMERICAN GEOLOGY: Early Writings / *Robert Hazen*

52 GEOCHEMISTRY OF ORGANIC MOLECULES / *Keith A. Kvenvolden*

53 TETHYS: The Ancestral Mediterranean / *Peter Sonnenfeld*

54 MAGNETIC STRATIGRAPHY OF SEDIMENTS / *James P. Kennett*

55 CATASTROPHIC FLOODING: The Origin of the Channeled Scabland / *Victor R. Baker*

56 SEAFLOOR SPREADING CENTERS: Hydrothermal Systems / *Peter A. Rona and Robert P. Lowell*

57 MEGACYCLES: Long-Term Episodicity in Earth and Planetary History / *G. E. Williams*

58 OVERWASH PROCESSES / *Stephen P. Leatherman*

59 KARST GEOMORPHOLOGY / *M. M. Sweeting*

60 RIFT VALLEYS: Afro-Arabian / *A. M. Quennell*

61 MODERN CONCEPTS OF OCEANOGRAPHY / *G. E. R. Deacon and Margaret B. Deacon*

62 OROGENY / *John G. Dennis*
63 EROSION AND SEDIMENT YIELD / *J. B. Laronne and M. P. Mosley*
64 GEOSYNCLINES: Concept and Place Within Plate Tectonics /
   F. L. Schwab
65 DOLOMITIZATION / *Donald H. Zenger and S. J. Mazzullo*
66 OPHIOLITIC AND RELATED MELANGES / *G. J. H. McCall*
67 ECONOMIC EVALUATION OF MINERAL PROPERTY / *Sam L.
   VanLandingham*
68 SUNSPOT CYCLES / *D. Justin Schove*
69 MINING GEOLOGY / *Willard C. Lacy*
70 MINERAL EXPLORATION / *Willard C. Lacy*
71 HUMAN IMPACT ON THE PHYSICAL ENVIRONMENT /
   Frederick Betz
72 PHYSICAL HYDROGEOLOGY / *R. Allan Freeze and William Back*
73 CHEMICAL HYDROGEOLOGY / *William Back and R. Allan Freeze*